Reservoir Fisheries Management:

Strategies For The 80's

RESERVOIR FISHERIES MANAGEMENT:
Strategies for the 80's

A National Symposium on Managing Reservoir Fishery Resources

Sponsors

Reservoir Committee, Southern Division
American Fisheries Society
United States Fish and Wildlife Service

Organizers

Reservoir Committee
Fred Harris, *Chairman*

Program Subcommittee
- Larry Aggus, *Chairman*
- David Combs
- Ronald Dent, Jr.
- Larry Olmsted
- Michael J. Van Den Avyle

Publicity Subcommittee
- Fred Harris, *Chairman*
- William Seawell
- William Neal
- Les Ager
- Bert Pierce
- Val Nash

Editorial Subcommittee
- Michael J. Van Den Avyle, *Chairman*
- Larry Aggus
- Larry Olmsted
- Michael Gibson
- David Combs

Local Arrangements & Finance Subcommittee
- Benjy Kinman, *Chairman*
- Robert Gasaway
- Tom Bankhead

Fish and Wildlife Service
Federal Assistance, Atlanta, Georgia
John Christian, Assistant Regional Director
John Hall, Chief, Federal Aid
Robert Gasaway, Project Officer

RESERVOIR FISHERIES MANAGEMENT
Strategies For The 80's

Gordon E. Hall
Technical Editor

and

Michael J. Van Den Avyle
Managing Editor

Proceedings of a Symposium
held in Lexington, Kentucky
June 13–16, 1983

Reservoir Committee
Southern Division
American Fisheries Society

1986

Publication of these Proceedings
was aided by a grant from the

United States Fish and Wildlife Service

Library of Congress Catalog Card Number: 84-062331
ISBN 0-913235-31-8

Orders should be addressed to

American Fisheries Society
5410 Grosvenor Lane
Bethesda, Maryland 21044 U.S.A.

Printed in the United States of America

CONTENTS

Preface

During the 1950's and 1960's, reservoir construction in the United States was carried out at a rapid rate. The impounded waters quickly became popular sport fishing areas that presented new challenges to fishery managers. The nature of reservoir systems, however, did not allow direct application of fishery management techniques that had been developed for natural lakes, rivers, or small ponds. As a result, the Southern Division of the American Fisheries Society established a Reservoir Committee in 1958 and charged it with conducting and coordinating investigations that would lead to more effective management of reservoir fishery resources.

The Reservoir Committee sponsored its first National Symposium on Reservoir Fishery Resources in 1967. The symposium allowed compilation of current reservoir research findings and provided a forum for the exchange of ideas concerning reservoir fisheries management. The proceedings were published and the resulting volume has been used extensively by reservoir biologists. Four years later, the President of the American Fisheries Society, Robert M. Jenkins, proposed the publication of a book that would consolidate the results of recently completed and ongoing reservoir research. The product of that effort was *Reservoir Fisheries and Limnology,* an American Fisheries Society special publication organized and edited by Gordon E. Hall. The special publication and the Reservoir Fishery Resources Symposium Proceedings provided fishery biologists with a state-of-the-science documentation of reservoir fisheries research and management.

Since these earlier compilations, construction of new reservoirs has slowed, and many reservoirs have experienced declines in sport fish populations. The demand for sport fishing recreation on reservoirs, however, has increased. These trends point to the need for development of new fishery management techniques to maintain desirable reservoir sport fisheries in aging reservoirs. In 1979, the Reservoir Committee decided that the development of such techniques would best be facilitated by a comprehensive assessment of current reservoir fishery management strategies and the identification of promising new procedures. Sponsoring a reservoir fishery management symposium and publication of a proceedings of the symposium were considered the appropriate means of achieving these goals.

Specific objectives of the symposium were (1) to provide a critical assessment of current practices for managing reservoir fisheries; (2) to provide a forum for the exchange of information and ideas among users and managers of reservoir fishery resources; (3) to stimulate new ideas and approaches for managing reservoir fishery resources; (4) to identify critical research and management needs; and (5) to clarify the role of fishery management relative to reservoir planning, operation, and other recreational uses.

To meet these objectives, the symposium was held in Lexington, Kentucky, on June 13–16, 1983. The program featured a general session with a keynote address, seven technical sessions, two poster sessions, an evening panel session, and a summarization of the symposium. The keynote speaker, Robert A. Jantzen, Director of the U.S. Fish and Wildlife Service, discussed the changing role of the Service in fisheries programs and identified areas of emphasis for the Service in future years. Each technical session addressed a selected theme by including review papers that synthesized past and potential approaches to reservoir fisheries management and case history papers that described specific management techniques. The poster presentations focused on the effectiveness of harvest regulations and fish introductions for managing reservoir fisheries; results from short-term or uncompleted studies that otherwise would not have been available to symposium participants were emphasized in these sessions. The evening panel session addressed the concerns of various reservoir fisheries user-groups and provided a forum for discussing conflicts that exist between groups. Dr. Robert Summerfelt presented an excellent summary of the technical sessions to conclude the symposium. Readers of the Proceedings may wish to review the summary first to obtain an overview of the technical papers' contents.

Many individuals worked industriously over a long period of time to develop, organize, and conduct this symposium. Members of the Reservoir Committee and employees of the U.S. Fish and Wildlife Service who performed these functions are named on page *ii* of these Proceedings. Early in the planning process, the critical need of financial support was met when the U.S. Fish and Wildlife Service agreed to function as co-sponsors. The Service financed most of the costs associated wtih the meeting and publication of the Proceedings. The Kentucky Department of Fish and Wildlife Resources hosted the symposium, and Commissioner Carl Kays and his staff contributed immeasurably to the symposium's success. About 50 fishery scientists in various governmental agencies, universities, and private organizations volunteered their expertise and reviewed one or more of the technical papers prior to acceptance for publication in the Proceedings.

Reservoirs now constitute and will undoubtedly continue to be one of our most valuable fishery resources—a resource that is exceedingly complex, poorly understood, and crudely managed, at best. As such, reservoirs continue to provide major fishery management problems and consequently offer tremendous opportunities for improvement. We hope this volume will provide a better understanding of the current state of reservoir fisheries and will stimulate the development of more effective management techniques.

GORDON E. HALL
Technical Editor
FRED A. HARRIS
Chairman, Reservoir Committee

Pages 1–4 *in* G. E. Hall and M. J. Van Den Avyle, editors.
Reservoir Fisheries Management: Strategies for the 80's.
Reservoir Committee, Southern Division American Fisheries Society, Bethesda, Maryland. 1986.

KEYNOTE ADDRESS

ROBERT A. JANTZEN
Director, U.S. Fish and Wildlife Service
Washington, D.C. 20240

Good morning. It's a pleasure to be here. I would like to thank the Reservoir Committee member States for this opportunity.

We all recognize that this has been a time of some concern and uncertainty among resource professionals. Today, I would like to discuss some of the changes in priorities, and goals in the Fish and Wildlife Service—changes which will likely affect you and areas of your work, if they have not already done so. And I would like to talk about the future direction of reservoir fishery management and some areas where improvement might be achieved.

First, I think it should be stated that fishery workers have made excellent progress in reservoir management in the past 10 years or so.

Since the Reservoir Symposium of 1967 there have been many significant achievements in areas that were then "uncharted waters." The remarkable increase in successful management of striped bass in reservoir fisheries is an excellent example. Likewise the use of rainbow trout in two-story reservoir fisheries and in tailwaters. In 1967, fish attractors were not even mentioned in the Symposium—today they are highly regarded and widely used. Sixteen years ago, predatory-prey dynamics in reservoir fisheries were little understood; today, a significant and expanding literature has been created, with management strategies to match.

To characterize the period since 1967, we can say it was one of growth, progress, and results for fishery managers and fishermen alike. These accomplishments speak very well indeed for the initiative and tenacity of State biologists—and they serve as an important and honorable testimonial to the past National Reservoir Research Program.

No one can question the fact that we have made tremendous progress in the scientific management of reservoir sport fisheries since the 1967 Symposium. The question now becomes "where do we go from here?" Let's look at the direction that the Fish and Wildlife Service is going.

Policy changes during the past 2 years illustrate that "business as usual" for the Fish and Wildlife Service won't be the way business is conducted in the future. These changes are a direct outgrowth of the ongoing redefinition of the roles and responsibilities of State and Federal governments. I want to emphasize up front that close cooperation with the States will continue to have a high priority with the Service. But the nature of that cooperation is changing. Changes in our fisheries' program stem largely from an examination of the Service fishery mandates and priorities relative to State responsibilities. The success of States that have assumed responsibility for reservoir management programs can be well illustrated using work by our hosts, the Kentucky Department of Fish and Wildlife Resources. Some very positive economic changes around Cave Run Lake, east of here, resulted from the establishment and maintenance of a trophy muskie fishery. A national striped bass tournament was held this year down at Lake Cumberland, illustrating the results of Kentucky's progressive striped bass management program. Many other State fish and wildlife agencies have similar success stories.

New emphases in our fisheries programs have provided some of the most vivid and controversial illustrations of the broad directional change in the Service. As you are aware, the work of Fishery Assistance offices has been redirected and in many cases discontinued. Many national fish hatcheries are being transferred to the States and a few will be closed. And, of course, our National Reservoir Research Program ended as of March 31, 1983.

The National Reservoir Research Program was the leading program in answering questions of common State interest concerning the management of fisheries in large reservoirs. It did a great job. Let me repeat that and emphasize: It did a great job. Our decision to change direction in this area comes from the following progression of events . . . which I will present only in brief:

The first event is the dramatic slowdown in reservoir construction throughout the United States. Through the 1960's and early 1970's construction of new reservoirs set a rapid pace. The Service's involvement with these reservoirs centered around the fact that many of these large reservoirs resulted from the action of another Federal agency—even though most such reservoirs are State managed. The National Reservoir Research Program met the challenge to develop predictive models that answered many questions. Now, however, the large construction programs have slowed dramatically while our reservoir management knowledge has skyrocketed.

Even more important than the slowdown in reservoir construction, though, was the tremendous improvement in technical capabilities within the States. When the National Reservoir Research Program was formed, only a small group of people were addressing the complex questions posed concerning reservoir fisheries management. You need only to review the literature on reservoir fisheries research to quickly see where the foundations were formed for today's management programs. The States have utilized the expertise and research provided by the National Reservoir Research Program. The Reservoir Committee that sponsored this symposium found much of its base and direction in our reservoir research group. Now, States with large numbers of reservoirs have developed their own successful programs.

Certainly anyone could argue that unified reservoir research is still needed. However, most species of fish found in reservoirs are legitimately the responsibilities of designated State agencies. The role of the Fish and Wildlife Service has changed to one of support on special problems, in areas such as fish disease studies and environmental contaminant monitoring. We feel that a regional consortium under State auspices could continue to provide coherence and unity to reservoir research. This may be a productive area to explore.

There are two other major changes: hatcheries and cooperative units. The reduction in the Fish and Wildlife Service National Fish Hatchery System, along with changes in the mission guiding our hatchery activities, will limit the number of fish available from Federal hatcheries. We believe that State hatchery operations should supply most future needs for reservoir fish stocks. But in those situations where we have mitigation responsibilities, they will be met fully, as the Secretary of Interior has indicated.

Reduction in the Cooperative Fishery Research Unit Program is another change that could impact activities on reservoirs, but to a lesser degree. State agencies or universities will likely assume research activities normally carried out under this program. In the short term, this may cause some problems. In the long run, however, it should provide States with greater control and more flexibility for their research monies.

Now let's look at our main fisheries-oriented programs and where new thrusts might appear. Two program goals of the Fish and Wildlife Service are centered in the Fisheries and Federal Assistance Programs. The goal of our Fisheries Program is to promote the wise use of the Nation's freshwater and anadromous fish populations for the benefit of people. Our goal for our Federal Aid Program is to strengthen the ability of State fish and wildlife agencies to meet effectively the consumptive and nonconsumptive needs of the public for fish and wildlife. These are the main program missions.

There will be increased emphasis in some areas, such as those pertaining to aquaculture. The Service has been working with the Departments of Commerce and Agriculture to implement the National Aquaculture Act of 1980. For many years, we've played a key role in the development of inland commercial aquaculture ventures; under the 1980 Act, our role in research, development, and scientific leadership is carried forward and, we hope, can be strengthened.

What else will our new role entail? This Symposium is one example—we will bring together and disseminate research and management information gathered by various States, making information readily available to field biologists. We may investigate high technology approaches for data transfer. Also, in the high tech realm, we will be exploring new areas in such fields as genetics and environmental concerns, such as acid rain, which might require more manpower and money than a given State is able to spend. We will address problems that transcend State boundaries.

Our Dingell-Johnson (D-J) Program remains a top priority. States are generally pleased with the way the program has been administered and managed in the past. D-J provides an important part of the foundation for most State fisheries programs and a means for documentation and reporting important management and research projects.

Many State fisheries administrators who now manage large State programs received their early training in the D-J Program. It is our strongest stabilizing force in fisheries and State-Federal relations. As you probably know, the Administration has given its support to the proposed D-J expansion which could add more than $60 million to the program annually. This is greatly needed funding, as you all know. It represents a major infusion of new fisheries monies. For many States, this could represent a three-fold increase in their D-J allotment.

Another of our priorities will be to make our Federal Aid/D-J efforts even more responsive and helpful to the States. In consultation with the International Association for Fish and Wildlife Agencies, we will identify what we see as emerging needs in reservoir fishery management that can be addressed via the D-J program. We will share these initiatives with the States. We will use the administrative portion of the D-J funds available to us to the maximum benefit for everyone. We will emphasize planning initiatives that have the broadest range of applicability and acceptability, both nationally and within each State. We believe that our expenditure of D-J administrative funds should address issues that affect many States and address issues of the highest priority. This Symposium is an excellent example of this type of Service expenditure of D-J monies. It's national in scope; it involves many States; and it will yield a tangible product that can be used by nearly every State fishery program.

Now let's look at the future direction of reservoir fishery management and the role of professional reservoir fishery managers and some areas where improvement might be achieved. The primary goal here should be to let sport fishermen know what we are doing and why it is important. Fishery biologists, or for that matter biologists in general, have largely ignored this need. Now, while we wade through tough economic times, everything comes under question, and program survival depends not only on technical need but also on public support.

How can we improve in this area? Fishery biologists and managers should strive to open new communication channels. Managers can't afford to be isolated. We must cultivate an *informed* public. Communicating technical information to the scientific community is important, to be sure. But we should be able to do likewise for the sportsmen, and other groups that directly or indirectly provide support for our activities. They pay the bills.

Another goal should be to economically quantify impacts of proposed reservoir management activities. The economic importance of actions we proposed, as well as predicted outcomes, should be highlighted before a decision is made to proceed with those actions. We live in a society heavily influenced by economics. Successful reservoir fishery managers in the future will have to make decisions utilizing comparative economic as well as biological data.

When we consider funding for reservoir fisheries management, rhetoric about National priorities has to be defined in real terms, as illustrated by the percentage of National income derived from reservoir fishing. Unfortunately, fishery management funds are far down the list. Such things as defense, education, and social services gobble up most of the Federal budget. When closely examining a State or Federal budget, you might ask "How have fishery programs managed to stay financially alive at all?"

In a very real sense, fishery management has achieved its current level because private enterprise supports those activities. Tackle manufacturers, boat builders, bait suppliers, and localized retail enterpreneurs actually pay a large portion of the bill. They give it priority because it is their livelihood! License sales from the angler provide another part of the support needed to run programs. This is what you expect in a society based on private enterprise.

Our National Hunting and Fishing Survey of 1980 revealed massive commercial expenditures by the public to go fishing. In the Southeast alone, estimated annual expenditures for fishing exceeded four billion dollars—a significant chunk of the GNP. The 1967 Symposium reported that the entire country spent less than three billion dollars annually to fish. Fishery management of reservoirs produces business with growing demand and rapidly expanding markets and expenditures.

To continue that growth in the future, we must assure that management activities yield definitive results. We must know what operations are really needed and what goals we strive to achieve. More than ever, we must produce *results*!

New direction may be needed here and I challenge you State fishery workers to take a hard look at your activities. What effect would the fishery, the sportsmen, or the economy feel if some management activities were reduced 50% or even eliminated?

A fishery biologist's time and mental energy should go toward increased creativity with a strong "product" orientation. It's time we quit hiding behind the "need to know" or "the public needs that information" and come face to face with the question of "what tangible product will this activity produce, who will benefit from it, how long will it take to produce it, and what will it cost?"

We think proper "action-oriented" planning has a key role here. Some States have been resistant to planning for a number of reasons. Some State administrators view planning as a bureaucratic weapon which transfers policymaking to the planner.

Another concern is that the process is cumbersome, time consuming, and costly. Some States in the past employed elaborate data collection and analysis schemes and expensive computer predictions. Critics say the means don't justify the end. Other States reorganized their natural resource functions and hired dozens of planners only to discover that this served only to build a larger administrative workload resulting in eventual disillusionment with planning and planners.

A third concern is that State agencies are "reaction oriented organizations," and planning could impose an inflexibility on an already established system which is responding to public needs.

All these concerns can be true, as the past well illustrates; however, planning can be simple, flexible, and productive. The key is to use planning as a tool—not a dominating force, but as a method to achieve identified goals.

I will do my best to simplify the Federal Aid planning process within the Fish and Wildlife Service. Policymaking will certainly remain within the States' established systems. Planning does not have to focus on a large volume of data collection, analysis, and computer projections. Planning can simply be the establishment of goals and the measurement of progress toward achieving those goals. When viewed from these perspectives, no additional staff may be needed since existing upper-level administration within the States can make basic directional planning decisions. Planning can be dynamic when viewed as a tool. Policymakers change plans every day and evaluate the impact of those changes. I assure you that our planning will focus on fishery resources and goals to be accomplished, and not on operating systems within the State agencies.

You are the first line decision makers for major reservoir fishery management decisions, which impact the fishermen, the economy, and the resource itself. We rely on your judgment to *manage* this very important part of our National fishery resources. Let's establish reservoir fishery goals and develop strategies to effectively meet those goals. Let the user know what you are doing and why it's important. Let them know what they can do to help you achieve your goals and assure productive results for all concerned. After all, your goals should be the public's goals.

Restoring productivity to America—whether we speak of economics or human effort—is not simply Federal rhetoric, it's a tough task we all must pursue. Although some decisions I make are unpopular, the benefits can strengthen the Service, the States, and the Nation, and increase productivity. We're doing our part. I am proud of the people in the Fish and Wildlife Service for the ability they have demonstrated in meeting the tough challenges and changes we are facing in a positive and productive manner. I have confidence that professionals within the States will also respond positively to our new efforts.

We are in a time of change . . . it is a time for us to reinforce and expand our partnership to address new resource goals.

The Fish and Wildlife Service's role in fisheries management and research has changed dramatically. But our role has *not* been abandoned.

The Service will work to complement and augment your efforts; not to duplicate them. The era of State leadership in many fisheries areas has been long in coming—but is now here. The States have long had the capability to lead; now they have the opportunity in many fishery endeavors. We want to aid and guide where and when we can, and we will.

With this approach, we believe, the Public and the Nation's fishery resources can be best served now and for the balance of this century.

Pages 5–10 *in* G. E. Hall and M. J. Van Den Avyle, editors.
Reservoir Fisheries Management: Strategies for the 80's.
Reservoir Committee, Southern Division American Fisheries Society, Bethesda, Maryland. 1986.

DEVELOPMENT OF MANAGEMENT PROGRAMS AND MEASUREMENT OF ECONOMIC VALUES

Economic Characteristics of Reservoir Fisheries

WARREN L. FISHER, J. J. CHARBONNEAU, AND MICHAEL J. HAY
Division of Program Plans
U.S. Fish and Wildlife Service
Washington, D.C. 20240

ABSTRACT

This paper assesses the relative importance of reservoir fishing to U.S. anglers. Using the results of the 1980 National Survey of Fishing, Hunting, and Wildlife-Associated Recreation, days of reservoir fishing and associated expenditures are compared with those for ponds, natural lakes and ponds (including the Great Lakes), and rivers and streams. Alternative methods are used for allocating days and expenditures for freshwater fishing by type of water fished. Reservoirs are shown to be a very important resource, accounting for nearly 272 million days of fishing and $5.4 billion of expenditures in 1980. Reservoir fishing accounts for 41% of total freshwater angling expenditures. Geographic differences in the relative importance of reservoir fishing appear to be related to the abundance of these manmade waters.

Reservoirs are the most-used freshwater fishery resource. Nearly 40% of all freshwater fishing activity occurs in reservoirs. This paper presents and discusses information from the 1980 National Survey of Fishing, Hunting, and Wildlife-Associated Recreation (FHWAR) survey as it relates to reservoir fisheries. The survey permits national and state-by-state projections of the number of participants who fished in freshwater in 1980, the number of days they fished, and their expenditures. Presented along with the estimates of effort and expenditures is a brief summary of how such information is helpful in understanding the economic value of reservoir fisheries, and suggested directions for further research that will better demonstrate this economic value.

DESIGN AND PROCEDURES

The FHWAR survey was designed by the U.S. Fish and Wildlife Service at the request of the International Association of Fish and Wildlife Agencies. The 1980 survey is the most comprehensive of the six surveys that have been conducted at five-year intervals since 1955. Data were collected and tabulated by the U.S. Bureau of the Census. Interviewed persons were asked about their wildlife-associated activities for the calendar year 1980. The nature of the sample, the three survey questionnaires, and their administration is described in USFWS (1982).

The first phase of the 1980 survey consisted of a telephone screen of a probability sample of 143,000 households in the United States. Results from the 116,000 completed interviews were used to project the total number of fishermen six years old and older in each state and their socio-economic characteristics. In the second phase, about 30,300 detailed, in-person interviews were completed with sportsmen 16 years old and older selected from among those screened. These interviews provide substantial information about participants' 1980 fishing and hunting expenditures and activities, including what, where, and how often they had fished or hunted.

The FHWAR survey resulted in a national report, individual reports for each of the 50 states, and a microdata file of individuals' survey responses. The tables and analysis in this paper rely upon those reports and the unpublished data. The estimates reported here are point estimates, i.e., the central point of a confidence interval whose width is determined by the relative size of the estimate's standard error. For example, an estimate of 87,000 fishermen may have a standard error of 9,000. A 68% confidence interval, one which would include the true number 68%

of the time, would then be 87,000 ± 9,000. For simplicity, standard errors are not reported here. They can be found using the statistical appendices of the FHWAR national and state reports. Coefficients of variation (CV's) for estimates of national totals that appear in this paper seldom exceed 5%. State-specific estimates have CV's that vary from 10% to over 100%, since each depends on sample size and the variability of the responses.

The discussion which follows does not include the estimated 11.8 million persons in the 6–15 year old age group who fished in 1980. While this group was screened, the detailed questionnaire was regarded as too complex for administration to persons less than 16 years old. To the extent that their participation in reservoir fishing follows the pattern of the older group, the comparative results will be appropriate for all age groups; in absolute terms, however, the tabulated results which follow are underestimates, since detailed information about this 6–15 year old age group is not available.

The FHWAR survey does not provide direct estimates of expenditures and days by type of water fished. Indirect estimates may be made using either of two survey questions. The first asked respondents to indicate whether they fished at all in 1980 in each of four water types: reservoirs, manmade ponds of less than 10 acres, natural bodies of water, and rivers and streams. The second question asked the respondent to indicate which one of the four types of water was used most often in the fisherman's resident state. In this paper results are reported using each criterion. Using "water fished in most often" may be biased to the extent that the average response to the question may differ from the actual distribution of the water types used by residents of each state. For example, it is possible in theory that all respondents from State X indicated that they fished most often in reservoirs, even though as little as 26% of their fishing effort was associated with reservoirs, with the remainder being equally divided among the other three water types. It is expected that errors from this source—while unknown in size—will not unduly bias the estimates.

The distribution of water type used most often by state residents may be biased in the national context only to the extent that water type use by non-residents differs from use in the resident state. Since 90% of freshwater fishing is done in the state of residence, the bias introduced from this source should be quite small. At the state reporting level, this method of allocation may be somewhat inaccurate if the average fisherman participated as a nonresident and sought a water type that was different from that used most often in the resident state. One such theoretical example would be the warmwater reservoir fishermen from Florida who visited another state to fish for trout in streams. To the extent that this kind of phenomenon is present, the estimates on this allocation criterion have an unknown bias.

Results and Discussion

Fishing in reservoirs and other freshwaters is a major activity. Freshwater fishing attracted 36.4 million of the total of 42.1 million fishermen in 1980. The FHWAR survey's national report shows that 40% of all freshwater fishermen fished in reservoirs, 23% fished in manmade ponds of less than 10 acres, 40% used natural bodies of water, and 43% did at least some fishing in rivers and streams. Persons from big cities show a slight preference for reservoirs compared with persons from small cities (who tend to prefer natural lakes) and those from towns and rural areas (who prefer rivers and streams).

Strong regional patterns of use emerge when freshwater fishing days per fisherman are projected for those fishermen who fished reservoirs most often in their state of residence. The contiguous states of Oklahoma, Arkansas, Texas, Kansas, and Nebraska had the highest per fisherman days of use (Table 1). As one might expect, states with relatively few reservoirs form the bottom of the distribution with the fewest days of fishing in reservoirs. The New England states, the Great Lakes states, Alaska, and Hawaii represent this group. One of the factors which appears to play a key role in determining the extent to which reservoirs are used for fishing is their availability. However, correlation analysis was inconclusive when it was used to test the hypothesis that reservoir use is a function of the percent of a state's area that is reservoir surface.

Relative use of reservoirs appears to be highest among persons in their middle years, compared with a greater interest in river use by younger persons who may be less likely to have boats. Males show a slight preference for rivers, while females tend to fish reservoirs and natural lakes more than rivers and ponds. Forty percent of the white anglers fished reservoirs, compared with

33% of blacks and 39% of fishermen of other races. However, 69% of all persons who fished in reservoirs were males and 92% of all reservoir fishermen were white. Regional differences in the availability and use of reservoirs and other freshwaters as well as the racial composition of the population are not fully reflected in these national summary statistics. No conspicuous patterns exist for participation in reservoir fishing as a function of income or education, if one recognizes the dominant role played by whites and males. That is, the patterns of participation in reservoir fishing with respect to income and education are similar to those for white males.

Freshwater fishermen, including Great Lakes fishermen, spent $13.2 billion for their sport in 1980. Travel-related expenditures, equipment and fees expenditures, and expenditures for special equipment were 42.8%, 16.4%, and 40.8% of the total, respectively. The first two categories are single-purpose items for freshwater fishing. The last, special equipment, represents estimates for equipment used primarily for freshwater fishing. Total freshwater expenditures of $7.8 billion shown in Table 26 of the national report of the FHWAR do not include those for special equipment.

Relative to the frequency distribution of water type fished in 1980, manmade reservoirs of 10 acres or more generated $3.7 billion, compared with $2.0 billion for manmade ponds of less than 10 acres, $3.7 billion for natural lakes and ponds, and $3.8 billion for rivers and streams. Following the same criterion, 26.8% of the 710.6 million freshwater fishing days may be allocated to reservoirs, 15.8% to ponds, 27.9% to natural waters, and 29.5% to rivers and streams. This criterion for allocation reflects accurate portions of expenditures and days only if the average fisherman in each state spent equal amounts and fished equal numbers of days for each water type used. On this measure, reservoirs rank second to rivers and streams as the water type responsible for freshwater fishing expenditures in 1980. Reservoir fishing ranks third as the water type on which days of fishing took place in 1980, with about 27% of all days being allocated to reservoirs.

The second criterion, the type of water fished most often in state of residence, appears to be the better of the two, and allocations of days and expenditures in Table 2 were made on this basis. Due to nonresponses, the projection of participants in Table 2 does not equal the total number

Table 1. Days of fishing and expenditures per freshwater fishermen (16 years and older), by state of residence, during 1980.

State	Days of fishing per freshwater fisherman in reservoirs	Expenditures for fishing in reservoirs per freshwater fisherman (U.S. dollars)
Oklahoma	14.56	287.70
Arkansas	12.15	179.00
Texas	12.08	345.03
Kansas	11.26	196.18
Nebraska	10.79	219.03
Nevada	10.76	468.78
Arizona	10.37	385.95
Ohio	10.24	182.33
Georgia	10.19	161.41
New Mexico	10.05	304.54
Indiana	9.75	140.68
South Carolina	9.52	114.60
West Virginia	9.33	141.47
Kentucky	9.22	145.26
Illinois	9.15	175.97
Pennsylvania	9.09	86.96
Tennessee	8.97	118.71
Missouri	8.80	142.69
California	8.07	267.65
Alabama	7.93	108.07
Mississippi	7.92	90.43
Idaho	7.76	117.60
North Dakota	7.42	149.87
Iowa	7.41	93.47
Virginia	7.21	142.56
North Carolina	7.13	83.64
Oregon	7.09	139.24
Utah	7.04	195.28
Wyoming	6.68	179.68
Colorado	6.43	151.80
Delaware	5.73	43.41
South Dakota	5.59	95.33
Maryland	5.58	114.79
Louisiana	4.99	75.66
Montana	4.82	81.73
Massachusetts	4.17	67.88
Hawaii	4.08	39.38
Washington	3.79	81.36
New Jersey	3.77	71.15
Wisconsin	3.73	57.75
Florida	3.41	63.08
Michigan	3.25	47.17
Connecticut	3.20	37.53
New York	2.27	39.65
Rhode Island	2.25	36.35
Vermont	1.50	13.82
New Hampshire	1.12	13.55
Minnesota	0.64	15.27
Maine	0.45	4.72
Alaska	0.26	9.11
District of Columbia	0.00	0.00

Table 2. Numbers of resident fishermen (16 years and older) using a water type most often in the resident state, and days of fishing and expenditures associated with reservoirs during 1980. All data are in thousands.

	Number of resident fishermen		Number of fishing days		Expenditures for fishing	
State	Total[a]	Reservoirs only[b]	Total	Reservoirs only[c]	Total	Reservoirs only[c]
Alabama	736.1	273.6	19,018	7,069	259,207	96,344
Alaska	103.3	1.5[d]	2,172	32[d]	75,617	1,098[d]
Arizona	319.4	221.2	6,042	4,185	224,867	155,732
Arkansas	449.1	228.5	13,840	7,042	203,945	103,766
California	2,203.5	1,156.1	47,712	25,033	1,582,162	830,106
Colorado	610.4	272.4	10,063	4,491	237,664	106,061
Connecticut	257.1	41.7[d]	6,122	993[d]	71,920	11,665[d]
Delaware	25.1	6.6[d]	848	223[d]	6,422[d]	1,689[d]
District of Columbia	8.1[d]	0.0[d]	572[d]	0[d]	5,785[d]	0[d]
Florida	906.9	125.7	33,232	4,606[d]	615,325	85,287[d]
Georgia	1,051.5	535.1	23,803	12,113	376,896	191,799
Hawaii	12.6[d]	3.1[d]	396[d]	97[d]	3,826[d]	941[d]
Idaho	214.1	98.4	4,857	2,232	73,618	33,835
Illinois	1,146.2	517.6	36,235	16,363	696,626	314,582
Indiana	794.2	354.6	24,146	10,781	348,360	155,538
Iowa	516.2	196.4	13,061	4,969	164,700	62,664
Kansas	466.9	242.7	12,220	6,352	212,822	110,627
Kentucky	678.5	291.1	16,682	7,157	262,767	112,736
Louisiana	629.0	151.8	16,264	3,925	246,740	59,547[d]
Maine	198.6	5.1[d]	3,933	101[d]	40,945	1,051[d]
Maryland	233.1	109.2	3,752	1,758[d]	77,115	36,126[d]
Massachusetts	304.0	64.7	8,381	1,784[d]	136,240	28,996[d]
Michigan	1,518.2	215.7	40,919	5,814	594,471	84,460[d]
Minnesota	1,064.4	43.0[d]	19,506	788[d]	468,367	18,921[d]
Mississippi	473.9	160.9	12,865	4,368	146,882	49,870
Missouri	889.6	421.3	19,810	9,382	321,042	152,040
Montana	196.3	56.4	3,422	983	58,033	16,674
Nebraska	271.5	161.3	5,933	3,525	120,409	71,536
Nevada	96.2	68.3	2,140	1,520	93,297	66,239
New Hampshire	106.8	5.8[d]	2,541	138[d]	30,788	1,672[d]
New Jersey	224.4	44.6[d]	7,366	1,464[d]	138,924	27,611[d]
New Mexico	180.8	113.0	3,462	2,164	104,908	65,567
New York	1,071.4	137.3	23,316	2,988[d]	406,371	52,077[d]
North Carolina	796.2	292.8	18,888	6,946	221,467	81,444
North Dakota	133.7	64.9	2,612	1,268	52,766	25,613
Ohio	1,435.5	638.3	43,538	19,359	775,031	344,620
Oklahoma	734.9	471.0	17,903	11,474	353,818	226,763
Oregon	542.4	198.6	12,357	4,525	242,733	88,877
Pennsylvania	979.3	407.3	27,063	11,256	258,932	107,692
Rhode Island	56.4	5.6[d]	1,513	150[d]	24,452	2,428[d]
South Carolina	423.7	173.4	11,991	4,907	144,323	59,064
South Dakota	143.3	47.6	2,639	876	44,971	14,938
Tennessee	789.4	347.6	19,691	8,670	260,710	114,800
Texas	1,845.2	1,334.1	40,301	29,138	1,151,025	832,204
Utah	244.8	130.9	3,658	1,956	101,417	54,230
Total, all states	28,736.0	11,140.6	710,571	271,726	13,177,053	5,365,989

[a] Number of fishermen extrapolated from those who responded to survey question asking the kind of water used most often in respondent's resident state.

[b] Extrapolated number of fishermen using reservoirs most often in resident state.

[c] Extrapolation based upon criterion of water type used most often in resident state.

[d] Coefficients of variation for these estimates exceeded 30%.

of 1980 freshwater fishermen. Specifically, 79% of the fishermen identified in the survey answered the question concerning the type of water used most often in their resident state. Therefore, while 36.4 million freshwater fishermen were projected from the survey, only 28.7 million of these could be associated with a particular water type used most often (Table 2).

Based on the frequency distribution of water type used most often in one's state of residence, reservoirs attracted more days of use and generated more expenditures than any of the other water types. Over 38.2% of freshwater fishing days, and about 40.7% of all freshwater fishing expenditures, may be attributed to reservoir fisheries; this compares with 15.1% of days and 14.1% of expenditures for ponds; 27.2% of days and 26.7% of expenditures for natural waters; and 19.5% of days and 18.5% of expenditures for rivers and streams. In absolute terms, this means that over 11 million freshwater fishermen, aged 16 or more, would have fished reservoirs most often in their state of residence, and that nearly 272 million of the total 710.6 million days of freshwater fishing were spent fishing in reservoirs. Direct expenditures for reservoir fishing were nearly $5.4 billion.

These expenditures generated additional economic activity in 1980 through the subsequent respending of income earned by the recipients of the direct expenditures—the "multiplier effect." The dollar value of their economic impact would be well in excess of $5.4 billion if the multiplier effect were known and if the direct spending of persons under 16 years of age could be identified. Obviously, reservoirs play a dominant role with their economic contribution, relative to other water types.

Economic Values

We could estimate the economic value of reservoir angling if the dollar value of an average day of fishing in reservoirs could be estimated. That would provide a particularly meaningful measure for decisions of resource allocation and use. Unfortunately, such an estimate of value is not currently available.

Much thought has been given to how to measure the economic values of non-market resources such as a public fishery. The easiest indicators to obtain are days of use and direct expenditures by participants. These do not measure economic value, although they do reflect the importance of reservoir fishing for state and regional economies. The theoretically appropriate measure of economic value is consumer surplus. Measuring consumer surplus (economic value) for a resource such as reservoirs is data-intensive. Further, the two methods which are available for such measurements—the travel cost method and the contingent valuation method—are relatively new in the sense that their practical applications are undergoing rapid development. Each is subject to criticism in application because available estimation techniques do not necessarily conform exactly to the underlying theory; Dwyer et al. (1977) and Weithman (1986) outlined these issues. Desvousges et al. (1983) presented a comprehensive discussion of methodologies.

In the absence of a good measure of the economic value for reservoir fishing, one may obtain an estimate by combining information for days of use with estimates of average consumer surplus per user day. Loomis and Sorg (1982) surveyed the economics literature and attempted to make the various recreational values therein comparable. Their adjustments include one that converts the values in the literature to 1982 dollars. The adjusted consumer surplus for anadromous fishing in Washington, Oregon, and Idaho ranges from $25.92 to $99.02 per day, with a median value of about $54.00 per day; their adjusted consumer surplus for a day's coldwater fishing ranges from $8.58 to $67.55, with a median value of about $16.50 per day; adjusted warmwater fishing estimates of consumer surplus range from $14.85 to $26.35 per day, with a $22.70 per day median.

The work of Loomis and Sorg (1982) is based upon studies undertaken at various points in time which use varying methodologies; therefore, this information cannot be used with precision in the assignment of economic values to freshwater fisheries. However, it may be used to show a probable order of magnitude for the consumer surplus of a reservoir fishing day of about $20.00. Given that persons age 16 years and older spent about 272 million days fishing in reservoirs in 1980, and using the approximate $20.00 per day value, the projected consumer surplus from reservoir use is about $5.44 billion (in 1982 dollars).

These results may illustrate the relative importance of the economic value of reservoirs, but a more precise measure of their value awaits further research which will be based on consumer surplus measures and encompass freshwater fishing and other reservoir-associated activities. Considering the extensive needs of such research, it may be some time before improvements over the second-best measures reported here are available.

Conclusions

Available information shows that reservoirs provide more days of fishing and trigger more spending by fishermen than any other water type. Based on the criterion of water type fished most often in one's state of residence, reservoirs accounted for 38.2% of fishing days and 40.7% of fishing expenditures in 1980. For fishermen aged 16 or more, this represented about 272 million fishing days and $5.4 billion of expenditures. As might be expected, much of this activity occurs in regions where reservoirs are most readily available to fishermen.

Although a precise measurement of the economic value (consumer surplus) of reservoir fishing is not now known, a rough estimate of $5.44 billion per year for persons 16 years old and older indicates the order of magnitude of this value. Coincidentally, consumer surplus of $5.44 billion is approximately equal to the $5.4 billion of expenditures in this case. Clearly, reservoirs are a valuable resource in this country's economy for the fishing opportunities they provide.

References

Desvousges, W. H., V. K. Smith, and M. P. McGivney. 1983. A comparison of alternative approaches for estimating recreation and related benefits of water quality improvements. Research Triangle Institute, Research Triangle Park, North Carolina, USA.

Dwyer, J. F., J. R. Kelly, and M. D. Bowes. 1977. Improved procedures for valuation of the contribution of recreation to national economic development. Water Resources Center, Research Report 128, University of Illinois, Champaign-Urbana, Illinois, USA.

Loomis, J., and C. Sorg. 1982. A critical summary of empirical estimates of the values of wildlife, wilderness and general recreation related to national forest regions. Report to U.S. Forest Service, Rocky Mountain Station, Fort Collins, Colorado, USA.

USFWS (U.S. Department of the Interior, Fish and Wildlife Service, and U.S. Department of Commerce, Bureau of the Census). 1982. 1980 national survey of fishing, hunting, and wildlife-associated recreation. U.S. Government Printing Office, Washington, D.C., USA.

Weithman, A. S. 1986. Measuring the value and benefits of reservoir fisheries programs. Pages 11–17 *in* G. E. Hall and M. J. Van Den Avyle, editors. Reservoir fisheries management: strategies for the 80's. Reservoir Committee, Southern Division American Fisheries Society, Bethesda, Maryland, USA.

Pages 11–17 *in* G. E. Hall and M. J. Van Den Avyle, editors.
Reservoir Fisheries Management: Strategies for the 80's.
Reservoir Committee, Southern Division American Fisheries Society, Bethesda, Maryland. 1986.

Measuring the Value and Benefits of Reservoir Fisheries Programs

A. Stephen Weithman

Fish and Wildlife Research Center
Missouri Department of Conservation
1110 College Avenue
Columbia, Missouri 65201

Abstract

Recreational fisheries at reservoirs need to be evaluated on an economic basis so their value can be compared to other potential uses for the water and to determine the value of specific management programs. Several methods have been used to determine the value of recreational fisheries based on the quantities of gross expenditures, unit-day values, property values, market value of fish, income multipliers, and consumers' surplus. The most widely accepted techniques are the travel-cost and contingent-valuation methods to estimate consumers' surplus—the value of a fishery to national economic development. The income-multiplier method is appropriate to determine the value of a fishery to the local economy. Future research should be directed at incorporating travel time and grouping participants in the travel-cost model; determining the importance of site quality, substitute sites, and option prices or existence values of reservoirs; and developing questioning techniques for the contingent-value method.

The purpose of this paper is to identify and compare methods for determining the economic value of reservoir fisheries and associated programs. We must learn how to measure these benefits so that the value of recreational fishing can be compared with the value of alternate resource uses. Knowledge of the value of recreational fisheries is also desirable for an evaluation of specific management programs. In this way a benefit:cost ratio can be calculated, with the economic benefits being compared to management expenses.

Reservoirs provide the opportunity for a tremendous amount of recreational fishing which can result in enjoyment for anglers and dollars for local economies (USFWS 1982). However, since market prices do not exist for this type of recreation, other methods must be employed to determine the value of the resource. Although some researchers do not believe values resulting from indirect assessments are appropriate (Cauvin 1980), some estimate of value is needed for the variety of economic considerations involved in managing reservoir fisheries.

Benefits from reservoir fisheries exist in a variety of forms. Determination of the economic value of the benefits can only be accomplished by specifying those benefits of interest. For example, are we interested in the intangible benefits derived by anglers (social well-being), or dollars provided to local or state economies (regional development), or the value of the fishery to the country (national economic development)?

Several methods for determining benefits from recreational fisheries have been used based on estimates of gross expenditures, unit-day values, property values, market value of fish, income multipliers, and consumers' surplus. All of these techniques are reviewed in this paper. Some of these methods are easy to apply, but others are complex. Some are founded on accepted economic theory, whereas others are generally considered unacceptable. A brief description of the theory, origin, assumptions, advantages, and disadvantages of each technique will be provided. Problems that could be addressed with the valuation techniques, literature with examples of applications of the techniques, and research that could lead to improvements in the techniques will also be discussed.

Methods for Measuring Recreation Values

Gross-Expenditure Method

This technique is based on the assumption that total gross expenditures of anglers are an indication of the minimum value of a recreational fishery. Total gross expenditures are calculated by multiplying an estimate of days of angler effort by the average expenditures of a representative sample of anglers. Gross expenditures were frequently reported for recreational fisheries throughout the country in the 1950's and 1960's (USFWS 1966). Crutchfield (1965) discussed the limitations of this method and provided several examples.

The major advantages of this value estimate are that it is comparatively easy to calculate and easy to interpret. One must assume that the dollars actually spent represent at least the minimum value of the fishery. However, this approach is not based on sound economic theory. The major weakness is that gross expenditures underestimate the true value of a fishery because any value received by the anglers over and above their capital outlay is ignored. Also, gross expenditures for different activities, such as agriculture, are not comparable to gross expenditures for recreational fishing; net values are appropriate in this instance. Gross expenditures should only be used to monitor trends or to assist in estimating the value of a fishery to regional development when used in conjunction with income multipliers (this application will be discussed in a subsequent section).

Unit-Day Values

Unit-day values were developed to assess the value of a variety of types of recreation in a consistent fashion. Given an estimate of angler effort, the value of a fishery is calculated by multiplying days of use by a specified dollar value per day. Unit-day values were promoted by federal agencies as a means to place a value on recreational fisheries. This technique has gradually been replaced by procedures for estimating the value of fisheries relative to national economic development.

Unit-day values are easy to apply, but a number of problems arise in the evaluation. One must assume a day of fishing has a particular value within a specified range. However, the approach is not reflective of variations in recreational fishing opportunities available locally, and it is unrelated to demand for that type of recreation. No economic basis exists for the range of values assigned for a day of fishing. Although variation within the range of unit-day values is permitted, an appropriate procedure for selecting an exact value has not been outlined. Unit-day values probably underestimate the true value of fishery resources. Dwyer et al. (1977) present a complete explanation of the drawbacks of using this technique.

Property Values

Presence of a reservoir or improvements in the condition of a reservoir, such as water quality or fishing, is assumed to increase local property values with this technique. The method of examining property values to determine the value of recreational fisheries was widely used in the 1960's and 1970's. Researchers involved in the development and application of this method include Knetsch (1964), David (1968, 1969), and Darling (1973).

This approach is usually based on a "cross-section" model of land value-benefit assessment and calculation of a "rent-gradient function." Land values are assumed to decrease as the distance from a body of water increases. However, difficulties arise in before-and-after evaluations because of changes in population or in local land values due to apparent or unrelated factors. Also, overall social values remain unchanged if the increase in land values around the lake causes a decrease in other local land prices. Comparisons with other techniques or other locations can also be difficult. A complete explanation of the drawbacks of using this technique was given by Binkley and Hanemann (1976).

Market Value of Fish

This technique is based on the assumption that the market value of the fish present should be an indication of the value of a recreational fishery. Given an estimate of the species, size, and number of fish available to anglers and a price per fish (either market value or replacement cost), the total value of the fishery can be calculated. Crutchfield (1962) originally equated the market value of the fish with the value of commercial fisheries and discussed the possible use of this technique as a minimum value for recreational fisheries. A similar type of valuation, based on replacement costs for fish, is frequently used to assess losses due to fish kills (Monetary Values of Freshwater Fish Committee, American Fisheries Society 1982).

This approach is easy to apply if a representative sample of fish can be collected. The market value or replacement cost of fish is taken to represent at least the minimum value of the fishery, because the costs are representative of prices one would have to pay to buy the fish in stores or at hatcheries. This technique takes into account that fish have value as a source of food; however, it does not take into account the value of the recreational aspects of sport fishing. Hence, the true value of the recreational fishery would be underestimated.

Income-Multiplier Method

The income-multiplier method was developed to determine the economic value of recreational fisheries on a regional basis. Given an estimate of total gross expenditures by anglers, the value of the fishery can be calculated by multiplying these values by appropriate input-output multipliers. State agencies are frequently interested in the effects of a recreational fishery on the local or regional economy. The income-multiplier method has been used to determine the value of a salmonid fishery in New York (Brown 1976), the effect of non-resident sportsmen in Vermont (Gilbert and Khayami 1973), the value of expenditures by hunters and fishermen in Colorado (Rohdy and Lovegrove 1970, as cited by Gilbert and Khayami 1973), and the value of a trout fishery in Missouri (Weithman and Haas 1982).

This approach is appropriate to determine the local or regional economic impacts of a recreational fishery. Some of the money spent in the area is sent outside the region to pay for goods and taxes, while that which remains continues to circulate (Ives and Youmans 1978). The effects of angler expenditures are studied with respect to income, employment, and taxes in the region of interest. The income-multiplier technique is not appropriate if an increase in visitation to one recreational fishery is the result of a decrease in participation at another fishery in the region.

Consumers' Surplus

The most widely recommended procedures for determining the value of recreational fisheries are based on estimates of consumers' surplus, which is the value of benefits that anglers receive from fishing in excess of their expenses. The two most suitable techniques are the travel-cost and contingent-valuation methods (Norling 1968; Binkley and Hanemann 1976; Dwyer et al. 1977). Both approaches are economically sound techniques suitable for determining the value of recreational fisheries to national economic development.

The travel-cost and contingent-value methods allow a more complete and accurate analysis of value and benefits provided by reservoir fisheries than the other methods reviewed. They take into account shifts in demand for reservoir recreation due to changes in the quantity and quality of recreation provided. This type of evaluation is the best measure of the value of management programs. At the same time, an economic evaluation allows comparisons with other fisheries, or other activities which compete with fisheries, for limited resources.

Both methods are based on the relationship between demand for and price of a particular fishery to anglers. The travel-cost method depends on observed days of fishing and actual angler expenses; the contingent-value method is based on an estimate of days fished and angler response to hypothetical questions about the value of the fishery.

Two important assumptions are involved in applying these methods. The travel-cost method depends on the assumption that anglers would respond the same to a fee for fishing as they would to an increase in their travel expenses; the basic assumption for the contingent-value method is that an anglers' response to a hypothetical question about the value of a fishery would actually fit his behavior in a real-market situation.

Several sources of information are available which outline the methodology to employ the travel-cost and contingent-value techniques. Early work was done by Hotelling (1949), Trice and Wood (1958), Clawson (1959), and Knetsch (1963) with the travel-cost method, and Davis (1963), Knetsch and Davis (1966), and Brown and Hammack (1972) with the contingent-value method.

The two methods have been used repeatedly to evaluate benefits from recreational fisheries and other types of outdoor recreation. Burt and Brewer (1974) tried to predict recreation benefits at a proposed reservoir in Missouri with a travel-cost model. Knetsch et al. (1976) developed a travel-cost model of use and benefits for a number of U.S. Army Corps of Engineers' reservoirs in California. Ullman and Volk (1962), Clawson and Knetsch (1966), Merewitz (1966), Stevens (1966), Burt and Brewer (1971), Michalson and Findeis (1975), Cesario and Knetsch (1976), Cicchetti et al. (1976), and Edwards et al. (1976) have applied travel-cost models to other outdoor recreation evaluations. The contingent-value method was used by Oster (1977) to determine willingness-to-pay to clean up the Merrimack River. Walsh et al. (1978) used the contingent-value approach to determine recreation benefits from high mountain reservoirs in Colorado. Horvath (1974) used this method to estimate the value of recreation benefits in the southeastern United States, and Hammack and Brown (1974) tried to determine benefits derived from a full season of duck hunting based on estimates of net willingness-to-pay and willingness-to-sell.

Research Needs and Future Direction

Use of the travel-cost and contingent-value methods should be emphasized in future valuation of reservoir fisheries. Studies already completed provide good examples for future applications, but there is room for improvement. Some areas currently under investigation include: incorporating travel time and grouping participants in the travel-cost model; determining the importance of site quality, substitute sites, and option prices or existence values of reservoirs; and questioning techniques for contingent valuation.

Incorporation of travel time into travel-cost models is an important consideration because this time is considered an expense of the trip (Cesario and Knetsch 1970; Wilman 1980). However, since travel time is usually correlated with distance traveled or total expense, inclusion in the model has been difficult (Johnston 1972; Dwyer et al. 1977). Travel time is taken into account most often by assuming a tradeoff between money and time based on some function of the average wage rate (Cesario 1976). McConnell and Strand (1981) developed a method to set the cost of travel time using an independent estimate of travel time and round-trip distance to the site. Two problems are apparent: (1) travel time might not be a liability in every instance, because some people derive benefits during the trip by sight-seeing; and (2) if trips for which the site of interest is not the only destination are included in the valuation, site benefits will be overestimated. Few attempts have been made to separate benefits provided by primary and secondary destinations (Dwyer et al. 1977).

The travel-cost model uses data on distance traveled from an angler's origin to the site under evaluation to produce a demand curve. Grouping fishermen into categories for consideration and comparison based on origin has become an art. Binkley and Hanemann (1976) recommended categorizing people according to the areas within concentric rings surrounding the site being evaluated. Merewitz (1966) and Palm and Malvestuto (1983), however, separated participants by county of origin, while Edwards et al. (1976) used geographic areas with homogeneous populations to make their evaluation. Another approach has been to consider individual responses and ignore aggregation over some area (Brown and Nawas 1973). This part of the evaluation is a matter of choice for individual researchers, but each type of grouping can provide valid estimates.

Site quality, including factors such as congestion, aesthetics, and fishing quality, can be an important variable for consideration in predicting visitation by fishermen. Much research has been conducted on the effects of congestion (Baron and Shecter 1973; Anderson and Bonsor 1974). Overcrowding is usually not a problem at reservoir fisheries, with the possible exception of sites near metropolitan areas. Cesario (1969) approached the question of site quality by observing actual behavior of recreationists. Fishing quality might be the most important site-quality variable, yet little consideration has been given to this topic. Stevens (1966) included catch rates as an indication of fishing quality. The overall quality of fishing index proposed by Weithman and Katti (1979) could be used as one indication of site quality. Further investigation of these variables is needed.

Visitation to one reservoir is affected by availability of other nearby fishing sites. Substitute sites can play a major role in the development of realistic models to predict reservoir visitation or benefits. Burt and Brewer (1974) developed a multiple-site model to evaluate benefits from a proposed reservoir in Missouri. Knetsch et al. (1976) considered the economic interactions of several reservoirs in California. Brown and Hansen (1974) also discuss a model with multiple sites. If the site being evaluated is unique, substitute sites are not a factor. Additional research should be directed at the comparability of a variety of sites for recreational fishing.

Option prices or existence values of a variety of natural resources or sites for recreational activity have been considered. These values also are accrued by non-participants solely from knowing a particular opportunity exists. Greenley et al. (1981) attempted to measure the option value of maintaining the water quality of the South Platte River Basin in Colorado. Brookshire and Randall (1978) estimated activity, option, and existence values for hunting and preserving wildlife resources in Wyoming. However, option prices or existence values of recreational fisheries at reservoirs have not been reported. Estimates of these values could be important in the application of the contingent-value methodology and therefore deserve further attention.

Many potential problems have been identified in regard to the questioning procedure or design of contingent-value surveys. If respondents cannot relate the questions to real-world situations, the survey could be biased (Ajzen and Fishbein 1977). Strategic bias is another problem. Survey results could be affected if people think they might benefit by giving a particular answer (Samuelson 1954), but evidence of strategic behavior usually has not been found (Brookshire et al. 1976; Rowe et al. 1980). Another area of concern is that the method selected for a hypothetical payment in the survey might be objectionable to a segment of the respondents. Care must also be taken not to infer that an actual fee for entry could change. In a bidding-game question to determine willingness-to-pay, a bias can be introduced by selecting a given starting point for dollars an angler is willing to spend to participate (Brookshire et al. 1979; Rowe et al. 1980). Thayer (1981) has developed a test and adjustment for this problem (starting-point bias).

The potential benefits from determining the economic value of recreational fishing at reservoirs are numerous. The value of these fisheries could be compared to other fisheries or other activities which compete for the use of water. As biologists, we need to overcome our hesitancy to place a dollar value on recreational fisheries for fear of being outbid. Reservoir fisheries yield dramatic economic values, such as the figure of $69 million reported for the Lake Mead fishery by Martin et al. (1982). Also, specific management programs could be evaluated on a more sound, economic basis. Benefits provided by an improvement in the quality of fishing could be translated into a dollar value—the value to anglers, national economic development, or local development. Development of comprehensive state-wide models based on the travel-cost and contingent-value methods will be instrumental in evaluating recreational fisheries and fisheries programs at reservoirs in the future.

Acknowledgments

I appreciate the assistance Glenn Covington provided in locating most of the literature for this paper.

References

Ajzen, I., and M. Fishbein. 1977. Attitude-behavior relations: a theoretical analysis and review of empirical research. Psychological Bulletin 84:888–918.

Anderson, F. J., and N. C. Bonsor. 1974. Allocation, congestion, and the valuation of recreational resources. Land Economics 50:51–57.

Baron, M., and M. Shecter. 1973. Simultaneous determination of visits to a system of outdoor recreation parks with capacity limitations. Regional and Urban Economics 3:327–359.

Binkley, C. S., and W. M. Hanemann. 1976. The recreation benefits of water quality improvements: analysis of day trips in an urban setting. U.S. Environmental Protection Agency PB-257719, National Technical Information Service, Springfield, Virginia, USA.

Brookshire, D. S., B. S. Ives, and W. D. Schulze. 1976. The valuation of aesthetic preferences. Journal of Environmental Economics and Management 3:325–346.

Brookshire, D. S., and A. Randall. 1978. Public policy alternatives, public goods, and contingent valuation mechanisms. Proceedings of the Western Economic Association 53:20–26.

Brookshire, D. S., R. C. d'Arge, W. D. Schulze, and M. A. Thayer. 1979. Methods development for assessing tradeoffs in environmental management. U.S. Environmental Protection Agency 600/6-79-0016, Washington, D.C., USA.

Brown, T. L. 1976. The 1973–75 salmon runs: New York's Salmon River sport fishery, angler activity, and economic impact. New York Sea Grant Institute NYSSGP-RS-76-025, Cornell University, Ithaca, New York, USA.

Brown, G. M., and J. Hammack. 1972. A preliminary investigation of the economics of migratory waterfowl. Pages 171–204 *in* J. V. Krutilla, editor. Natural environments studies in theoretical and applied analysis. The Johns Hopkins University Press, Baltimore, Maryland, USA.

Brown, R. E., and W. J. Hansen. 1974. A preliminary analysis of day use recreation and benefit estimation models for selected reservoirs. U.S. Army Engineer Institute for Water Resources, Fort Belvoir, Virginia, USA.

Brown, W. G., and F. Nawas. 1973. Impact of aggregation on the estimation of outdoor recreation demand functions. American Journal of Agricultural Economics 55:246–249.

Burt, O. R., and D. Brewer. 1971. Estimation of net social benefits from outdoor recreation. Econometrica 39:813–827.

BURT, O. R., AND D. BREWER. 1974. Evaluation of recreational benefits associated with the Pattonsburg Dam Reservoir. Office of Industrial Development Studies, University of Missouri, Columbia, Missouri, USA.

CAUVIN, D. 1980. The valuation of recreational fisheries. Canadian Journal of Fisheries and Aquatic Sciences 37:1321–1327.

CESARIO, F. J. 1969. Operations research in outdoor recreation. Journal of Leisure Research 1:33–51.

CESARIO, F. J. 1976. Value of time in recreation benefit studies. Land Economics 52:32–41.

CESARIO, F. J., AND J. L. KNETSCH. 1970. Time bias in recreation benefit estimates. Water Resources Research 6:700–704.

CESARIO, F. J., AND J. L. KNETSCH. 1976. A recreation site demand and benefit estimation model. Regional Studies 10:97–104.

CICCHETTI, C. J., A. C. FISHER, AND V. K. SMITH. 1976. An economic evaluation of a generalized consumer surplus measure: the Mineral King controversy. Econometrica 44:1259–1276.

CLAWSON, M. 1959. Methods of measuring the demand for and value of outdoor recreation. Reprint Number 10, Resources for the Future, Washington, D.C., USA.

CLAWSON, M., AND J. L. KNETSCH. 1966. Economics of outdoor recreation. The Johns Hopkins University Press, Baltimore, Maryland, USA.

CRUTCHFIELD, J. A. 1962. Valuation of fishery resources. Land Economics 38:145–154.

CRUTCHFIELD, J. A. 1965. Can we put an economic value on fish and wildlife? Colorado Outdoors 14: 1–5.

DARLING, A. H. 1973. Measuring the benefits generated by urban water parks. Land Economics 49: 327–338.

DAVID, E. 1968. Lakeshore property values: a guide to public investment in recreation. Water Resources Research 4:697–707.

DAVID, E. 1969. The exploding demand for recreational property. Land Economics 45:206–217.

DAVIS, R. K. 1963. Recreation planning as an economic problem. Natural Resources Journal 3:239–249.

DWYER, J. F., J. R. KELLY, AND M. D. BOWES. 1977. Improved procedures for valuation of the contribution of recreation to national economic development. Water Resources Center, Research Report 128, University of Illinois, Champaign-Urbana, Illinois, USA.

EDWARDS, J. A., K. C. GIBBS, L. J. GUEDRY, AND H. H. STOEVENER. 1976. The demand for non-unique outdoor recreational services: methodological issues. Oregon State University Agricultural Experiment Station Technical Bulletin 133, Corvallis, Oregon, USA.

GILBERT, A. H., AND S. M. KHAYAMI. 1973. Expenditure patterns of nonresident sportsmen in Vermont, 1970. University of Vermont Agricultural Experiment Station Research Report MP78, Burlington, Vermont, USA.

GREENLEY, D. A., R. B. WALSH, AND R. A. YOUNG. 1981. Option value: empirical evidence from a case study of recreation and water quality. Quarterly Journal of Economics 96:657–674.

HAMMACK, J., AND G. M. BROWN. 1974. Waterfowl and wetlands: toward bioeconomic analysis. The Johns Hopkins University Press, Baltimore, Maryland, USA.

HORVATH, J. C. 1974. Economic survey of southeastern wildlife and wildlife-oriented recreation. Transactions of the North American Wildlife Conference 39:187–194.

HOTELLING, H. 1949. A letter reply to a United States National Park Service request. *In* R. A. Prewitt, editor. The economics of public recreation: an economic study of the monetary evaluation of recreation in the national parks. National Park Service, U.S. Department of the Interior, Land and Recreational Planning Division, Washington, D.C., USA.

IVES, E., AND R. YOUMANS. 1978. Business interrelationships of the Tillamook County economy: a study for analysis of economic change. Oregon State University Agricultural Experiment Station Circular of Information 672, Corvallis, Oregon, USA.

JOHNSTON, J. 1972. Econometric methods. McGraw Hill, New York, New York, USA.

KNETSCH, J. L. 1963. Outdoor recreation demands and benefits. Land Economics 39:387–396.

KNETSCH, J. L. 1964. Economics of including recreation as a purpose of eastern water projects. Journal of Farm Economics 46:1148–1157.

KNETSCH, J. L., AND R. K. DAVIS. 1966. Comparisons of methods for recreation evaluation. Pages 125–142 *in* A. V. Kneese and S. C. Smith, editors. Water Research. The Johns Hopkins University Press, Baltimore, Maryland, USA.

KNETSCH, J. L., R. E. BROWN, AND W. J. HANSEN. 1976. Estimating expected use and value of recreation sites. Pages 103–115 *in* C. E. Gearing, W. W. Swart, and T. Var, editors. Planning for tourism development: quantitative approaches. Praeger Publishers, New York, New York, USA.

MARTIN, W. E., F. H. BOLLMAN, AND R. L. GUM. 1982. Economic value of the Lake Mead fishery. Fisheries (Bethesda, Maryland) 7(6):20–24.

MCCONNELL, K. E., AND I. STRAND. 1981. Measuring the cost of time in recreation demand analysis: an application to sport fishing. American Journal of Agricultural Economics 63:153–156.

MEREWITZ, L. 1966. Recreational benefits of water resource development. Water Resources Research 2:625–640.

MICHALSON, E. L., AND J. FINDEIS. 1975. Economic impact of mountain pine beetle on outdoor recreation. Southern Journal of Agricultural Economics 7:43–50.

MONETARY VALUES OF FRESHWATER FISH COMMITTEE,

American Fisheries Society. 1982. Monetary values of freshwater fish. American Fisheries Society Special Publication 13:1–26.

Norling, I. 1968. Economic evaluation of inland sport fishing. European Inland Fisheries Advisory Commission Technical Paper Number 7, Rome, Italy.

Oster, S. 1977. Survey results on the benefit of water pollution abatement in the Merrimack River basin. Water Resources Research 13:882–884.

Palm, R. C., and S. P. Malvestuto. 1983. Relationships between economic benefit and sportfishing effort on West Point Reservoir, Alabama-Georgia. Transactions of the American Fisheries Society 112:71–78.

Rohdy, D. D., and R. E. Lovegrove. 1970. Economic impact of hunting and fishing expenditures in Grand County, Colorado. Colorado Experiment Station GS916.

Rowe, R. D., R. C. d'Arge, and D. S. Brookshire. 1980. An experiment on the economic value of visibility. Journal of Environmental Economics and Management 7:1–19.

Samuelson, P. 1954. The pure theory of public expenditure. Review of Economics and Statistics 36: 387–389.

Stevens, J. B. 1966. Recreation benefits from water pollution control. Water Resources Research 2: 167–181.

Thayer, M. A. 1981. Contingent valuation techniques for assessing environmental impacts: further evidence. Journal of Environmental Economics and Management 8:27–44.

Trice, A. H., and S. E. Wood. 1958. Measurement of recreation benefits. Land Economics 34:195–207.

Ullman, E. L., and D. J. Volk. 1962. An operational model for predicting reservoir attendance and benefits. Papers of the Michigan Academy of Science, Arts, and Letters 47:473–484.

USFWS (U.S. Department of Interior, Fish and Wildlife Service). 1966. 1965 national survey of hunting and fishing. U.S. Fish and Wildlife Service, Washington, D.C., USA.

USFWS (U.S. Department of Interior, Fish and Wildlife Service, and U.S. Department of Commerce, Bureau of the Census). 1982. 1980 national survey of fishing, hunting, and wildlife-associated recreation. U.S. Government Printing Office, Washington, D.C., USA.

Walsh, G., D. A. Greenley, R. A. Young, J. R. McKean, and A. A. Prato. 1978. Option values, preservation values, and recreational benefits of improved water quality. U.S. Environmental Protection Agency EPA-600/5-78-001, Washington, D.C., USA.

Weithman, A. S., and M. A. Haas. 1982. Socioeconomic value of the trout fishery in Lake Taneycomo, Missouri. Transactions of the American Fisheries Society 111:223–230.

Weithman, A. S., and S. K. Katti. 1979. Testing of fishing quality indices. Transactions of the American Fisheries Society 108:320–325.

Wilman, E. A. 1980. The value of time in recreation benefit studies. Journal of Environmental Economics and Management 7:272–286.

Pages 18–27 *in* G. E. Hall and M. J. Van Den Avyle, editors.
Reservoir Fisheries Management: Strategies for the 80's.
Reservoir Committee, Southern Division American Fisheries Society, Bethesda, Maryland. 1986.

A Simulation Model for Managing Fisheries in Reservoirs on the Rio Grande of New Mexico

RICHARD COLE

Department of Fishery and Wildlife Sciences
New Mexico State University
Las Cruces, New Mexico 88003

TIMOTHY WARD

Department of Civil Engineering
New Mexico State University
Las Cruces, New Mexico 88003

FRANK WARD

Department of Agricultural Economics and Agricultural Business
New Mexico State University
Las Cruces, New Mexico 88003

ROBERT DEITNER

Department of Fishery and Wildlife Sciences
New Mexico State University
Las Cruces, New Mexico 88003

ABSTRACT

A mathematical model of the upper Rio Grande system above El Paso, Texas is being constructed to predict biological and economic consequences of alternatives for managing river flow through six reservoirs and their connecting waters. The hydrologic component of the model simulates inputs of water, suspended sediments, and limiting plant nutrients to reservoirs which, in turn, alter reservoir water level, depth of light penetration, and concentration of limiting nutrients. The biological component uses the simulated fluctuations of water levels and material loadings as well as meterological data, including solar radiation, to determine flows of energy and materials through aquatic plants, intermediate consumers, and fish. Fish production is partitioned into groups that feed primarily on zooplankton, zoobenthos, and fish, and occupy five different habitats based on levels of illumination. The economic component requires predicted estimates of fish production and water-level changes as well as user travel costs, income, fishing success rates, and population distribution. A recreational use-projection and dollar-benefit prediction model has been developed, which includes an empirically determined relationship between fish yield, fishing effort, and economic benefits. Predictive capabilities, applications, and limitations of the model are also discussed.

Fisheries management strategies for the 80's in reservoirs of New Mexico must recognize that water rights in the Rio Grande basin are property rights. Virtually all of the surface water in the basin is appropriated, and without purchase to retire existing rights, development or use of a particular water right must not interfere with any other appropriation. Within this structure, appropriated water in Rio Grande reservoirs may be used for recreation, including sport fishing, without cost to the public as long as it in no way diminishes the value to water owners. Since modification of any existing water storage and delivery scheme may change evaporative losses, and, therefore, water consumption, changes requested for recreational benefit must be accompanied by evidence that downstream water values will remain intact or improve. Of course, water may be purchased and managed specifically for recreational use by private or public interests after recreational value and appropriate price are established.

Economic valuation of water depends on natural and anthropogenic factors. Annual fluctuations in surface-water production are large and random and therefore predictable only in terms of mean and extreme production expected over decades. The Rio Grande consistently receives 80 to 90% of its water from snowmelt, and water demand for irrigation and other uses downstream is predictable; therefore, changes in the total water volume available for storage may be predicted. The exact timing of changes in water volumes and levels is less predictable than the total volume because of the uncertainty of the spring weather that controls snowmelt. Superimposed on natural fluctuations are the effects of compacts, treaties, and other management obligations in the six mainstream reservoirs of New Mexico.

A survey of four of the six mainstream reservoirs in 1981 showed that fishing was the single most important recreational use, 38% of total recreational activity. Fisheries vary from predominantly cold-water at the higher reservoirs to warm-water at the lower reservoirs. Projected yields of fish based on two empirical models, the morphoedaphic index (Ryder 1965) and the growing-season index (Leidy and Jenkins 1977), indicate that fish yield should increase substantially from the highest elevations to the lowest elevations. Most recreational users of the reservoirs are concentrated in two areas: centrally near Albuquerque, New Mexico, and near the lower end of the system at El Paso, Texas (Cole et al. 1983). To help contend with such habitat instability and spatial variation in production and resource use, the New Mexico Department of Game and Fish sought a predictive tool for evaluating the effects of water-management alternatives on the distribution, stability, and quality of reservoir fisheries and the cost effectiveness associated with their management. Mathematical simulation of the Rio Grande basin was thought to be the most flexible approach and therefore most useful because of the wide variety of potential applications.

In 1980, an interdisciplinary team of engineers, biologists, and economists was assembled at New Mexico State University to begin development of a model to simulate 1) the movement of water, energy, and other material through Rio Grande reservoirs, and 2) reservoir fish production and its associated economic value. Initially funded by the New Mexico Department of Game and Fish and the New Mexico Water Resources Research Institute, model development has progressed about half way toward completion in 1985. Upon completion in 1985, the sponsors hope that recreational value of various water-management strategies can be contrasted with values of other uses to optimize management and overall economic benefit from all water consumed in reservoirs of the Rio Grande basin of New Mexico. This paper describes major aspects of model formulation, predictive capabilities, and possible applications.

The Model

The Rio Grande basin model is being constructed from three subunits which simulate hydrologic, biologic, and economic processes (Fig. 1). The hydrologic component processes data on discharge of water, phosphorus, nitrogen, suspended matter, and heat to predict time-related outputs of water volume, surface area, water elevation, water temperature, and concentrations of phosphorus, nitrogen, and suspended solids in the reservoirs. Other submodels (Lorenzen et al. 1972) link loading rates to sedimentation, and predict mean concentrations of suspended matter and nutrients for each reservoir. One important requirement is the need to rely on readily available data. Therefore, the model uses physical-chemical inputs measured at stations operated by the U.S. Geologic Survey and the National Oceanographic and Atmospheric Administration.

Predicted changes in water quality and quantity serve as inputs for the biologic component which also requires climatic inputs, particularly solar radiation, derived from routinely monitored weather stations. From these inputs, we predict fish production or yield with a relatively simple process model developed by us or through a more complex model developed during the International Biological Program (Scavia and Park 1976). With estimates of yield per unit of fishing effort, fluctuations in production and yield are predicted from the variations in water and material generated by the hydrologic component.

Inputs required for the economic model include fish production from the biologic model, water fluctuation from the hydrologic model, and fishing success, travel costs, user incomes, and population distributions based on economic surveys. From these inputs, recreational values associated with fishing and other water-based activities are generated to serve management agencies and user groups.

Simulation of flow through the Rio Grande can address long- or short-term modeling needs.

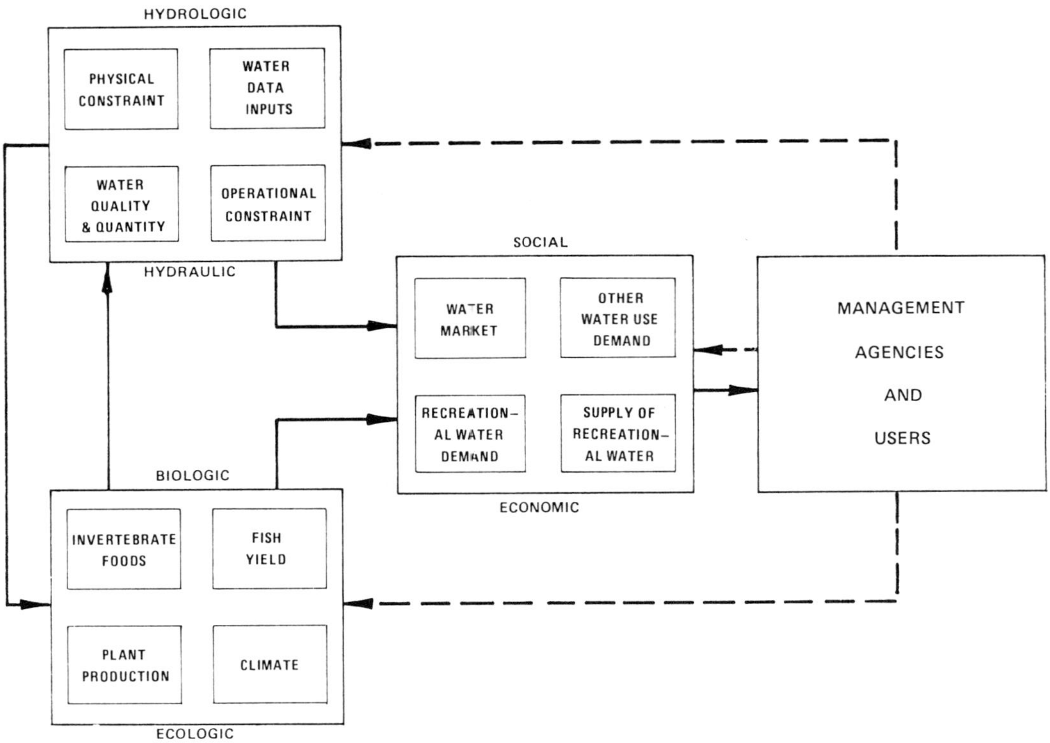

Figure 1. Overview of model subunits illustrating the primary interactions. Hydrologic submodels generate inputs to the biologic submodel with feedback effects on water quality generated by primary productivity. Fish production and yield as well as water quality outputs provide input for economic submodels. Values generated by the economic submodel provide input for management decisions which can, in turn, influence the system's structure.

Over the long-term (decades), stochastically generated water and material fluctuation will be used to predict mean and extreme conditions. For the short-term (months), a deterministic approach based on predictions of water and material changes from snow-pack information is used to answer more immediate questions, such as when or how to stock fish or regulate harvest.

Hydrologic Component

We have incorporated different temporal and spatial resolutions into the hydrologic component based on need and availability of data. The number of locations monitored along the Rio Grande is limited (Fig. 2), but the sites adequately represent a larger associated segment of the river system. Each of those hydrologic segments can be treated separately from the other segments in the system or collectively with other segments as a complete system.

The temporal resolution available and chosen for water discharge is usually more precise than for loadings of nutrients and suspended solids. Since water discharge and reservoir content are monitored daily, the short-term effects of water-level fluctuation can be readily incorporated in the model. On the other hand, data on nutrients and suspended solids usually are obtained less frequently. Therefore, the periods of time the model can realistically be expected to predict biological production based on those inputs will be limited more by data availability than the effects of water-level fluctuation on, for example, spawning success. Biological production is estimated seasonally while rates of water-level stranding of eggs and larvae may be estimated daily. Wherever data are available at a frequency greater than needed, the model builds to the required time period by summation of the shorter-term data.

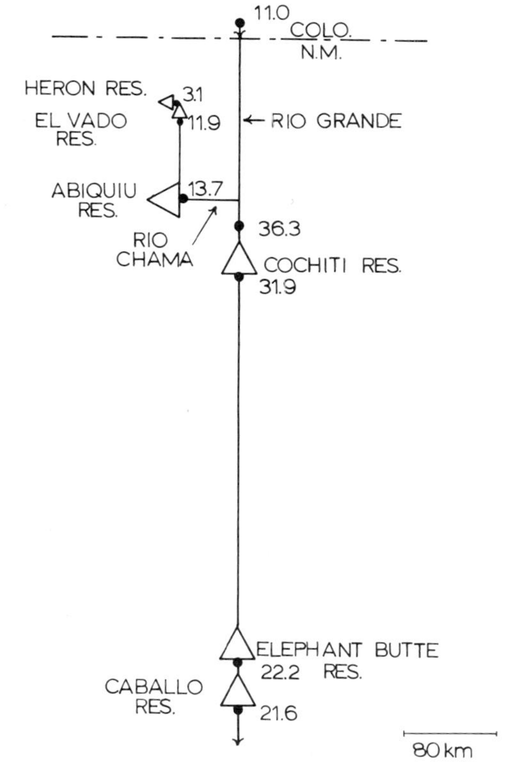

Figure 2. A schematic map of reservoir and hydrologic monitoring (black dots) locations in the Rio Grande basin of New Mexico. Reservoirs are not to scale. The mean annual discharge rate (m³/second) is given for each monitoring station.

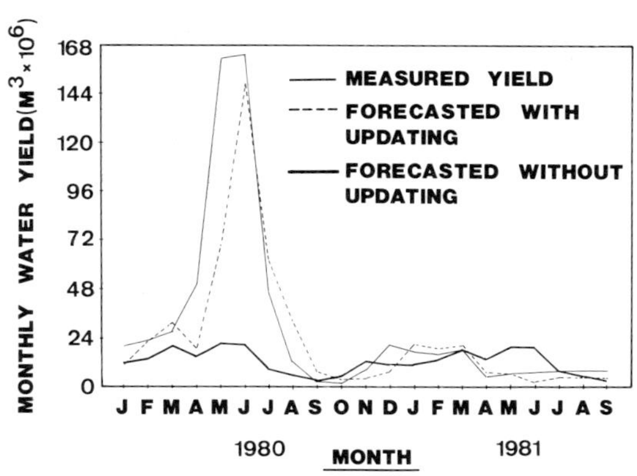

Figure 3. A comparison of measured water yield with forecasts of models which include or disregard updating of yield estimations with the most recent monthly yield measurements.

For longer periods (weeks or a season), an inflow-outflow, mass-balance approach is used based on USACE (1974), but for shorter time steps of 1 hr up to 1 day we have developed a flow-routing approach. We are currently using Muskingum-type routing models (e.g., Linsley et al. 1982) to simulate flows in connecting waters. Because water inputs are stochastic, a series of randomly generated inputs have been developed to analyze long-term projections. Flow inputs in the Rio Grande have been greatly modified during recent decades by altered withdrawal practices, interbasin water-transfer, and operation of Cochiti Reservoir, which has significantly altered the river downstream since it was completed in 1975. Since the long-term records incorporate a history no longer applicable to the Rio Grande system, we have relied on more recent records, usually a decade or less in length, for predicting flows.

For short-term predictions, we are developing an approach that uses stochastic prediction of monthly yields up to 12 months from time-series coefficients and water yield during the previous month. A comparison between the measured time series and updated (previous flow) time series in Fig. 3 shows that updating flows is more appropriate than a static estimate based on the long-term record.

The hydrologic component is structured to accept data on river flow, reservoir volume, and climatological characteristics including daily rainfall, wind, and pan evaporation. This component also uses empirical regression equations to estimate loadings of sediment and nutrients which serve as input for the biological component. Future modifications of the model will include a refined routine for simulating reservoir water chemistry, detailed simulation of flow routing between reservoirs, and incorporation of intermittent contributions from tributary streams.

Biologic Component

Our biologic component is intermediate in complexity between the simple morphoedaphic index (Ryder 1965) and more data-demanding ecosystem models (Scavia and Park 1976; Scavia 1979). The morphoedaphic model predicts annual yield of all fish in aggregate while the ecosystem models predict instantaneous production of three generalized fish groups. Our model pre-

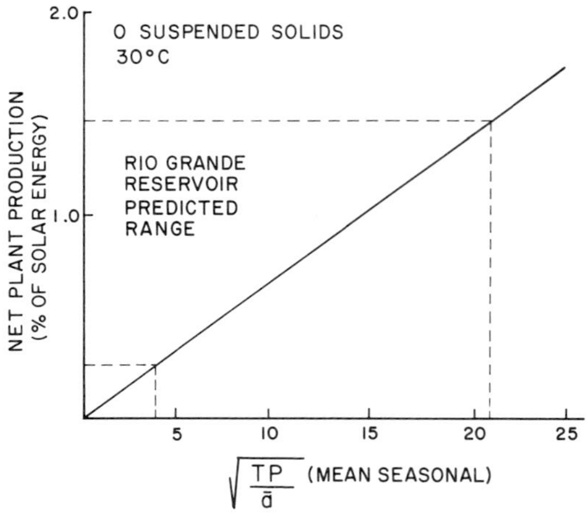

Figure 4. Modeled relationship between a modified morphoedaphic model (TP = total phosphorus and $\bar{d}$ = mean depth) and the percentage of seasonal solar energy converted to net plant production when water transparency and temperature are optimum for plant conversion of solar energy.

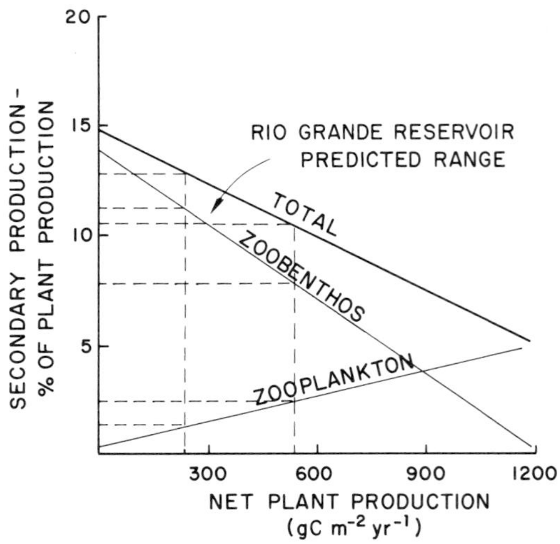

Figure 5. Modeled relationships of the percentages of secondary herbivorous production converted from net plant production.

dicts seasonal production of up to 11 groups of fish based on estimated primary production, conversion efficiencies of intermediate trophic groups, and resource partitioning of food and habitat.

Basic assumptions of the model are that the maximum conversion of light energy to plant net production occurs at an optimum combination of light, nutrient, and temperature, and that conversion efficiency of light by plants decreases as environmental conditions vary from optimum. Based on reported efficiencies from nutrient-rich lakes in the tropics (Westlake et al. 1980), we have chosen as optimum a 2% conversion efficiency of light by phytoplankton at 30 C and mean concentrations of total phosphorus exceeding 0.75 mg/liter in shallow, well-mixed water with no inorganic suspended solids. Total nutrient, rather than dissolved nutrient, was used because it is theoretically a better predictor of long-term production (Golterman and Kouwe 1980). Phosphorus is considered limiting when the N:P ratio is more than 8.0; nitrogen is limiting otherwise.

We incorporated the effect of nutrient limitation, represented by the mean seasonal concentration of phosphorus or nitrogen depending on which was indicated to be limiting, in the form of the morphoedaphic expression shown in Fig. 4. A first approximation of the relationship between the morphoedaphic expression and the light conversion efficiency of phytoplankton has been derived from published data reported in Likens (1975) and unpublished data collected from New Mexico reservoirs. This variable alone is likely to generate a range of 6 or 7 to 1 in phytoplankton production among the Rio Grande reservoirs over several years of water-level fluctuation. We have included a modifying temperature term (Q_{10} of 1.5) based on research in New Mexico and similar relationships found elsewhere (Westlake et al. 1980). Absorption and reflection of light by suspended matter is a third term included that varies particularly at times of high loading of suspended solids when low-water years are followed by high spring runoff. Turbidity generated by wind erosion in these reservoirs appears to be much less important and is not included in the model.

The prediction of net primary production, as described above, serves as input for estimating herbivore production as zooplankton and zoobenthos. Blazka et al. (1980) and Parsons (1980) estimated that about 15% of the primary production is converted to secondary production in nutrient-poor fresh and marine waters, and declines to about 5% in the most productive waters. Parsons (1980) indicated that in marine environments the conversion efficiency progressively declines as primary production increases, suggesting a relationship as shown in Fig. 5. Hillbricht-Ilkowska et al. (1972) thought efficiencies in lakes declined based on evidence from European lakes. Work by Makarewicz and Likens (1979) on zooplankton, and Hargrave (1973) on benthos provide evidence for the direction of trends, although the actual relationships shown in Fig. 5 are speculative.

From our structure, an estimate of total non-piscivorous fish production is made based on conversion efficiencies at different levels of net primary productivity varying between 5% and 15% of the herbivore production. One end result is that benthos feeders will be more efficient in converting food to fish production than planktovores. Although zooplanktonic and zoobenthic predators undoubtedly exist, for purposes of this initial model we assume their role is negligible. Production of fishes that feed on plankton and benthos are used to calculate the total production of piscivorous fish which are also assumed to have a conversion efficiency between 5% and 15% as primary production increases. A habitat partitioning coefficient is superimposed on the food partitioning coefficient to develop more specific categories of fish production. Reservoirs in the Rio Grande basin are usually well-mixed, therefore, illumination is a better measure of habitat diversification than temperature. Reservoirs are divided into five onshore and offshore habitats defined by light intensity. Detailed equations and methodology are given in Cole et al. (1983).

Once the volumes of habitats are determined for a reservoir from predictions of light penetration from loadings of suspended solids, the relative production of fish that feed on zooplankton, zoobenthos, and other fish can be estimated for each habitat in proportion to the volumes or areas of habitat present. We assume that the overall production of the reservoir is distributed evenly. Since zooplankton are prey mostly for sight-feeders in illuminated waters, we assume none of them are consumed in the profundal zone. Also, since the littoral and limnetic are illuminated longer than the sublittoral and the pelagic habitats, we assume that proportionally more zooplankton will be eaten in the surface habitats. Piscivorous fish are also divided up among all those habitats which are sometimes illuminated. Zoobenthos are consumed only in the littoral, sublittoral, and profundal zones where bottom habitat occurs.

Although our model now enables generalized predictions of fish production from routinely monitored inputs associated with seasonal and annual changes in water levels, light, and nutrients, management agencies in the Rio Grande basin need a model which is more sensitive to the specific effects of water-level fluctuation on recruitment and growth of fish, particularly in the diverse littoral habitat. We are presently constructing submodels which predict partitioning of resources, recruitment, growth, and yield of individual species. Particularly important in the Rio Grande reservoirs is the influence of water-level fluctuation on reproductive success, our present modeling focus.

Economic Component

We chose the travel-cost method (Gum and Martin 1975; Ward 1982) to determine the recreational benefits of the site and improvements in site quality (such as fish caught per unit of effort) because of the non-market nature of outdoor recreational resources provided by the public sector. Travel costs included both the money and time spent traveling to the site. Water-level fluctuations and angling success rates per unit of effort derived from hydrologic and biologic components of the model also serve as inputs to the economic component.

Per capita user-demand schedules were estimated statistically from data collected in 1981 at 14 study sites in the Rio Grande basin of New Mexico. The purpose of this pilot study was to develop the modeling approach and determine sampling intensity needed for future research. These sites included tributaries, connecting waters, and recreational reservoirs to develop demand relationships among different kinds of habitats. Because no entry fees are charged for water-based recreation at any of the sites, benefits per trip are calculated as the consumer surplus area under the recreationists' demand function (relates trips taken as a function of travel cost per trip and other major determinants). Consumer surplus for each recreationist estimates the individual's willingness to pay extra travel costs per trip in excess of observed travel cost. With this approach, we can isolate the gain in economic value due to an extra fish caught per trip by increasing the number of fish caught per trip while holding other determinants constant.

To illustrate the method, demand schedules

Table 1. Annual visit rates per capita (relative to aggregate county-of-origin-population) by New Mexico residents for cold-water fishing at Heron Lake depending on fish caught per visit and other determinants (ordinary least squares regression).

Independent variable	Units	Parameter estimate[a]	t value	Probability of a greater t
Intercept		−0.2010		
Travel cost per trip	1981 dollars	−0.0013	−2.71	0.0303
Travel time per trip	Hours	−0.0285	−6.06	0.0005
(Coldwater fish)$^{1/2}$ per trip	(Number caught)$^{1/2}$	0.1115	3.06	0.0182
Total days per year of outdoor recreation	Days per year	0.0177	3.35	0.0015
Size of home city	Units	0.00006	3.18	0.0004
Overall F ratio = 10.60				
$R^2 = 0.91$				

[a] Estimated increase in trips per capita from a single unit change in the variable at the left.

that relate visit rates to site "prices" and fishing success rates were developed in order to estimate recreational benefits (consumer surplus) at a cold-water reservoir, Heron Lake, in the Rio Grande basin. The travel-cost method is used to construct the desired demand schedules for the site. Use of the method assumes that observed visits would fall equally as a result of an unobserved higher entry price as they would from an equal increase in observed travel costs per visit, i.e., households coming from greater distances that paid higher travel costs would visit at lower rates. Benefits are calculated (1) per visitor, (2) per unit of zone (county) population, (3) per average visitor trip, and (4) per gain in benefits for each trip for each additional fish caught.

Least-squares regression methods were used to estimate the household demand functions for the different sites as follows:

$$V_{ij} = B_0 + B_1TC_{ij} + B_2TT_{ij} + B_3F_{ij}^{1/2} + B_4S_{ij} + e_{ij} \quad (1)$$

where:

V_{ij} = the ith household's visits per year to the jth site, weighted by total participation rates from that household's zone of origin (Brown 1983);
TC = out-of-pocket travel cost per trip;
TT = travel time per trip in hours;
$F^{1/2}$ = square root of fish caught per trip;
S = a vector of socioeconomic variables;
e = an error term, with assumed zero mean and constant variance; and
B = parameter to be estimated.

The data in Table 1 indicate that for cold-water anglers at Heron Reservoir, travel cost per trip was a significantly negative factor which explained the variance in household trips per year (weighted by total sample trips per year from the ith household's home county), and had an estimated coefficient of −0.0013 which was significantly non-zero ($P < 0.04$). This coefficient implies that each one-dollar increase in travel cost per trip reduces annual trips per capita (of base county population) by 0.0014 (mean value of 0.05). The parameter estimate of "fish caught per trip," 0.3188, was of the expected sign and significantly non-zero ($P < 0.02$). A square-root term was included for the fish-caught-per-trip variable, assuming that fish caught per trip would increase annual trips per capita (average value, 2.6 days per trip) up to a point, but further increases in fish caught, above 7.1 (due to stocking or water-level fluctuations), would actually reduce the rate of increase in visits.

Table 2 shows the results of the general recreational demand function. Since there were more observations (568) on which to estimate functions, more determinants could be included as explanatory variables with maintenance of acceptable statistical precision. Of the 14 study sites, 10 variables significantly entered the demand equation (as 0–1 dummy variables). Relative to the base site at Elephant Butte Reservoir, all 10 of the other sites entered with a negative sign, implying that relative to Elephant Butte, each of these other 10 showed significantly less visitation on a per capita basis. Again the critical travel cost variable entered significantly ($P < 0.02$) with the expected negative sign. The overall F statistic (12.63) was significant ($P < 0.01$), although the percentage of explained variation in trips per capita ($R^2 = 0.30$) was lower than for anglers alone

Table 2. Annual visit rates per capita (relative to aggregate county-of-origin-population) by New Mexico residents for general water-based recreation, depending on major determinants (ordinary least squares regression).

Independent variable	Units	Parameter estimate[a]	*t* value	Probability of a greater *t*
Intercept		−0.0136		
Travel cost per trip (including travel time)	1981 dollars	−0.0025	−2.39	0.0171
Site (relative to Elephant Butte Reservoir)				
Jemez Springs	1 if Yes; 0 if No	−0.8868	−3.80	0.0002
Fenton Lake	1 if Yes; 0 if No	−0.7725	−3.17	0.0016
Heron Lake	1 if Yes; 0 if No	−0.7179	−3.23	0.0013
Upper Chama	1 if Yes; 0 if No	−0.9255	−3.55	0.0004
Lower Chama	1 if Yes; 0 if No	−1.0509	−2.27	0.0237
Lower Rio Grande	1 if Yes; 0 if No	−0.8281	−2.96	0.0032
Cochiti Lake	1 if Yes; 0 if No	−1.0954	−4.82	0.0001
Caballo Lake	1 if Yes; 0 if No	−1.1300	−3.35	0.0009
Belen Drain	1 if Yes; 0 if No	−1.4649	−2.88	0.0042
Tingley Drain	1 if Yes; 0 if No	−0.9021	−2.23	0.0260
Activity (relative to RV camping)				
Swimming	1 if Yes; 0 if No	0.7085	3.04	0.0025
Rafting	1 if Yes; 0 if No	3.5454	4.10	0.0001
Number of teens in group	Units	0.0667	2.03	0.0425
Outdoor recreation days per year	Units	0.0034	3.53	0.0004
Vacation days per year	Units	0.0081	2.21	0.0273
Size of home city	1 = farm 6 = over 1 million	0.3829	5.92	0.0001
Overall F = 12.63				
$R^2 = 0.30$				

[a] Estimated increase in trips per capita from a single unit change in the variable at the left.

because of the much larger sample size for all recreationists (Table 1).

Preliminary estimates of benefits per trip were $35.68 for cold-water anglers, $27.50 for warm-water anglers, and $51.25 for general recreationists. Benefits calculated per extra fish caught (i.e., value of each extra fish) for cold-water fish was $6.58, and for warm-water fish, $2.55. For each class of recreationist, these benefits are based on (1) the travel cost parameter estimate, (2) the "fish-caught" per trip parameter estimate (anglers only), and (3) actual annual trips taken by the recreationist surveyed (further details are found in Ward 1982). These results are generally consistent with other research on the value of benefits provided by public recreational facilities. Although the U.S. Water Resources Council's (1979) published unit-day values are in the range of $2.30–$2.85 per visitor day of general water-based recreation, when the present study's $35.68 per cold-water angler is deflated by our study sample's average of 2.6 days per trip at an inflation rate of 1.21 (10% per year from 1979 to 1981), this study's equivalent 1979 value per day would be $10.14, a reasonable first estimation.

Model Application

In its completed version, our model will be useful for several purposes including (1) improved predictions of fluctuations that cause reproductive failure, changes in productivity, or imbalances among predators and prey which may be managed through stocking and harvest regulation; (2) development of alternative water management plans that increase the probability of fish reproductive success and reliable fish yields without interfering with rights of those who own the water in the reservoirs; (3) assessing the value of water for improved fisheries and, when justified and affordable, purchasing water to stabilize levels in mainstream reservoirs or transferring storage to smaller, more manageable reservoirs somewhere off the mainstream; and

(4) influencing policy or program decisions that affect the water quality or structure of the river system. Application of any of these options will benefit from the use of a predictive model capable of testing the effects of alternative water-management strategies on recreational values.

The most far-reaching potential use of our model would be analyzing where best to store water for optimum overall benefit from the water consumed as it passes through the system (including construction of new storage reservoirs and retirement of existing ones). Such use of the model requires additional inputs of industrial and agricultural market values (generally available), legal constraints imposed by compacts and treaties, and linkages with flood-control models used by the U.S. Army Corp of Engineers and other agencies. With linkage to watershed models, our model could be used to evaluate downstream effects of changes in watersheds, including land development or altered agricultural practices. With completion of the biological model, short-term deterministic and long-term stochastic projections of the effects of water-level fluctuations on recruitment, growth, and yield of fish are possible. To some extent, modification of existing water storage and routing practices may benefit fisheries management without diminishing other user's water value.

What are the critical needs for our model to work accurately? First, the functions used in the model must reasonably represent reality for projections to be meaningful. Certain variables are more likely to have an important impact on outputs of interest. One of the most valuable uses of the model initially will be to test the sensitivity of outputs to alternative functions. Those functions having the most impact on output and the least confidence in their accuracy need research attention first. Since any error in input values tends to be compounded through the simulations, certain parameters should be continuously monitored at key locations within the system to check and reset the simulations. It will require routine monitoring at key locations of such inputs as discharge; concentrations of nutrients and suspended solids; solar radiation; light transmission; fish biomass, growth, recruitment, and yield to anglers; and willingness to pay to travel to available sites for recreation, preferably at randomly chosen localities. These are for the most part data now routinely monitored by various agencies, although some modification of present research practices may be needed for best model application.

We believe that the limitations of using simulation modeling as a management technique are a function of user attitudes. For water-managers with pressing needs, there is too often a tendency to assume that once a model is "completed," projections are reliable even though accuracy and precision of internal functions and input data may be questionable. In one of its best applications, sensitivity analysis, a model helps identify critical research needs and potential management conflicts. This application, however, requires the time demanded of close cooperation among model developers and users because, without cooperation, models are too often viewed as valid only when they predict that your special interest is in everyone's best interest. We have tried to avoid this inherent limitation by using a steering committee of potential users (agencies) to provide guidance for model development.

Our future efforts in completing the model will focus on developing linkages with existing watershed models and fish-population models. We will also develop more accurate mathematical functions relating loadings of materials and energy to resulting reservoir concentrations, and habitat quality to recreationists' willingness to pay for angling and other recreational uses.

With modification dependent on the site, our model could be used in any system of reservoirs where input data are available. In other basins, inputs of allochthonous matter, wind-agitated turbidity, commercial fishing, and thermal stratification may have to be added. The model is likely to be most useful where water availability and aquatic habitat are scarce and unpredictable and therefore management is greatly needed. Where aquatic habitat is predictable and abundant, recruitment-yield population models could be more cost-effective alternatives.

Acknowledgments

We wish to acknowledge financial support and philosophical encouragement for model development to date from the New Mexico Department of Game and Fish, New Mexico Water Resources Research Institute, New Mexico State University Agricultural Experiment Station, and the New Mexico Interstate Stream Commission. This is publication number SP 193 of the New Mexico State University Agricultural Experiment Station.

REFERENCES

BLAZKA, P., T. BACKIEL, AND F. B. TAUB. 1980. Trophic relationships and efficiencies. Pages 393–410 *in* E. D. LeCren and R. H. Lowe-McConnell, editors. The functioning of freshwater ecosystems. Cambridge University Press, New York, New York, USA.

BROWN, W. G. 1983. Using individual observations to estimate recreation demand functions: a caution. American Journal of Agricultural Economics 65:155–159.

COLE, R., F. WARD, T. WARD, R. DEITNER, AND P. TURNER. 1983. Model for optimization of socioeconomic values on the Rio Grande. Performance Report, Federal Aid Project F-52-R-4, New Mexico Department of Game and Fish, Santa Fe, New Mexico, USA.

GOLTERMAN, H. L., AND F. A. KOUWE. 1980. Chemical budgets and nutrient pathways. Pages 85–141 *in* E. D. LeCren and R. H. Lowe-McConnell, editors. The functioning of freshwater ecosystems. Cambridge University Press, New York, New York, USA.

GUM, R. L., AND W. E. MARTIN. 1975. Problems and solutions in estimating the demand for and value of outdoor recreation. American Journal of Agricultural Economics 57:558–566.

HARGRAVE, B. T. 1973. Coupling carbon flow through some pelagic and benthic communities. Journal of the Fisheries Research Board of Canada 30: 1317–1326.

HILLBRICHT-ILKOWSKA, A., I. SPODNIEWSKA, T. WEGLENSKA, AND A. KARABIN. 1972. The seasonal variation in some ecological efficiencies for production rates in the plankton community for several Polish lakes of different trophy. Pages 111–270 *in* Z. Kajak and A. Hillbricht-Ilkowska, editors. Productivity problems of freshwater. Polish Scientific Editors, Warsaw, Poland.

LEIDY, G. R., AND R. M. JENKINS. 1977. The development of fishery compartments and population coefficients for use in reservoir ecosystem modeling. Office, Chief of Engineers, U.S. Army, Contract Report 4-77-1. Washington, D.C., USA.

LIKENS, G. E. 1975. Primary production of inland aquatic systems. Pages 185–202 *in* H. Lieth and R. H. Whitaker, editors. Primary productivity of the biosphere, Springer-Verlag, New York, New York, USA.

LINSLEY, R. K. JR., M. A. KOHLER, AND J. L. H. PAULHUS. 1982. Hydrology for engineers, 3rd edition. McGraw-Hill Book Company, New York, New York, USA.

LORENZEN, M. W. 1972. Predicting the effects of nutrient diversion on lake recovery. Pages 205–211 *in* E. J. Middlebrook, D. H. Falkenborg, and T. E. Maloney, editors. Modeling the eutrophication process. Ann Arbor Science Publications, Inc., Ann Arbor, Michigan, USA.

MAKAREWICZ, J. C., AND G. E. LIKENS. 1979. Structure and function of the zooplankton community of Mirror Lake, New Hampshire. Ecological Monographs 49:109–127.

PARSONS, T. R. 1980. Zooplanktonic production. Pages 46–66 *in* R. S. K. Barnes and K. H. Mann, editors. Fundamentals of aquatic ecosystems. Blackwell Scientific Publications, Boston, Massachusetts, USA.

RYDER, R. A. 1965. A method for estimating the potential fish production of north-temperate lakes. Transactions of the American Fisheries Society 94:214–218.

SCAVIA, D. 1979. The use of ecological models of lakes in synthesizing available information and identifying research needs. Pages 109–168 *in* D. Scavia and R. Robertson, editors. Perspectives on lake ecosystem modeling. Ann Arbor Science Publications, Inc., Ann Arbor, Michigan, USA.

SCAVIA, D., AND R. A. PARK. 1976. Documentation of selected constructs and parameter values in the aquatic model CLEANER. Ecological Modeling 2:33–58.

USACE (U.S. ARMY CORPS OF ENGINEERS). 1974. HEC-3 Reservoir System Analysis for Conservation, Computer Program 723-X6-L2030 User's Manual. The Hydrologic Engineering Center, Davis, California, USA.

U.S. WATER RESOURCES COUNCIL. 1979. Procedures for evaluation of national economic development (NED). Page 72892 *in* Benefits and costs in water research planning (Level C). U.S. Government Printing Office, Superintendent of Documents, Washington, D.C., USA.

WARD, F. A. 1982. The demand for and value of recreational use of water in southeastern New Mexico. Research Report 465, New Mexico Agricultural Experiment Station, New Mexico State University, Las Cruces, New Mexico, USA.

WESTLAKE, D. F., M. S. ADAMS, M. E. BINDLOSS, G. G. GANF, G. C. GERLOFF, U. T. HAMMER, P. JAVORNICKY, J. F. KOONCE, A. F. H. MARKER, M. D. MCCRACKEN, B. MOSS, A. NAUWERK, I. L. PYRINA, J. A. P. STEEL, M. TILZER, AND C. J. WALTERS. 1980. Primary production. Pages 141–246 *in* E. D. LeCren and R. H. Lowe-McConnell, editors. The functioning of freshwater ecosystems. Cambridge University Press, New York, New York, USA.

Pages 28–38 *in* G. E. Hall and M. J. Van Den Avyle, editors.
Reservoir Fisheries Management: Strategies for the 80's.
Reservoir Committee, Southern Division American Fisheries Society, Bethesda, Maryland. 1986.

Management of Fisheries on Large African Reservoirs—An Overview

J. M. KAPETSKY

Inland Water Resources and Aquaculture Service
Fishery Resources and Environment Division
Food and Agriculture Organization of the United Nations
00100 Rome, Italy

ABSTRACT

Reservoirs provide about 10% of the 1.4 million tons of fish harvested annually from Africa's inland waters. The cumulative surface area of African reservoirs exceeds 40,400 km². The predominant resources in these reservoirs are tilapias. Yields in the larger reservoirs under moderate to heavy fishing pressure currently vary from about 27 to 65 kg/hectare/year.

Many African countries are food-poor, and much of the available labor force is underemployed; therefore, African reservoir fisheries are managed primarily to produce food fishes and to create employment. Fisheries are labor intensive rather than mechanized and are little regulated. Introduction of non-indigenous species has been a successful management technique for the development of littoral and pelagic fisheries. Development of reservoir fisheries has included a wide variety of activities—brush and tree clearing; boatbuilding; training in fishing methods; improvement of processing techniques; establishment of water and road transport systems; market facilities; and organization of fishermen's cooperatives and credit schemes.

Inland waters are an important source of food in Africa. They provide most of the available fish supply in about one-half of the countries, and one-third of the countries are entirely dependent on inland waters for their fish needs. Among the inland water systems of Africa, reservoirs contribute about 10% of the 1.4 million ton annual inland fish harvest. The objective of this paper is to present a brief overview of African reservoirs and their management and development for fisheries. Emphasis is placed on the largest of the African reservoirs—Volta in Ghana; Nasser/Nubia in Egypt and Sudan; Kariba in Zambia and Zimbabwe; Cahora Bassa in Mozambique; Kossou in Ivory Coast; Kainji in Nigeria; and Gebel Aulia in Sudan (Fig. 1). These larger reservoirs have received most of the research and development attention, so more information is available about them. Also, many of the larger reservoirs date from the 1950's and 1960's, thus providing a good historical cross-section of the evolution of the ecological and socio-economic factors which have shaped fishery management and development in Africa.

RESERVOIR CHARACTERISTICS

African reservoirs are among the largest in the world. Volta Lake in Ghana is pre-eminent with a surface area of 8,482 km² at full supply level (FSL). Six others exceed 1,000 km² in area at FSL (Table 1). Actual surface areas are, however, quite variable from year to year. Surface areas at FSL have been taken from Coche (1974) and from Petr (1978) who also provide additional comparative information on physical and chemical aspects of African reservoirs. Mean depths range from 2.3 m for Gebel Aulia reservoir to 29 m for Lake Kariba (Table 1). The cumulative surface area of 86 African reservoirs larger than 10 km² is 40,400 km² (Bernacsek 1984).

All of the large reservoirs incorporate the generation of electricity as a purpose, but some, like Kossou and particularly Nasser, also provide water for irrigation. The oldest of the large African reservoirs is Gebel Aulia (dam closed in 1937), and the youngest is Cahora Bassa (1974) (Table 1); all are well past their "trophic upsurges."

African reservoirs occur in various biomes (Table 1) including savanna, desert, and rainforest. The type of biome in which a reservoir occurs not only has important implications for aquatic biological potential but also affects development activities. An example is brush and tree clearing to facilitate fishing; clearing is more expensive in dense forest (Kossou) than in sa-

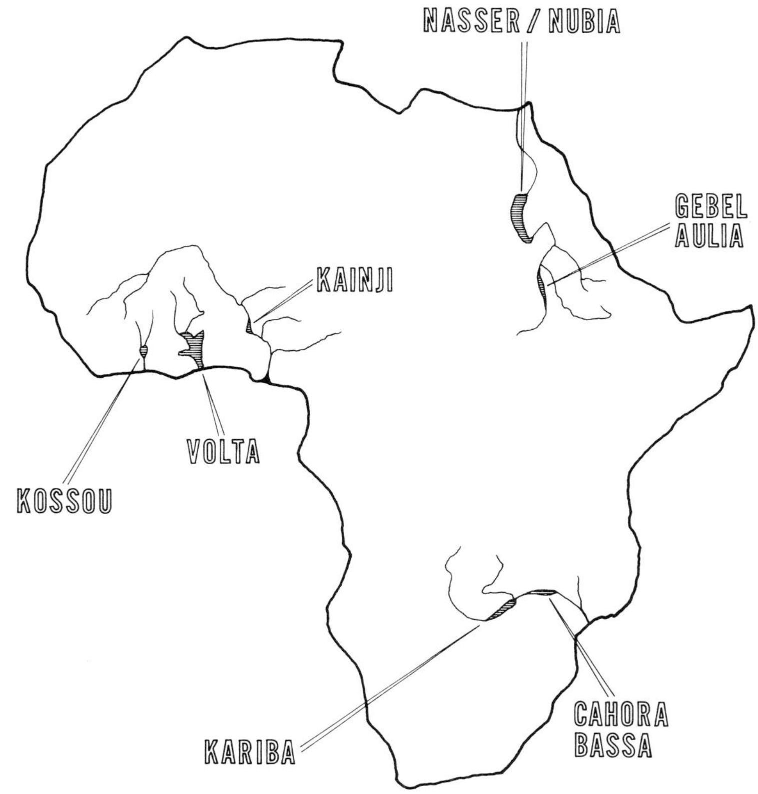

Figure 1. Locations of Africa's largest reservoirs.

vanna (e.g., Kariba), and not needed at all in a treeless desert (Nasser/Nubia).

Fish Populations

The predominant fishes of African reservoirs are the tilapias. The prominent species are the Nile tilapia, *Oreochromis niloticus,* and the Galilaean tilapia, *Sarotherodon galilaeus* (Table 2). Other important fish groups are clarid and schilbeid catfishes, and, in the upper, more river-like parts of the reservoirs, cyprinids. Top predators are the tigerfish, *Hydrocynus vittatus,* and the

Table 1. Characteristics of large African reservoirs.

Reservoir	Country	Year completed	Surface area (km^2) Full level	Surface area (km^2) Mean[a]	Mean depth[b]	Biome
Volta	Ghana	1964	8,482	7,394 (1968–79)	19.0	Savanna
Kariba	Zambia/Zimbabwe	1958	5,364	5,384 (1968–81)	29.2	Savanna
Nasser/Nubia	Egypt/Sudan	1964	6,216	4,889 (1976–81)	25.0	Desert
Cahora Bassa	Mozambique	1974	2,665	2,316 (1977–81)	20.9	Savanna
Kossou	Ivory Coast	1971	1,600	600 (1972–74)	14.3	Savanna/ rainforest
Gebel Aulia	Sudan	1937	1,500	—	2.3	Desert
Kainji	Nigeria	1968	1,270	—	11.0	Savanna

[a] Mean annual surface area for the years indicated.
[b] In meters, at full level.

Table 2. Fishery characteristics of large African reservoirs.

Reservoir	Principal exploited species	Annual fishery yield (kg/hectare) Observed[a]	Predicted[b]	Number of fishermen	Present level of exploitation: Littoral species	Pelagic species
Volta	*Tilapia zilli* *Sarotherodon galilaeus* *Lates niloticus*	53 (1969–79)	33	19,487 (1970–75)	Heavy	Light
Kariba	*Limnothrissa miodon* *Oreochromis mortimeri* *Tilapia rendalli*	6 (littoral) (1966–72) 21 (pelagic) (1980–82)	23	No data	Heavy	Moderate
Nasser/Nubia	*Sarotherodon galilaeus* *Oreochromis niloticus*	65 (1979–81)	40	9,260 (1982)	Heavy	Light
Cahora Bassa	*Hydrocynus vittatus* *Distichodus schenga* *Labeo* sp.	16 (1982)	32	1,200 (1982)	Moderate	None
Kossou	*Sarotherodon galilaeus* *Oreochromis niloticus* *Citharinus eburneensis*	88 (1972–74)	32	2,171 (1972–74)	Light	None
Gebel Aulia	*Sarotherodon galilaeus* *Oreochromis niloticus* *Lates niloticus*	55 (1981–82)	104	1,594 (1981–82)	Light	Light
Kainji	*Oreochromis niloticus* *Sarotherodon galilaeus*	55 (1971–78)	35	4,867	Heavy	Light

[a] Values are means for years shown in parentheses. Estimates for Volta and Nasser/Nubia were computed using actual surface areas for each year; full-level surface areas were used for other cases.

[b] Predicted from a relationship in Henderson and Welcomme (1974), where yield = $14.3136(\text{conductivity/mean depth})^{0.4681}$.

Nile perch, *Lates niloticus.* Introductions, mainly of tilapias, the freshwater sardine, *Limnothrissa miodon,* and cyprinids, have altered reservoir species compositions; these changes are discussed in detail in the later section on management.

As the reservoirs aged there have been changes in the relative abundance of species and species groups, but, except in the case of Lac Kossou, which is now very much under-fished, it is difficult to separate the effect of fishing from the changes in the physical, chemical, and biological environment associated with ageing. Such changes have important economic as well as biological implications. Some species are more desirable than others from the point of view of ease of harvest, adaptability to processing and preservation, and consumer preference. Thus, the ability to anticipate irreversible changes or to manipulate species composition is an important aspect of the management of fisheries in African reservoirs.

In Lake Nasser there has been a change in the relative abundance in the catch of both the Nile and Gailaean tilapias. The former has been the backbone of the fishery while the latter, of smaller size and slower growth, is less desirable but has now become the most abundant fish in the catch (Latif 1984). The tilapias, as compared with other species, have come to occupy a position of prominence in the overall catch, presently making up about 95% by weight of the fresh fish output whereas in the early years of the reservoir other species were more important.

Ita (1982) pointed out that *Citharinus* spp., reservoir phytoplanktivores, changed dramatically in abundance since the early years of the Kainji reservoir. Apparently one large and successful year class, spawned as the reservoir was expanding in 1968, sustained the fishery and made up 90% of the weight of the catch in 1970/71. But by 1972, these species provided only 34% of the catch.

Types of Fisheries

In contrast to inland fisheries in North America, African fisheries are almost exclusively for food. Reservoir fisheries are predominantly small-scale—carried out by individuals working alone, in pairs, or in small groups, usually using nylon multi-filament gillnets from dugout canoes or other small craft powered by paddles or by outboards. Some fishing is done at the subsistence level and fishermen also may be part-time farmers. Fishermen on the large reservoirs catch from about 1.4 tons/year to 5.2 tons/year and densities range from 0.5 to 3.8 fishermen/km^2. The organization of catching, transporting, and distributing fish varies from an almost completely laissez-faire situation on Volta Lake in Ghana to a state-owned company responsible for fishing, fish transporting, processing, and marketing on Lake Nasser.

Sport fishing has not developed to any appreciable extent on African reservoirs. Africa generally is a food-poor continent and freshwater fisheries are an important food source. Thus, African reservoirs are not likely to see any substantial development of sport fisheries until there is a food surplus combined with a much greater economic and physical mobility of the population than presently exists. In fact, any attempt to develop large-scale sport fisheries on an African reservoir could be looked on by the food fishermen as undue competition, just as commercial fishing on reservoirs is often unacceptable to sport fishermen in North America. On the Zimbabwean side of Lake Kariba an annual tigerfish tournament has been held for more than 19 years; however, the sport catch never amounted to more than a few tons among the literally thousands of tons of fish annually harvested from the reservoir for food.

Gillnets are the predominant fishing gear on Volta Lake (Bazigos 1970) and on Kainji Lake (Bazigos 1979); gill and trammel nets are the usual gears on Lake Nasser (Latif 1984) and on Lake Nubia (Ali 1984). Seines and gillnets are the principal gears on Gebel Aulia reservoir, but cast nets are also common (Henderson 1975).

Yield Characteristics and Status of Reservoir Resources

Yield performances of African reservoirs have been only partly related to their biological capacities to produce fish. Rather, much has depended on such diverse factors as the degree of physical isolation of the reservoir, availability of fish from other sources, and the supply of appropriate gears, materials, and equipment for fishing and for preserving and transporting the catch, as well as on internal or external political and economic stability.

In order to compare yield performances among African reservoirs, it is useful to place yield on a unit area basis. Therefore, in describing reservoir fishery yield characteristics, allowance has been made for varying surface areas among years for reservoirs with large variations, or where actual surface areas have been markedly different from surface areas calculated at full supply level (Table 1). In such cases annual total yield was divided by mean annual reservoir surface area to give yield in kilograms per hectare of average surface area.

Volta Lake

The year of peak fishery output was 1969 with nearly 62,000 tons caught. Over a recent 11-year period yield has averaged 53 kg/hectare. The principal fisheries of Volta Lake are for *Tilapia zillii, Sarotherodon galilaeus,* and *Lates niloticus.* Fishing is carried out mostly in the littoral area. Although experimental fishing has shown that there is an ample pelagic resource in the form of the small clupeids, *Sierrathrissa* spp. and *Pellonula* spp., this resource is little exploited. Littoral resources are fairly intensively exploited, but there is no evidence of overexploitation in recent yield trends.

Lake Nasser and Lake Nubia

Aswan dam on the Nile forms a 500 km long reservoir; in Egypt it is called Lake Nasser and the part in the Sudan is called Lake Nubia.

Resources of Lake Nubia are very much underexploited; it makes up about 16% of the surface area of the entire reservoir but produces only about 2% of the total catch. The low yield from Lake Nubia is due to its geographic isolation, the desert environment, and competition from river and reservoir fisheries closer to Sudan's population centers. Nasser/Nubia has yet to fill completely and recent average annual surface areas have varied from about 4,700 km^2 to 5,000 km^2, compared with an FSL area of 6,218 km^2. The fishery output of Lake Nasser has been steadily increasing each year since its impoundment in 1964. For 1981, the latest year for which data are available, combined Nasser/Nubia yield had reached 73 kg/hectare; it has increased at an average annual rate of 8 kg/hectare since 1976. Actual yield per unit area fished is much higher than indicated, because nearly all of the catch, about 90% tilapias by weight, comes from the littoral zone within the 20 m depth contour; this zone makes up only about 20% of the total reservoir area. Earlier assessments of yield potential by Ryder and Henderson (1974) and Bazigos

(1972) were 38 kg/hectare and 43 kg/hectare, respectively.

Results from experimental fishing and preliminary findings from hydroacoustical surveys indicate that pelagic resources in Lake Nasser are sparse. Its littoral resources do not appear to be overexploited; however, heavy fishing pressure on spawning grounds is believed to be responsible for the gradual replacement of the Nile tilapia by the Galilaean tilapia.

Cahora Bassa Reservoir

Cahora Bassa Reservoir in Mozambique presently yields only about 16 kg/hectare (Bernacsek and Lopes 1984); only the littoral areas are exploited. The principal reasons for the relatively low yield are political turmoil in the area of the reservoir and general economic problems that result in an inadequate supply of fishing materials and insufficient fish transport and marketing facilities, which provide few incentives for fishing.

First results from biological and limnological studies now underway and preliminary data from a hydroacoustical survey (Lindem 1983) indicate that there are substantial pelagic resources consisting primarily of *Limnothrissa miodon,* with a small characid, *Alestes* sp., and a schilbeid catfish, *Eutropius* sp., as secondary components. Efforts are underway to find appropriate methods for the harvest of these pelagic resources. Freshwater mussels also appear to be abundant here and are another potential resource.

Lac Kossou

Lac Kossou in Ivory Coast is at present little exploited. The only available yield data are from 1972–74, but they indicate good yields, averaging 88 kg/hectare, during the early history of the reservoir. More recently, Kouassi (1979) observed that the large-mesh gillnets used on the reservoir were harvesting only about 5% by weight of the fishes available to the multi-mesh experimental fleet.

The present underexploitation appears to be largely a matter of economics and culture. Fish from other sources is available and the peoples of the reservoir area, farmers by tradition, have not taken well to fishing. Expatriate fishermen from neighbouring Mali, who once accounted for about 45% of the yield, were later not welcome in the area (Bazigos 1979).

Kainji Lake

Resources of Kainji Lake in Nigeria are heavily exploited (Ita 1982) and at least one species among the citharinids is overexploited in the single-species sense of overexploitation. Nevertheless, average annual yield for the period 1971–78 was 54 kg/hectare as compared with a predicted potential of about 35 kg/hectare based on the Henderson and Welcomme (1974) MEI-yield relationship (Table 2). Peak yield was attained in the second and third years after impoundment; since then yield has been gradually decreasing.

Lake Kariba

Lake Kariba, like Lake Nasser/Nubia, is an international reservoir; however, in this case each of the two riparian countries, Zimbabwe and Zambia, have about one-half of the reservoir area along its total length. Yield performance of this reservoir has been much influenced by the countries' varying capacities for fishery development and by the recent Rhodesian war, which affected each country differently. For the period 1966–72, when the littoral fisheries on both shores of the reservoir were active, yield amounted to about 6 kg/hectare. The recent combined yield from the Zimbabwean and Zambian pelagic sardine (clupeid) fisheries is about 21 kg/hectare. Zambia has only recently begun to license a pelagic fishery for sardines and the small-scale littoral fishery there is slowly coming back to pre-war levels. Predicted yield (MEI) is 25 kg/hectare.

Lake Kariba may be the best example of efficient utilization of African reservoir fishery resources because there are both littoral and pelagic fisheries, the latter made possible by the introduction of sardines from Lake Tanganyika. Marshall et al. (1982) point out that the best prospects for increasing the fishery output of Kariba are through the exploitation of presently little-used resources. Some fishes, like *Tilapia rendalli,* are not very susceptible to gillnets and thus not all of their potential is being realized. Freshwater mussels have an estimated biomass of 167,000 tons on Kariba (Kenmuir 1980), but only 1 to 2 tons/month are utilized there.

Fishery Management Activities

Stock Assessment

All of the large African reservoirs had internationally-funded technical assistance projects associated with them, but in the case of fisheries, the start-up usually was after impoundment. Nevertheless, such projects provided a technical, financial, administrative, and logistic framework through which resource information could be acquired in a timely manner and at a reasonable cost when considered in relation to the value of the fisheries and the sizes of the systems under investigation. The core group for resource evaluation generally consisted of an expatriate fishery biologist and limnologist together with national counterparts, technicians, and support personnel. The Food and Agriculture Organization of the United Nations (FAO) has been involved in these efforts on all of the large African reservoirs, but bilateral assistance agencies such as the United States Agency for International Development, Japanese International Cooperation Agency, and Swedish International Development Authority, also have sponsored reservoir fishery projects.

Not surprisingly, periodic fishing of standard gillnet fleets at fixed stations has been a universal stock assessment tool for African reservoirs. The assessment program on Volta Lake was the most ambitious. It used a fleet of nine mesh sizes that was set two or three times per year at 30 stations (Vanderpuye 1984). This type of sampling offers the advantages that results can be made available readily, and analyses are straightforward. Moreover, data can be obtained regularly and inexpensively by a small field staff, and the equipment is inexpensive, simple, and easily maintained. Disadvantages are that the gear is selective and provides only relative estimates of abundance, which are not serious problems where gillnets are the most important fishing method.

Other more sophisticated sampling techniques have been tried. Among them was the use of toxicants in blocked-off areas on Lakes Kariba (Balon 1974), Volta (Loiselle 1972), and Kainji (Ita, 1984), and sampling with an explosive grid on Lake Kariba (Marshall 1984a). Major drawbacks of these methods in the African context are that they are relatively demanding of time, equipment, supplies, manpower and finances, and they cannot be effectively applied in all habitats.

Acoustic techniques have been used along with test fishing to survey pelagic resources on Lake Kariba (Woodward 1974), Lake Nasser, and most recently on Cahora Bassa reservoir (Lindem 1983). Although this is a useful tool, the field equipment is complex and fairly delicate and repairs must usually be made abroad. Also, processing, analysis and interpretation of the magnetic tapes must be done at a specialized laboratory abroad.

Catch and effort data along with socio-economic statistics on the fishery, although recognized by fishery managers as fundamental for resource evaluation, have not always been collected in a consistent or timely manner. One reason for this is that it has been difficult to make the operational and administrative shift from full-time, labor-intensive, census-based programs to the streamlined, stratified sampling programs which use coordinated aircraft and ground surveys to provide the essential data on a part-time basis. Bazigos and collaborators have been instrumental in instituting the latter on large African reservoirs and in revising and correcting estimates obtained with earlier methodologies (Bazigos 1972, 1973, and 1979; Bazigos et al. 1975; Coppola and Agadzi 1977).

When data have been available from experimental fishing or from fishery statistics programs, they have been used in various ways to evaluate the status of resources. For example, on Volta Lake, data on size range and abundance from the experimental gillnet fleet are compared with the same parameters from the commercial gillnet catch. In this way an indication of the effects of fishing on the commercial species of the reservoir is obtained. On Kainji Lake, experimental gillnets were used to evaluate the status of citharinids by Ita (1980), using the mean-size approach developed by Henderson (1972). On Lake Kariba, the availability of complete and reliable data on the pelagic sardine fishery has enabled the use of the various, well-known catch-effort models to be applied for resource evaluation (Marshall et al. 1982).

Development of a morpho-edaphic index (MEI) for African waters (Henderson and Welcomme 1974) has facilitated rapid predictions of potential fishery yield for new reservoirs based on conductivity and mean depth (e.g., Cahora Bassa; Bernacsek and Lopes 1984). As a means of judging yield performance, the MEI yield estimate has been used as a yardstick against actual yield from Lake Nasser (Ryder and Henderson 1974) and Kariba (Marshall et al. 1982). As an aid to planning for the management and development of new African reservoir fisheries, Marshall (1984b) has developed several indices useful for predicting future reservoir conditions. The

Table 3. Species of fish introduced into African reservoirs.

Reservoir	Area (km²)	Country	Species	Years introduced	Economic success	References
Kariba	5,364	Zambia/ Zimbabwe	*Tilapia rendalli*	1951–62	Poor	Marshall et al., 1982
			Oreochromis macrochir	1959–62	Poor	
			Limnothrissa miodon	1967–68	Good	Bell-Cross and Bell-Cross 1971
McIlwaine	26	Zimbabwe	*Oreochromis macrochir*	mid-1950's	Fair	Van der Lingen 1981
Robertson	81	Zimbabwe	*Serranochromis condringtoni*	early-1980's	Poor	Marshall, pers. comm.
Ayamé	140	Ivory Coast	*Oreochromis niloticus*	1962–63	Good	Lessent 1973; Kouassi 1980
			Heterotis niloticus	1964–65	Poor	
Kossou	1,600	Ivory Coast	*Heterotis niloticus*	1971–72	Poor	CTFT 1973; Kouassi 1979
			Oreochromis niloticus	1971–72	Poor	
Nyumba ya Mungu	180	Tanzania	*Tilapia rendalli*[a]	1974	Poor	Bailey 1978; Bernacsek 1980
			Sarotherodon esculentus[a]	1973	Good	
Mantasoa	15	Madagascar	*Tilapia rendalli*	1955	Good	Moreau (1984)
			Micropterus salmoides	1963	Fair	
			Oreochromis niloticus	1970	Poor	
			Cyprinus carpio[a]	unknown	Fair	

[a] Probably introduced in the basin before impoundment rather than directly into the reservoir.

indices are based on physico-chemical data that can be easily acquired before impoundment. Finally, Pauly (1984) has made a compilation of recently developed stock assessment methods adaptable to African reservoirs.

Constraints on Management

Management in the classical sense of regulation of fisheries—such as limits on fishermen numbers, fishing gears, gear characteristics, closed areas or closed seasons—is little practiced on African reservoirs at present, mainly because such regulations would be politically and economically unacceptable. For example, many of the African countries suffer from underemployment and high rates of unemployment; so African reservoir fisheries are important first as producers of food and secondly as providers of employment. Any type of regulation that would limit employment, either directly or indirectly, would not find favor. Another reason is that most African fishery departments would be hard pressed to find sufficient funds to adequately enforce regulations, particularly on the larger reservoirs. Furthermore, several large reservoirs remain underharvested, and they would benefit from development activities that would stimulate fishing rather than from restrictive regulations. Finally, African reservoir fisheries appear to be relatively unaffected by overfishing; species compositions of the catch change, but total fishery output remains little affected, as for example on Kainji Lake. Additionally, where regulation might be beneficial, efforts to translate resource and economic information into management policy, and management policy into the implementation of management programs, have not always been successful.

Another aspect of fisheries management on African reservoirs is their international character. Two, Lake Nasser/Nubia and Lake Kariba, are shared between two countries, and another in Mauritania and Senegal is being developed. Thus, coordination and cooperation between riparian countries are necessary, particularly with regard to the management of shared pelagic resources and the introduction of exotic species.

Fish Introductions

The most utilized reservoir management technique in Africa has been the introduction of nonindigenous species, most frequently tilapias (Table 3), but undoubtedly most successfully the sardine, *Limnothrissa miodon,* into Lake Kariba (Bell-Cross and Bell-Cross 1971). It is the latter introduction which is the focus of the remainder of this section.

Pre-impoundment studies indicated that no fish species indigenous to the Zambezi River would be likely to flourish in the pelagic zone of the new Kariba reservoir. In 1963 Zambia began researching the practical aspects of introducing clupeids there and these were continued with FAO assistance. In 1967 and 1968 a total of about 720,000 *L. miodon* were airlifted from Lake Tanganyika by light plane. Fishing trials, first undertaken by Rhodesia (now Zimbabwe) and Zambia in 1969 and 1970, showed that the sardine had become well established but initial evaluations of the introduction were negative because stock densities were believed too low to support a fishery (Balon 1974). Nevertheless, experimental fishing to identify effective means for harvest were continued and a commercial fishery on the Zimbabwean side was established in 1973. The principal fishing methods are purse seining, now losing favor because of high costs, and lift netting, both of which use light attraction.

Since 1973 the sardine yield from Zimbabwe waters increased annually with a maximum yield of about 12,000 tons in 1981 followed by a decrease to about 9,000 tons in 1982. Yield, dependent on nutrient inflows from the Zambezi and other smaller rivers (Marshall 1982), is likely to vary in the future. However, with Zambia's sardine fishery beginning to expand (1981 harvest about 1,000 tons and 3,000 tons in 1982), total sardine yield from Lake Kariba should be on the increase.

There are several notable points about this introduction. First, the introduction permitted the establishment of a fishery where none had existed previously; secondly, the pelagic fishery has resulted in a more than three-fold increase in output from the reservoir compared with its littoral potential of only about 2,500–3,000 tons. Also, the sardine fishing industry—catching and processing—has contributed to local employment. Finally, from the economic point of view of a developing country the sardine is an ideal species. Processing is simple and the costs are correspondingly low. The fish can be sun-dried whole in from 24 to 48 hours and then sealed in plastic bags in which the product has a long shelf life. Alternatively, sardines can be frozen whole in blocks, or canned. The tigerfish caught in sardine nets are also canned.

As a consequence of the planned introduction of *L. miodon* into Lake Kariba, this species has moved downstream to Cahora Bassa reservoir in Mozambique where the results of biological investigations, a preliminary acoustical survey, and test fishing indicate that it is well established (Bernacsek and Lopes 1984; Lindem 1983).

Other Management Techniques

On Lake Nasser an "artificial nurseries" management scheme is now in its pilot stage. The scheme is aimed at overcoming two related problems. The first is heavy fishing on Nile tilapia spawners (40% of the annual tilapia catch is made during the 3-month spawning period). The second problem is the increasing proportion of the smaller, slower-growing Galilaean tilapia in the catch. It now makes up more than 50% as compared with less than 20% ten years ago. Because nearly all of the tilapia are sold as fillets, the larger 1 kg-size tilapia are desired.

Basically, the management techniques of the nurseries scheme are predator control and regulation of fishing. The physical layout of the first scheme consists of a 1,000 hectare arm blocked off from the remainder of the reservoir with a fine mesh net. Within this larger blocked-off area much smaller secondary enclosures were placed along the littoral edge. After predators had been removed from these small enclosures, they were stocked with Nile tilapia breeders. In this first pilot scheme, ducks were being farmed as a stimulus to aquatic production. Early results from the pilot scheme were said to be good in that the desired species, the Nile tilapia, made up 80% of the catch in the blocked-off arm; however, with the data available it has not been possible to determine if overall yield had increased above levels from the open, unmanaged areas of the reservoir.

Value of African Reservoir Fisheries

It is difficult to assign economic values to African reservoir fisheries, first because comparative economic data are scarce, and second because different living standards and exchange and inflation rates among African countries complicate comparisons. Finally, employment generated by reservoir fisheries, even if of low remuneration, has a social value in its own right. This can be reflected, but not adequately measured, in terms of fishermen numbers.

The sardine fishery in Zimbabwe's portion of Lake Kariba had a landed value of \$5.2 million (U.S.) and a retail value reaching about \$12.2 million in 1980 for the nearly 8,000 tons landed. through 1980 capital investment in the industry was about \$7.6 million, and about 2,000 people were employed where there were previously few other job opportunities. In contrast, the tigerfish sport fishery on the Zimbabwe side of Lake Kariba, the largest on African reservoirs, generates about $60,000 in fishermen's expenditures per ton of tigerfish caught in the annual tournament. The tournament catch reached a peak of 3.2 tons in 1975 and 1979 and has averaged 2.2 tons through its 19-year history. Commercial tigerfish catches, in comparison, have ranged from about 40 to 60 tons/year in recent times and the fish is much in demand for canning.

The direct annual increase in national income due to the Aswan High Dam in Egypt has been placed at $263 million mainly from water storage for irrigation and electricity generation (Latif 1984). The landed value of the catch from Lake Nasser in 1981 was about $4.6 million based on a yield of 33,900 tons. The cost to consumers was about double that amount. Employment of fishermen in reservoir fishing in 1982 was about 9,000. Perhaps half again as many people find work in transporting and processing the catch.

On Volta Lake the landed value of the catch in 1978 exceeded the income surplus from the generation of electricity by a factor of two (Vanderpuye 1984). Fishermen in 1975 numbered about 20,600. Nearly all of them fished on a full-time basis.

Fishery Development Activities

Fishery management, as discussed above, has been taken to include only those activities that contribute to increased fishery yield through increasing biological production. Fishery development, in comparison, is used here to define those activities that increase fishery output through more efficient fishing, processing, transporting, and marketing. Fishery development activities undertaken in most reservoirs include fisherman training, boat design and construction, and fish transport, processing, and marketing. Other less common development activities include aquaculture, brush and tree clearing to facilitate fishing, organization of fisherman cooperatives, and provision of credit to purchase equipment. These have been undertaken in a variety of ways—by the private sector, with government funds, with technical assistance from international organizations, or with financing from religious or other non-governmental agencies. From whatever source, the purpose has usually been to get development started; later, when established, it can be taken over by private industry, state companies, or by fishermen's cooperatives.

Development activities on Volta Lake are a good example of most large impoundments. As described by Vanderpuye (1984), the fishing villages on Volta Lake sprang up as the reservoir formed; they were founded by fishermen who had emigrated from downstream on the Volta River. Government-planned development included the establishment of fishing complexes at each important fish landing point. Each complex was to include landing facilities, a store for fishermen's supplies, a market, fish smoking houses, and training facilities for fishermen, boatbuilding, and boat motor repair. Only one such complex has been built thus far.

Individually less ambitious than the fishery complex, but also contributing to fishery development on Volta Lake, have been a mobile fishery school; a fish processing demonstration program; an extension service to introduce more efficient nets and improved smoking methods; establishment of a revolving fund for the construction of boats and the sale of outboard motors to fishermen; the organization of fishermen into cooperatives; and the placing of floating jetties to facilitate fish landings.

Other more general development efforts that have served to stimulate reservoir fisheries by providing improved living conditions have been rural health schemes, assistance with agriculture, and the establishment of schools and communications. These have been an important impetus to fisheries, especially where living conditions are harsh. For example, one of the major constraints on fishery output from Lake Nubia has been the difficulty in attracting fishermen to live and work in isolated desert fishing camps (Ali 1984), and this also was a factor affecting the Lake Nasser fishery until recent times. There, however, government and private efforts have been instrumental in providing small pumps so that fishermen could irrigate land for subsistence farming, and houses were designed to be comfortable during hot desert days and cold nights.

Conclusions

African reservoir fisheries are comparatively important in providing food and employment on a continent which periodically suffers from food shortages and underutilization of its available labor. Finfish resources of most of the large reservoirs are now moderately to heavily exploited. Nevertheless, there are good opportunities for increasing fishery harvests by implementing additional management and development activities. There are enormous littoral freshwater mussel resources on some reservoirs and others have pelagic fish stocks which are little exploited. Where no pelagic resources presently exist, well-researched, well-planned fish introductions offer a possibility to increase fishery outputs. Improvements in existing methods of processing and in systems of fish transport could make reservoir fisheries more productive without sacrificing employment.

Acknowledgments

I am grateful to H. F. Henderson, T. Petr, A. Coche, G. Bernacsek, B. E. Marshall, and Z. Song for their critical review of the manuscript.

References

Ali, M. T. 1984. Nubia (Sudan). *In* J. M. Kapetsky and T. Petr, editors. Status of African Reservoir Fisheries. FAO. Committee for Inland Fisheries of Africa, Technical Paper No. 10, Rome, Italy.

Bailey, R. G. 1978. The ecology of the fishes in Nyumba ya Mungu reservoir, Tanzania. Biological Journal of the Linnean Society 10:109–131.

Balon, E. K. 1974. Fish production of a tropical ecosystem. Pages 249–647 *in* E. K. Balon and A. G. Coche, editors. Lake Kariba: a man-made tropical ecosystem in central Africa. W. Junk. The Hague, The Netherlands.

Bazigos, G. P. 1970. Sampling techniques in inland fisheries with special reference to Volta Lake. FAO. Statistical Studies, UNDP/SF/GHA/10, St.S./1, Rome, Italy.

Bazigos, G. P. 1972. The yield pattern at Lake Nasser (Arab Republic of Egypt). FAO. UNDP Lake Nasser Development Center, Statistical Studies, UNDP/SF/EGY/66/558, St.S./1, Rome, Italy.

Bazigos, G. P. 1973. The improvement of the fisheries statistical system at Lake Kossou (Ivory Coast). FAO. Lake Kossou Fisheries Development Project Statistical Studies, UNDP/SF/IVC/526, St.S./1, Rome, Italy.

Bazigos, G. P. 1979. The yield patterns and their implications for fisheries development of some selected man-made lakes and river basins in Africa. Pages 246–315 *in* Proceedings of the International Conference on Kainji Lake and River Basin Development in Africa, Vol. 2, Kainji Lake Research Institute, New Bussa, Nigeria.

Bazigos, G. P., C. Grant, and C. Williams. 1975. Revised fisheries time series of the natural waters of Zambia (1966–1974). FAO. Fisheries Training Development Project Statistical Studies, FAO/ZAM/73/009/St.S./3, Rome, Italy.

Bell-Cross, G., and B. Bell-Cross. 1971. Introduction of *Limnothrissa miodon* and *Limnocaridina tanganicae* from Lake Tanganyika into Lake Kariba. Fisheries Research Bulletin of Zambia 5: 207–214.

Bernacsek, G. M. 1980. Introduction to the freshwater fishes of Tanzania. Department of Zoology, University of Dar-es-Salaam, Dar-es-Salaam, Tanzania (mimeographed).

Bernacsek, G. M. 1984. Guidelines for dam design and operation to optimize fish production in impounded river basins (based on a review of the ecological effects of large dams in Africa). FAO. Committee for Island Fisheries of Africa, Technical Paper No. 11, Rome, Italy.

Bernacsek, G. M., and S. Lopes. 1984. *In* J. M. Kapetsky and T. Petr, editors. Status of African reservoir fisheries. FAO. Committee for Inland Fisheries of Africa, Technical Paper No. 10, Rome, Italy.

CTFT (Centre Technique Forestier Tropical). 1973. Introduction de nouvelles especes de poissons. Campagnes 1971–1972 et 1972–1973. Rapport Final. Centre Technique Forestier Tropical, Nogent-sur-Marne, France.

Coche, A. G. 1974. Limnological study of a tropical reservoir. Pages 1–246 *in* E. K. Balon and A. G. Coche, editors. Lake Kariba: a man-made tropical ecosystem in Central Africa. W. Junk. The Hague, The Netherlands.

Coppola, S. R., and K. Agadzi. 1977. Evolution of the fishing industry over time at Volta Lake 1970–1976. FAO. Volta Lake Research and Development Project Statistical Studies GHA/71/553/St.S./6, Rome, Italy.

Henderson, H. F. 1972. The dynamics of mean-size statistics in a changing fishery. FAO. Fisheries Technical Paper No. 116, Rome, Italy.

Henderson, H. F. 1975. Report to the Government of the Democratic Republic of the Sudan on the fisheries of the reservoirs of central Sudan. FAO Near East Regional Office, Cairo, Egypt.

Henderson, H. F., and R. L. Welcomme. 1974. The relationship of yield to morpho-edaphic index and numbers of fishermen in African inland fisheries. FAO. Committee for Inland Fisheries of Africa, Occasional Paper No. 1, Rome, Italy.

Ita, E. O. 1980. Contributions to the dynamics of mean size statistics in a changing fishery. 1. Family Citharinidae in Kainji Lake. Hydrobiologia 68: 269–277.

Ita, E. O. 1982. Biological indices of overfishing in Kainji Lake and the management proposal for the lake fishery. Kainji Lake Research Institute Technical Report Series No. 8, New Bussa, Nigeria.

Ita, E. O. 1984. Kainji (Nigeria). *In* J. M. Kapetsky and T. Petr, editors. Status of African reservoir fisheries. FAO. Committee for Inland Fisheries of Africa, Technical Paper No. 10, Rome, Italy.

KENMUIR, D. H. S. 1980. The mussel resources of Lake Kariba. Transactions of the Zimbabwe Scientific Association 60:7–10.

KOUASSI, N. 1979. Installation et evolution des populations piscicoles dans le lac de barrage de Kossou (Cote d'Ivoire) entre 1972 et 1977. Annales de l'Universite d'Abidjan, Serie E (Ecologie) 12:159–174.

KOUASSI, N. 1980. Donnees sur l'effort de peche et la production piscicole du lac d'Ayame (Cote d'Ivoire) en 1978. Annales de l'Universite d'Abidjan, Serie E (Ecologie) 13:155–181.

LATIF, A. A. 1984. Nasser (Egypt). *In* J. M. Kapetsky and T. Petr, editors. Status of African reservoir fisheries. FAO. Committee for Inland Fisheries of Africa, Technical Paper No. 10, Rome, Italy.

LESSENT, P. 1973. Man-made lakes in Ivory Coast. Pages 74–76 *in* W. C. Ackerman et al., editors. Man-made lakes: their problems and environmental effects. American Geophysical Union Geophysical Monograph 17, Washington, D.C., USA.

LINDEM, T. 1983. A preliminary analysis of acoustic data from Maputo Bay, Lake Niassa and Cahora Bassa. FAO. Research and Development of Inland Fisheries Project Field Document 6, FAO/GCP/MOZ/006/SWE, Rome, Italy.

LOISELLE, P. V. 1972. Preliminary survey of inshore habitats in the Volta Lake. FAO. Volta Lake Fisheries Research Project, FI/DP/GHA/67/510/2, Rome, Italy.

MARSHALL, B. E. 1982. The influence of river flow on pelagic sardine catches in Lake Kariba. Journal of Fish Biology 20:465–469.

MARSHALL, B. E. 1984a. Kariba (Zimbabwe/Zambia). *In* J. M. Kapetsky and T. Petr, editors. Status of African reservoir fisheries. FAO. Committee for Inland Fisheries of Africa, Technical Paper No. 10, Rome, Italy.

MARSHALL, B. E. 1984b. Predicting ecology and fish yields in African reservoirs from pre-impoundment physico-chemical data. FAO. Committee for Inland Fisheries of Africa, Technical Paper No. 14, Rome, Italy.

MARSHALL, B. E., F. J. R. JUNOR, AND J. D. LANGERMAN. 1982. Fisheries and fish production on the Zimbabwean side of Lake Kariba. Kariba Studies Papers 10:175–231.

MOREAU, J. 1984. Mantasoa (Madagascar). *In* J. M. Kapetsky and T. Petr, editors. Status of African reservoir fisheries. FAO. Committee for Inland Fisheries of Africa, Technical Paper No. 10, Rome, Italy.

PAULY, D. 1984. Recent developments in the methodology available for the assessment of exploited fish stocks of reservoirs. *In* J. M. Kapetsky and T. Petr, editors. Status of African reservoir fisheries. FAO. Committee for Inland Fisheries of Africa, Technical Paper No. 10, Rome, Italy.

PETR, T. 1978. Tropical man-made lakes—their ecological impact. Archiv fur Hydrobiologie 81:368–385.

RYDER, R. A., AND H. F. HENDERSON. 1974. Fish yield projections on the Nasser Reservoir (including Lake Nubia, the Sudan). FAO. Lake Nasser Development Centre Aswan Technical Report, FI: DP/EGY/66/558:5, Rome, Italy.

VAN DER LINGEN, M. I. 1981. Potential, planning and problems in aquatic resource management in Zimbabwe. Pages 42–51 *in* J. M. Kapetsky, editor. Seminar on river basin management and development. FAO. Committee for Inland Fisheries of Africa, Technical Paper No. 8, Rome, Italy.

VANDERPUYE, J. 1984. Volta (Ghana). *In* J. M. Kapetsky and T. Petr, editors. Status of African reservoir fisheries. FAO. Committee for Inland Fisheries of Africa, Technical Paper No. 10, Rome, Italy.

WOODWARD, J. 1974. Successful reproduction and distribution in Lake Kariba. Pages 526–536. *In* E. K. Balon and A. G. Coche, editors. Lake Kariba: a man-made tropical ecosystem in central Africa. W. Junk. The Hague, The Netherlands.

Pages 39–45 *in* G. E. Hall and M. J. Van Den Avyle, editors.
Reservoir Fisheries Management: Strategies for the 80's.
Reservoir Committee, Southern Division American Fisheries Society, Bethesda, Maryland. 1986.

Why Are They Here and How Much Do They Spend? A Survey of Muskellunge Angler Characteristics, Expenditures, and Benefits

LAWRENCE C. BELUSZ AND DANIEL J. WITTER

Missouri Department of Conservation
P.O. Box 180
Jefferson City, Missouri 65102

ABSTRACT

A survey of tournament entrants at the 1981 Pomme de Terre Lake muskellunge tournament was undertaken to identify and evaluate the social and economic benefits associated with the tournament and to estimate the value of the muskellunge fishery to regional and state economies. The typical respondent was a Missouri male between 30 and 50 years of age who traveled an average of 160 miles one-way and devoted 4 days to the tournament. Participants indicated their involvement afforded them opportunities to socialize with other anglers while testing angling skills for muskellunge.

Total tournament expenditures averaged $193.54 per respondent for the 2.5-day tournament and were adjusted by regional-income multipliers to express benefits to the local lake community and state economy. Total worth of the tournament to the state economy was estimated at $100,519. Daily costs for muskellunge fishing were estimated at $48.94 and $21.90 for non-local and local anglers respectively. The annual value of the muskellunge fishery to the state economy in 1981 was estimated at $479,000.

These data indicate that muskellunge stocking programs can provide a high quality, trophy-fishing opportunity, in spite of characteristically low angler catch rates, and produce substantial economic benefits. Fishery management agencies might well consider establishing similar trophy-fishery programs to help meet anglers' ever increasing desires for new and challenging opportunities.

Tradition dictates that fisheries management programs be evaluated biologically even though it is assumed that human socioeconomic benefits accrue as a result of habitat or species manipulations. One method by which state fisheries agencies have attempted to provide benefits to anglers has been to introduce non-native species, particularly those with reputations as trophy fish. Such programs are not without cost, and many fail to develop as predicted or must be maintained by constant restocking of the desired species. For those that are successful, continued program support often depends upon an "acceptable" harvest rate of the stocked fish. However, methods which define the value of fisheries management only in terms of numbers of fish caught are inadequate when applied to trophy fisheries which often provide unique angling experiences. Missouri's muskellunge fishery is one such program whose benefits cannot accurately be measured strictly in terms of angler harvest or other standard biological assessments.

Muskellunge were first released in Pomme de Terre Lake, located in southwestern Missouri, in 1966 by the Missouri Department of Conservation. The objective was to increase fishing diversity at this 7,800-acre, U.S. Army Corps of Engineers flood-control reservoir, and to provide anglers an added dimension in trophy fishing. The muskellunge population resulting from this and subsequent stocking has been studied to measure angler success and to develop methods of increasing survival of stocked fingerlings (Belusz 1975, 1978). The muskellunge fishery in this lake is maintained entirely by hatchery-produced fish; about 2,000 fingerlings (10–12 in) are released each year. Currently Pomme de Terre is Missouri's only muskellunge lake.

METHODS

Since 1975, the local chapter of Muskies, Inc. (MI), a national organization of muskellunge fishing enthusiasts, has sponsored an annual muskie tournament at Pomme de Terre Lake. Entrants in the 1979, 1980 and 1981 tournaments were asked to complete questionnaires for the purpose of gathering socioeconomic data. Methods of distributing and collecting questionnaires varied slightly over the three years, as did questionnaire items. Expenditure data obtained from the 1980 survey were used to refine or clarify the method of assigning expenditures to specific cost categories in the 1981 survey. Unless specified otherwise, reported data are from the most recent survey in 1981. Of the 269 entrants in 1981, 192, or 71%, returned their questionnaires after the initial mailing and one reminder mailing. Further information regarding questionnaire design and procedures is available from the authors.

To help secure reliable expenditure data, a pretournament meeting was held at which participants were reminded to keep track of expenses. To encourage attendance at this meeting, local muskellunge anglers and previous tournament winners presented a fishing seminar. Also, the majority of tournament entrants were repeat participants, familiar with the survey procedure, and thus expected to provide reliable information. Comparisons with preliminary expenditure data collected in 1980 indicated that record-keeping by participants was reasonably accurate and did not show wide ranges of fluctuations in expenditures from year to year. Participants were instructed to allocate only those costs specifically generated by the tournament, and very little adjustment of reported costs was necessary.

Expenditure data for anglers (N = 192) responding to the 1981 tournament questionnaire were placed in nine categories: lodging (motel, campground, and resort fees); groceries (snacks, and food purchased by entrants who prepared meals); restaurant meals; boat/motor (operational expenses, rental fees, or specific purchase for tournament); fishing equipment; travel (automobile operational expenses); entertainment (socializing in local establishments); general merchandise (all other non-food and non-fishing items); and tournament fees.

All expenditure categories except tournament fees were considered in calculating the daily cost of a muskellunge fishing trip for non-local and local anglers. Expenditures were divided by 3.8, the average number of days spent to fish the tournament, to calculate a daily fishing cost for all anglers in each group. The daily total was then divided by the number of anglers in the group to arrive at the daily cost per angler. Only expenditures generated by each angler group were considered in calculating costs for that particular group.

Results

Participant Characteristics, Fishing Success, and Evaluation

Background and motives.—Most entrants were Missouri males, 30 to 50 years old; about 60% resided in the Kansas City or St. Louis areas. Non-residents numbered 48, or 25% of the total response. Most of the non-residents were from Kansas (70%), with representatives from Arkansas, Oklahoma, Nebraska, Iowa, Illinois, and California.

Tournament participants devoted an average of 3.8 days to preparation for and participation in the event; the tournament itself lasted 2.5 days. Respondents were asked what their tournament involvement represented of the annual time they devoted to muskie fishing. One-third replied it accounted for most or all of their time; another third indicated about half the time; and the remaining third said the tournament was a rather small part of their total time spent fishing for muskie.

Fifty-two percent of all participants belonged to MI. Characteristics of participants were compared on the basis of group membership. Both MI members and non-members, on the average, devoted the same amount of time to fishing the event, about 21.5 hours. Differences were observed in (1) the number of years attending past competitions—about three tournaments for members to two for non-members; and (2) years experience fishing for muskies—6 versus about 3.5, respectively. Moreover, when asked to indicate on a scale from 1 to 10 the importance muskie fishing ranks in their angling interests, MI members averaged a higher interest rating ($\bar{x} = 7.2$) than non-members ($\bar{x} = 4.8$).

Participants were presented seven possible reasons or motives for participating in the tournament and asked to indicate whether each was very important, somewhat important, or not important in explaining their participation. Motive ranking was determined using a two-step scoring procedure (Gilbert 1977) to determine whether specific benefits were either *very important, somewhat important,* or *not important* reasons for participating in the tournament.

"Being with friends," "catching a muskie," and "meeting people" were *very important* to the re-

Table 1. Importance responses of 186 participants in the 1981 Pomme de Terre Lake muskellunge tournament for seven potential benefits.

Reason	Percent of respondents (N = 186)		
	Very important	Somewhat important	Not important
Very important			
Being with friends	68	27	5
Catching a muskie	66	28	6
Testing angling skills	47	42	11
Meeting people	46	43	11
Somewhat important			
Getting away on vacation	52	25	23
Competition aspect	22	54	24
Prospect of winning a prize	18	56	26

spondents (Table 1). When compared on the basis of membership in MI, however, non-members reported "testing angling skills" and "meeting people" as only *somewhat important* in explaining their involvement. MI members indicated that "meeting people" was a *very important* aspect of their participation, supporting the idea that relating to other anglers about muskie fishing is an important part of their interest in this gathering.

As a group, respondents saw "getting away on vacation," "competition aspect," and "prospect of winning a prize" as *somewhat important* in explaining their participation. Again, there was a difference when considering angler membership. Participants who were not members of MI said getting away on vacation was a *very important* factor. The fact that the competition and prize aspects are only *somewhat important* supports the notion that being a winner and getting recognition at the awards ceremony is on participants' minds, but they do not dwell on it to where they cannot enjoy their 2.5 days at the event. In fact, in this kind of tournament, it may be that the majority of anglers are no more occupied with the idea of doing well than during most fishing trips.

Additional insights into reasons for attending the muskellunge tournament were gained in the 1979 survey of participants. Entrants that year were asked to rate the importance of eight possible reasons for participating. These reasons were generally different from the motives reported. Reasons given as *very important* were "for the thrill of it," "to enjoy the outdoors," and "for the challenge." Reasons rated *somewhat important* were "to catch a trophy," for "family togetherness," and for "the exercise." Reasons *not important* in explaining participation were, "to show others my fishing skills" and "for food."

Fishing success.—Ten percent of the respondents caught at least one legal muskie, a fish 30 in or larger; 5% caught at least one sub-legal fish. Overall, angling success for muskie was 15%. Forty-one percent of respondents said they had at least one muskie "follow." Follows are easily observed characteristics of muskie; they follow a lure, often up to the boat, without striking. Follows, strikes, and swirls are significant aspects of muskie fishing, and, in the view of muskie anglers, are almost as highly regarded as catching a muskie. Considering these as well, 47% of the respondents reported some contact with a muskie during the tournament.

In all, 36 muskies were caught by survey respondents during 4,028 hours of fishing, for a catch rate of 0.009, or one fish for each 112 fishing hours. This tournament rate compares favorably with the catch rate of 0.01 (1 fish for each 100 fishing hours) common on northern lakes. As in all types of trophy fishing, catch-and-release does much toward sustaining trophy fish, and MI encourages this by offering harvest substitutes—called release patches. Of the 27 legal muskies caught during the tournament, 16 were released. Considering the success rate by membership in MI, 13% of the non-members caught at least one legal or sub-legal muskie, compared with 16% for members. This catch rate was 0.01 for MI members, and 0.007 for non-members.

Tournament evaluation.—When asked to evaluate muskie fishing in the tournament, 10% rated the fishing poor. Of those, none had experienced any contact with muskie during the tournament. Interestingly though, of the respondents who rated the fishing good or excellent (69%), almost one-half also had no muskie contact during the tournament. Possibly, for these participants seeing others catch muskies was sufficient for them to feel that the fishing was good to excellent, or that the prospects for catching a muskie were good to excellent.

Almost all participants (94%) were satisfied

Table 2. Expenditures reported by 192 participants in the 1981 Pomme de Terre Lake muskellunge tournament.

	Reported expenditures (dollars)			Average cost (dollars) per angler
	Local anglers (N = 27)	Non-local anglers (N = 165)	Total (N = 192)	
Lodging	150	3,198	3,348	17.44
Groceries	259	3,559	3,818	19.88
Restaurant meals	243	3,100	3,343	17.41
Boat/motor	407	3,521	3,928	20.46
Fishing equipment	730	8,715	9,445	49.19
Travel	266	6,246	6,512	33.92
Entertainment	162	1,047	1,209	6.30
General merchandise	30	1,302	1,332	6.94
Tournament fees	594	3,630	4,224	22.00
	2,841	34,318	37,159	193.54

with the tournament; 95% said they would participate again next year.

Economic Benefits

To avoid the constraints imposed by biological assessments, fishery managers should assume that the ultimate purpose in management of a fishery is to provide public benefits. A further assumption is that these benefits acquire a value equal to that which the public is willing to spend in obtaining them. For the anglers participating in the 1981 muskie tournament at Pomme de Terre Lake, the value of their experience could not be judged solely on the basis of expending 4,028 hours to catch 36 muskellunge. A more realistic and accurate measure, however, is the value of dollars spent in participating in that opportunity. These fishing opportunity dollars have a positive impact on the local and state economy and can be used to evaluate the significance of alternative management strategies (e.g., benefits of not stocking muskellunge or of stocking more).

Actual expenditures for all respondents totaled $37,159, or an average of $193.54 per respondent (Table 2). If expenditures are extrapolated to reflect all entrants (N = 269), total dollars spent by tournament participants were estimated to be $52,062. Since total expenditures were listed under specific categories, they could be allocated on the basis of where they were spent. Local economy benefits included expenditures in four of the nine expense categories (lodging, restaurant meals, boat/motor, and entertainment) totaling $13,172. Costs for travel (mainly the cost of gasoline), purchase of fishing equipment, groceries (used by entrants who camped), general merchandise purchases, and tournament fees were deleted since it was not possible to determine what percentage could be allocated locally. Realistically it could be assumed that gasoline was purchased for the return trip home, or that local marinas and tackle suppliers had the best inventory of muskellunge fishing tackle; however, it was deemed more prudent to present an admittedly conservative estimate of dollars spent locally. When extrapolated to include all entrants (N = 269), total dollars reportedly spent of benefit to the local economy were estimated at $18,455.

Dollars spent in a community, however, are not merely spent one time. They are used to purchase other goods or services and pay taxes, producing a multiplying effect. The income multiplier method has been used in a number of studies (Gilbert and Khayami 1973; Brown 1976; Weithman and Haas 1982) to account for direct and indirect (multiplier) income derived from expenditures on a regional basis. Income-multiplier values for southwest Missouri were applied to angler expenditures at Pomme de Terre Lake (S. Weithman, Missouri Department of Conservation, Columbia, Missouri, USA [personal communication]).

Multiplier values vary for different expenditure categories of goods and services. For example, lodging costs are increased by a factor of 2.03 for each dollar spent. Restaurant meals have the lowest multiplier value (1.71), indicating that a dollar spent in this category has a more limited effect on the economy. For the four categories of expenditures included in the conservative estimate of local community benefits, the income multiplier method suggested that the real value of tournament angler expenditures increased from $13,172 to $25,148 (Table 3). When extrapolated to include all entrants (N = 269), local community benefit dollars were estimated to be $35,233. In terms of benefit to the state economy, when all entrants and all categories of expenditures were considered and adjusted by appropriate multiplier values, the 1981 muskellunge tournament was worth an estimated $100,519 to the economy of the state of Missouri.

Table 3. Community benefit value of muskellunge tournament angler expenditures (N = 192) when calculated on the basis of regional value multipliers.

	Reported expenditures		Value multiplier		Estimated community benefits
Lodging	$ 3,348	×	2.03	=	$ 6,796
Restaurant meals	4,687	×	1.71	=	8,015
Boat/motor	3,928	×	1.97	=	7,738
Entertainment	1,209	×	2.15	=	2,599
	$13,172				$25,148

Annual Value of the Muskellunge Fishery

About 78% of all muskellunge fishing trips (3,542 of an estimated 4,570 fishing days) at Pomme de Terre Lake are made by non-local anglers principally from the St. Louis, Kansas City, and Springfield areas (R. J. Dent, Missouri Department of Conservation, Sedalia, Missouri, USA [personal communication]). Their expenses are not unlike those generated by participants at the annual tournament, since 60% of tournament entrants are from metropolitan areas. Similarly, expenses for most local muskellunge anglers (residents of counties adjacent to the lake) would not be appreciably different from those reported by local residents in the tournament survey. Consequently, the categorization of expenditure data would allow calculation of the daily cost of a muskellunge fishing trip for each angling group (local and non-local) not associated with a special event, such as a tournament.

Estimates of daily costs for muskellunge anglers at Pomme de Terre Lake were $48.94 for non-local anglers and $21.90 for local anglers. Inasmuch as a cost of $17 per day was estimated for Missouri anglers during 1980 (USFWS 1982), trip costs reported here may seem high. However, because muskellunge represent a specialized trophy fishery, with limited availability and no substitute sites in Missouri, it does not seem unreasonable that anglers would assume the greater cost and effort required for participation. Additionally, since the expenditures data reported here were collected within weeks of the event (as compared to over a year for the U.S. Fish and Wildlife Service survey), recall of expenditures data by tournament respondents was assumed to be more reliable.

Estimates of daily fishing costs for muskellunge at individual lakes are not readily available from the literature. However, the estimated value of the muskellunge fishery at Lake Chautauqua, New York, provided a useful comparison. A postcard census of muskellunge anglers in 1965 estimated the value of the fishery at $1.8 million dollars annually (Mooradian and Shepherd 1973), or about $14.83 per fishing day. If this value is adjusted by the Consumers Price Index in terms of 1981 dollars, the cost per fishing day is about $28.84. In comparison, the income multiplier value of expenditures by trout anglers at Lake Taneycomo, Missouri, in 1979 was $22.75 per day (Weithman and Haas 1982), while Martin et al. (1982) estimated a mean consumers surplus value of $86 per day for bass club members at Lake Mead, Nevada.

Assuming that the values generated for the cost of a muskellunge fishing trip for local and non-local anglers are reliable and somewhat conservative, the annual worth (excluding tournament value) of the muskellunge fishery to the state economy in 1981, adjusted by value multipliers, was estimated at about $379,000. This estimate was made by multiplying the adjusted tournament daily expenditure values for local ($42.52) and non-local ($94.88) anglers by the corresponding estimated number of days fished (1,028 local, 3,542 non-local, R. J. Dent, personal communication) and adding the products. When the estimated value of the tournament is included, total worth of the muskellunge fishery in 1981 was about $479,000.

On a per fish basis, the stocking cost per fingerling at Pomme de Terre Lake in 1981 was $17.84 (4 year average, 1977–82; range $8.59–$36.97), which includes rearing, distribution, and overhead costs. Based on what muskie anglers were willing to spend in 1981, each muskie caught was worth $503—approximately a 30-fold increase in benefits over cost. This value was obtained by dividing the annual muskellunge fishery value ($479,000) by the estimated harvest of muskellunge (952; R. J. Dent, personal communication). At Pomme de Terre Lake, the average return of stocked muskellunge (through Age VII) to the creel from 1966 to 1972 was 39% (Belusz 1978). More recent estimates show an average return of 31% (range 23%–44% through Age X, R. J. Dent, personal communication). Adjusting for survival to harvestable size and using a return of 31% increases production costs to $57.52 per legal fish and reduces benefits from 30:1 to 9:1.

Management Implications

Though the task of state fish and wildlife agencies is resource management, the ultimate products of their management activities are opportunities for appreciation of fish and wildlife. As evidenced by the high degree of satisfaction expressed over the present quality of muskellunge fishing and anglers' willingness to spend substantial amounts of money in pursuit of this opportunity, there is little doubt that the fishery provides anglers with a high-quality outdoor experience. Indeed, the information gathered in this survey supports the notion that muskellunge fishing in Missouri is analogous to hunting big game like deer and turkey. In these activities, the thrill of pursuing the quarry often provides enough satisfaction.

Because the muskellunge grows large, is elusive and a fighter, the taking of even a sub-legal muskie is acknowledged as a challenge which can be met only infrequently—catch rates at Pomme de Terre Lake average 0.027 fish/hr (R. J. Dent, personal communication). Muskellunge anglers thus come to equate success with a muskie "follow" or "raise," finding excitement in speculating on the size of the fish. For them, the "one that got away" is the rule, rather than the exception. All of this, which is referred to as the muskie mystique, contributes to the development of a "big-game frame of mind." MI chapter members at Pomme de Terre Lake take this idea even further by releasing virtually all legal muskellunge caught (93% release rate in 1982; K. Richards, Missouri Department of Conservation, Sedalia, Missouri, USA [personal communication]), helping to insure stocks of trophy-size fish.

Unlike regulations, which artificially set "quality" size and catch limits in attempts to satisfy diverse angler needs, the natural lure of trophy fishing can do much to assist agencies in meeting ever increasing angler demand for something new and challenging. Highly regarded species like muskellunge produce perceptions of quality simply by their presence, and so require only limited availability to yield positive angler feedback. Managers should recognize that today's anglers are very aware of the new opportunities made possible through introductions of "exotic" species, and that they welcome attempts to establish such fisheries even on a limited basis, especially for a trophy experience.

The introduction of non-native game fish has often been criticized, and, perhaps, in some instances with good cause. However, where such introductions do not dramatically or irreversibly alter the character of the native fishery, and where sportsmen are afforded new and unique fishing opportunities, it appears that the value of the resource will be enhanced by such diversity. Moreover, when managers begin to scrutinize these special trophy fisheries with other than standard biological techniques, they may discover overlooked economic values which can more than compensate for seemingly low program benefits. However, few state agencies, particularly those receiving only license fees and/or legislative appropriations, benefit directly from an economy stimulated by a particular fishery program. For those agencies, perhaps economic information, such as provided here, would help justify increased appropriations, or gain more public support when they seek increased license fees.

In Missouri, anglers who previously had to travel to northern states for muskellunge (and thereby spend money outside Missouri) can now fish for them within their own state, and thus would also have some positive impact on the state economy. There is danger, however, in attempting to increase fishing diversity in all waters capable of supporting a particular sportfish. Part of the unique attraction of muskie in Missouri stems from their being confined to one reservoir. Additional fishing waters could dilute the feeling of uniqueness associated with muskie fishing and adversely impact those communities which currently have come to depend on muskellunge-generated dollars. Possible adverse economic impacts or negative public reactions should be weighed, along with the likely benefits, in any decision to provide additional trophy-fishing waters.

The social and economic information collected by the tournament survey has provided an assessment of the value which anglers placed on a muskellunge-fishing opportunity. Additionally, if the need arises, these data can assist in decision making when biological realities and fiscal constraints must be balanced.

Acknowledgments

The authors wish to thank the Pomme de Terre Chapter of Muskie's Inc. for assistance and support with the tournament surveys. Thanks also to Rich Wehnes and Stan Michaelson for reviewing the manuscript.

References

BELUSZ, L. C. 1975. The use of isolation coves in assessing muskellunge stocking mortality. Proceeding of the Annual Conference of the Southeastern Association of Game and Fish Commissioners 29:251–253.

BELUSZ, L. C. 1978. An evaluation of the muskellunge fishery of Lake Pomme de Terre and efforts to improve stocking success. Pages 292–297 *in* R. L. Kendall, editor. Selected coolwater fishes of North America. Special Publication 11, American Fisheries Society, Bethesda, Maryland, USA.

BROWN, T. L. 1976. The 1973–75 salmon runs: New York's Salmon River sport fishery, angler activity, and economic impact. New York Sea Grant Institute publication NYSSGP-RS-76-025.

GILBERT, A. H. 1977. Vermont hunters: characteristics, attitudes and levels of participation. Agricultural Experiment Station Miscellaneous Publication 92, University of Vermont, Burlington, Vermont, USA.

GILBERT, A. H., AND S. M. KHAYAMI. 1973. Expenditure patterns of non-resident sportsmen in Vermont, 1970. Agricultural Experiment Station Miscellaneous Publication 78, University of Vermont, Burlington, Vermont, USA.

MARTIN, W. E., F. H. BALLMAN, AND R. L. GUM. 1982. Economic value of Lake Mead fishery. Fisheries 7(6):20–24.

MOORADIAN, S. R., AND W. F. SHEPHERD. 1973. Management of muskellunge in Chautauqua Lake. New York Fish and Game Journal 20:152–157.

USFWS (U.S. DEPARTMENT OF INTERIOR, FISH AND WILDLIFE SERVICE, AND U.S. DEPARTMENT OF COMMERCE, BUREAU OF THE CENSUS). 1982. 1980 national survey of fishing, hunting, and wildlife-associated recreation. U.S. Government Printing Office, Washington, D.C., USA.

WEITHMAN, A. S., AND M. A. HAAS. 1982. Socioeconomic value of the trout fishery in Lake Taneycomo, Missouri. Transactions of the American Fisheries Society 111:223–230.

Pages 46–56 *in* G. E. Hall and M. J. Van Den Avyle, editors.
Reservoir Fisheries Management: Strategies for the 80's.
Reservoir Committee, Southern Division American Fisheries Society, Bethesda, Maryland. 1986.

ASSESSMENT OF FISH POPULATIONS AND MEASUREMENT OF ANGLER HARVEST

Indices as Predictors of Fish Community Traits

ROBERT F. CARLINE
Ohio Cooperative Fishery Research Unit[1]
The Ohio State University
Columbus, Ohio 43210

ABSTRACT

Objectives of this paper are to review the use of indices as predictors of fish standing crops, yields, and community structure and to outline some alternative approaches to reservoir research. Considerable progress has been made in predicting standing crops and yields in lakes and reservoirs from relatively simple indices. Earlier work relied mostly on abiotic variables, but more recent studies have incorporated biotic parameters into predictive indices. Present models may account for over 90% of the variability in yield or standing crops from a subset of lakes. In contrast, little progress has been made in relating fish community structure to environmental variables. Different ways of describing community structure should be explored. Besides taxonomic groupings, fishes can be grouped by trophic level, by habitat use, and by reproductive mode. Relating community characteristics to environmental variables may be best accomplished by first classifying reservoirs according to major habitat features and then proceeding with analyses of individual classes.

In this paper I review some uses of indices as predictors of fish community traits, specifically standing crops, yields, and community structure, and suggest approaches to reservoir biology that merit further attention. Community is used in the broadest sense throughout the paper to include all game and nongame fish species.

Indices can be labeled descriptive or predictive. Fisheries biologists have long used descriptive indices to characterize various attributes of fish populations. For example, catches of young-of-year (YOY) have been used as indices of recruitment and gonad to body weight ratios have been employed as indices of maturity. Such descriptive indices can be particularly useful when the attribute of interest cannot be easily measured. At the community level, indices have been used to describe species diversity in a variety of ways (Pielou 1977). While such tools have undoubtedly been useful in describing populations or communities, perhaps the most valuable use of indices has been as predictors of community traits, such as standing crops and yields. I use index to denote a simple or transformed variable or a small set of variables expressed as a formula.

Over the past 15 to 20 years considerable progress has been made in predicting fish standing crops and yields from a variety of aquatic habitats. Early attempts at correlating fish standing crops to environmental variables were motivated by the need to predict carrying capacities of proposed reservoirs (Rounsefell 1946). More recently, development of this predictive capability has been fervently pursued, because it provides biologists with a basis for estimating yields and thus formulating management plans. Ideally, indices derived from one or a few readily obtainable measurements would be all that are needed for making predictions about individual systems. Use of readily measured variables is particularly important to reservoir biologists, because the size and complexity of impoundments make intensive assessment surveys expensive.

Development of predictive indices does not require an understanding of how fish communities function nor are large amounts of data essential. Alternatively, reliance on complex mathematical models for predictions requires quantitative descriptions of important biotic and

---

[1] The Unit is jointly sponsored by the U.S. Fish and Wildlife Service, Ohio Department of Natural Resources, and The Ohio State University.

Table 1. Summary of percentages of variation explained by single-variable and multivariate models used to predict fish yield and standing crop in lakes and reservoirs from abiotic and biotic variables.

Independent variables tested[a]	Model type[b]	Percent of variation explained	Source
	Yield from natural lakes		
Mean depth	S	28–92	Rawson (1952)
Mean depth; MEI	S	54–73	Ryder (1965)
Phytoplankton production	S	57–82	Melack (1976)
Phytoplankton standing crop; phytoplankton production	S	74–84	Oglesby (1977)
Benthic biomass; MEI	S	62–83	Matuszek (1978)
Surface area	S	94	Youngs and Heimbuch (1982)
Length of growing season	S	74	Schlesinger and Regier (1982)
Total phosphorus; TDS; MEI; benthic biomass; mean depth	S	54–73	Hanson and Leggett (1982)
Mean depth; MEI	S	63–92	Prepas (1983)
Mean depth; TDS	M	46–76	Ryder (1965)
Mean depth; TDS	M	63–68	Matuszek (1978)
Surface area; mean depth; TDS	M	95–97	Youngs and Heimbuch (1982)
Length of growing season; MEI	M	81–83	Schlesinger and Regier (1982)
Surface area; mean depth; TDS; total phosphorus; benthic biomass	M	40–97	Hanson and Leggett (1982)
	Yield from reservoirs		
Chlorophyll *a*; total phosphorus	S	52–83	Jones and Hoyer (1982)[c]
Mean depth; MEI	S	8–20	Jenkins (1982)
Length of growing season; surface area; reservoir age	M	17	Jenkins and Morais (1971)
	Standing crop in natural lakes		
Alkalinity	S	27–41	Carlander (1955)
Benthic biomass ÷ mean depth; mean depth; benthic biomass	S	20–83	Hanson and Leggett (1982)
Surface area; mean depth; benthic biomass; benthic biomass ÷ mean depth	M	35–83	Hanson and Leggett (1982)
Length of growing season; surface area; Secchi-disc depth; macrophyte abundance; panfish index; rough fish index	M	44–56	Schneider (1978)[d]
	Standing crop in reservoirs		
Alkalinity	S	69	Carlander (1955)
Retention time	S	72	Jenkins (1976)
Annual outflow volume; retention time; surface area	S	9–31	Aggus and Lewis (1978)
MEI	S	21–72	Jenkins (1982)
Total phosphorus; Secchi-disc depth; chlorophyll *a*	S	16–41	G. R. Ploskey (Aquatic Ecosystems Analysts, Fayetteville, Arkansas, USA, personal communication)
Length of growing season; annual outflow volume; shoreline development index; TDS	M	52	Aggus and Lewis (1978)

[a] Abbreviated variables are MEI = morphoedaphic index, TDS = total dissolved solids.
[b] S = single-variable model; M = multivariate model. Where more than one single-variable model was tested, the independent variable giving the best predictions is listed first.
[c] Some data from lakes were included.
[d] Some data from reservoirs were included.

abiotic interactions. While this approach has received a great deal of attention, no fishery has been successfully managed solely on results from such models (Walters 1980). Our understanding of reservoir dynamics is still too limited for the construction of detailed models, although simplified models may adequately describe reservoir dynamics and generate testable hypotheses (Ploskey and Jenkins 1982). Until realistic simulation models become available, further refinement and development of simple indices offer the most promise for enhancing our ability to predict standing crops, yields, and species composition.

Indices of Standing Crops and Yields

In a summary of published studies (Table 1), I have attempted to provide a brief overview of environmental variables related to standing crops and yields and to show ranges in the amount of variability that has been explained. Authors have employed both simple and multiple regression techniques and have tried a variety of transformations (usually logarithmic) and variable combinations to derive the highest possible coefficients of determination. I have not incorporated these details into this summary. Rather, I have simply indicated the individual variables or combination of variables that produced significant correlations.

Early attempts at relating yields and standing crops to environmental variables were restricted to simple regressions. Rounsefell (1946) correlated yields to lake surface area, Rawson (1952) used mean depth, Carlander (1955) used alkalinity, and Moyle (1956) used phosphorus. Ryder (1965) then combined total dissolved solids and mean depth in his morphoedaphic index (MEI) to successfully predict fish yields. The usefulness of MEI as a predictor of yields has been re-examined a number of times and has been compared to other simple indices (Hanson and Leggett 1982; Youngs and Heimbuch 1982; Prepas 1983). While some disagreement remains among authors as to which single variable best predicts fish yield, the MEI clearly ranks among the best for both natural lakes and reservoirs (Jenkins 1982). Recent studies have used additional edaphic factors, such as total phosphorus and biotic variables, including chlorophyll *a*, phytoplankton biomass and production, and benthos to develop rather impressive correlations with standing crops and yields (Table 1). Biotic variables were oftentimes the best predictor of yield or standing crops, even though they have not been used as frequently as abiotic variables.

In both natural lakes and reservoirs, the same variables produced significant correlations with yields and standing crops. The one exception was mean annual retention time, which is important in reservoirs, but is of minor significance in lakes where retention times are often rather long (i.e., several years). The best correlations for yield and standing crops with MEI in reservoirs were derived when reservoirs were subdivided into similar operational or chemical groups—hydropower mainstream, hydropower storage, and chemical types of nonhydropower reservoirs (Jenkins 1982). Retention times were similar within groups; hence the influence of retention time on productivity was taken into account.

With continued refinement of available regression models and accumulation of more data, particularly biotic variables, we can look forward to some increase in predictive capabilities. However, it may be unrealistic to hope for indices that will account for more than 80–90% of the variability in standing crops or yields from reservoirs. Year-to-year variability in fish populations and climatic factors, together with sampling difficulties, will always contribute some unaccountable error to regression models. Estimates of yield may be improved by incorporating fishing pressure into predictive models (Jenkins and Morais 1971; Schlesinger and Regier 1982), but precise standing crop estimates will continue to be difficult to obtain in large reservoirs. Despite sampling problems and natural variability, present capabilities to predict standing crops and yields in reservoirs are quite good, and further improvement in the future is likely to be modest.

Developing Indices of Community Structure

Biologists dealing with lentic systems have devoted much effort to predicting correlates of productivity (mainly standing crops and yields), yet they have seemingly ignored developing predictive indices of fish community structure. This is somewhat curious because managers cannot readily influence the basic productivity of large lakes and reservoirs, yet they can modify fish community structure. Primary productivity has been altered in small lakes (Cooke and Kennedy 1981) and infrequently in large reservoirs. However, fish community structures in reservoirs have been modified by regularly stocking predators, introducing forage fishes, manipulating water levels to enhance spawning, etc. Despite these management successes, there are few generalizations in the literature regarding interactions of reservoir productivity, habitat features, and fish community structure. I suggest that these interactions merit more attention and that efforts toward developing predictive indices of fish community structure will provide immediately usable results and foster increased understanding of reservoir dynamics. In this section I first suggest different ways to describe community structure and then examine reservoir communities in relation to habitat features.

Describing Community Structure

I use the phrase community structure to mean the relative proportions of species or species groups according to their numbers or weight. There seems to be no consistent use of numerical or biomass units as community descriptors. Fish densities will vary more than biomass, particularly on a seasonal basis, because of the dynamics of YOY fishes. Hence, biomass will most often be the unit of choice.

A more challenging problem is how to group species into meaningful units. Groupings can be based on at least four criteria: 1) taxonomic, 2) trophic position, 3) habitat use, and 4) reproductive mode.

Fish communities are frequently described on the basis of presence of each species or relative numerical proportions. McDonough and Barr (1978) described reservoir species associations in relation to elevation in the Tennessee and Cumberland river systems. Numbers of species increased from headwater to mainstream impoundments. Similarly, location of natural lakes in the watershed and presence of connecting waterways are important in affecting species assemblages and diversity (Tonn and Magnuson 1982). However, in making generalizations about reservoirs across drainages, taxonomic grouping above the species level seems appropriate, because morphologically and ecologically similar species frequently occur in different watersheds. Taxonomic descriptions of communities will often be complicated by introductions of exotic freshwater and marine species in addition to various hybrids. In such instances, alternate modes of characterizing communities may be warranted.

Descriptions of fish communities based on trophic position or foods eaten by species groups have been used for streams (Grossman et al. 1982) and reservoirs (Ploskey and Jenkins 1982). The difficulty with this classification is that fish change trophic position as they grow; a typical life history sequence for a top predator would be from planktivore to insectivore to piscivore. The situation is further complicated, because fish tend to be food generalists (Keast 1978), and can simultaneously exploit two or more trophic levels. Ploskey and Jenkins (1982) tried to overcome this difficulty by apportioning food consumption of different size groups of fish according to empirical data. For computer simulation modeling this procedure has merit, but for static descriptors of community structure it has obvious limitations. A complex classification scheme may be needed to reflect this plasticity in feeding habits. Thus, some categories will span two or even three trophic levels such as 1) zooplanktivore-piscivore, 2) invertebrate omnivore-piscivore, 3) benthivore-planktivore, 4) detritivore-benthivore, etc.

Fishes may be grouped according to habitat use. In natural lakes the littoral area (Werner et al. 1977) and presence of macrophytes (Tonn and Magnuson 1982) have marked effects on species assemblages. In reservoirs, destruction of littoral areas and embayments can result in drastic shifts in community structure (Walburg 1977). In addition to structural features, one might also consider thermal characteristics in describing habitats. Thermal heterogeneity allows coexistence of fishes with substantially different temperature tolerances. Thus, the degree of thermal heterogeneity, particularly during summer stratification, will influence the array of species a reservoir can support. A thermally restrictive environment may substantially curtail fish movement, as Schaich and Coutant (1980) found for striped bass (*Morone saxatilis*). Thermal preferences and habitat use may be combined into categories such as warmwater-littoral, warmwater-pelagic, coldwater-pelagic, etc. Given the importance of habitat in influencing species assemblages, its inclusion in describing communities should be worth pursuing.

The fourth approach to grouping fishes is through their reproductive biology. Balon (1975) classified fish according to reproductive guilds, which were used as a basis for explaining success of percids in different environments (Balon et al. 1977). Clady (1978) had little success in relating reproductive guilds to community structures of lakes with percids. In reservoirs, particularly ones that have short retention times and large fluctuations in water levels, modes of reproduction would seem important in determining which species are successful. In addition, timing of reproduction should be considered. Proximity of peak spawning times among predator and prey species determines sizes of YOY fishes and consequently how well YOY predators can exploit YOY forage. For example, walleyes (*Stizostedion vitreum vitreum*) spawn early and their young are usually large enough to eat YOY gizzard shad (*Dorosoma cepedianum*) throughout the growing season. However, where spawning of black bass and gizzard shad overlap in time, few YOY bass are large enough to prey on young gizzard shad. The effect of piscivory on growth of some fish is marked. In Pleasant Hill Reservoir, Ohio, gizzard shad are the major food item for YOY walleye, which average 207 mm in mid-September. In contrast, most YOY smallmouth (*Micropterus dolomieui*) and largemouth bass (*M. salmoides*) average only 86 mm in mid-September and are not large enough to prey on gizzard shad. Timing and modes of reproduction will no doubt influence the success of a species in a given environment and I suggest these characteristics might be profitably used to describe community structure.

In the above examples I proposed some different ways of describing fish communities; there may well be other equally effective modes of description. Once community structures have been characterized, the second phase of developing predictive indices is that of adopting a strategy for relating community structure to habitat features.

Reservoirs Within the Lake-Stream Continuum

In developing a framework for understanding factors affecting community structure in reservoirs, it may be helpful to view reservoirs as an intermediate habitat type between natural lakes and streams (Regier and Henderson 1973). Perhaps one of the most divergent features of these habitats is annual rate of water movement through the system, i.e., hydraulic retention time. Differences in retention times between small streams and large lakes are six orders of magnitude (G. Ploskey, Aquatic Ecosystem Analysts, Fayetteville, Arkansas, USA [personal communication]). Large deep natural lakes have long retention times, e.g., 180 years for Lake Superior. At the opposite extreme a first order stream with an average velocity of 0.5 m/second and 3 km long would have a retention of about 1.7 hours or 0.0002 years. From a sample of 180 reservoirs in the United States, retention times ranged from 0.004 to 7.5 years, a span of about three orders of magnitude (G. Ploskey, personal communication). Clearly, reservoirs represent habitats with widely divergent physical features and their similarity to lakes or streams ought to be a function of their location within this continuum. Examination of properties of communities in both lakes and streams should then provide some insight into reservoir communities.

Ryder and Kerr (1978) provided an intriguing representation of fish communities in lakes of Canada's boreal forest zone (Fig. 1). They proposed that certain communities, mainly salmonid, percid, and centrarchid, exhibit a high degree of persistence. Ryder and Kerr (1978) labeled these assemblages of coevolved species "harmonic communities." Intermediate assemblages, they suggest, are astatic and are less persistent than harmonic communities. Thus, one can infer that certain mixtures of percids and centrarchids, although they may represent natural assemblages, will be less stable than the harmonic communities. While conceptually attractive, these ideas remain to be tested with empirical data. Relatively constant commercial yields from a large number of percid lakes (Adams and Olver 1977) provide circumstantial evidence that entire communities are persistent. However, long term studies of entire fish communities in natural lakes appear to be nonexistent. Ecologists studying fishes in lakes generally focus on processes such as resource allocation and predation among certain species, but not on entire communities. For the moment we shall have to rely on circumstantial evidence to infer a high degree of persistence for fish communities in natural lakes.

While there is a paucity of good descriptive studies of entire fish communities in natural lakes, a large data set exists for reservoirs, because of extensive use of cove-rotenone sampling to estimate standing crops. Ironically, there are few published studies of fish community structures in reservoirs. A good example of how fish communities in reservoirs can be related to environmental features is provided by Jenkins (1976). He used regression techniques on data from 20 Oklahoma reservoirs. Standing crops of all species combined and that of all bottom feeders were negatively related to retention time. Data for standing crops of other species and retention time were best described by curvilinear or polynomial regressions. Reasons for these changes in biomass were not evident, but the effect of retention time was apparent.

Small streams have been extensively studied by ecologists, probably because they can be readily sampled with inexpensive gear. A good example is a recent study by Grossman et al. (1982) in which fish populations were monitored for 12 years. They used taxonomic and trophic descrip-

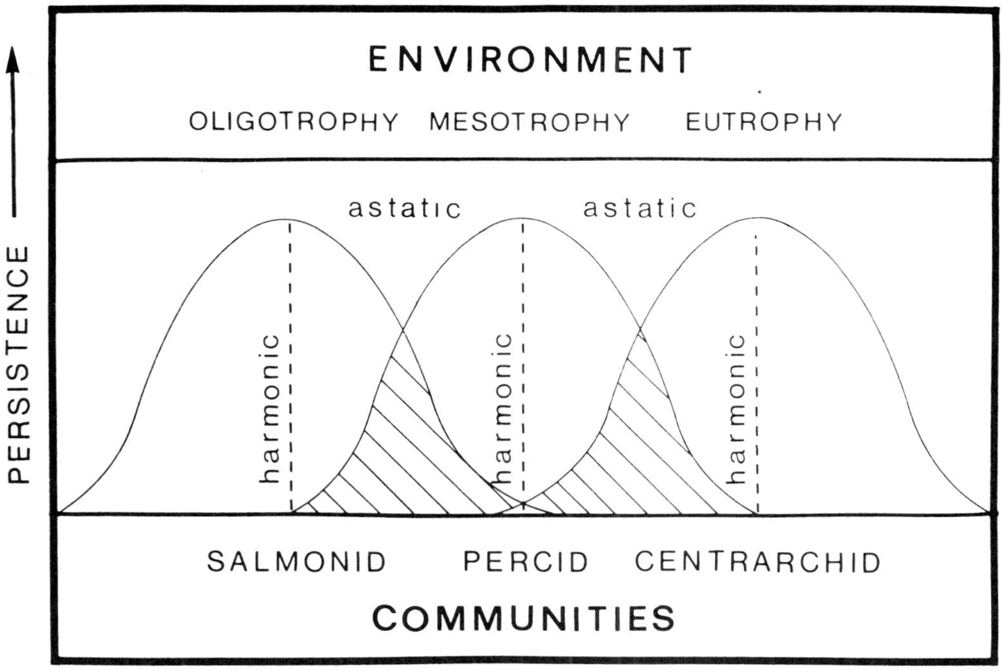

Figure 1. Schematic representations of there fish communities of the boreal forest zone. The vertical dotted lines show the medians of central tendency, producing harmonic species associations complementary to prevailing environmental conditions. This, in turn, results in high levels of community persistence and integrity. The crosshatched areas indicate astatic species associations of low persistence and variable levels of community integrity. (From Ryder and Kerr 1978.)

tors to assess persistence of community structure and found no relationships among species assemblages from year to year. They concluded that stochastic processes brought about by environmental instability were responsible for this lack of persistence. Rapid changes in stream flow were thought to disrupt reproduction. Variation in flow appeared to be a major influence in other stream studies. Based on Grossman et al. (1982) and other studies they reviewed, one might conclude that at the end of the retention time gradient, small streams characteristically support unstable fish communities in contrast to natural lakes where persistent communities are common.

I believe there is another class of streams that support persistent communities. Lawrence Creek, Wisconsin, is an example of a typical trout stream in the glacial till region of the northcentral United States in which groundwater inputs maintain stable stream flows throughout the year. Hunt (1974) found remarkable consistency in annual production (instantaneous growth × mean biomass) of brook trout (*Salvelinus fontinalis*); over an 11-year period production varied by less than 20%. Data on mottled sculpins (*Cottus bairdi*), the second most abundant species, were limited to four years of population estimates; biomass ranged from about 1.1 to 3.8 g/m^2, which was more variable than brook trout yet still fairly stable. No data are available for other species in the stream, but their combined biomass would probably not exceed 5% of the total fish community. In another long-term study of brook trout in a stable stream in Michigan, McFadden et al. (1967) found that autumn densities of all age groups varied by a maximum of 68% over a 14-year period. In contrast, Grossman et al. (1982) found that total catches of the most numerous species had annual ranges as large as 0 to 1,700 and that 6 of the 10 most numerous species had abundance ranks ranging from 1 to 10 over the 12-year period. Although data from these brook trout streams are incomplete, I believe it is safe to conclude that the communities in these stable streams are vastly more persistent than that described by Grossman et al. (1982). Differences

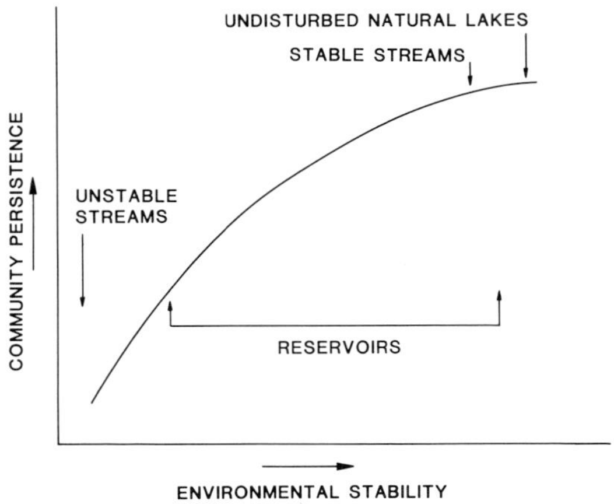

Figure 2. Hypothesized relationship between fish community persistence and environmental stability, indicating relative positions of different habitat types.

in community persistence can be attributed to environmental instability, which Grossman et al. (1982) suggested was the major factor influencing changes in species abundances.

From the above examples it is apparent that community structure is influenced by productivity (Ryder and Kerr 1978; Jenkins 1976) and habitat features (Walburg 1977; Tonn and Magnuson 1982). The other community attribute of significance is persistence, which appears to be a function of environmental stability. As with retention time, reservoirs span a broad range on the environmental stability axis (Fig. 2). Thus, one could infer that reservoirs with little year-to-year fluctuation in water level should support relatively persistent communities while reservoirs that have large annual fluctuations in water level would likely support communities with variable structures. It is not clear, however, if environmental stability affects community structure per se, or just persistence. In either case it would appear useful to group reservoirs according to a few important environmental variables before proceeding with detailed analyses of community structure.

Resource Classification

In the following examples of fisheries resource classification from Iowa and Wisconsin lakes, different approaches were used to group lakes and then compare lake characteristics within and among groups. Because these lakes are from limited geographical areas, the first classification step was in fact a regionalization. The main difference in approaches was choice of criteria for initial classification. Iowa lakes were first classified according to habitat variables and the Wisconsin lakes were initially grouped by fish community characteristics.

Iowa Lakes

In an ongoing study, Hill (1982) used mean basin slope and an index of siltation to classify 57 public fishing lakes into five categories. The siltation index reflected each watershed's susceptibility to erosion. Fish species composition and standing crops were determined from cove-rotenone samples for three lakes in each category.

Initial results suggested that lakes with the steepest basin slope (>12%) and the lowest siltation index supported the lowest total standing crops, but the largest proportion of acceptable size game fish. This group represented the best fishing lakes. At the other end of the scale, lakes with a small basin slope (<7%) and a high siltation index supported the highest standing crops of fish, but the lowest proportion of acceptable size game fish. Lakes with the intermediate slopes and/or siltation indices appeared to support fish communities with intermediate characteristics. Thus, two environmental factors were related to both productivity and size structure of game fish populations. Providing additional data corroborate initial findings, this straightforward classification system should offer managers a simple way to judge which lakes have the most potential for producing desirable game fish populations and to predict fisheries characteristics in planned impoundments.

Northwestern Wisconsin Lakes

Walleye abundance varies greatly among northern Wisconsin lakes. Walleyes may range from a few percent to over 90% of the total game fish biomass. In general, biomass proportions of walleye and centrarchids are inversely related. I analyzed survey data from 71 lakes in an attempt to answer the following questions: "What conditions favor domination of the fish community by walleye or centrarchids?" and "Is it possible to identify lakes that do not support walleye but have a high potential for doing so if they are stocked regularly?" Recruitment of walleyes was not a limiting factor in lakes selected for this

Table 2. Watershed and lake characteristics (mean and standard deviation in parentheses) for northwest Wisconsin lakes having relatively low, moderate, and high walleye abundance expressed as a proportion of total gamefish biomass.

Characteristic	Relative walleye abundance		
	Low	Moderate	High
Walleye biomass (% of all gamefish)			
Mean	12	44	78
Range	0–30	31–60	61–94
Watershed area (km^2)[a]	84 (158)	46 (53)	340 (1,108)
Lake type			
Drainage (%)	74	70	53
Seepage (%)	24	30	41
Outlet flow (m^3/second)[a]	0.87 (1.8)	0.34 (0.47)	3.1 (11)
Alkalinity (ppm as $CaCO_3$)	50 (25)	41 (21)	30 (18)
Gravel area (%)[a]	22 (17)	20 (13)	17 (12)
Rubble area (%)	6 (12)	13 (15)	9 (15)
Macrophyte area (%)[a]	18 (17)	10 (6)	7 (5)
Dimictic (%)[a]	50	60	18

[a] Included in discriminant analyses.

analysis. All lakes either supported naturally reproducing populations, were stocked regularly, or both. Numbers of game fish collected by electrofishing or trapnets were converted to weights and community structure was expressed as biomass proportions.

Simple and multiple regression analyses indicated that walleye biomass proportions were related to maximum depth, outlet flow, lake type, amount of rubble and gravel bottom, stratification pattern, and amount of submersed macrophytes. Coefficients of determination were often significant, but too low for making reliable predictions about community structures.

When lakes were grouped according to walleye biomass as a percentage of total game fish biomass, there were apparent differences in the means of several environmental variables, but variances were large (Table 2). No single variable or combination of variables could be used to reliably classify lakes to appropriate groups. I then used discriminant analysis to determine which combination of variables would best classify lakes. A discriminant function that used six variables (Table 2) correctly classified 57% of the lakes into three categories. Lakes with low or medium proportions of walleyes were most frequently misclassified. These groups were then combined into a single group so that the two new categories were 0–60% walleye and >60%. Another discriminant function was derived and classification accuracy increased to 82%.

Two sets of results emerged from this analysis. The first was a general description of lakes with low or high proportions of walleyes. For example, walleye dominant lakes had relatively large drainage areas and outlet flows, low alkalinity, few submersed macrophytes, and usually did not stratify during summer. This description does not explain why walleyes dominate these lakes, but it does provide a basis for further research. The second result is that discriminant analyses provide a means of identifying lakes that currently do not have walleyes but have a high potential for supporting walleye fisheries if they are stocked regularly.

In the above examples the initial classification step was based on environmental factors or community attributes. Choice of initial classification criteria can depend upon established (or perceived) relationships or the kind of question being asked. In either case the important result is that generalities emerge. It would be convenient if successful classification could be achieved with just one or two simple indices as in Hill's (1982) work. However, where communities are not strongly influenced by a few environmental factors, more complicated procedures may be needed.

Conclusions

Considerable progress has been made in formulating predictive indices of fish standing crops and yields from natural lakes and reservoirs. Further refinements of these indices are likely to be relatively small, because of inherent biases in sampling and natural year-to-year variability. I suggest that the next step should be development of predictive indices for fish community structure. In a synopsis of the morphoedaphic index, Oglesby (1982) urged further research on relationships among MEI, yield, and fish community structure in various types of lakes and fisheries. By combining predictions of standing crops and community structure, expected densities of individual species or groups of species can be determined for individual systems. Along with densities, size structures of populations are im-

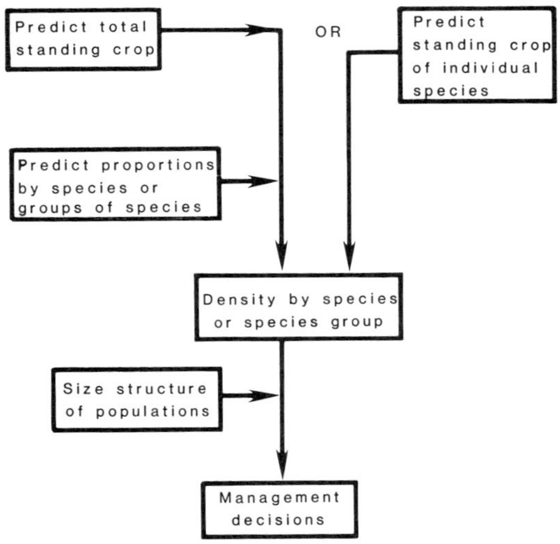

Figure 3. Alternative steps for arriving at management decisions given predictive relationships for standing crops and community structure.

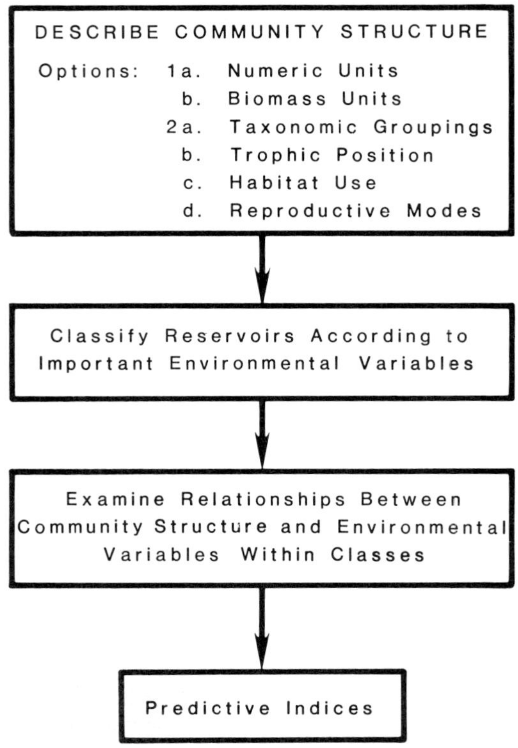

Figure 4. A sequence of phases for development of indices to predict fish community structure in reservoirs.

portant to making management decisions. Anderson's (1976) Proportional Stock Density is a useful descriptor of population size structure, but densities are equally important (Carline et al. 1984), because sport fishing catch rates are directly related to densities.

An alternative procedure for predicting densities of individual species is to develop species-specific indices (Fig. 3). Several researchers have had good success in developing such predictions (Jenkins and Morais 1971; Aggus and Bivin 1982). While species-specific indices can be quite valuable for certain purposes, I believe the community approach has greater merit. Management actions (e.g., stocking, length limits) are often species-directed, but the effects may be felt by a number of species. The same may be said for natural perturbations, such as increased inflows, prolonged winters, etc. Thus for many natural or human perturbations, there will likely be a community response.

Attempts at making generalizations about reservoir communities and environmental features may be viewed as a three-step process (Fig. 4). Alternate ways of describing communities should be explored. Habitat use or reproductive modes may prove to be more effective than taxonomic or trophic position schemes for characterizing species assemblages. Regardless of the descriptor, grouping community data and treating them as discrete variables may be most appropriate. Standing stocks and yields are continuous variables that are readily analyzed with regression techniques. Descriptors of community structure, when grouped, may require different techniques. The second phase of the process would be to classify reservoirs on the basis of important habitat features. The final phase is one of relating community structure to environmental variables within classification groups. From such an analysis, predictive indices would hopefully emerge.

The process of making predictions about reservoir communities based on relatively simple indices represents an approach that Rigler (1982) called empirical ecology. He contrasts empirical ecology with analytical or explanatory ecology in which predictions emanate from quantitative descriptions of how the entire system functions. In a related work, Kerr (1982) explains external analysis of systems as a process of describing system behavior in relation to a few important variables. For example, elucidation of relationships such as MEI and yield is a first step in understanding the fish production process. The challenge to reservoir biologists is to extend this approach to fish community structure and attempt to identify those variables related to species assemblages. Such indices should not only provide predictive capability, but enhance our understanding of community dynamics.

References

Adams, G. F., and C. H. Olver. 1977. Yield properties and structure of boreal percid communities in Ontario. Journal of the Fisheries Research Board of Canada 34:1613–1625.

Aggus, L. R., and W. M. Bivin. 1982. Habitat suitability index models: regression models based on harvest of coolwater and coldwater fishes in reservoirs. U.S. Fish and Wildlife Service, FWS/OBS-82/10.25, Washington, D.C., USA.

Aggus, L. R., and S. A. Lewis. 1978. Environmental conditions and standing crops of fishes in predator-stocking-evaluation reservoirs. Proceedings of the Annual Conference of the Southeastern Association of Fish and Wildlife Agencies 30:131–140.

Anderson, R. O. 1976. Management of small warm water impoundments. Fisheries (Bethesda, Maryland) 1(6):5–7, 26–27.

Balon, E. K. 1975. Reproductive guilds of fishes: a proposal and definition. Journal of the Fisheries Research Board of Canada 32:821–864.

Balon, E. K., W. T. Momot, and H. A. Regier. 1977. Reproductive guilds of percids: results of paleogeographical history and ecological succession. Journal of the Fisheries Research Board of Canada 34:1910–1921.

Carlander, K. D. 1955. The standing crop of fish in lakes. Journal of the Fisheries Research Board of Canada 12:543–570.

Carline, R. F., B. L. Johnson, and T. J. Hall. 1984. Estimation and interpretation of Proportional Stock Density for fish populations in Ohio impoundments. North American Journal of Fisheries Management 4:139–154.

Clady, M. D. 1978. Structure of fish communities in lakes that contain yellow perch, sauger, and walleye populations. Pages 100–108 *in* R. Kendall, editor. Selected coolwater fishes of North America. Special Publication 11, American Fisheries Society, Bethesda, Maryland, USA.

Cooke, G. D., and R. H. Kennedy. 1981. Precipitation and inactivation of phosphorus as a lake restoration technique. Environmental Protection Agency, EPA-600/3-81-012, Corvallis, Oregon, USA.

Grossman, G. D., P. B. Moyle, and J. O. Whitaker, Jr. 1982. Stochasticity in structural and functional characteristics of an Indiana stream fish assemblage: a test of community theory. American Naturalist 120:423–454.

Hanson, J. M., and W. C. Leggett. 1982. Empirical prediction of fish biomass and yield. Canadian Journal of Fisheries and Aquatic Science 39:257–263.

Hill, K. R. 1982. Classification of Iowa lakes and their fish standing stock. Progress Report, Federal Aid Project F-90-R, Iowa Conservation Commission, Des Moines, Iowa, USA.

Hunt, R. L. 1974. Annual production by brook trout in Lawrence Creek during eleven successive years. Technical Bulletin 82, Wisconsin Department of Natural Resources, Madison, Wisconsin, USA.

Jenkins, R. M. 1976. Prediction of fish production in Oklahoma reservoirs on the basis of environmental variables. Annals of the Oklahoma Academy of Science 5:11–20.

Jenkins, R. M. 1982. The morphoedaphic index and reservoir fish production. Transactions of the American Fisheries Society 111:133–140.

Jenkins, R. M., and D. I. Morais. 1971. Reservoir sport fishery effort and harvest in relation to environmental variables. Pages 381–384 *in* G. E. Hall, editor. Reservoir fisheries and limnology. Special Publication Number 8, American Fisheries Society, Bethesda, Maryland, USA.

Jones, J. R., and M. V. Hoyer. 1982. Sportfish harvest predicted by summer chlorophyll-a concentration in midwestern lakes and reservoirs. Transactions of the American Fisheries Society 111:176–179.

Keast, A. 1978. Trophic and spatial interrelationships in the fish species of an Ontario temperate lake. Environmental Biology of Fishes 3:7–31.

Kerr, S. R. 1982. The role of external analysis in fisheries science. Transactions of the American Fisheries Society 111:165–170.

Matuszek, J. E. 1978. Empirical predictions of fish yields of large North American lakes. Transactions of the American Fisheries Society 107:385–394.

McDonough, T. A., and W. C. Barr. 1978. An analysis of fish associations in Tennessee and Cumberland drainage impoundments. Proceedings of the Annual Conference of the Southeastern Association of Fish and Wildlife Agencies 31:555–563.

McFadden, J. T., G. R. Alexander, and D. S. Shetter. 1967. Numerical changes and population regulation in brook trout *Salvelinus fontinalis*. Journal of the Fisheries Research Board of Canada 24:1425–1459.

Melack, J. M. 1976. Primary productivity and fish yields in tropical lakes. Transactions of the American Fisheries Society 105:575–580.

Moyle, J. B. 1956. Relationships between the chemistry of Minnesota surface waters and wildlife management. Journal of Wildlife Management 20: 303–320.

OGLESBY, R. T. 1977. Relationships of fish yield to lake phytoplankton standing crop production, and morphoedaphic factors. Journal of the Fisheries Research Board of Canada 34:2271–2279.

OGLESBY, R. T. 1982. The MEI symposium—overview and observations. Transactions of the American Fisheries Society 111:171–175.

PIELOU, E. C. 1977. Mathematical ecology. John Wiley and Sons, New York, New York, USA.

PLOSKEY, G. R., AND R. M. JENKINS. 1982. Biomass model of reservoir fish and fish-food interactions, with implications for management. North American Journal of Fisheries Management 2:105–121.

PREPAS, E. E. 1983. Total dissolved solids as a predictor of lake biomass and productivity. Canadian Journal of Fisheries and Aquatic Science 40:92–95.

RAWSON, D. S. 1952. Mean depth and the fish production of large lakes. Ecology 33:513–521.

REGIER, H. A., AND H. F. HENDERSON. 1973. Towards a broad ecological model of fish communities. Transactions of the American Fisheries Society 102:56–72.

RIGLER, F. H. 1982. The relation between fisheries management and limnology. Transactions of the American Fisheries Society 111:121–132.

ROUNSEFELL, G. A. 1946. Fish production in lakes as a guide for estimating production in proposed reservoirs. Copeia 1946:29–40.

RYDER, R. A. 1965. A method for estimating the potential fish production of north-temperate lakes. Transactions of the American Fisheries Society 94:214–218.

RYDER, R. A., AND S. R. KERR. 1978. The adult walleye in the percid community—a niche definition based on feeding behavior and food specificity. Pages 39–51 *in* R. Kendall, editor. Selected coolwater fishes of North America. Special Publication 11, American Fisheries Society, Bethesda, Maryland, USA.

SCHAICH, B. A., AND C. C. COUTANT. 1980. A biotelemetry study of spring and summer habitat selection by striped bass in Cherokee Reservoir, Tennessee, 1978. Publication No. 144, Oak Ridge National Laboratories, Environmental Sciences Division, Oak Ridge, Tennessee, USA.

SCHLESINGER, D. A., AND H. A. REGIER. 1982. Climatic and morphoedaphic indices of fish yields from natural lakes. Transactions of the American Fisheries Society 111:1–150.

SCHNEIDER, J. C. 1978. Predicting the standing crop of fish in Michigan lakes. Fisheries Research Report Number 1860, Michigan Department of Natural Resources, Lansing, Michigan, USA.

TONN, W. M., AND J. J. MAGNUSON. 1982. Patterns in the species composition and richness of fish assemblages in the northern Wisconsin lakes. Ecology 63:1149–1166.

WALBURG, C. H. 1977. Lake Francis Case, a Missouri River reservoir: changes in the fish population in 1954–1975, and suggestions for management. Technical Paper 95, U.S. Fish and Wildlife Service, Washington, D.C., USA.

WALTERS, C. J. 1980. Systems principles in fisheries management. Pages 167–183 *in* R. T. Lackey and L. A. Nielsen, editors. Fisheries management. Halsted Press, New York, New York, USA.

WERNER, E. E., D. J. HALL, D. R. LAUGHLIN, D. J. WAGNER, L. A. WILSMANN, AND F. C. FUNK. 1977. Habitat partitioning in a freshwater fish community. Journal of the Fisheries Research Board of Canada 34:360–370.

YOUNGS, W. D., AND D. G. HEIMBUCH. 1982. Another consideration of the morphoedaphic index. Transactions of the American Fisheries Society 111:151–153.

Pages 57–64 *in* G. E. Hall and M. J. Van Den Avyle, editors.
Reservoir Fisheries Management: Strategies for the 80's.
Reservoir Committee, Southern Division American Fisheries Society, Bethesda, Maryland. 1986.

Measuring Angler Effort, Success, and Harvest

MICHAEL J. VAN DEN AVYLE
Georgia Cooperative Fish and Wildlife Research Unit[1]
School of Forest Resources
University of Georgia
Athens, Georgia 30602

ABSTRACT

Information on angler effort, success, and harvest is needed to monitor resource trends, evaluate management tools, and document fisherman satisfaction associated with specific fishery resources. The large size, ease of access, and multiple use of reservoirs make collection of suitable survey information difficult, yet extremely important. A review of published information revealed that few publications give adequate "how to" details for designing creel surveys and analyzing the resulting data. Furthermore, details of sampling designs and precision of estimates have not routinely been included in the published results of creel survey investigations. Little research has been done on alternative methods or on ways that accuracy and precision of commonly-used procedures could be improved. Further research should concentrate on refinement of current techniques rather than development of entirely new approaches. Aspects that warrant further attention include (1) allocation of survey effort by criteria other than expected fishing effort, (2) development of highly-specialized surveys, (3) optimization of routine surveys, and (4) development of survey methods and interpretation procedures that will be useful to reservoir managers.

Information on fishing pressure, success, and harvest is used to monitor resource trends, evaluate management programs, and document fisherman satisfaction associated with sport fisheries. Quantification of these attributes often is the means by which ecological aspects of a fishery, such as fish population and habitat characteristics, are related to its economic aspects so that benefit-cost ratios can be used to define efficient management alternatives (Gordon et al. 1973; Copes and Knetsch 1981; Weithman and Haas 1982). Collection of data on angling effort and harvest is especially important in managing reservoirs that are built and operated primarily for purposes other than recreational fishing (Fisher et al. 1986).

Management agencies have recognized the need for quantitative information and have used a variety of approaches for obtaining data. Applications have included evaluations of a variety of factors: species introductions (Hanson and Dillard 1976; Campbell et al. 1978; Crandall 1978); heated power plant discharges (Hanson 1974; Schneider et al. 1977); tailwater fisheries (Fry 1965; Aggus et al. 1977; Hanson 1977); benefits of standing timber (Burress 1961); and fisherman and fishery traits (Jarman et al. 1968; Von Geldern 1972; Lingerfelter and Summerfelt 1973; Palm and Malvestuto 1983).

Despite the need for quantification of fishing pressure and success, development of methods for obtaining reliable estimates has not received the attention historically devoted to techniques for measuring other types of vital statistics such as population size, growth rate, and mortality. Until recently (Malvestuto 1983), fishery science textbooks have not included methods for angler surveys, and little research has been devoted to the development and refinement of angler survey techniques.

The purposes of this paper are to review published information relevant to measuring angling effort, success, and harvest in reservoirs, to briefly summarize the methods now in common use, and to identify aspects for which further technique development is warranted. Coverage is limited to survey methods (as opposed to complete censuses) for large reservoirs. Fishing effort is used here as the hours of fishing expended during a period of time (e.g., hours per month); success (CPUE) is defined as the catch per unit (hour) of fishing effort; and harvest (or catch) is expressed as the number or weight of fish caught

[1] The Unit is jointly sponsored by the U.S. Fish and Wildlife Service, the University of Georgia, the Georgia Department of Natural Resources, and the Wildlife Management Institute.

by anglers during a period of time (e.g., pounds per month).

Survey Methods

Background

Carlander et al. (1958) classified angler surveys into eight groups according to the nature and extent of sampling and the type of person providing information (fishermen versus biologists or clerks). They stated that many survey programs were "indiscriminant" and prone to inaccuracy, imprecision, and inefficiency, and they recommended that greater attention be given to statistical designs that could improve precision and accuracy, allow quantification of reliability, and reduce costs. They specifically recommended stratified random sampling designs and cited some of the earliest applications to creel surveys (Tarzwell and Miller 1943; Kathrein 1953; Murphy 1955). Present sampling designs generally satisfy the recommendations of Carlander et al. (1958) and are fundamentally similar to those used by the authors cited above; however, significant modifications and improvements have been made by many investigators, including Neuhold and Li (1957), Lambou (1961, 1966), and Von Geldern and Tomlinson (1973).

Most survey techniques published since the early 1960's can be placed in the eighth category of Carlander et al. (1958) in that they provide "statistically-selected" coverage with data obtained by trained clerks or biologists. Because of the variety of site-specific factors, as well as agency priorities, designs differ with respect to the degree of randomness used, allocation of sampling effort in space and time, locations where clerks interview fishermen, the manner in which estimates of total fishing effort are obtained, and the types of ancillary information obtained from anglers. Unique aspects of individual reservoirs and the frequent need for specialized information has led to the use of "custom-fit" survey designs by most agencies. Design modifications are generally obtained from consulting statisticians. In the southeastern United States, consulting services relative to reservoir creel survey design have been provided by the Southeastern Cooperative Fish and Game Statistics Project (within the Institute of Statistics at North Carolina State University), which is jointly supported by several state resource management agencies. General aspects of the Project's designs and procedures for conducting roving-clerk and access-point surveys have been reported in mimeo form (Anonymous 1976). Malvestuto et al. (1978) summarized the use of a fundamentally similar roving-clerk design on West Point Reservoir, Georgia and Alabama.

Five elements are common to most current statistically based creel survey designs: (1) the entire study period is divided into time blocks for which separate estimates are desired (usually 2-week periods, months, or quarters), (2) each time block is divided into non-overlapping sampling intervals (often morning versus afternoon or week-end versus week days), (3) the reservoir is divided into distinct spatial units, sections, or access points, each of which can be surveyed adequately by a clerk during one sampling interval, (4) sampling units (one of the possible combinations of sampling intervals and sections) to be surveyed during each time block are selected according to predetermined probabilities, and (5) sampling is conducted to obtain counts of fishermen and to determine the catch and hours fished for anglers interviewed. Many additional types of information, such as species fished for, type of gear used, county of residence or other economically oriented data, and numbers of fish caught and released, can be obtained when conducting surveys, but detailed descriptions of these are beyond the scope of this review. Malvestuto (1983) described ways to use survey data to quantify economic benefits of recreational fisheries.

Selecting Sampling Units

The manner in which sampling probabilities are assigned and sampling effort is allocated among the sampling units establishes the statistical design that provides a "label" for a creel survey. If all sampling units within a time block are assigned equal (uniform) probabilities of being sampled, the design is equivalent to simple random sampling within each time block. Usually, however, unequal (nonuniform) probabilities of selection are assigned to the sampling units in proportion to an expected distribution of fishing pressure; this practice increases the chance that units having relatively high pressure (e.g., weekends, mornings, or one arm of a reservoir) will be included in the sample. The intended result is to improve survey efficiency by increasing the number of anglers contacted and increasing the likelihood that each person fishing during the specified time block has an equal chance of being sampled (Lambou 1961). For this approach to be unbiased, every sampling unit must be assigned a non-zero probability of selection, and all unit probabilities must sum to 1.0. Frequently, the lack of information needed to accurately predict fishing pressure or constraints on creel clerk work schedules limit the flexibility of assigning selection probabilities.

In reservoirs where fishing pressure is known or expected to shift seasonally, sampling probabilities for the units can be altered accordingly. Malvestuto et al. (1978) used data from monthly airplane overflights to adjust sample selection probabilities used for each subsequent month in their survey of West Point Reservoir.

If it is desirable to obtain estimates that will allow statistical comparisons of success, catch, or effort among portions of the reservoir, or between periods within a time block, the sample units must be subdivided (stratified) into two or more groups and assigned separate sampling probabilities; unit selection probabilities must then total 1.0 in each stratum. If every unit within a stratum is assigned the same probability of selection for sampling, the design is stratified random sampling (Abramson and Tolladay 1959); otherwise, the design is termed stratified nonuniform probability sampling (Malvestuto et al. 1978).

Stratification has been used primarily to improve precision of estimates for all strata combined. In this application, stratification is similar in intent to nonuniform probability sampling; the major difference is that stratification guarantees sampling of a predetermined number of units in each stratum, whereas a strictly nonuniform probability design only increases the chance that certain units will be sampled. Stratification requires specification, in advance, of the number of units to be sampled in each stratum during each time block. Although many criteria can be used to allocate this effort, sample sizes generally are set in proportion to expected fishing effort. For example, Malvestuto et al. (1978) stratified the sample units in each time block (month) as weekend or weekday and worked five units in each stratum during summer months. Such allocation in proportion to fishing pressure was expected to improve the precision of effort estimates relative to those that would have been obtained by using simple random sampling of 10 units. To be efficient, any stratified design should allocate at least two samples to each stratum; otherwise, no gain in precision or ability to compare stratum results can be realized.

In practice, few survey designs have used probability sampling exclusively. Most modern approaches combine some systematic features with probability sampling due to practical constraints imposed by budgets and personnel schedules. Since most surveys are conducted by one clerk, only one unit can be sampled per day (or perhaps, more realistically, the length of the sample interval is adjusted to fit the clerk's work day), which often leads to a two- or three-stage process for randomly selecting the units sampled. The first stage generally is to select the sampling day, the second would be to select the time interval within the day, and the third is to select the lake section or access point to be sampled (Malvestuto 1983).

Angler Contacts

Gathering data on fishing pressure, effort, and success has traditionally been conducted at access points or by a roving clerk who works in a boat. In an access point survey, the clerk works a selected sampling unit (e.g., a launching ramp or boat dock) and interviews all departing fishermen to determine their catch and time spent fishing. Effort is calculated by adding the trip lengths reported by each angler. Lakewide (or stratum) effort and catch for the sampling day are estimated by direct expansion, and the daily CPUE is calculated by dividing total catch by total time fished (Anonymous 1976). CPUE could also be computed as the average of the catch-per-hour values observed for individual fishermen (Malvestuto 1983).

A roving clerk interviews anglers to determine their catch and the length of time fished up to the interview (incomplete trip); fishing pressure is usually determined by making a separate circuit of the sample section and counting the number of fishermen (termed "instantaneous" count). Less frequently, no separate pressure count is made; rather, anglers are counted as they are encountered during the interview circuit (termed a "progressive" count by Lambou 1961). The number of fishermen seen during a pressure count is assumed to be an unbiased estimate of the average number of anglers fishing during all hours in the sampling interval, and lakewide effort is estimated by expansion of the product of the fisherman count times the length of the sampling interval (Lambou 1961; Malvestuto et al. 1978). Lakewide catch for the sampling day is estimated by multiplying lakewide effort (in hours) by the CPUE determined from the sample interviews (Anonymous 1976).

The major advantage of the roving method is that the clerk actively seeks out fishermen for interviews, thereby increasing the fisherman sample size, whereas the access clerk is at a fixed point and the number of interviews is subject to fishermen returning to the sampled access point. A roving clerk often can effectively count certain types of fishermen, such as bank or dock anglers, who might be missed by access point methods. Low numbers of interviews may be obtained from access point surveys on reservoirs that have many launching areas (DiCostanzo 1956). The major reported concern about the adequacy of the roving survey has been the use of incompleted trips to estimate CPUE (Carlander et al. 1958; Malvestuto et al. 1978). Robson (1960, 1961) claimed, on theoretical grounds, that unbiased CPUE data could be obtained only from completed trips. Investigators that have compared CPUE from incomplete and completed trips have not detected bias in using CPUE from interviews based on incompleted trips (DiCostanzo 1956; Brown 1970; Malvestuto et al. 1978), but additional evaluations of bias over a broader range of survey conditions need to be done before any general conclusions can be made.

Aerial counts of fishermen can be used with access or roving clerk surveys; when catch is estimated, the counts made in the sampling unit are treated like the pressure counts of a roving clerk. Data on CPUE are obtained either from incomplete trips (roving clerk) or complete trips (access point), and in both approaches, total effort is estimated by multiplying the aerial fisherman count by the length of the sampling interval. In cases where aerial counts encompass the entire reservoir, the pressure estimate requires no expansion over space, and the CPUE determined for the section sampled by the clerk is assumed to be representative of all sections within a stratum. Disadvantages of aerial counts include difficulties with distinguishing anglers from other boaters, with counting bank and dock fishermen, and with missing sampling schedules due to inclement weather (Anonymous 1976; Malvestuto 1983).

Relative Effectiveness of Designs

Few comparisons of creel survey efficiency, precision, or accuracy have been reported. Evaluations published by Best and Boles (1956), Johnson and Wroblewski (1962), and Pfeiffer (1966) have shown that surveys in which 20–25% of the possible fishing time was sampled provided acceptably accurate estimates when compared with complete censuses. Comparisons of survey designs have shown (1) probabilistic surveys to be more efficient than systematic ones and (2) stratified designs to be more precise than unstratified ones (Best and Boles 1956; Abramson and Tolladay 1959). In these studies, however, complete census information was available to the investigators to establish strata and define sample selection probabilities before they simulated the survey results. The paucity of research in this area is probably due, in part, to routine use of generalized survey designs that typically require only minor modifications to provide results adequate to satisfy nonspecific, monitoring objectives.

Discussion and Recommendations

Statistically designed angler surveys used by most agencies are fundamentally similar yet they differ according to the amount of emphasis placed on estimating certain variables. For example, one agency might regard changes in the amount of angling effort as the best measure of a length-limit's effect, whereas another agency might emphasize trends in CPUE. The precision and overall cost efficiency of a survey designed to meet either of these agencies' needs would be affected by the sampling design, the criteria used to assign selection probabilities to the sampling units, and the methods by which anglers are contacted.

Allocation of Sampling Effort

Most stratified and nonuniform probability designs place the greatest sampling emphasis into intervals or sections expected to have the greatest fishing pressure. This placement can improve efficiency for estimating effort, but it may sacrifice precision and/or accuracy of CPUE estimates, especially if CPUE and effort are correlated. Such a trade-off probably would be acceptable in situations where the major concern is to monitor trends in resource use but unacceptable where major goals are to measure trends in stock abundance or angling quality by using CPUE as an index.

Although the need for consideration of criteria other than fishing pressure when designing angler surveys has been expressed repeatedly (Carlander et al. 1958; Brown 1970), little research appears to have been directed toward meeting this need. A stratified design that increases sampling intensity in strata having high internal variation in CPUE could improve precision of success estimates over those obtained by allocating sampling in proportion to effort. Von Geldern and Tomlinson (1973) chose to emphasize refinement of CPUE estimates at the expense of precise effort estimates because they were mainly interested in developing an index of fish population abundance.

Design Options

Criteria for choosing stratified versus unstratified or uniform versus nonuniform probability sampling have not been clearly defined in relation to conducting creel surveys; general aspects, however, have been treated extensively in statistical texts, such as the one published by Cochran (1977). One limitation of stratified and nonuniform probability designs is that both require before-hand knowledge of patterns in effort or success through time (e.g., AM versus PM, weekday versus weekend) or space (e.g., different arms of the reservoir). Unavailability of these types of information may require use of "best-guess" approximations for initial phases of a survey, with refinements in design for subsequent time blocks. Further research should be conducted to compare the relative precision and cost-efficiency of various survey approaches, especially for situations where only easily obtained data are available for establishing strata or setting sampling-unit selection probabilities.

When there is no interest in testing for differences in success or effort among strata, the use of simple random sampling or nonuniform probability sampling could produce unbiased estimates with fewer observations than required by stratified sampling, but precision could suffer. Stratification improves precision relative to unstratified designs only if sampling effort is allocated in proportion to within-stratum variation of the variable of principal interest. Stratification would probably also be better when nonuniform designs would require the use of extremely divergent sample selection probabilities. Although nonuniform probability sampling is unbiased, the chance that a given set of samples within a time block will produce inaccurate estimates increases as the sampling unit probabilities diverge.

Unmodified, general purpose surveys of effort and success often will not provide adequate evaluations of specific management techniques regardless of design. For example, surveys to measure the effect of stocking, or of a restrictive length limit, on CPUE of one species might be better conducted by establishing sampling units and probabilities in relation to the variability and magnitude of "fished-for" effort for the target species, rather than in relation to total effort or catch. Evaluations of management programs should be done in a direct manner, rather than "making do" with the best available (yet often inadequate) data that might be extracted from a generalized survey that was designed to satisfy other objectives.

Specialized creel surveys are particularly needed for fisheries that have extreme spatial or temporal variations in effort or success (Malvestuto et al. 1979). In some cases, surveys could be specialized to an extent that other information, such as records maintained by fishing clubs (Alexander and Holbrook 1978), could be used as part of the design. Separation of CPUE according to "fished for" effort generally is a better measure of fishing quality for a particular species than is CPUE based on effort for "any species" (Lambou and Stern 1959; Lambou 1966).

Improvements in survey design appear to have been hampered by a general lack of agreement (or understanding) about the best ways to compute some of the quantities needed to expand the sampling data. For example, CPUE for a given sampling unit is sometimes calculated as the ratio of total observed catch divided by total observed effort; in other applications, CPUE is computed as the average of the ratios observed for each fisherman (Malvestuto 1983). The equations and information used to compute the variance of CPUE differ with the computational technique; thus, any other estimate based on CPUE, such as total catch, will have a variance (and a variance equation) dependent on the method used to estimate daily CPUE. Since optimization of most sample designs and tests of statistical significance depend on quantification of variation, investigators should clearly define computational procedures used in their surveys and report variances of estimated parameters.

Angler Contacts

The effects of different angler contact methods on accuracy and precision of surveys are probably not as pronounced as the effects of selecting sampling units. For surveys that emphasize information on total effort and harvest, roving clerk contacts are probably best; however, where accurate CPUE estimates are of principal interest, bias should either be evaluated or reduced by ensuring the inclusion of relatively large numbers of completed trip interviews in the samples. In surveys where aerial pressure counts can be made, assignment of high sampling selection probabilities to periods when many anglers are completing their trips could be a way of obtaining more accurate and precise estimates of CPUE from access point interviews.

Creel clerk consistency and questionnaire design can also affect precision and accuracy of estimates. Hudgins and Malvestuto (1985) recommended more intensive clerk training programs and simpler response formats for questionnaires in order to reduce discrepancies among survey clerks.

General Needs

Further improvements of survey techniques will require consideration of variance components associated with specific designs and greater emphasis on procedures that can be used to control precision. More careful planning, establishment of priorities for types of estimates needed from each survey, and recognition of trade-offs associated with options inherent in conventional survey approaches will probably be more productive than attempts to invent fundamentally new designs. The most useful lines of research will probably include ways to improve the use of aerial pressure counts, procedures for decreasing labor requirements and other costs while maintaining adequate precision of routine surveys, and refining methods for conducting precise, specialized surveys to test specific management techniques. Computer simulations and sensitivity analyses will probably be the most effective approaches for developing and evaluating designs.

The ability of many fishery managers to design, execute, and analyze data for a reservoir creel survey is severely hampered by the absence of "how-to" publications. Malvestuto's (1983) chapter in *Fisheries Techniques* is an excellent treatment of sampling designs and use of survey data in measuring economic values of fisheries; however, no papers have adequately described computational methods and data interpretation at a level of detail to assist fishery biologists in understanding and applying the techniques to reservoirs. Some agency or professional organization, such as the sponsors of this symposium, should identify and support an interdisciplinary team of qualified fishery scientists and statisticians to organize, write, and publish such manuals.

References

Abramson, N., and J. Tolladay. 1959. The use of probability sampling for estimating annual numbers of angler days. California Fish and Game 45: 303–311.

Aggus, L. R., D. I. Morais, and R. F. Baker. 1977. Evaluation of the trout fishery in the tailwater of Bull Shoals Reservoir, Arkansas, 1971–73. Proceedings of the Annual Conference of the Southeastern Association of Fish and Wildlife Agencies 31:565–573.

Alexander, C. M., and J. A. Holbrook, II. 1978. Comparison of bass club records with creel censuses. Proceedings of the Annual Conference of the Southeastern Association of Fish and Wildlife Agencies 32:446–448.

Anonymous. 1976. Field forms for fishing, hunting, and other recreational surveys. North Carolina State University Institute of Statistics, Raleigh, North Carolina, USA. (Mimeo.)

Best, E. A., and H. D. Boles. 1956. An evaluation of creel census methods. California Fish and Game 42:109–115.

Brown, B. E. 1970. Implications from the Oklahoma state lake creel survey to improve creel survey design. Proceedings of the Annual Conference of the Southeastern Association of Game and Fish Commissioners 24:577–590.

Burress, R. M. 1961. Fishing pressure and success in areas of flooded standing timber in Bull Shoals Reservoir, Missouri. Proceedings of the Annual Conference of the Southeastern Association of Game and Fish Commissioners 15:296–298.

Campbell, W. J., E. Hayes, W. R. Chapman, and W. Seawell. 1978. Angling pressure and sport fish harvest in the predator-stocking-evaluation reservoirs. Proceedings of the Annual Conference of the Southeastern Association of Game and Fish Commissioners 30:114–119.

Carlander, K. D., C. J. DiCostanzo, and R. J. Jessen. 1958. Sampling problems in creel census. The Progressive Fish-Culturist 20:73–81.

Cochran, W. G. 1977. Sampling techniques, 3rd edition. John Wiley & Sons, New York, New York, USA.

Copes, P., and J. L. Knetsch. 1981. Recreational fisheries analysis: management modes and benefit implications. Canadian Journal of Fisheries and Aquatic Sciences 38:559–570.

Crandall, P. S. 1978. Evaluation of striped bass × white bass hybrids in a heated Texas reservoir. Proceedings of the Annual Conference of the Southeastern Association of Fish and Wildlife Agencies 32:588–598.

DiCostanzo, C. J. 1956. Creel census techniques and harvest of fishes in Clear Lake, Iowa. Ph.D. Dissertation, Iowa State University, Ames, Iowa, USA.

Fisher, W. L., J. J. Charbonneau, and M. J. Hay. 1986. Economic characteristics of reservoir fisheries. Pages 5–10 *in* G. E. Hall and M. J. Van Den Avyle, editors. Reservoir fisheries management: strategies for the 80's. Reservoir Committee, Southern Division American Fisheries Society, Bethesda, Maryland, USA.

Fry, J. P. 1965. Harvest of fish from tailwaters of three large impoundments in Missouri. Proceedings of the Annual Conference of the Southeastern Association of Game and Fish Commissioners 16: 405–411.

Gordon, D., D. W. Chapman, and T. C. Bjornn. 1973. Economic evaluation of sport fisheries—what do they mean? Transactions of the American Fisheries Society 102:293–311.

Hanson, W. D. 1974. The fishery of a Missouri reservoir receiving thermal effluent. Proceedings of the Annual Conference of the Southeastern Association of Game and Fish Commissioners 27: 722–731.

Hanson, W. D. 1977. Tailwater fisheries of Lake of the Ozarks and Pomme de Terre Lake, Missouri. Proceedings of the Annual Conference of the Southeastern Association of Fish and Wildlife Agencies 31:505–513.

Hanson, W. D, and J. G. Dillard. 1976. Contribution of striped bass to the fishery of Lake of the Ozarks, Missouri. Proceedings of the Annual Conference of the Southeastern Association of Game and Fish Commissioners 29:162–167.

Hudgins, M. D., and S. P. Malvestuto. 1985. An evaluation of factors affecting creel clerk performance. Proceedings of the Annual Conference of the Southeastern Association of Fish and Wildlife Agencies 36:252–263.

Jarman, R., C. Bennett, C. Collins, and B. E. Brown. 1968. Angling success and recreational use on twelve state-owned lakes in Oklahoma. Proceedings of the Annual Conference of the Southeastern Association of Game and Fish Commissioners 21: 484–495.

Johnson, M. W., and L. Wroblewski. 1962. Errors associated with a systematic sampling creel census. Transactions of the American Fisheries Society 91:201–207.

Kathrein, J. A. 1953. An intensive creel census on Clearwater Lake, Missouri, during its first four years of impoundment, 1949–1952. Transactions of the North American Wildlife Conference 18:282–295.

Lambou, V. W. 1961. Determination of fishing pressure from fisherman or party counts with a discussion of sampling problems. Proceedings of the Annual Conference of the Southeastern Association of Game and Fish Commissioners 15:380–401.

Lambou, V. W. 1966. Recommended method of reporting creel survey data for reservoirs. Oklahoma Fishery Research Laboratory Bulletin No. 4, Norman, Oklahoma, USA.

Lambou, V. W., and H. Stern, Jr. 1959. Creel census methods used on Clear Lake, Richland Parrish, Louisiana. Proceedings of the Annual Conference of the Southeastern Association of Game and Fish Commissioners 12:169–175.

Lingerfelter, D. P., and R. C. Summerfelt. 1973. Angler harvest in heated fishing docks on an Oklahoma reservoir. Proceedings of the Annual Conference of the Southeastern Association of Game and Fish Commissioners 26:611–621.

Malvestuto, S P. 1983. Sampling the recreational fishery. Pages 397–419 *in* L. A. Nielsen and D. L. Johnson, editors. Fisheries techniques. American Fisheries Society, Bethesda, Maryland, USA.

Malvestuto, S. P., W. D. Davies, and W. L. Shelton. 1978. An evaluation of the roving creel survey with nonuniform probability sampling. Transactions of the American Fisheries Society 107:255–262.

Malvestuto, S. P., W. D. Davies, and W. L. Shelton. 1979. Predicting the precision of creel survey estimates of fishing effort by use of climatic variables. Transactions of the American Fisheries Society 108:43–55.

Murphy, L. W. 1955. Two-year creel census of the Little Salmon River, Idaho. Transactions of the American Fisheries Society 84:200–206.

Neuhold, J. M., and K. H. Li. 1957. Creel census method. Utah Department of Fish and Game Publication 8, Salt Lake City, Utah, USA.

Palm, R. C., Jr., and S. P. Malvestuto. 1983. Relationships between economic benefit and sportfishing effort on West Point Reservoir, Alabama-Georgia. Transactions of the American Fisheries Society 112:71–78.

Pfeiffer, P. W. 1966. The results of a non-uniform probability creel survey on a small state-owned lake. Proceedings of the Annual Conference of the Southeastern Association of Game and Fish Commissioners 20:409–412.

Robson, D. S. 1960. An unbiased sampling and estimation procedure for creel census of fishermen. Biometrics 16:261–277.

Robson, D. S. 1961. On the statistical theory of a roving creel census of fishermen. Biometrics 17: 415–437.

Schneider, R. W., A. O. Smith, and V. P. Mitchell, Jr. 1977. Characteristics of sport fishing activity in three warmwater discharges. Proceedings of the

Annual Conference of the Southeastern Association of Fish and Wildlife Agencies 31:546–554.

TARZWELL, C. M., AND L. F. MILLER. 1943. The measurement of fishing intensity on the lower T.V.A. reservoirs. Transactions of the American Fisheries Society 72:246–256.

VON GELDERN, C. E., JR. 1972. Angling quality at Folsom Lake, California, as determined by a roving creel census. California Fish and Game 58:75–93.

VON GELDERN, C. E., JR., AND P. K. TOMLINSON. 1973. On the analysis of angler catch rate data from warmwater reservoirs. California Fish and Game 59:281–292.

WEITHMAN, A. S., AND M. A. HAAS. 1982. Socioeconomic value of the trout fishery in Lake Taneycomo, Missouri. Transactions of the American Fisheries Society 111:223–230.

Pages 65–72 *in* G. E. Hall and M. J. Van Den Avyle, editors.
Reservoir Fisheries Management: Strategies for the 80's.
Reservoir Committee, Southern Division American Fisheries Society, Bethesda, Maryland. 1986.

Methods and Parameters Used in Evaluating Bass Length Limits on Pomme de Terre Lake, Missouri

RONALD J. DENT, JR.
Missouri Department of Conservation
1104 S. Grand St.
Sedalia, Missouri 65301

ABSTRACT

An evaluation of a 15-in minimum length limit on black basses (*Micropterus* spp.) was conducted from 1976 to 1982 on Pomme de Terre Lake, Lake of the Ozarks, and Table Rock Lake in Missouri. One objective of these studies was the determination and assessment of sampling methods and parameters used in evaluating harvest regulations. Electrofishing, cove and shoreline rotenone sampling, and creel data from Pomme de Terre Lake were used to compare the effectiveness and cost efficiency of different sampling procedures.

Spring nighttime electrofishing of at least 1,000 largemouth bass (*Micropterus salmoides*) provided trend information on length-frequencies and indices of relative abundance. Scale analyses of 150–300 largemouth bass yielded reliable data for separation of age classes and calculation of annual mortality estimates. Reward tagging data were used to estimate angler exploitation rates. An angler opinion survey also provided useful information on the importance of bass catch versus harvest.

Growth, mortality, exploitation rates, and relative abundance of the bass population and bass angler catch rates and harvest were important parameters for assessing bass length limits on large Missouri reservoirs. Electrofishing required the least manpower for the amount of data collected. Definitive objectives for catch and harvest, based on the ability of the bass population to respond to a size-limit regulation, are an important consideration in evaluating the effectiveness of any imposed size limit. Electrofishing and creel survey data are particularly helpful in assessing results of size-limit regulations.

Length limits have been used extensively to restrict bass angler harvest on small impoundments (Hickman and Congdon 1972; Hoey and Redmond 1972; Ball 1980; Paragamian 1982). Interest in using size limits to manage bass populations in large reservoirs has increased substantially in recent years, as shown by an unpublished survey conducted by the Reservoir Committee of the Southern Division of the American Fisheries Society in 1981. The survey revealed that 11 of 14 southern states had some type of size-limit regulation on bass in reservoirs. Documentation of the effects of bass size limits is needed to refine management techniques for large reservoirs.

Responses of fish populations to management activities are traditionally measured by monitoring changes in growth, mortality, recruitment, size structure, and population size as well as angler catch and harvest, success rate, and effort. To effectively manage large reservoirs, it it necessary to determine which of these parameters are most important and which sampling methods are most useful for evaluating size limits.

On January 1, 1976, a 15-in minimum length limit was imposed for black basses (*Micropterus* spp.) in Pomme de Terre Lake, Lake of the Ozarks, and Table Rock Lake. Since these were the first established bass size limits on large reservoirs in Missouri, considerable effort was devoted to selecting sampling methods and parameters that would quantify the effectiveness of this regulation. The Pomme de Terre Lake study is presented in this paper as an example of parameters and sampling methods needed to describe changes in the bass population and fishery resulting from the imposition of a size limit. Included is a brief discussion of manpower requirements for collection of pre- and post-size limit data.

Data on largemouth bass (*Micropterus salmoides*) are presented; the same techniques, however, have been used to assess size limit effects on spotted bass (*M. punctulatus*) and smallmouth bass (*M. dolomieui*) in other Missouri reservoirs.

STUDY SITE

Pomme de Terre Lake is a 7,800-acre flood control and recreational impoundment built by

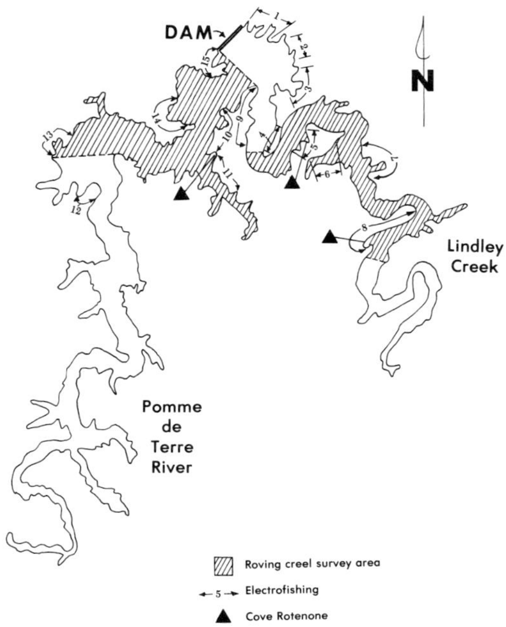

Figure 1. Creel survey areas (shaded), electrofishing sites (numbered shoreline transects), and cove rotenone sampling areas (solid triangles) used on Pomme de Terre Lake, Missouri, 1976-1982.

the U.S. Army Corps of Engineers in 1961. The lake is dimictic and stratified from June through September. Secchi-disc transparency ranges from 22 to 116 in and averages 54 in. Specific conductance averages 281 micromhos/cm² and ranges from 236 to 440 micromhos/cm². During the study period, average water temperatures ranged 47–60°F in the spring and 65–75°F in the fall; average spring and fall secchi-disc transparencies were similar (49 and 53 in, respectively).

Methods

Relative abundance, expressed as catch per unit of effort (CPUE), size composition, and rates of growth, mortality, and recruitment were calculated from cove and shoreline rotenone data, electrofishing samples, tagging, and creel census information. Methods of data collection were based on the least amount of manpower required to obtain the maximum amount of representative data. Preferences of parameters are based on their effectiveness to describe changes in the bass population resulting from imposition of a size limit. Brief descriptions of the various sampling and survey methods follow.

Electrofishing

Nighttime electrofishing (direct current) was used from 1976 through 1982 to collect largemouth bass for age and growth determinations and defining length-frequency distributions. Shoreline samples were collected along 15 selected coves, bluffs, and points in April and October each year (Fig. 1). Attempts were made to collect at least 1,000 largemouth bass each spring. All bass were measured to the nearest 0.1 in, and those from which scales were collected were weighed to the nearest 0.5 oz. The number of bass collected per hour of electrofishing was used to describe relative abundance. Length-frequency distributions were evaluated using the Proportional Stock Density (PSD) index described by Anderson and Weithman (1978), and fish condition was described by relative weight values (Wege and Anderson 1978).

Cove Rotenone Sampling

Cove rotenone sampling was used to estimate standing crop and size composition of populations of black bass and other fishes in August 1977. Three 1-acre coves, each blocked with a 0.5-in square mesh net, were sampled with 0.5 ppm concentration of rotenone. The coves were situated in the three regions of the reservoir to correspond with the spring and fall electrofishing sample areas (Fig. 1). All fish collected over a 3-day period were measured to the nearest inch. Methods used were similar to those of Grinstead et al. (1978).

Shoreline Rotenone Sampling

A 150 × 6 ft block net (0.25-in mesh) was used to enclose ten 100-ft sections of shoreline during October 1976 to 1979 and May 1978 to 1980. Each sample area approximated a rectangle of 0.06 acre and was treated with 0.5 ppm rotenone. Maximum depth near the outer edge of the net was 5 ft. All sites were selected to obtain representative samples of various habitats and were within shoreline areas sampled by electrofishing.

Creel Survey and Angler Opinion

Two creel survey areas, each about 2,000 acres, were established in the arms of the reservoir (Fig. 1). A roving creel survey was conducted during daylight hours in each area (11 days/month) from

March–November 1976–1982. The daily work schedule, selection of sampling days and time periods, and calculations of total fishing hours, trips, catch rates, and total estimated catch were similar to the methods described by Hanson (1977). Estimated catch and harvest by species were calculated by multiplying the total estimated catch by the percent composition of the observed catch. The number and approximate lengths of sublegal bass were included.

A survey was conducted March–June 1982 to obtain the opinions of anglers about the 15-in size limit and to provide future management direction on bass harvest regulations. Two questions were asked by the creel clerk in his interviews: 1. What is your preference for bass length limits at Pomme de Terre Lake? (potential answers: no change, change to a lower minimum length limit, change to a slot length limit, no opinion); 2. Have you ever been asked this question before? (Yes or No). Explanations of each option were included before the angler's response was recorded.

Age and Growth

Scales from ten bass in each 0.5-in group above 7 in were collected each spring in 1976–1978 to determine age composition. Sample size each year ranged from 50 to 100 bass. In subsequent years, scales were taken from all bass during the first few hours of sampling, and the total number used in age determinations was increased to 150–300.

Tagging Procedures and Rewards

To determine exploitation rates, largemouth bass were tagged with Floy FD68B tags and released between March 20 and April 10 in 1975–1978. The number of tags voluntarily returned from fish caught by anglers within one year of tagging divided by the total number tagged provided estimates of exploitation rates. In 1975, 157 reward tags, worth $1–100, and 157 non-reward tags were placed on fish longer than 8.9 in. In 1976–77, only non-reward tags were used, but it soon was apparent that a strictly voluntary program with no reward incentive provided severe underestimates of exploitation. Reward tags worth $1–10 were again used in 1978, when 119 largemouth bass longer than 14.9 in were tagged.

RESULTS

Parameters originally considered important for evaluating length limits on black bass were rates of survival, growth, exploitation, recruitment, angler catch and harvest, fishing pressure, and measures of population abundance, size structure, and relative weight. Following examination of preliminary results, parameters now considered to be the most important for evaluating effects of a size limit on bass populations are growth, survival, and exploitation rates, and population abundance and size structure. Creel data parameters essential to describe changes in the bass fishery include legal and sublegal bass catch and harvest and fishing pressure. Relative weight indices were not essential, and estimates of recruitment rates were judged impractical due to our inability to find a reliable and cost-effective method of estimation. The following sections illustrate methods of interpretation, results, and problems encountered with estimating important parameters.

Population Abundance and Size Structure

Electrofishing.—CPUE data for four size groups of bass were compared for fall and subsequent spring collections for 1976 to 1982 (Table 1). Electrofishing effort to collect the samples averaged 16 hours (range 7.5–27.4 hours) in the spring and 6 hours (range 2.3 to 8.2 hours) in the fall. The average number of bass collected was 1,586 in the spring (range 942 to 3,436) and 496 (range 130 to 703) in the fall. The CPUE for the spring months for bass over 8 in was usually greater than the previous fall and especially so for 15-in and larger bass. Bass shorter than 8 in were usually collected at a higher rate in the fall. Hence, the relative proportions of bass within the 8.0 to 14.9-in range were similar in spring and fall samples whereas percentages for bass shorter than 8.0 in (primarily Age 0) were greater in fall. Proportions of bass exceeding 15.0 in often were higher in spring. PSD varied considerably between the two sample periods, primarily due to variability in sampling efficiency for the larger bass.

Cove sampling.—Pomme de Terre Lake supported an estimated standing crop of 24.3 lb/acre of black bass (94% largemouth and 6% spotted bass) in 1977. This estimate was probably above average because the CPUE for the spring 1977 electrofishing sample was the highest recorded during the 7-year study (Table 1).

The PSD of largemouth bass collected in the August cove sample was similar to the 40% value obtained by spring electrofishing and somewhat

Table 1. Comparisons of spring (S) and fall (F) electrofishing catch rates, sampling effort, and Proportional Stock Density (PSD) for largemouth bass in Pomme de Terre Lake, 1976–1982.

Size group (in)	Number collected per hour												
	1976		1977		1978		1979		1980		1981		1982
	S	F	S	F	S	F	S	F	S	F	S	F	S
<8.0	4	21	24	21	15	51	13	42	3	19	5	30	71
8.0–11.9	55	26	80	16	32	29	24	30	40	48	76	7	38
12.0–14.9	17	16	50	16	33	24	18	19	21	18	46	14	42
≥15.0	7	1	7	2	5	8	9	5	4	1	6	2	6
Effort (hours)	16.0	6.3	20.0	7.0	27.4	6.5	17.5	6.0	13.9	8.2	7.5	2.3	8
PSD	28	41	40	53	55	53	62	44	39	28	41	68	56
Total number collected	1,376	403	3,436	499	1,938	674	1,164	569	942	703	995	130	1,254

lower than the 53% PSD observed in the fall. Similar size distributions of largemouth bass collected by cove rotenone sampling and electrofishing were also reported by Harris et al. (1980). Even though both methods gave similar PSD data in Pomme de Terre Lake, the number of largemouth bass collected by cove sampling averaged only 0.97 bass over 8 in for each man-day of effort whereas electrofishing produced about 31 bass per man-day of effort. Because of the relative inefficiency of cove rotenone sampling for collecting largemouth bass, this procedure was discontinued.

Survival

Annual survival estimates were made using age-specific CPUE data from electrofishing in a model that allowed non-uniform recruitment. This method computes survival between successive years for age-specific data and accounts for variable recruitment (Ricker 1975). CPUE values were determined by assigning bass collected during springtime electrofishing to age classes on the basis of the results of scale analyses made for each 0.5-in length group.

Annual survival estimates calculated for each age interval in successive years are presented in Table 2. Only the spring electrofishing collections were used in estimating survival because of the low numbers of bass 8 in and larger collected in the fall. Age groups 1 and 2 were frequently under-represented in the samples because of selectivity of electrofishing for bass longer than 12.0 in. Hence, survival estimates between Ages 1 and 2 were not calculated, and some of the estimates from Age 2 to Age 3 exceeded 100% (Table 2). Size selectivity of electrofishing gear for longer bass has also been documented by Simpson (1978) and Harris et al. (1980). Survival beyond Age 6 was not calculated because these bass were rare in the electrofishing samples. Collection of older age groups by electrofishing is generally unpredictable because of their low abundance and preference for deep water (Cooper and Schafer 1954).

Exploitation Rates

Results of the tagging studies illustrate the importance of using a reward system to encourage anglers to report catches of tagged fish. The annual return rate of fish tagged in 1975 was 44% for reward tags and 15% for non-reward tags. A non-response rate calculated according to Chad-

Table 2. Age-specific catch per unit effort (number/hour) and estimated annual survival rates (%) for largemouth bass in Pomme de Terre Lake spring electrofishing collections for 1976–1982. Survival rates exceeding 100% were due to gear selectivity.

Sample year	Number/hour at age 1	2	3	4	5	6
1976	5.8	48.9	17.3	9.0	3.7	1.4
1977	35.8	68.2	48.0	13.1	3.2	2.1
1978	12.9	26.1	21.7	6.7	2.0	0.7
1979	11.0	17.6	14.9	9.8	4.1	3.1
1980	3.6	35.0	19.5	5.9	2.5	1.0
1981	4.7	52.8	55.9	11.5	4.0	2.0
1982	79.9	24.8	26.3	19.2	5.5	1.4

Interval	Annual survival (%) for ages 2–3	3–4	4–5	5–6
1976–77	98	76	36	57
1977–78	32	14	15	22
1978–79	57	45	61	155
1979–80	111	40	26	24
1980–81	160	59	68	80
1981–82	50	34	48	35

wick (1968) equaled 0.63, which is comparable to previously published estimates of 0.66 (Chadwick 1972), 0.46 (Rawstron 1971), and 0.66 (Folmar et al. 1980) for largemouth bass. Although it was planned to use the 1975 data to adjust non-reward tagging data in 1976 and 1977, it became apparent that anglers returned the tags applied these years at even a substantially lower rate than for non-reward tags in 1975. The adjusted exploitation rates of 19% in 1976 and 6% in 1977 were considered severe underestimates and unreliable; therefore, subsequent procedures for estimating exploitation used a reward system. The exploitation rate for bass longer than 14.9 in was 34% in 1978.

The rates of return for reward-tagged bass in the Pomme de Terre Lake study are intermediate to exploitation rates of 13–77% reported by other authors (Rawstron 1971; Chadwick 1972; Ager 1979; Yeager and Van Den Avyle 1979; Folmar et al. 1980). Variability of return rates among reservoirs can be attributed to several factors including climate, size of lake, fishing pressure, number and sizes of bass tagged, rate of tag loss, and variations in response rate.

Growth Rates

Age determinations of largemouth bass from Pomme de Terre Lake revealed considerable

Table 3. Indices of largemouth bass recruitment obtained from electrofishing and shoreline rotenone sampling in Pomme de Terre Lake, Missouri.

Year class	Age 1: Springtime electro-fishing (number/hour)	Age 1: May shore-line rotenone (number/0.06 acre)	Age 0: October shoreline rotenone (number/0.06 acre)
1976	36		8
1977	13	5	37
1978	11	4	15
1979	4	6	15
1980	5	5	19
1981	46		

overlap of size ranges for different age groups each year. Ball (1980) also reported this phenomenon. The wide range in sizes apparently was affected by the 15-in size limit. Fast growing individuals of a year class ordinarily would have been subjected to angler harvest at an earlier age than slow growing individuals, making the size-at-age data for the survivors more uniform.

Recruitment Rates

Efficiencies of springtime electrofishing and shoreline rotenone sampling as indicators of largemouth bass recruitment were evaluated. Electrofishing indicated considerable variation in abundance of Age 1 bass among the spring samples (Table 3), but it is not clear if this accurately reflects variable recruitment or is an artifact of electrofishing's bias against young bass as discussed earlier. Catch rates of Age 1 bass in May shoreline rotenone samples were relatively constant and not correlated with electrofishing results (Table 3).

The only clear advantage of one technique over the other was that electrofishing produced more specimens per man-day than did shoreline rotenoning. About 15 Age 0 bass were collected for each man-day of effort by electrofishing whereas only 2.1 bass were collected per man-day of rotenoning. Hightower et al. (1982) collected an average of 2.5 Age 0 bass per man-day of shoreline rotenoning, and they estimated that it would take 9–18 man-days of sampling to provide statistically valid estimates of Age 0 and Age 1 bass abundance. Shoreline rotenone sampling was discontinued in Pomme de Terre Lake in 1980 because it was felt that the inconsistent bias in

Table 4. Estimated catch rates (number per hour) and fishing effort (hours per acre per year) for bass anglers and all anglers in Pomme de Terre Lake, Missouri, 1976-1982.

	1976	1977	1978	1979	1980	1981	1982
Bass anglers							
Effort	14	11	11	14	11	11	14
Sublegal catch (<15 in)	0.47	0.69	0.41	0.34	0.39	0.50	0.79
Legal catch (≧15 in)	0.03	0.04	0.07	0.08	0.04	0.03	0.04
All anglers							
Effort	49	32	34	41	36	39	43
Sublegal catch (<15 in)	0.15	0.29	0.15	0.13	0.14	0.15	0.30
Legal catch (≧15 in)	0.01	0.02	0.03	0.03	0.02	0.01	0.02

collection of Age 0 or Age 1 largemouth bass was too great for the manpower expended.

Creel Survey and Angler Opinion

Quantification of catch-and-release rates of bass shorter than the minimum length limit in addition to monitoring harvest of legal-sized bass is important because one objective of many size limits is to increase overall catch rates rather than increase total harvest. This importance is illustrated by the creel survey results for Pomme de Terre Lake, where estimated catch rates of sublegal bass were about 4–20 times greater than rates for bass longer than 14.9 in (Table 4). Angler preference information also is needed to assess changes in bass fishing pressure or total fishing pressure due to imposition of size limits. On Pomme de Terre Lake, the 15-in limit had no apparent impact on the amount of effort expended by bass anglers or other fishermen (Table 4). Bass anglers in Pomme de Terre Lake caught legal and sublegal bass at a rate two to three times faster than anglers fishing for any species.

Of 2,440 anglers contacted during the creel survey in 1982, 2,166 expressed an opinion relative to bass harvest regulations. Most preferred a shorter, less restrictive length limit (79% of all anglers and 58% of the bass anglers). Comments made by anglers during interviews revealed that most would prefer to harvest more bass provided that adequate numbers of fish were still available for catch-and-release fishing. Many anglers considered size limits important to protect bass and did not approve of their elimination.

Discussion and Recommendations

Considerable effort was made in the Pomme de Terre Lake study to evaluate the effectiveness of sampling methods and important parameters that could be used in bass size-limit studies. In the original study plan, PSD was considered the best parameter to use for assessing changes in bass populations; however, it was apparent after 1976 that information on PSD alone was not sufficient to determine why bass populations responded differently to size limits. Therefore, electrofishing data collected in 1976–1982 included CPUE information as an index of bass abundance.

The need for estimates of survival rates became apparent when the yearly CPUE indices and age and growth data showed annual reductions in the numbers of bass over 8 in. Age determinations were initially aimed at identifying changes in growth rates; however, after survival esimates were recognized as important, more scale samples were collected to accurately separate bass length-frequencies into age groups.

Results of the Pomme de Terre Lake investigation lead to several recommendations that could improve evaluations of bass harvest regulations in other large reservoirs. The parameters and methods outlined below are not a complete listing, but they can provide a basis for expansion as specific cases warrant.

Electrofishing is recommended as the best method for bass collection. Both major arms of Pomme de Terre Lake were included in the size limit evaluation. From this and size-limit studies on other Missouri reservoirs up to 55,000 acres, it is recommended that at least two major arms of a lake be sampled since large reservoirs can have extremes in habitat or angler use that can produce differences in bass populations. Sampling effort should be adjusted to provide a minimum of 1,000 bass from each study area. This will provide a sample large enough to quantify CPUE, define length-frequencies, and to adequately describe relative abundance for most age groups based on scale analyses. Spring nighttime electrofishing is recommended because of difficulties with obtaining adequate numbers of bass longer than 12 in during fall. Collection of data from the same sites and at similar times of the year should provide data that adequately reflect annual trends of abundance and survival. The amount of electrofishing effort needed to collect 1,000 bass in spring was 8–20 hours; this level of effort probably would be adequate on most large reservoirs because the standing stock of bass in Pomme de Terre Lake was similar to estimates for several reservoirs in the Predator Stocking Evaluation (Grinstead et al. 1978).

A comprehensive analysis of the effectiveness of a size limit must include assessment of mortality rates. When bass are protected from harvest (by a length limit or other restriction), slow growth and increased natural mortality may result and compensate for decreased fishing mortality. Comparisons of survival rates of protected and unprotected age classes of bass before and after the imposition of a size limit would help define changes in survival rates due to compensatory mechanisms. Because of wide variability of growth within year classes, scale samples from 150 to 300 largemouth bass representative of the entire population should be collected to reduce errors in the conversion of CPUE data to survival rate estimates. Exploitation estimates should use variable-reward tags to improve accuracy because of the poor response rate for non-reward tags.

Manpower required to evaluate the size limit on Pomme de Terre Lake averaged 312 man-days per year for 1976–1982. Man-day requirements for each method, including data collection and analysis, were: electrofishing 32, creel survey 208, cove sampling 65 (1977 only), and bass tagging 7. Levels of effort would have to increase somewhat for larger reservoirs. Electrofishing effort needed to obtain a reliable bass sample is more cost-efficient than is cove rotenoning. In 1977, effort devoted to electrofishing was half that of cove rotenoning, but electrofishing provided about 32 times as many bass for analysis.

Although recruitment rates of black bass populations are needed for some evaluations, the amount of manpower required to determine Age 0 or Age 1 bass abundance was considered too great to warrant their inclusion as a part of size-limit evaluations. Recruitments rates should correspond to abundance at a certain minimum size or age; this size or age may vary among bass populations due to differences in fishery characteristics.

A creel survey is recommended for size-limit evaluations. Information on bass angler catch rates and effort should be included because anglers fishing for a particular species are usually more adept at catching this fish than those anglers whose efforts are not specific. Definitive objectives for desired catch and harvest rates should be specified, and several years of creel data should be collected prior to the imposition of the regulation to help define these objectives and provide a basis for future comparisons.

Angler acceptance and compliance with harvest regulations is necessary for the establishment and success of size limits to produce desirable results. Although either catch or harvest can be used to measure management success, an angler-opinion survey can help define angler preferences for specific fisheries.

Acknowledgments

I appreciate the assistance of James P. Fry, Lee C. Redmond, Ken R. Perry, Kevin R. Richards, Bill M. Turner, Gary D. Novinger, John A. Goddard, and Lawrence C. Belusz for their critical review of the manuscript. John E. McPherson, Kevin R. Richards, Bill M. Turner, and Lawrence C. Belusz assisted with data collections. I thank Melanie A. Benitz for her patience and understanding in typing the manuscript.

References

Ager, L. M. 1979. Exploitation and mortality of largemouth bass in Lake Tobesofkee, Georgia. Proceedings of the Annual Conference of the Southeastern Association of Fish and Wildlife Agencies 32:429–436.

Anderson, R. O., and S. A. Weithman. 1978. Concept of balance of coolwater fish populations. Pages 371–381 *in* R. Kendall, editor. Selected coolwater fishes of North America. Special Publication 11, American Fisheries Society, Bethesda, Maryland, USA.

Ball, R. L. 1980. Evaluation of a 14-inch minimum size limit on largemouth bass and other sportfishing changes in Monroe Reservoir. Final Report, Indiana Department of Natural Resources, Indianapolis, Indiana, USA.

Chadwick, H. K. 1968. Mortality rates in the California striped bass population. California Fish and Game 54:228–246.

Chadwick, H. K. 1972. Non-reporting of tagged

largemouth bass 1966–1969. Califronia Fish and Game 58:145–147.

COOPER, G. P., AND R. N. SCHAFER. 1954. Studies on the population of legal-size fish in Whitmore Lake, Washtenaw and Livingston Counties, Michigan. Transactions North American Wildlife Conference 19:239–257.

FOLMAR, H. G., W. D. DAVIES, AND W. L. SHELTON. 1980. Factors affecting estimates of fishing mortality of largemouth bass in a Southeastern Reservoir. Proceedings of the Annual Conference of the Southeastern Association of Fish and Wildlife Agencies 33:402–407.

GRINSTEAD, B. G., R. M. GENNINGS, G. R. HOOPER, C. A. SCHULTZ, AND D. A. WHORTON. 1978. Estimation of standing crop of fishes in the Predator-Stocking-Evaluation reservoirs. Proceedings of the Annual Conference of the Southeastern Association of Fish and Wildlife Agencies 30:120–130.

HANSON, W. D. 1977. The tailwater fisheries of Lake of the Ozarks and Pomme de Terre Lake, Missouri. Missouri Department of Conservation, Final report, Federal Aid Project F-1-R-25, Jefferson City, Missouri, USA.

HARRIS, F. A., L. M. AGER, AND E. HAYES. 1980. Comparisons of various mark-recapture techniques for estimating abundance of largemouth bass in Barkley Lake, Kentucky. Proceedings of the Annual Conference of the Southeastern Association of Fish and Wildlife Agencies 33:703–709.

HICKMAN, G. D., AND J. C. CONGDON. 1972. Effects of length limits on fish populations of five North Missouri lakes. Pages 84–94 *in* J. Funk, editor. Symposium on overharvest and management of largemouth bass in small impoundments. Special Publication 3, North Central Division American Fisheries Society, Bethesda, Maryland, USA.

HIGHTOWER, J. E., R. J. GILBERT, AND T. B. HESS. 1982. Shoreline and cove sampling to estimate survival of young largemouth bass in Lake Oconee, Georgia. North American Journal of Fisheries Management 2:257–261.

HOEY, J. W., AND L. C. REDMOND. 1972. Evaluation of opening Binder Lake with a length limit for bass. Pages 100–105 *in* J. Funk, editor. Symposium on overharvest and management of largemouth bass in small impoundments. Special Publication 3, North Central Division American Fisheries Society, Bethesda, Maryland, USA.

PARAGAMIAN, V. L. 1982. Catch rates and harvest results under a 14.0 inch minimum length limit for largemouth bass in a new Iowa impoundment. North American Journal of Fisheries Management 2:224–231.

RAWSTRON, R. R. 1971. Non-reporting of tagged white catfish, largemouth bass, and bluegills by anglers at Folsom Lake, California. California Fish and Game 57:246–252.

RICKER, W. E. 1975. Computation and interpretation of biological statistics of fish populations. Bulletin 191, Fisheries Research Board of Canada, Ottawa, Canada.

SIMPSON, D. E. 1978. Evaluation of electrofishing efficiency for largemouth bass and bluegill populations. Master's thesis, University of Missouri, Columbia, Missouri, USA.

WEGE, G. J., AND R. O. ANDERSON. 1978. Relative weight (W_r): a new index of condition for largemouth bass. Pages 79–91 *in* G. D. Novinger and J. G. Dillard, editors. New approaches to the management of small impoundments. Special Publication 5, North Central Division American Fisheries Society, Bethesda, Maryland, USA.

YEAGER, D. M., AND M. J. VAN DEN AVYLE. 1979. Rates of angler exploitation of largemouth, smallmouth, and spotted bass in Center Hill Reservoir, Tennessee. Proceedings of the Annual Conference of the Southeastern Association of Fish and Wildlife Agencies 32:449–458.

Pages 73–78 *in* G. E. Hall and M. J. Van Den Avyle, editors.
Reservoir Fisheries Management: Strategies for the 80's.
Reservoir Committee, Southern Division American Fisheries Society, Bethesda, Maryland. 1986.

Comparison of Rotenone and Trawling Methodology for Estimating Threadfin Shad Populations

J. ROBERT SILER
Duke Power Company
Production Environmental Services
Route 4, Box 531
Huntersville, North Carolina 28078

ABSTRACT

A technique was developed to estimate the standing crop and total population size of threadfin shad (*Dorosoma petenense*) in 13,156-hectare Lake Norman, North Carolina. Eighty-six oblique or stair-stepped samples from the surface to near bottom were taken at night with a bow-fished, midwater Tucker Trawl. Population estimates (in millions) and 95% confidence intervals for threadfin shad during October 1979-1981 were 304.4 (±57.3), 589.6 (±165.6), and 560.9 (±111.2), respectively. Average standing crops based on trawling samples ranged from 16.5 kg/hectare in 1979 to 26.3 kg/hectare in 1980, with an 11% coefficient of variation for each estimate. Cove rotenone samples taken at six locations produced average standing crop estimates that ranged from 8.4 kg/hectare in 1979 to 33.0 kg/hectare in 1980, with coefficients of variation of 91 to 197%. The better precision of estimates from trawl samples was attributed to use of a stratified random sampling design, uniform horizontal distribution of shad at night, and the expression of trawl catches on an areal basis. The extreme variability in cove rotenone data made estimates unacceptable for management decisions. Trawling can provide precise estimates of threadfin shad abundance in reservoirs at about one-third of the annual manpower costs of the cove rotenone program.

Threadfin shad (*Dorosoma petenense*) have been widely introduced into south-temperate reservoirs of the United States to provide additional prey for piscivorous fish. Piscivorous species clearly utilize threadfin shad as prey (Olmsted and Kilambi 1971; Aggus 1973; Ball and Kilambi 1973; May et al. 1975; McGee et al. 1978); however, effects of threadfin shad introductions on growth of piscivorous fishes have not been clearly documented (Miller 1971; Range 1973; Hepworth and Gloss 1976; Fast et al. 1982). The uncertainty of the effects of threadfin shad introductions on piscivorous fish populations is compounded both by the lack of information concerning changes in growth and abundance of piscivorous fish and, perhaps more importantly, the lack of quantitative information concerning the success of prey introductions. Because of the threadfin shad's sensitivity to low water temperatures (Strawn 1965) and its high reproductive potential (Swingle 1970), its biomass may fluctuate drastically within and among years. Quantitative estimates of threadfin shad biomass are important for evaluating the success of introductions, and are essential for forage fish management (Ney 1981) and evaluating predator-stocking strategies. This paper outlines a methodology for estimating threadfin shad abundance using results from trawl samples and compares the manpower cost and results of this methodology to a conventional cove rotenone sampling program.

METHODS

A bow-fished, midwater Tucker Trawl (Siler 1983) was used to sample threadfin shad at night on 13,156-hectare Lake Norman, North Carolina. Sampling was done with a 1.5 × 2-m trawl having a mesh size of 3.2 mm in early October 1979, 1980, and 1981. Sample locations were chosen based on a stratification scheme whereby the lake was partitioned into nine sampling zones (Fig. 1). Six of the nine zones were further separated into littoral and limnetic strata at the 6-m depth contour. Approximately 15 random samples were taken from each of Zones 1, 2, and 3, 14 samples from each of Zones 4 and 5, three samples from Zone 6, six samples from Zone 7, and two samples from each of Zones 8 and 9. Sampling in Zones 1–5 and Zone 7 was proportioned between littoral and limnetic strata based on the surface areas of the strata. A map with numbered grids was used to randomly select sample locations and depths each year. Trawl samples in limnetic strata were taken obliquely, and littoral samples were taken by stair-stepping

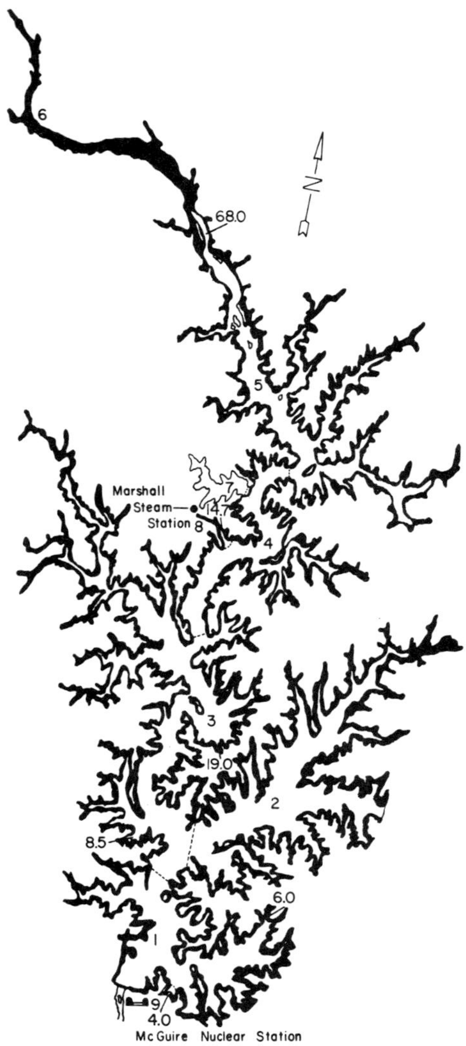

Figure 1. Map of Lake Norman, North Carolina, showing cove rotenone sampling locations (4.0, 6.0, 8.5, 14.7, 19.0, and 68.0) and midwater trawl sampling zones (1-9) used in 1979-1981.

from the surface to the randomly chosen depth. Trawl catches were expressed in numbers or weight of fish per unit surface area (Siler 1985) and were directly comparable to rotenone catches.

Population estimates and variances were calculated separately for trawl data taken from each zone and stratum of the reservoir, using the direct enumeration methodology of Regier and Robson (1967) and the stratified random sampling design of Cochran (1963). Mean catch rates from trawling were expanded to total population size based on the area of each zone and stratum. The surface area of each zone and stratum at the time of sampling was determined using area : elevation equations based on topographic maps. Analysis of variance was used to determine if statistical differences in threadfin shad standing crop estimates from trawl samples existed among zones, strata, and years. The predetermined significance level for all statistical tests was $P =$ 0.05.

Fish sampling with rotenone was conducted during the first two weeks of August in 1979, 1980, and 1981. Six sampling locations, each approximately 1.2 hectare in area, were located in uplake, midlake, and downlake areas of Lake Norman (Fig. 1). All coves were surveyed, and permanent markers were placed to define sample areas. Maximum depth of the coves ranged from 3.4 to 6.7 m. Prior to applying rotenone, a 9.5-mm mesh block net was positioned across the mouth of the sample cove and checked by divers. Rotenone was applied to achieve a concentration of 1 mg/liter in the sample area. A special effort was made to collect stressed threadfin shad as soon as they surfaced. A bow-mounted scoop, as well as individual netters, collected fish for two consecutive days following rotenone application. On the first day of sampling, all species were sorted into 2-cm size classes, counted, and weighed. Fish from the second day were similarly measured and counted, but weights were based on specimens collected during the first day.

Results

Rotenone

Threadfin shad standing crop estimates from cove rotenone samples were imprecise despite a relatively high level of effort devoted to obtain the data. Estimates for cove samples ranged from 0.02 to 165 kg/hectare, with coefficients of variation (standard deviation divided by the mean) of 159, 197, and 91% among samples taken in 1979, 1980, and 1981, respectively. Manpower required for sampling and data summarization was about 175 man-days/year.

Variability among coves was so great that mean standing crop estimates for 1979 and 1980 were not significantly different from zero, and annual differences in standing crop were not statistically detectable despite a four-fold range among the

Table 1. Average threadfin shad standing crop estimates (kg/hectare), 95% confidence intervals (CI), and sample size (N) for trawling and cove rotenone samples in Lake Norman, North Carolina.

	Trawling estimates											
	Littoral strata			Limnetic strata			Reservoir-wide			Cove rotenone estimates		
Year	N	kg/hectare	CI	N	kg/hectare	CI	N	kg/hectare	CI	N	kg/hectare	CI
1979	39	8.4	±3.1	47	21.3	±5.2	86	16.5	±3.5	6	8.4	±10.7
1980	39	8.7	±7.5	46	37.0	±8.1	85	26.3	±5.8	6	33.0	±51.9
1981	36	6.5	±2.4	51	33.6	±7.8	87	23.3	±4.9	6	15.2	±11.1

means (Table 1). To have obtained desirable levels of precision (coefficients of variation of 20% or less), data from at least 61, 93, and 20 coves would have been required in 1979, 1980, and 1981, respectively. Length-frequency distributions of shad also varied considerably among coves (Fig. 2).

Trawling

Trawling provided more precise estimates of shad standing crop at considerably less effort than devoted to rotenone sampling. Estimated standing crops for 258 midwater trawl samples collected during the three years ranged from 0 to 217 kg/hectare. Coefficients of variation for reservoir-wide standing crop estimates were about 11% each year using the stratified design. Manpower required for sampling with the trawl, sample processing, and data summarization was about 65 man-days/year.

Much of the gain in precision was due to the use of a stratified sampling design, made feasible by an ability to collect a large number of trawl samples. If the data collected each year were treated as a simple random sample, the coefficients of variation would have been about 120% each year. Length-frequency distributions of shad collected by trawling were similar among samples and usually had a modal size of 4 cm. Distributions for six trawl samples collected near the coves sampled with rotenone were nearly homogeneous (Fig. 2).

The use of trawling and the randomized sampling design allowed spatial comparisons of threadfin shad density along the main axis of Lake Norman (Siler et al. 1986) and between littoral and limnetic zones. Standing crops were significantly greater in limnetic areas (Table 1). This methodology also allows a more statistically valid basis for expanding the sample data to whole-lake estimates because sample sites can be selected at random (within zones and strata) whereas rotenone data come from sites with physical features conducive to sampling. Total threadfin shad biomass estimates (kg) from trawling data and 95% confidence intervals for Lake Norman during October 1979, 1980, and 1981 were 2.1 (±0.4) × 10^5, 3.2 (±0.7) × 10^5, and 2.7 (±0.6) × 10^5, respectively.

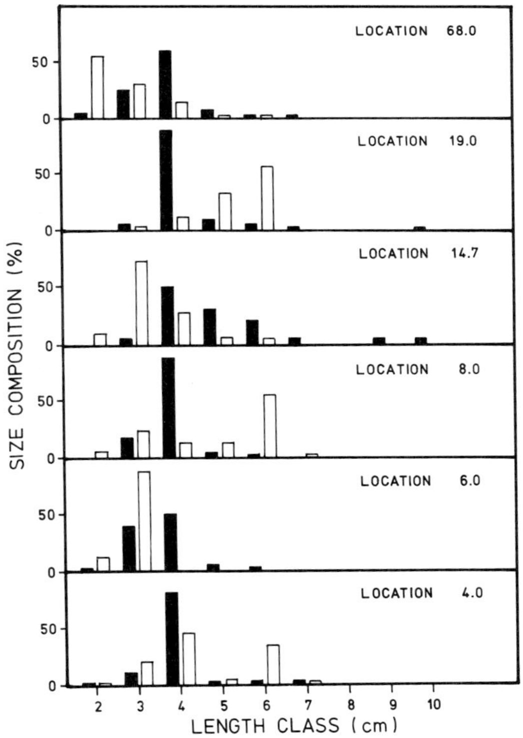

Figure 2. Length-frequency distributions of threadfin shad from six cove rotenone samples (open bars), and six limnetic midwater trawl samples (shaded bars) collected near the cove samples in Lake Norman, North Carolina, during 1979.

Discussion

The precision of estimates made by cove rotenoning and trawling affects the utility of these techniques for making management decisions concerning threadfin shad. Ranges of standing crop estimates were generally similar; however, the more variable data and a general lack of suitable rotenone sample sites illustrate the limitations of cove rotenoning as a technique for monitoring threadfin shad populations. Cove rotenone sampling locations were selected to represent particular areas of the reservoir and were similar in size and morphometry. Selection of the six locations was made after visually inspecting all alternative locations. On the other hand, the 86 trawling locations were randomly selected from all possible areas of the reservoir.

Coefficients of variation and confidence intervals indicate the lack of precision of threadfin shad standing crop estimates derived from rotenone sampling; standard deviations were usually greater than 100% of the mean estimate from rotenone samples but were only 11% of the mean from trawling. The extreme variation of the rotenone data was partially attributable to the small number of samples. Roughly 240 trawl samples could be taken with the same manpower necessary to collect six rotenone samples. It is cost prohibitive to increase the cove rotenone sample size to a level that would provide data necessary for management decisions or for detecting changes in threadfin shad populations. While cove rotenone sampling may provide adequate data for management decisions for some species, these results indicate that it has limited applicability for threadfin shad.

The trawling program provided more precise abundance and biomass estimates of threadfin shad than did rotenone sampling. The improved precision in trawl estimates is attributed to sample stratification, uniform horizontal distribution of threadfin shad at night, and the expression of individual trawl catches on an areal basis. Sample stratification into high and low density areas, as compared to samples pooled across zones and strata, accounted for a decrease in coefficients of variation from approximately 120 to 11%. Stratification of the reservoir into littoral and limnetic strata accounted for most of the increased precision; however, uplake zones supported higher standing crops than downlake zones (Siler et al. 1986). Provided that some knowledge of the distribution of threadfin shad within a reservoir is available, stratification will likely improve the precision and cost-effectiveness of a trawling program.

Observations by Houser and Dunn (1967) provided information critical for the development of my trawling methodology. They observed that threadfin shad school during the day, but horizontal distribution of shad at night was basically uniform in Bull Shoals Reservoir, Arkansas. They also observed decreasing density with depth and noted that threadfin shad seldom occupied waters below the thermocline. Similar distribution patterns occurred in Lake Norman, and these observations provided the basis for the stratification scheme.

Schooling and clumped distribution patterns of threadfin shad during the day probably contributed to the variability of estimated threadfin shad standing crops observed in cove rotenone samples because the block net was positioned after threadfin shad schools formed. This is also suggested by the wide variability in length-frequency distributions of threadfin shad collected with rotenone. The precision of threadfin shad standing crop estimates might be improved by positioning the block net at night, but the trawling data for Lake Norman indicate that these littoral estimates would severely underestimate the overall biomass because of the disparity between littoral and limnetic strata. Aggus et al. (1980) also observed this disparity in Lake Barkley, Kentucky, where threadfin shad standing crops were 40 times higher in open water areas than in coves.

The area-based sampling units (Siler 1985) made it possible to express catch rates from individual samples in conventional "kilograms per hectare" units instead of volumetric units. Area-based estimates and oblique sampling made it possible to take advantage of Houser and Dunn's (1967) observation that horizontal distribution patterns of threadfin shad were remarkably uniform. Using oblique sampling, I was able to avoid the intensive sampling required by Houser and Dunn to estimate densities in the vertical depth strata. The area-based sampling units made it possible to sample variable amounts of water below the thermocline, which had very few or no threadfin shad, without affecting the actual standing crop estimates.

The trawling program used at Lake Norman offers a cost-effective method of estimating threadfin shad abundance in reservoirs. These data can also provide a basis for considering supplemental stockings or introductions of piscivorous species, as well as providing information on one of the most important prey species in southern reservoirs; however, the trawling program could not be justified without a special interest in small limnetic species, especially threadfin shad. Efforts to enhance energy flow in reservoirs should be directed at forage-fish management (Ney 1981); therefore, precise threadfin shad abundance data are desirable to make valid management decisions and to evaluate management strategies with respect to this species and its major predators.

Acknowledgments

I thank Kim Baker, Rich Hansen, Tim Leonard, and Danny Sherrill for their assistance in field collections. Steve Johnson, Don Degan, and Larry Olmsted provided valuable suggestions on the manuscript. Other personnel from Duke Power Company's Production Environmental Services Group provided assistance in cove rotenone sampling and laboratory processing of fish. Al Houser and Dr. Jim Dunn provided helpful suggestions in the design of the sampling program.

References

AGGUS, L. R. 1973. Food of angler harvested largemouth, spotted and smallmouth bass in Bull Shoals Reservoir. Proceedings of the Annual Conference of the Southeastern Association of Game and Fish Commissioners 26:519–529.

AGGUS, L. R., D. C. CARVER, L. L. OLMSTED, L. L. RYDER, AND G. L. SUMMERS. 1980. Evaluation of standing crops of fishes in Crooked Creek Bay, Barkley Lake, Kentucky. Proceedings of the Annual Conference of the Southeastern Association of Fish and Wildlife Agencies 33:710–722.

BALL, R. L., AND R. V. KILAMBI. 1973. The feeding ecology of the black and white crappies in Beaver Reservoir, Arkansas, and its effect on the relative abundance of the crappie species. Proceedings of the Annual Conference of the Southeastern Association of Game and Fish Commissioners 26: 577–589.

COCHRAN, W. G. 1963. Sampling techniques. John Wiley and Sons, New York, New York, USA.

FAST, A. W., L. H. BOTTROFF, AND R. L. MILLER. 1982. Largemouth bass, *Micropterus salmoides,* and bluegill, *Lepomis macrochirus,* growth rates associated with artificial destratification and threadfin shad, *Dorosoma petenense,* introductions at El Capitan Reservoir, California. California Fish and Game 68:4–20.

HEPWORTH, D. K., AND S. P. GLOSS. 1976. Food habits and age-growth of walleye in Lake Powell, Utah-Arizona, with reference to introduction of theadfin shad. Utah State Division Wildlife Resources Publication 76-15, Salt Lake City, Utah, USA.

HOUSER, A., AND J. E. DUNN. 1967. Estimating the size of the threadfin shad population in Bull Shoals Reservoir from midwater trawl catches. Transactions of the American Fisheries Society 96:176–184.

MAY, B. F., C. THOMPSON, AND S. P. GLOSS. 1975. Impact of threadfin shad (*Dorosoma petenense*) introduction on food habits of four centrarchids. Utah State Division Wildlife Resources Publication 75-4, Salt Lake City, Utah, USA.

MCGEE, M. V., J. S. GRIFFITH, AND R. B. MCLEAN. 1978. Prey selection by sauger in Watts Bar Reservoir, Tennessee, as affected by cold-induced mortality of threadfin shad. Proceedings of the Annual Conference of the Southeastern Association of Fish and Wildlife Agencies 31:404–411.

MILLER, E. E. 1971. The age and growth of centrarchid fishes in Millerton and Pine Flat Reservoirs, California. California Department Fish and Game. Inland Fisheries Administration Report 71-4, Sacramento, California, USA.

NEY, J. J. 1981. Evolution of forage-fish management in lakes and reservoirs. Transactions of the American Fisheries Society 110:725–728.

OLMSTED, L. L., AND R. V. KILAMBI. 1971. Interrelationships between environmental factors and feeding biology of white bass in Beaver Reservoir, Ark. Pages 397–409 *in* G. E. Hall, editor. Reservoir fisheries and limnology. Special Publication 8, American Fisheries Society, Bethesda, Maryland, USA.

RANGE, J. D. 1973. Growth of five species of game fishes before and after introduction of threadfin shad into Dale Hollow Reservoir. Proceedings of the Annual Conference of the Southeastern Association of Game and Fish Commissioners 26: 510–518.

REGIER, H. A., AND D. S. ROBSON. 1967. Estimating population numbers and mortality rates. Pages 31–66 *in* S. D. Gerking, editor. The biological bases of freshwater fish production. Blackwell Scientific Publications, Oxford, England.

SILER, J. R. 1983. A description of trawl handling structures for a bow-fished midwater trawl. Progressive Fish-Culturist 45:217–220.

SILER, J. R. 1985. A comparison of area- and volume-based population estimates from midwater trawl catch statistics. Research report PES/85-16, Duke Power Company Production Environmental Services, Huntersville, North Carolina, USA.

SILER, J. R., W. J. FORIS, AND M. C. MCINERNY. 1986. Spatial heterogeneity in fish parameters within a reservoir. Pages 122–136 *in* G. E. Hall and M. J. Van Den Avyle, editors. Reservoir fisheries management: strategies for the 80's. Reservoir Committee, Southern Division American Fisheries Society, Bethesda, Maryland, USA.

STRAWN, K. 1965. Resistance of threadfin shad to low temperatures. Proceedings of the Annual Conference of the Southeastern Association of Game and Fish Commissioners 17:290–293.

SWINGLE, H. A. 1970. Production of the threadfin shad, *Dorosoma petenense* (Günther). Proceedings of the Annual Conference of the Southeastern Association of Game and Fish Commissioners 23: 407–421.

Pages 79–85 *in* G. E. Hall and M. J. Van Den Avyle, editors.
Reservoir Fisheries Management: Strategies for the 80's.
Reservoir Committee, Southern Division American Fisheries Society, Bethesda, Maryland. 1986.

A Method of Qualitatively Assessing White Crappie Populations in Missouri Reservoirs

MICHAEL A. COLVIN AND FRED W. VASEY

Missouri Department of Conservation
Fish and Wildlife Research Center
1110 College Avenue
Columbia, Missouri 65201

ABSTRACT

White crappie (*Pomoxis annularis*) populations in Missouri's large reservoirs are assessed by rating population density, growth rate, age structure, size structure, and recruitment. The assessment is based on October trap-net samples. Each of the five parameters is assigned a point value from 0 to 10, and an overall rating is made by summing the points. Individual parameter values less than 5 were determined to be unsatisfactory. Point values were adjusted for dependent parameters, for example, slower growth was accepted when densities were high. Assessments for population density and size structure were related to angler harvest the following spring. This assessment enables biologists to determine which parameters may be preventing a population from achieving a satisfactory rating and helps determine if management or research efforts are needed.

A population is considered excellent if Age 1 and older white crappie are captured at a rate of 40 or more per trap-net day; 10–19% are Age 4 or older; and 40–60% are 9 in and longer. In addition, crappie should average at least 9 in long after three growing seasons and Age 0 should be captured at a rate of four to 10 per trap-net day.

Fishery workers have been confronted with the difficult task of identifying causes of problems with fish populations so that correct management practices can be implemented. Important questions the biologist should be able to answer are: "Which combinations of recruitment, growth, and mortality result in populations with desirable densities and size structures?" and "What is the problem if a population is not satisfactory?"

In the past few years, considerable attention has been given to the use of structural indices to describe populations. The concept of proportional stock density, or PSD (Anderson 1976, 1980), has been widely used because it is easy to calculate and is influenced by the three primary rate functions of recruitment, growth, and mortality. However, the use of PSD alone to assess fish populations has some potential drawbacks in that it is not possible to identify the causes of problems when a population does not meet the objective. It also is possible for a population to have a desirable PSD and still not be satisfactory because of poor recruitment, low population density, undesirable growth rates, or other factors. As a result, many biologists now use several indices in conjunction with PSD in order to assess fish populations. While relative weight indices (W*r*) and relative stock density, RSD (Wege and Anderson 1978), aid in describing populations, they do not adequately define causes of problems.

We have studied white crappie (*Pomoxis annularis*) in Missouri's large reservoirs since 1968 and have developed a method of assessing populations by rating five parameters measured from October trap-net catches. We chose the parameters of population density, growth rate, age structure, size structure, and recruitment because they are either assessments of, or are determined by, interactions of the three primary rate functions; they also relate in some way to sport fishing success. By rating these parameters we are better able to identify specific problems with white crappie populations and define needed research and management activities. The method also provides a consistent and objective way of determining the status of white crappie populations. Results of the assessment can easily be communicated to supervisors, other biologists, and anglers. Accurate assessments can be made without detailed data analysis because most of the parameters can usually be rated by examining length-frequency data.

Table 1. Length-at-age and total annual mortality rates used to model white crappie length-frequency distributions in Missouri reservoirs.

Age (years)	Total length (in)		Total annual mortality rates (%)		
	Mean	Standard deviation	High levels	Medium levels	Low levels
0	3.6	0.3			
1	6.9	0.7	50	50	30
2	9.3	0.7	80	50	30
3	10.7	0.6	80	50	30
4	11.8	0.6	80	50	30
5	12.5	0.5	85	50	30
6	13.0	0.5	85	50	30
7	13.4	0.5	85	50	50
8	13.7	0.5			

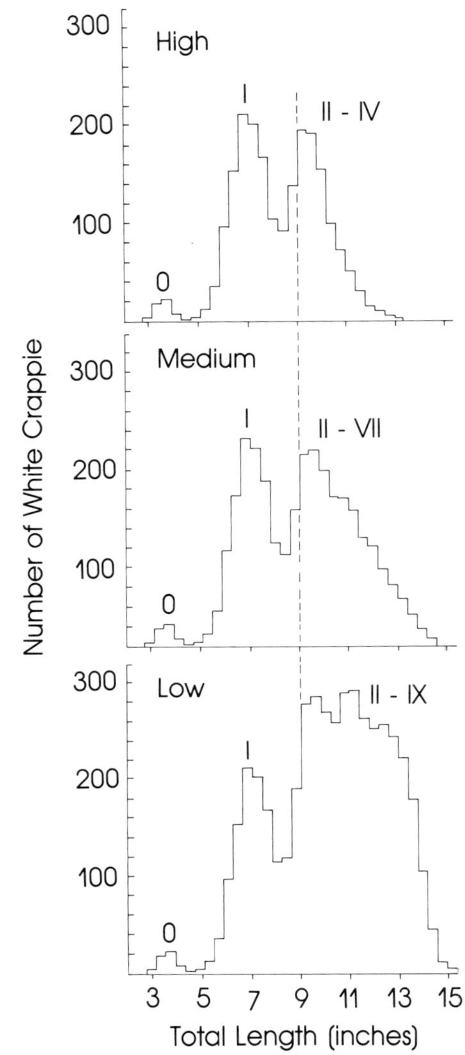

Figure 1. Length frequencies constructed from simulated white crappie populations using high, medium, and low total annual mortality levels. The numbers of Age 0 and Age 1 crappie were adjusted downward to reflect size selectivity that typically occurs in October trap-net samples.

Methods and General Approach

Our method is designed to assess white crappie populations sampled with trap nets during October when surface water temperatures are 55 to 68 F. We sample in October because all sizes of white crappie are captured and the corresponding trap-net catch rates have been found to be positively related to angler harvest the following year. Our trap nets have 0.5-in square, knotless nylon mesh and are constructed with two 3 × 6-ft frames with center braces and four 2.5-ft diameter hoops. Each net has a single lead that is either 55, 65, or 85 feet long, 3.5 ft deep, and of 0.625-in square mesh.

We sampled from one to seven reservoirs every year since 1967 and have collected more than 70 reservoir-years of white crappie data. Each reservoir typically was sampled with 20 trap nets until 1,500 Age 1 and older white crappie were captured, or for a maximum of four nights. Fish were removed from the nets once each day and measured to the nearest millimeter. Scale samples were taken from 10 fish per 10-millimeter length interval. Data from scale-sampled fish were extrapolated into growth rate and age structure for the entire sample.

The population assessment was begun by assigning a point value from 0 to 10 to the data for each of the five parameters. The point value was determined from a set of standard values based on optimum conditions expected from populations with desirable size structures and densities. High point values of 9 or 10 are assigned to optimum measurements of each parameter, whereas lesser or greater measurements are given fewer points. Any parameter receiving four or fewer points is considered to be unsatisfactory.

We determined optimum levels for each parameter using two methods. First, we examined length-frequency distributions from populations that we felt provided a good fishery and made

Table 2. Point values assigned to the five population parameters used to assess Missouri reservoir white crappie populations sampled with trap nets in October.

	Assigned values							
	Fish per trap-net day[a]							
Parameters	0–4	5–9	10–19	20–29	30–39	40–49	50–59	>59
Population density	0	1	3	6	8	9	7	4
(with good growth)[b]	0	0	3	6	8	10	10	10
	Mean length at end of third growing season (in)							
	6.0–6.9	7.0–7.9	8.0–8.9	9.0–9.9	10.0–10.9	>10.9		
Growth rate	0	1	5	9	8	4		
(with good density)[c]	1	4	8	10	10	10		
	Percent Age 4 and older[a]							
	<1.0	1.0–2.9	3.0–5.9	6.0–9.9	10–19	20–29	30–39	>39
Age structure	0	2	4	7	10	7	4	0
(with good density)[c]	0	3	6	9	10	10	8	4
	Percent over 9.0 in[a]							
	0–9	10–19	20–29	30–39	40–49	50–59	60–69	>69
Size structure	1	4	7	9	10	7	3	1
(with good density)[c]	1	4	9	10	10	10	9	6
(with good age structure)[d]	0	3	6	9	10	10	10	8
	Number of Age 0 per trap-net day							
	<0.2	0.2–0.5	0.5–0.9	1.0–1.9	2.0–3.9	4.0–9.9	10–19	>19
Recruitment	0	1	2	4	6	8	4	0
(with good growth)[b]	0	0	1	3	6	10	8	6

[a] Excludes young-of-the-year crappie.
[b] Mean length of at least 9.0 in after three growing seasons.
[c] At least 25 Age 1 and older per trap-net day.
[d] At least 6% Age 4 and older.

subjective decisions as to the satisfactory and optimum parameter measurements. Satisfactory measurements were those that we believed did not need further management adjustments. Second, because many of the populations we observed were not considered good, we modeled structural relationships of populations similar to the method used by Reynolds and Babb (1978). Length-frequency distributions were constructed using growth rates similar to most Missouri reservoirs and three levels of total annual mortality (Table 1 and Fig. 1). Mortality levels classified as high were similar to those calculated from trap-net data for most Missouri reservoirs; therefore, predicted population length frequencies were similar to actual samples.

Size of white crappie was important in our analysis. We considered 9 in to be the minimum size that should be harvested by anglers. This decision was based on length-weight relationships which showed that white crappie began adding proportionately more weight per unit length when they are 8 to 9 in long. In terms of a quality fishery, we believe 9-in and larger crappie provide considerably more benefits to anglers than do smaller ones.

Population Parameters

Population Density

Density is a result of recruitment and mortality. We used the catch per trap-net day of Age 1 and older fish in October as an index of density. Our basic premise was that higher densities were better for anglers, which generally holds true until densities become so high that growth declines to an unacceptable level. We judged that satisfactory densities occurred when the catch rate of Age 1 and older white crappie was 20 to 59 fish per trap-net day; optimum catch rates were 40 or more fish per net day if growth rate was good (Table 2).

Population density influences the ratings of three other parameters. When densities of white crappie were good, broader ranges of growth rate, age structure, and size structure were accepted as desirable. Standard values assigned to these parameters were increased when catch rates were at least 25 white crappie per trap-net day.

Growth Rate

Growth influences the size structure of a population and affects sizes of fish present for anglers. In a desirable population, fish should reach a minimum harvestable size in a reasonable period of time. We believe white crappie are growing at an optimum rate if they average 9 to 10 in after three growing seasons; hence, we quantify growth as the mean length of Age 2 white crappie from October netting samples. Satisfactory values were assigned to growth when Age 2 crappie averaged 8.0 to 10.9 in long. Higher points were assigned over a wider range when densities were good (Table 2).

Age Structure

Age structure is a function of recruitment and mortality. We measured it as the percent of Age 4 and older white crappie with young-of-the-year excluded. Age structure gives an indication of the size of crappie that are present in a population because a higher proportion of older fish generally means a higher proportion of larger fish if growth is good. Exploitation rate may also be reflected by the age structure; for example, high exploitation can cause a lower percentage of older fish.

We used population models to determine satisfactory and optimum levels of age structure because most white crappie populations in Missouri reservoirs have relatively high total annual mortality levels and similar age structures. Percent Age 4 and older white crappie calculated from the population models (Table 1 and Fig. 1) were 2, 12, and 30 for the high, medium, and low mortality levels, respectively. We wanted our approach to result in the assignment of high points to populations providing opportunities to catch large crappies.

We knew white crappie were capable of living longer than the three or four years observed in most reservoirs because of information collected from the Long Creek Arm of Table Rock Lake. This arm, which is very difficult to sample with trap nets, consistently produced older white crappie in the creel. Age 5 and older white crappie usually provided about 30% (range of 6 to 60%) of the springtime harvest. Although we were not able to calculate mortality rates, they were probably similar to the low level used in the population models (Table 1). Mortality in the Long Creek Arm was probably as low as could be expected for a population subjected to angling, but it did indicate the species' ability to survive to older ages. We therefore used the medium mortality level as a guide when rating the age structure parameter. The satisfactory range for population age structure (percent Age 4 and older) was determined to be 6–29%, while the optimum range was 10–19%; a broader range was accepted when densities were good (Table 2).

Size Structure

As an index of the size of white crappie that are present for anglers, size structure is the only parameter in our assessment that is influenced directly by all three population rates of recruitment, growth, and mortality. Our measurement of size structure is similar to PSD and is the percent of the sample that is 9 in and larger (excluding young-of-the-year). Satisfactory points were assigned to size structure when the proportion of 9 in and larger white crappie in the trap nets ranged from 20 to 59% (Table 2). Optimum points were assigned over a wider range of size structure when densities were good and to higher percentages when the age structure was good (at least 6% Age 4 and older).

Recruitment

The recruitment parameter is a measure of annual reproductive success; therefore it affects population levels and angler success in subsequent years. Recruitment is measured as the catch per trap-net day of young-of-the-year white crappie.

Age 0 white crappie are not as vulnerable as older and larger individuals when sampled with trap nets in October; however, their catch rates do give an indication of year-class strength. Our data show that good year classes of white crappie were usually produced when the catch of young-of-the-year was at least two fish per trap-net day. Generally, higher catch rates were assigned higher points, but since excessive recruitment theoretically causes slow growth, exceptionally high catch rates received lower points. Optimum recruitment occurred when growth was good and catch

Table 3. Assessments of the Pomme de Terre Lake white crappie population sampled with trap nets in October 1975–1982.

		Point values for						
Year		Population density	Growth rate	Age structure	Size structure	Recruitment	Total value	Population rating
1975	Actual measurement	31	9.2	0.05	46	0.1		
	Assigned value	8	10	0	10	0	28	Good
1976	Actual measurement	16	10.0	0.3	46	0.3		
	Assigned value	3	8	0	10	0	21	Fair
1977	Actual measurement	14	8.6	7.7	29	2.2		
	Assigned value	3	5	7	6	6	30[a]	Good
1978	Actual measurement	26	10.2	0.8	20	<0.1		
	Assigned value	6	10	0	9	0	25	Good
1979	Actual measurement	15	9.7	0.6	65	2.1		
	Assigned value	3	9	0	3	6	21	Fair
1980	Actual measurement	20	9.3	0.2	20	0.1		
	Assigned value	6	9	0	7	0	22	Fair
1981	Actual measurement	25	9.4	0.3	69	1.3		
	Assigned value	6	10	0	9	3	28	Good
1982	Actual measurement	12	9.8	0.2	36	0.3		
	Assigned value	3	9	0	9	0	21	Fair

[a] Includes three additional points because four parameters were satisfactory or better.

rates ranged from 4.0 to 9.9 young-of-the-year per trap-net day (Table 2).

Overall Rating

An overall population assessment is made by summing the points obtained for each of the five parameters. We added three points to this total when a population received a satisfactory rating (five or more points) from four of the five parameters. We felt that a population structured this way would provide more benefits to anglers than one that received higher points from just two or three parameters, but a similar overall rating.

We established five population ratings ranging from poor to excellent as reflected by the various point totals obtained from the overall assessment:

Point Total	*Population Rating*
≤18	Poor
19–24	Fair
25–34	Good
35–40	Very Good
≥41	Excellent

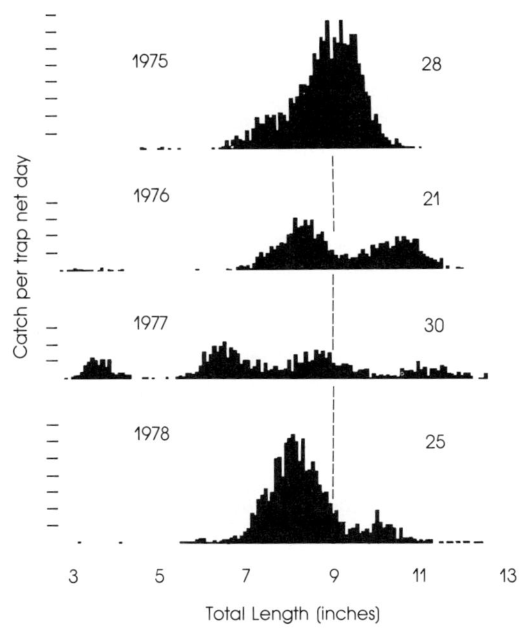

Figure 2. Length frequencies of white crappie captured with trap nets in October from Pomme de Terre Lake, Missouri, 1975–1978. Numbers on the right of the graph are point totals obtained from the qualitative assessment.

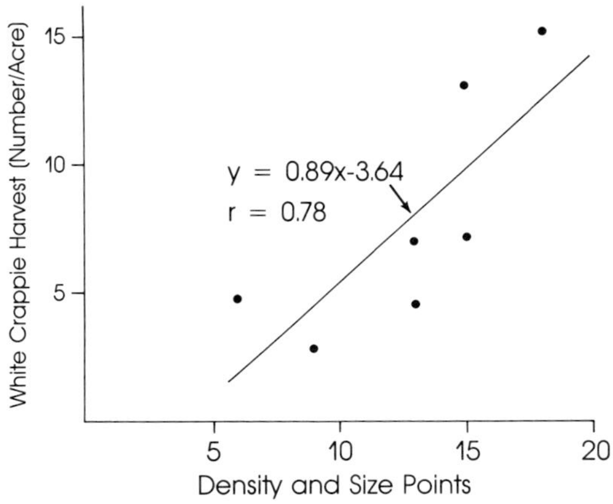

Figure 3. Relationship between population density and size structure assessment points from October samples and angler harvest of white crappie the following spring for Pomme de Terre Lake, Missouri, 1975-1981.

Application and Discussion

Our method of assessing white crappie populations by rating five population parameters has advantages over other methods such as PSD that are based on only one parameter. For instance, one or two years of poor recruitment may cause age or size structure to appear satisfactory or optimum and therefore receive high point values. The overall assessment, however, would reflect the poor recruitment because another parameter, such as population density, would receive a low rating. Also, assessments from previous years would reflect problems in recruitment or other parameters.

Results of the assessment show that different population structures can have similar overall ratings. White crappie populations at 7,800-acre Pomme de Terre Lake in 1975, 1977, and 1978 received similar overall assessments and total points, but the length-frequency distributions revealed considerably different population structures (Table 3 and Fig. 2). Samples from 1975 and 1978 received high ratings for population density, growth rate, and size structure, but low ratings for age structure and recruitment. The 1977 sample received a low rating for population density, and satisfactory ratings for the other four parameters. These assessments corresponded well with angler success the following spring. Estimated harvests in spring 1976, 1979, and 1978 were 15, 7, and 3 white crappie per acre, respectively, but the harvested crappie averaged approximately 1 in larger in 1978 (9.9 in versus about 9.0 in).

We believe this qualitative method provides a consistent and objective means of assessing white crappie populations and helps establish priorities for crappie research and management. By using similar yearly assessments, strong and weak aspects of population dynamics can be identified. Populations that consistently rate "poor" or "fair" and individual parameters that are consistently rated unsatisfactory should receive higher priorities for management efforts than those that receive better ratings. Assessments for eight years at Pomme de Terre Lake indicate that variable recruitment and poor age structure consistently affected the white crappie population even though growth rate and size structure were usually rated satisfactory or better (Table 3). In other Missouri reservoirs age structure has consistently been unsatisfactory in five of six lakes and recruitment has been variable in four. Even though some of these populations are regularly rated "good," opportunities for improvement still exist. As a result, our current research is designed to address these major problems. When corrective management actions are implemented the assessment procedure can be used to monitor the resultant changes in the population.

An effective assessment method should be related in some way to angler success; populations assessed as "good" should provide more benefits to anglers than those rated "poor" or "fair." The population density (numbers) plus size structure (size) assessments at Pomme de Terre Lake from 1975 through 1981 were positively related to angler harvest the following spring (Table 3 and Fig. 3). The relationship was significant at the 0.05 level ($r = 0.78$). This kind of positive relationship would be expected from populations where density is variable because of inconsistent recruitment and age structure is generally poor. We suspect that populations receiving higher ratings in age structure would result in increases in the size, but not necessarily the number, of white crappie harvested.

This method of assessing white crappie populations in large reservoirs needs to be tested more on populations that consistently rate "very good" or "excellent," particularly in relation to angler benefits. We believe, however, that it adequately assesses white crappie populations over the range of conditions that exist in Missouri's large reservoirs.

Acknowledgments

We appreciate the assistance given by Gary Novinger and Tom Russell in their review of the manuscript. This paper was a result of Dingell Johnson Project F-1-R, Study I-7, Missouri.

References

ANDERSON, R. O. 1976. Management of small warm-water impoundments. Fisheries 1(6):5–7.

ANDERSON, R. O. 1980. Proportional stock density (PSD) and relative weight (Wr): interpretive indices for fish populations and communities. Pages 27–33 *in* S. Gloss and B. Shupp, editors. Practical fisheries management: more with less in the 1980's. Proceedings of the 1st annual workshop of the New York Chapter of the American Fisheries Society, Ithaca, New York, USA.

REYNOLDS, J. B., AND L. R. BABB. 1978. Structure and dynamics of largemouth bass populations. Pages 50–61 *in* G. Novinger and J. Dillard, editors. New approaches to the management of small impoundments. Special Publication No. 5, North Central Division of the American Fisheries Society, Bethesda, Maryland, USA.

WEGE, G. J., AND R. O. ANDERSON. 1978. Relative weight (Wr): a new index of condition for largemouth bass. Pages 79–91 *in* G. Novinger and J. Dillard, editors. New approaches to the management of small impoundments. Special Publication No. 5, North Central Division of the American Fisheries Society, Bethesda, Maryland, USA.

Pages 86–97 *in* G. E. Hall and M. J. Van Den Avyle, editors.
Reservoir Fisheries Management: Strategies for the 80's.
Reservoir Committee, Southern Division American Fisheries Society, Bethesda, Maryland. 1986.

MANAGEMENT OF THE PHYSICAL AND CHEMICAL ENVIRONMENT

Effects of Water-Level Changes on Reservoir Ecosystems, with Implications for Fisheries Management

GENE R. PLOSKEY
Aquatic Ecosystem Analysts
P.O. Box 4188
Fayetteville, Arkansas 72702

ABSTRACT

Over 350 papers published since 1930 compose the extensive literature about effects of water-level changes on aquatic biota. Most of the literature details single pre- and post-treatment studies evaluating effects on species of plants, invertebrates, or fish. Although some conflicting results have been obtained, fishery managers have developed successful management plans by relying on the consensus of published observations. Most managers seek to (1) draw down water levels in late summer or fall, (2) establish terrestrial vegetation by seeding or allowing for recolonization, (3) flood vegetation in spring, and (4) maintain high water for as much of the growing season as possible. Variations of this general plan, with regard to magnitude, duration, and timing have been used, but the literature suggests that only broad manipulations on seasonal or annual time scales yield significant benefits. Long-term data replicating several management and recruitment events are needed to develop predictive models. Comprehensive models probably are lacking today because manipulation and evaluation are expensive multi-year efforts. However, sound predictions will become increasingly important to justify future requests for modified reservoir operations.

Inasmuch as the literature on water-level management is very extensive, this paper is only an overview of the history, state of the art, and literature. Many potentially valuable papers were not cited because of space limitations. The literature on reservoir water-level management has been summarized in annotated bibliographies by Triplett et al. (1980) and Ploskey (1982) and discussed in papers by Wood (1951), Hulsey (1959), Wood and Pfitzer (1960), Keith (1975), Groen and Schroeder (1978), and Ploskey (1983). Common names of fishes mentioned in the text correspond to those accepted by the American Fisheries Society (Robins et al. 1980).

HISTORICAL PERSPECTIVE

In early studies, Ellis (1937) considered water-level fluctuations to be detrimental to fish because of impaired spawning, although others recognized the potential for enhancing fertility (Hasler and Einsele 1948). Deleterious effects of uncontrolled fluctuations on fish reproduction and survival were later demonstrated (Heman et al. 1969; June 1970; Walburg 1976). Additionally, water-level manipulation for fishery management purposes was virtually abandoned in reservoirs with salmonids except to provide stable levels for spawning and nurseries (Martin 1955; Aass 1960; Calhoun 1966). Nevertheless, management of water levels showed promise for improving fish populations in large warm-water reservoirs.

Work in culture ponds, small warm-water impoundments, and the first large reservoirs laid the groundwork for comprehensive plans to manage water levels for fish. Wood (1951) was the first to synthesize information pertaining to effects of water-level changes on nutrients, productivity, and fish abundance. Findings of Wood and Pfitzer (1960), who relied upon work by Hulsey (1959), provided a central axiom for managing water levels: Effects of water-level fluctuations depend more on time, area, and duration of flooding than the amount of fluctuation as measured in vertical feet. High stages of long duration usually produce the greatest fish crops, and effects may be accentuated by topography, soil conditions, and vegetation.

In the early 1950's, the study of effects of fluctuating water levels entered a "testing phase." During this period, questions of researchers often included the following: Was the change detrimental? Which species were affected? What was the degree of change in phytoplankton, periphyton, benthos, zooplankton, or fish? To a lesser extent: What were the mechanisms causing change? Over 350 papers have been published on the subject (Triplett et al. 1980; Ploskey 1982). Most evaluations of management plans were designed as pre- and post-treatment studies, and statistical analyses were hence limited to paired tests of significance. Consequently, much of the literature details positive and negative results for different species, and conflicting findings have been obtained for some. Opportunistic fishery managers such as Hulsey (1959), Keith (1975), and Groen and Schroeder (1978) followed what appeared to be the majority of evidence. They developed successful management plans even before the complex reasons for success were completely understood.

State of the Art

A panel of 35 experts in ecology, hydrology, and hydropower optimization attended a workshop on research needs to include ecological issues in basin-level hydropower planning (Hildebrand and Goss 1981). They concluded that the capability to predict water-level changes was adequate but models for predicting biological consequences were inadequate. They also concluded that (1) knowledge of bank and riverbed stability was poor, (2) the capability to predict effects of water-level changes on aquatic biota in lower trophic levels was poor, and (3) effects of water-level changes on fish had not been quantified except to determine requirements for spawning. The consensus was that the ability to address ecosystem-level effects was qualitative at best.

After 50 years of studying reservoirs, biologists have accumulated many observations of physicochemical and biological changes resulting from changes in water levels. Although observations are sometimes disjunct or contradictory, enough information is available to develop conceptual models of effects on fisheries. However, standardized quantitative data are needed for rigorous statistical analyses and further development of predictive models.

Literature Highlights

Lower Trophic Levels

Aquatic plants.—Because plants support bacteria, zooplankton, benthos, and fish, effects of water-level changes on primary production can greatly influence responses at higher trophic levels. Phytoplankton, periphyton, and macrophytes are directly or indirectly important as food for fish, and plants provide habitat for invertebrates and fish (Cowell and Hudson 1967; Claflin 1968; Johnson and Stein 1979). Macrophytes may influence productivity and predator-prey relations (Cooper and Crowder 1979), and aquatic or terrestrial vegetation is important for spawning of many species of fish (Carlander 1969, 1977).

Effects of water-level changes on phytoplankton have received little attention (Mitchell 1975). Phytoplankton production probably is affected indirectly by altered nutrient concentrations, light, temperature, and grazing pressure that result from water-level changes. Direct effects include the physical removal of phytoplankton by release of water from reservoirs (Benson and Cowell 1967).

The abundance or biomass of phytoplankton can usually be explained by changes in nutrient levels or light (Jones and Bachmann 1978a, 1978b), as modified by factors such as temperature or turbidity (Guseva 1958; Markosyan 1969). Thus, changes in nutrient levels resulting from water-level changes should produce predictable changes in phytoplankton abundance provided that light is not limiting. Effects of changes in water levels probably have more impact on primary production during the growing season than in winter because production is regulated largely by solar radiation and temperature in temperate waters (Wetzel 1975). Mitchell (1971) was able to correlate phytoplankton production with rising water levels that continually inundated nutrient-rich pasture areas in Lake Mahinerangi, New Zealand. After water levels stabilized, Mitchell (1975) found that measured phytoplankton production was higher than predicted, suggesting that a delayed release of nutrients maintained high levels of productivity although no new areas were being flooded.

Periphyton can be directly affected by changes in water levels. When exposed to air for several days, periphyton dessicates and dies (Benson and Cowell 1967; Claflin 1968). Barman and Baarda (1978) found that periphyton biomass and production was significantly reduced for almost a year after subtrates of Lake Laurel, Georgia, were exposed for 178 days and reflooded.

The presence of some macrophytes is desirable for many fisheries because of increased productivity and diversity in littoral areas. However, at high densities, macrophytes may cause fish kills due to oxygen depletion or limit surface area available for fishing or boating. Control of overabundant macrophytes usually has been of more concern to fisheries managers than the absence or scarcity of such vegetation.

Success in regulating macrophytes by manipulating water levels has been inconsistent (Bennett 1962; Lantz et al. 1965), but it offers a viable alternative to chemical control (Davis 1967). Desiccation, freezing, and soil compaction after drawdowns reduce densities of some species (Dunst et al. 1974), many of which have been categorized according to preferences for fluctuating or stable water levels (Nichols 1974).

Zooplankton. —Zooplankton, like phytoplankton, is rarely directly affected by changes in water levels. Direct effects are limited to displacement or removal due to shortened water retention time (Benson and Cowell 1967). Rodhe (1964) observed that seasonal regulation of water levels that resulted in damming from spring to late autumn and discharge in winter favored zooplankton production more than operations that used high flushing rates in spring and summer.

In reservoirs where water-retention time exceeds 0.2 years, production of zooplankton probably is regulated by grazing pressure and food availability, both of which can be influenced by water levels. Increased abundance of some zooplankters has been associated with increased food availability after water inundated terrestrial areas (Lötmarker 1964). Two- to three-fold increases in zooplankton volume in Lake Ransaran, Swedish Lappland, resulted from increased nutrient and detrital inputs from flooded terrestrial areas (Rodhe 1964). Benson (1968) attributed increased abundance of rotifers in two Missouri River reservoirs to rising water levels that continually inundated grassy areas. As waters rose and flooded pastures in Lake Mahinerangi, densities of *Daphnia carinata* and cyclopoid copepods increased significantly (Mitchell 1975).

Benthos.—Benthic invertebrates are directly and indirectly affected by changes in water levels. Direct effects include (1) exposure and mortality of species with poor mobility or without resting mechanisms (Aass 1960), and (2) entrainment and loss of planktonic stages from reservoirs during periods of rapid discharge. Indirect effects result from changes in habitat, food resources, or the chemical environment.

Of the direct effects, discharge of benthic invertebrates has been studied the least, though it can be significant in reservoirs with occasionally or continuously high rates of water exchange (Benson 1973). In Lewis and Clark Lake, 24 metric tons of *Hexagenia* spp. and 20 metric tons of Diptera passed through the turbines of the dam in spring 1965 (Cowell and Hudson 1967; Swanson 1967).

The most obvious direct effect of water-level changes on benthos is exposure and desiccation after drawdown. In Lake Francis Case, a 4-m drawdown in August 1966 stranded up to 6,146 organisms/m^2 (Cowell and Hudson 1967). Kaster and Jacobi (1978) observed that annual fluctuations of 7.7 m in a central Wisconsin reservoir exposed 1.8 g/m^2 (dry weight) of benthos. Mortality of exposed organisms undoubtedly reduces populations within the fluctuation zone and may partly explain the inverted vertical distributions of benthos observed in fluctuating reservoirs (Grimås 1961; Fillion 1967; Kaster and Jacobi 1978). The abundance of benthos in nonfluctuating reservoirs usually is greater in shallow than in deep areas, but in fluctuating reservoirs, inverted distributions may result from the lack of a littoral community or concentration of mobile species at or just below the drawdown limit (Davis and Hughes 1966; Cowell and Hudson 1967).

Regardless of the mechanism, water-level fluctuations have a negative impact on littoral benthos. The impact is modified by basin morphometry and the frequency, timing, and duration of fluctuations. Water-level changes in shallow, gently sloping reservoirs affect more benthic habitat per unit of change than they do in steep-sided reservoirs. Frequency and duration of changes determine potential for recovery, annual loss of production, and impacts on littoral habitats (Grimås 1964). Littoral benthos communities in regulated lakes of Norway and Sweden were destroyed by water-level fluctuations of 13 m. Grimås (1962) reported that densities were reduced 50% during 10 years of 6-m fluctuations.

After fall drawdowns in Lake Francis Case were reduced from about 11 to 7 m, the density of benthos in May increased by a factor of five (Benson and Hudson 1975). Apparently, reduced drawdowns deposited sediments at shallower depths and reduced shoreline scouring, thereby enhancing the habitat of burrowing species.

Changes from nonfluctuating to fluctuating regimes may be followed by shifts in species composition from diverse littoral communities (mayflies, stoneflies, odonates, beetles, and snails) to communities dominated by chironomids and oligochaetes (Hunt and Jones 1972a, 1972b). Because littoral invertebrates have distinct preferences for habitats such as soft mud (Benson 1973), macrophytes (Wegener et al. 1975), or inundated terrestrial vegetation (Aggus 1971; Jones and Selgeby 1974), the loss or addition of such habitat by changes in water levels may greatly alter species composition. Because some species of benthos are more productive or important as fish food than others, productivity and trophic efficiency may also be altered.

Fish Populations and Fisheries

Trophic relations and growth.—Inflows and resultant water-level changes alter trophic relations and growth of fish by regulating input of allochthonous nutrients and foods, productivity or species composition of fish-food biota, or prey availability and vulnerability. Changes in water levels that significantly affect trophic relations have several characteristics: they are large, last several months, occur during the growing season, and inundate or eliminate productive areas of littoral or terrestrial vegetation. Small, short-term fluctuations have little positive effect on nutrients, plants, or invertebrates. Winter fluctuations generally do not increase productivity because of low water temperatures. If areas of the fluctuation zone are barren, even large changes in water level will have little effect on productivity (Grimås 1962).

Rapidly rising waters that inundate terrestrial areas—especially areas supporting lush vegetation—temporarily increase supplies of terrestrial invertebrate foods for opportunistic fish such as white catfish (Goodson 1965), black bullheads (Applegate and Mullan 1967), and sunfishes and black basses (Mullan and Applegate 1968). Changes in trophic conditions after waters flood vegetation may be inferred from growth rates as well as from changes in diet. In Lake Oahe, Nelson (1974) associated the inundation of areas with terrestrial vegetation with increased growth of 13 species of fish (goldeye, northern pike, common carp, river carpsucker, smallmouth buffalo, bigmouth buffalo, white bass, white crappie, black crappie, yellow perch, sauger, walleye, and freshwater drum). Similar observations have been made elsewhere for largemouth bass, blue sucker, shorthead redhorse, gizzard shad, spotted sucker, white crappie, bluegill, yellow perch, and channel catfish (Wright 1950; Jackson 1958; Elrod and Hassler 1971; Aggus and Elliott 1975).

Production of fish-food biota also is increased in newly inundated areas (Aggus 1971; McLachlan 1977; Strange et al. 1982). However, terrestrial areas must be inundated long enough to be colonized. According to Benson (1973), 25 days of inundation were required for periphyton development on plant stems in Missouri River reservoirs and 40 days for the colonization of periphyton by aquatic insects.

Large drawdowns that concentrate prey fish for at least 2–3 months (Jenkins 1970) at temperatures above 13 C (Bennett 1962) can increase predator foraging and growth (Aggus 1979). Increased feeding activity or growth by piscivores during or immediately after drawdown has been reported for northern pike (Beard and Snow 1970), smallmouth bass (Heisey et al. 1980), largemouth bass (Heman et al. 1969), and white crappie and flathead catfish (Johnson and Andrews 1974). However, after prolonged drawdown, growth of fish may decrease as concentrations of prey diminish and the production of most invertebrates and small fish declines. Johnson (1974) reported that declining water levels over 5 years reduced the growth of young white crappie, channel catfish, and common carp in Lake Carl Blackwell, Oklahoma, by prolonging the loss of littoral benthos. Eschmeyer (1949) suggested that food limitations experienced by benthophagous fishes may account for their low abundance in widely fluctuating storage impoundments and high densities in stable mainstream reservoirs.

Predator-prey relations may also be altered by changes in habitat complexity as water levels change. Not only are prey concentrated by large drawdowns, but they also may be forced to abandon refuge in littoral areas. Consequently, prey become more vulnerable to predators (Bennett 1962; Keith 1975). Complex structure not only provides refuge for prey but also reduces the foraging efficiency of predators (Murdock and Oaten 1975; Savino and Stein 1982). Heman et al. (1969) observed an inverse relation between amounts of inundated vegetation and feeding and growth of largemouth bass.

Reproduction. —Reproduction of fish that spawn near shore is influenced by changes in water levels. Adverse effects on reproduction are related to loss of habitat by drawdown or shoreline modification or to mortality of eggs of young-of-year (YOY) fish after exposure or suffocation by eroded sediments (Hassler 1970). Mortality of eggs or YOY fish stranded by drawdown has been documented for many species, including salmonids (Aass 1960), sunfishes (Heman et al. 1969), walleyes (Priegel 1970), and black basses (Estes 1971). Species of fish that frequently spawn in tributaries (e.g., white bass, walleye, and saugers) or in open-water areas (e.g., goldeye and freshwater drum) generally are not adversely affected (Benson 1973).

In addition to stranding eggs or young fish, rapidly receding waters may result in desertion of nests, poor egg survival, and disrupted spawning for species that spawn in shallow water (centrarchids, yellow perch, northern pike, common carp, buffalofishes, and gizzard shad). Low and variable spring water levels adversely affected the spawning success of gizzard shad, emerald shiner, white bass, white crappie, and yellow perch in Lewis and Clark Lake (Walburg 1976). Nest desertion permits sunfishes to prey intensively on eggs (Vogele 1975). Intraovarian mortality of eggs and disrupted spawning in northern pike was caused by fluctuating water levels (June 1970).

Nevertheless, reproductive success of many fishes probably is ultimately regulated by post-spawning survival which can be affected by temperature, wind-induced turbulence (Summerfelt 1975), predation (Aggus and Elliott 1975), and food availability (Hassler 1970; Eipper 1975). This may explain why drawdowns to control spawning of common carp (Shields 1958) did not always suppress strong year classes (Gasaway 1970). Conversely, provision of ideal spawning conditions may not produce a strong year class. Fourt (1978) observed large numbers of YOY black basses and crappies after 7,000 acres of terrestrial vegetation were flooded during the spawning season in Beaver Lake, Arkansas. However, poor survival of YOY fish after August resulted from inadequate prey abundance, and the year class was not exceptional.

Spawning success does not insure a strong year class, but it may increase the chances of producing one, given environmental and trophic conditions favorable for survival of YOY fish. Therefore, knowledge of spawning and nursery requirements of fishes is important for improving spawning by manipulating water-levels. Benson (1976) listed spawning and nursery habitats for 16 species, and Carlander (1969, 1977) recorded spawning temperatures and habits of many sport and commercial fishes. Because time of spawning varies with temperature, temperature is more reliable than date as an index to spawning, given geographical variations in temperature. Increased spawning success has often been associated with the inundation of appropriate spawning substrates, including vegetation or gravel for yellow perch (Beckman and Elrod 1971; Nelson and Walburg 1977) or terrestrial vegetation for northern pike (Benson 1968; Hassler 1970), buffalofishes (Moen 1974), and common carp (Gabel 1974).

Strong year classes of many freshwater fish have been correlated with rising or high water during and for several months after the spawning season. Examples include the largemouth bass (Summerfelt and Shirley 1978); black basses (Rainwater and Houser 1975); northern pike (Hassler 1970); sauger (Walburg 1972); and common carp, river carpsucker, smallmouth buffalo, and bigmouth buffalo (Gasaway 1970; Elrod and Hassler 1971). Rainwater and Houser (1975) found that the reproductive success of black basses in Bull Shoals Lake, Arkansas, from 1966 to 1973 was negatively correlated with fluctuations of water levels during each 3-month spawning season. High rates of survival of young-of-year black bass were attributed to accelerated growth and increased survival due to refuge provided by flooded shoreline vegetation.

Reproductive success of fish that spawn near shore in reservoirs is influenced by the time and duration of flooding and the type of substrate inundated (Aggus 1979). Water levels determine the amount of nursery area available by inundating or receding from vegetation. The density of YOY largemouth bass in August in Bull Shoals Lake was directly related to acre-days of flooding of terrestrial vegetation (Aggus and Elliott 1975). Survival of young fish of many species apparently is increased when cover is abundant, but lack of habitat exposes YOY fish to increased predation. Decreasing water levels reportedly reduced cover and refuge for larval and juvenile stages of spotted bass (Vogele 1975). The abundance of young walleyes has been directly correlated with water levels (Nelson and Walburg 1977).

Because year-class strength of most fishes in reservoirs varies greatly among years depending on environmental conditions and predator-prey relations, consecutive annual manipulation of water levels may not be essential or productive. Strong year classes of one species may suppress future year classes of its own or other species for several years by competing for food (von Geldern 1971) or by preying on eggs, embryos, larvae (Bennett 1962), or other YOY fish (Aggus and Elliott 1975). Jenkins (1975) concluded that cannibalism and a scarcity of submerged vegetation in some years made it virtually impossible to produce strong year classes of largemouth bass. Swingle and Swingle (1967) noted that weak year classes of crappies occurred when a strong year class of largemouth bass had developed the previous year.

Manipulation of water levels for spawning is the most practical and inexpensive method of producing littoral-spawned YOY. Other methods have included (1) provision of artificial spawning sites (Martin 1955), (2) construction of nonfluctuating "inlet impoundments" adjoining reservoirs (Ellis 1937; Grimås 1965), and (3) artificial propagation of important fish in hatcheries. Artificial spawning sites are expensive and impractical for large reservoirs, and inlet impoundments do not make up for the loss of littoral areas, at least in steep-sided, deep reservoirs (Grimås 1965). Although important sport and forage fishes are stocked extensively in large reservoirs by State fishery agencies, fish stocks in most large reservoirs are still produced and sustained by natural reproduction.

Standing Crop and Harvest.—Effects of changes in water levels or rates of water exchange on trophic relations, reproduction, and survival of fish ultimately are reflected in fish standing crop. Jenkins (1967) found a negative correlation between the total standing crop of fish and yearly vertical fluctuations in water levels in 70 reservoirs. Jenkins (1970) found positive correlations between annual water-level fluctuations and standing crops of spotted gar, flathead catfish, black basses, and white crappie. Annual water-level fluctuation was negatively correlated with the biomass of gizzard shad, pikes, carpsuckers, and sunfishes. Aggus and Lewis (1977) found that standing crops of fish were more variable in reservoirs with large seasonal changes in water levels (flood control or storage reservoirs) than in mainstream impoundments.

Although fish-carrying capacity may be reduced by seasonally fluctuating water levels, large changes that occur every 3 or 4 years can be beneficial. Productivity and carrying capacity can be significantly increased by large prolonged changes in water level that are infrequent enough to permit the growth of terrestrial vegetation in exposed areas. Controlled annual drawdown and reflooding of three Louisiana lakes resulted in a gradual increase in fish standing crop and rapid increases in the biomass of harvestable-size fish (Lantz 1974). In Council Grove Reservoir, Kansas, the percent (by weight) of harvestable-size fish increased from 30.3 to 44.1 after water levels were managed (Groen and Schroeder 1978). Wegener and Williams (1975) observed a large increase in total standing crop (from 214 to 510 kg/hectare) within 3 years after water levels of Lake Tohopekaliga, Florida, were manipulated. The biomass of black basses increased from 39 to 67 kg/hectare. Drawdown exposed 50% of the bottom for 6 months, and refilling to normal pool required another 6 months. Afterward, high inflows caused water levels to rise and remain above normal pool for the next year, and significant changes in aquatic vegetation and fish standing crop were observed.

Harvest of fish is affected by many factors, some of which are influenced by changes in water level. Jenkins (1967) found that sport-fish harvest per unit area was directly correlated with total dissolved solids, storage ratio, and shoreline development ratio and inversely related to reservoir age, area, and mean depth. Commercial fish harvest was inversely related to storage ratio, mean depth, shoreline development ratio, and water-level fluctuations; it was directly related to reservoir age. Chevalier (1977) found positive correlations between the commercial catch of walleyes from Rainy Lake, Minnesota, and water levels 4, 5, and 6 years earlier. Drawdown followed by extensive flooding in Lake Tohopekaliga nearly quadrupled the harvest of black basses (Wegener and Williams 1975).

Management.—Manipulation of water levels should be regarded as only one of several procedures used to manage the fluctuation zone. Because substrates exposed by drawdown are vulnerable to erosion, early establishment of herbaceous vegetation after drawdown is important for erosion control and aesthetics (Benson 1976). Erosion and resuspension of sediments may be an important temporary source of nutrients for biota, but over a number of years, erosion and sedimentation usually reduce the productive capacity of littoral areas by removing sediments, especially in steep-sided reservoirs (Grimås 1961; Hunt and Jones 1972a). Inundated terrestrial plants provide substrates for algae and release nutrients upon decomposition. They furnish substrates, refuge, or food for many invertebrates and fish. In Lake Nottely, Georgia, significantly more zooplankton was present in June in flooded shoreline areas that had been seeded with grasses the previous fall than in unseeded areas (Strange et al. 1982). The number of aquatic insects present in seeded areas also was significantly higher than in control areas, as was the density of YOY black basses (from mid-June to mid-July), YOY sunfish, YOY white crappie, and YOY yellow perch.

Drawdown must occur during the growing season for successful seeding of herbaceous terrestrial plants in fluctuation zones and should allow for substantial growth before winter. Hulsey (1959) recommended drawdown by 15 September in Arkansas for rye. Groen and Schroeder (1978) discussed planting rapidly growing ryegrass, wheat, or rye during September or October after drawdown in Kansas. Ryegrass was seeded at about 11 kg/hectare and wheat and rye at 34 to 68 kg/hectare. For drawdowns before August, Japanese millet and hybrid sudan-sorghum also were seeded successfully in fluctuation zones of reservoirs in Arkansas and Kansas. Strange et al. (1982) broadcast seed (49 kg/hectare) and fertilizer (25 kg/hectare) from July to September in plots in the drawdown zone of Lake Nottely, Georgia. A drought caused most of the sudan hybrids sown in July and August to die after attaining 10 cm height. Seeded during September, fescue and especially rye were more successful than other grasses because of fall rains. The rye, seeded at 35 kg/hectare, established a strong root system over winter and grew rapidly in early spring.

Drawdown provides fishery managers with time to improve submerged structure in drained areas. In new impoundments, where habitat in the form of submerged timber or brush is present, construction of artificial structures probably is unjustified. However, as reservoirs age and timber in littoral areas deteriorates, the use of artificial shelters may become an economically-viable management option. For further information and references on artificial structures, see papers by Johnson and Stein (1979), Pierce and Hooper (1979), and Brown (1986).

Based on a consensus of available literature, managers should attempt to (1) draw down water in late summer or fall, (2) establish herbaceous vegetation by natural colonization or seeding, (3) flood terrestrial vegetation in spring, and (4) maintain high water for as much of the growing season as possible. Variations of this basic plan, with regard to magnitude, duration, and time of water-level changes, generally meet specific needs of most conservation agencies. The Kansas Fish and Game Commission, for example, often limits the extent of drawdown to 10–20% of the original area, seeds vegetation extensively, and raises water levels slightly in fall to flood vegetation for waterfowl (Groen and Schroeder 1978). Management of flood-control reservoirs in Arkansas (Hulsey 1959) incorporated extensive drawdowns (80% of surface area). In most reservoirs, drawdowns are best scheduled for late summer or fall because water temperatures are above 13 C and warm-water piscivores are still feeding and growing. Earlier drawdowns may not be favorable to survival of YOY fishes, and drawdown in winter does not permit the establishment of terrestrial vegetation.

Operational flexibility to provide appropriate seasonal water levels increases where a system of reservoirs exists. Flexibility generally is related to the number of reservoirs in the drainage basin, at least for the downstream impoundments. Every 2 years one reservoir in a series can be selected and managed to improve fisheries. With appropriate scheduling, power and water demands often can be met by releases from other reservoirs. To maximize effectiveness, biologists should monitor the age structure of fish populations and relations between prey and predator biomass (see methods of Swingle 1950; Jenkins and Morais 1977) in candidate reservoirs. In this way, reservoirs that would benefit most by intensive management could be selected. Impoundments with below average populations consisting of mostly older sport fish would be priority candidates for management.

In reservoirs where other essential uses prohibit manipulation of water levels to improve fisheries, an alternative is to construct inlet or subimpoundments. This approach has not been used extensively; subimpoundments were discussed by Ellis (1937), and inlet impoundments were studied by Grimås (1965) in Norway. Subimpoundments formed by water-control structures across large coves or arms of reservoirs could provide small, highly productive fishing lakes in parts of large reservoirs where fisheries are severely impaired by extensive fluctuations. Water levels in subimpoundments could be managed to benefit fish (or waterfowl), thus offering mitigation for losses of fishery resources.

Needs for the Future

The literature illustrates why water-level management is a powerful tool for managing large reservoirs. Although exact response mechanisms vary somewhat, reservoir fisheries can be dramatically improved by manipulating water levels. However, plans to fine-tune water-level changes or alter fish populations by making small adjustments probably will fail. Fishery managers still can greatly increase their chances of success by monitoring fish communities annually and identifying community and reservoir characteristics that favor production of strong year classes.

The scarcity of quantitative data collected from one reservoir for four or more consecutive years is perhaps the greatest hindrance to statistical testing of management hypotheses. Previous research on effects of water-level changes was essential to the development of the management plans and testable hypotheses we have today. But, only by collecting long-term data can we replicate management operations, statistically determine which elements are important, pinpoint cause-effect relations, and develop predictive models. Sound predictions will become increasingly important to justify requests for altered reservoir operations.

Informed decisions concerning reservoir operations to benefit fish require extensive data to model effects of water-level changes on fisheries. Scheduling models have been developed to help reservoir operators optimize reservoir use and resolve conflicts among competing users (Shane 1981). Unfortunately, current scheduling models rarely address fishery concerns because appropriate quantitative data suitable for predicting effects on fish, fisheries, and limnology are scarce. One vehicle for conducting future research would be to form cooperative research teams of fishery biologists from several States, private industry, or universities, perhaps organized by the Reservoir Committee of the Southern Division, American Fisheries Society. Predictive capabilities are needed to have water-level plans seriously considered as part of multi-purpose operations of large Federal reservoirs.

References

Aass, P. 1960. The effects of impoundment on inland fisheries. Technical Meeting, International Union for Conservation of Nature and Natural Resources, Athens, Greece 4:69–76.

Aggus, L. R. 1971. Summer benthos in newly flooded areas of Beaver Reservoir during the second and third years of filling 1965–1966. Pages 139–152 *in* G. E. Hall, editor. Reservoir fisheries and limnology. Special Publication 8, American Fisheries Society, Bethesda, Maryland, USA.

Aggus, L. R. 1979. Effects of weather on freshwater fish predator-prey dynamics. Pages 47–56 *in* R. H. Stroud and H. Clepper, editors. Predator-prey systems in fisheries management. Sport Fishing Institute, Washington, D.C., USA.

Aggus, L. R., and G. V. Elliott. 1975. Effects of cover and food on year-class strength of largemouth bass. Pages 317–322 *in* R. H. Stroud and H. Clepper, editors. Black bass biology and management. Sport Fishing Institute, Washington, D.C., USA.

Aggus, L. R., and S. A. Lewis. 1977. Environmental conditions and standing crops of fishes in predator-stocking-evaluation reservoirs. Proceedings of the Annual Conference of the Southeastern Association of Fish and Wildlife Agencies 30:131–140.

Applegate, R. L., and J. W. Mullan. 1967. Food of the black bullhead (*Ictalurus melas*) in a new reservoir. Proceedings of the Annual Conference of the Southeastern Association of Fish and Game Commissioners 20:288–292.

Barman, E. H., Jr., and D. G. Baarda. 1978. An evaluation of the effect of drawdown on the trophic status of a small reservoir. Environmental Resources Center (ERC 01-78), Georgia Tech University, Atlanta, Georgia, USA.

Beard, T. D., and H. E. Snow. 1970. Impact of winter drawdown on a slow-growing panfish population and associated species. Bureau Research Report, Wisconson Department of Natural Resources, Madison, Wisconsin, USA.

Beckman, L. G., and J. H. Elrod. 1971. Apparent abundance and distribution of young-of-year fishes in Lake Oahe, 1965–69. Pages 333–347 *in* G. E. Hall, editor. Reservoir fisheries and limnology. Special Publication 8, American Fisheries Society, Bethesda, Maryland, USA.

Bennett, G. W. 1962. Management of artificial lakes and ponds. Reinhold Publishing Company, New York, USA.

Benson, N. G. 1968. Review of fishery studies on Missouri River main stem impoundments. Research Report 71, U.S. Fish and Wildlife Service, Washington, D.C., USA.

Benson, N. G. 1973. Evaluating the effects of discharge rates, water levels, and peaking on fish populations in Missouri River mainstem impoundments. Pages 683–689 *in* W. C. Ackermann, G. F. White, and E. B. Worthington, editors. Man-made lakes: their problems and environmental effects. Geophysical Monograph 17, American Geophysical Union, Washington, D.C., USA.

Benson, N. G. 1976. Water management and fish production in Missouri River main stem reservoirs. Pages 141–147 *in* J. F. Osborn and C. H. Allman, editors. Instream flow needs, volume II. American Fisheries Society, Bethesda, Maryland, USA.

Benson, N. G., and B. C. Cowell. 1967. The environment and plankton density in Missouri River reservoirs. Pages 358–373 *in* Reservoir fishery resources symposium. Southern Division American Fisheries Society, Bethesda, Maryland, USA.

Benson, N. G., and P. L. Hudson. 1975. Effects of a reduced fall drawdown on benthos abundance in Lake Francis Case. Transactions of the American Fisheries Society 104:526–528.

Brown, A. M. 1986. Modifying reservoir fish habitat with artificial structures. Pages 98–102 *in* G. H. Hall and M. J. Van Den Avyle, editors. Reservoir fisheries management: strategies for the 80's. Reservoir Committee, Southern Division American Fisheries Society, Bethesda, Maryland, USA.

Calhoun, A. 1966. Habitat protection and improvement. Pages 40–48 *in* A. Calhoun, editor. Inland fisheries management. California Department of Fish and Game, Sacramento, California, USA.

Carlander, K. D. 1969. Handbook of freshwater fishery biology, volume 1. Iowa State University Press, Ames, Iowa, USA.

Carlander, K. D. 1977. Handbook of freshwater fishery biology, volume 2. Iowa State University Press, Ames, Iowa, USA.

Chevalier, J. R. 1977. Changes in walleye (*Stizostedion vitreum vitreum*) population in Rainy Lake and factors in abundance, 1924–75. Journal of the Fisheries Research Board of Canada 34:1696–1702.

Claflin, T. O. 1968. Reservoir aufwuchs on inundated trees. Transactions of the American Microscopical Society 87:97–104.

Cooper, W. E., and L. B. Crowder. 1979. Patterns of predation in simple and complex environments. Pages 257–267 *in* R. H. Stroud and H. Clepper, editors. Predator-prey systems in fisheries management. Sport Fishing Institute, Washington, D.C., USA.

Cowell, B. C., and P. L. Hudson. 1967. Some environmental factors influencing benthic invertebrates in two Missouri River reservoirs. Pages 541–555 *in* Reservoir fishery resources symposium. Southern Division American Fisheries Society, Bethesda, Maryland, USA.

Davis, J. T. 1967. Water fluctuation. Louisiana Conservationist (Jan–Feb):5–7.

Davis, J. T., and J. S. Hughes. 1966. Effects of impoundment on the benthic population of Bayou D'Arbonne, Louisiana. Proceedings of the Annual Conference of the Southeastern Association of Game and Fish Commissioners 19:364–374.

Dunst, R.C., S. M. Born, P. D. Uttormark, S. A. Smith, S. A. Nichols, J. O. Peterson, D. R. Knauer, S. L. Serns, D. R. Winter, and T. L. Wirth. 1974. Survey of lake rehabilitation techniques and experiences. Technical Bulletin 75, Wisconsin Department of Natural Resources, Madison, Wisconsin, USA.

Eipper, A. W. 1975. Environmental influences on the majority of bass embryos and larvae. Pages 295–305 *in* R. H. Stroud and H. Clepper, editors. Black bass biology and management. Sport Fishing Institute, Washington, D.C., USA.

Ellis, M. M. 1937. Some fishery problems in impounded waters. Transactions of the American Fisheries Society 66:63–71.

Elrod, J. H., and T. J. Hassler. 1971. Vital statistics of seven fish species in Lake Sharpe, South Dakota 1964–69. Pages 27–40 *in* G. E. Hall, editor. Reservoir fisheries and limnology. Special Publication 8, American Fisheries Society, Bethesda, Maryland, USA.

Eschmeyer, R. W. 1949. The fisheries picture—with special reference to TVA impoundments. The Progressive Fish-Culturist 11:267–271.

Estes, R. D. 1971. The effects of the Smith Mountain pump storage project on the fishery of the lower reservoir, Leesville, Virginia. Doctoral dissertation, Virginia Polytechnic Institute and State University, Blacksburg, Virginia, USA.

Fillion, D. B. 1967. The abundance and distribution of benthic fauna of three mountain reservoirs on the Kananaskis River in Alberta. Journal of Applied Ecology 4:1–11.

Fourt, R. A. 1978. The effects of a two year water-level management plan on the production of sport fish in Beaver Reservoir. Unpublished report, Arkansas Game and Fish Commission, Little Rock, Arkansas, USA.

Gabel, J. A. 1974. Species and age composition of trap net catches in Lake Oahe, South Dakota, 1963–67. Technical Paper 75, U.S. Fish and Wildlife Service, Washington, D.C., USA.

Gasaway, C. R. 1970. Changes in the fish population in Lake Francis Case in South Dakota in the first 16 years of impoundment. Technical Paper 56, U.S. Fish and Wildlife Service, Washington, D.C., USA.

Goodson, L. F., Jr. 1965. Diets of four warmwater game fishes in a fluctuating, steep-sided, California reservoir. California Fish and Game 51:259–269.

Grimås, U. 1961. The bottom fauna of natural and impounded lakes in northern Sweden (Ankarvattnet and Blasjon). Institute of Freshwater Research, Drottningholm 42:183–237.

Grimås, U. 1962. The effect of increased water level fluctuation upon the bottom fauna in Lake Blasjon, northern Sweden. Institute of Freshwater Research, Drottningholm 44:14–41.

GRIMÅS, U. 1964. Studies on the bottom fauna of impounded lakes in southern Norway (Tunnhovdfjord, Paalsbufjord, and Rodungen). Institute of Freshwater Research, Drottningholm 45:94–104.

GRIMÅS, U. 1965. Inlet impoundments. An attempt to preserve littoral animals in regulated subarctic lakes. Institute of Freshwater Research, Drottningholm 46:22–30.

GROEN, C. L., AND T. A. SCHROEDER. 1978. Effects of water level management on walleye and other coolwater fishes in Kansas reservoirs. Pages 278–283 *in* R. L. Kendall, editor. Selected coolwater fishes of North America. Special Publication 11, American Fisheries Society, Bethesda, Maryland, USA.

GUSEVA, K. A. 1958. The influence of water level regime of the Rybinsk Reservoir on the development of phytoplankton. Trudy Biologicheskoi Stantsii "Borok" 3:112–124. (In Russian; Biological Abstracts 47:65732).

HASLER, A. D., AND W. G. EINSELE. 1948. Fertilization for increasing productivity of inland waters. Transactions of the North American Wildlife Conference 13:527–554.

HASSLER, T. J. 1970. Environmental influences on early development and year-class strength of northern pike in Lakes Oahe and Sharpe, South Dakota. Transactions of the American Fisheries Society 99:369–380.

HEISEY, P.G., D. MATHUR, AND N. C. MAGNUSSON. 1980. Accelerated growth of smallmouth bass in a pumped storage system. Transactions of the American Fisheries Society 109:371–377.

HEMAN, M. L., R. S. CAMPBELL, AND L. C. REDMOND. 1969. Manipulation of fish populations through reservoir drawdown. Transactions of the American Fisheries Society 98:293–304.

HILDEBRAND, S. G., AND L. B. GOSS. 1981. Hydroelectric operation at the river basin level: Research needs to include ecological issues in basin-level hydropower planning. Electric Power Research Institute Report EPRI WS-80-155, Palo Alto, California, USA.

HULSEY, A. H. 1959. A proposal for the management of reservoirs for fisheries. Proceedings of the Annual Conference of the Southeastern Association of Game and Fish Commissioners 12:132–143.

HUNT, P. C., AND J. W. JONES. 1972a. The effect of water level fluctuations on littoral fauna. Journal of Fish Biology 4:385–394.

HUNT, P. C., AND J. W. JONES. 1972b. The profundal fauna of Llyn Tegid, North Wales. Journal of Zoology 168:9–49.

JACKSON, S. W., JR. 1958. Comparison of the age and growth of four fishes from Lower and Upper Spavinaw Lakes, Oklahoma. Proceedings of the Annual Conference of the Southeastern Association of Game and Fish Commissioners 11:232–249.

JENKINS, R. M. 1967. The influence of some environmental factors on standing crop and harvest of fishes in U.S. reservoirs. Pages 298–321 *in* Reservoir fishery resources symposium. Southern Division American Fisheries Society, Bethesda, Maryland, USA.

JENKINS, R. M. 1970. The influence of engineering design and operation and other environmental factors on reservoir fishery resources. Water Resources Bulletin 6:110–119.

JENKINS, R. M. 1975. Black bass crops and species associations in reservoirs. Pages 114–124 *in* R. H. Stroud and H. Clepper, editors. Black bass biology and management. Sport Fishing Institute, Washington, D.C., USA.

JENKINS, R. M., AND D. I. MORAIS. 1977. Predator-prey relations in the predator-stocking-evaluation reservoirs. Proceedings of the Annual Conference of the Southeastern Association of Game and Fish Commissioners 30:14–157.

JOHNSON, D. L., AND R. A. STEIN, editors. 1979. Response of fish to habitat structure in standing water. Special Publication 6, North Central Division American Fisheries Society, Bethesda, Maryland, USA.

JOHNSON, J. N. 1974. The effects of water level fluctuations on the growth, relative abundance, mortality, and standing crop of fishes in Lake Carl Blackwell, Oklahoma. Masters thesis, Oklahoma State University, Stillwater, Oklahoma, USA.

JOHNSON, J. N., AND A. K. ANDREWS. 1974. Growth of white crappie and channel catfish in relation to variations in mean annual water level of Lake Carl Blackwell, Oklahoma. Proceedings of the Annual Conference of the Southeastern Association of Game and Fish Commissioners 27:767–776.

JONES, J. R., AND R. W. BACHMANN. 1978a. Prediction of phosphorus and chlorophyll *a* in lakes. Journal of the Water Pollution Control Federation 48:2176–2182.

JONES, J. R., AND R. W. BACHMANN. 1978b. Trophic status of Iowa lakes in relation to origin and glacial geology. Hydrobiologia 57:267–273.

JONES, W. E., AND J. H. SELGEBY. 1974. Invertebrate macrobenthos of Lake Oahe, 1968–69. Technical Paper 73, U.S. Fish and Wildlife Service, Washington, D.C., USA.

JUNE, F. C. 1970. Atresia and year-class abundance of northern pike, *Esox lucius,* in two Missouri River impoundments. Journal of the Fisheries Research Board of Canada 37:487–591.

KASTER, J. L., AND G. Z. JACOBI. 1978. Benthic macroinvertebrates of a fluctuating reservoir. Freshwater Biology 8:283–290.

KEITH, W. E. 1975. Management by water level manipulation. Pages 489–497 *in* R. H. Stroud and H. Clepper, editors. Black bass biology and management. Sport Fishing Institute, Washington, D.C., USA.

LANTZ, K. E. 1974. Natural and controlled water level fluctuations in a backwater lake and three Louisiana impoundments. Fishery Bulletin 11, Louisiana Wildlife and Fisheries Commission, Baton Rouge, Louisiana, USA.

LANTZ, K. E., J. T. DAVIS, J. S. HUGHES, AND H. E. SCHAFER, JR. 1965. Water level fluctuation—its effect on vegetation control and fish population management. Proceedings of the Annual Conference of the Southeastern Association of Game and Fish Commissioners 18:483–494.

LÖTMARKER, T. 1964. Studies on planktonic crustacea in thirteen lakes in northern Sweden. Institute of Freshwater Research, Drottningholm 45: 113–189.

MARKOSYAN, A. K. 1969. Benthic productivity of Lake Sevan. Pages 146–152 *in* B. Golek, editor. Transactions of the 6th conference on the biology of inland waters. Israel Program for Scientific Translations, Catalog No. 5136, U.S. Department of Commerce, Washington, D.C., USA. (Translated from Russian).

MARTIN, N. V. 1955. The effect of drawdowns on lake trout reproduction and the use of artificial spawning beds. Transactions of the North American Wildlife Conference 20:263–271.

MCLACHLAN, A. J. 1977. The changing role of terrestrial and autochthonous organic matter in newly flooded lakes. Hydrobiologia 54:215–217.

MITCHELL, S. F. 1971. Phytoplankton productivity in Tomahawk Lagoon, Lake Waipori, and Lake Mahinerangi. Fishery Research Bulletin (New Zealand) 3:1–87.

MITCHELL, S. F. 1975. Some effects of agricultural development and fluctuations in water level on the phytoplankton productivity and zooplankton of a New Zealand reservoir. Freshwater Biology 5:547–562.

MOEN, T. E. 1974. Population trends, growth, and movement of bigmouth buffalo, *Ictiobus cyprinellus,* in Lake Oahe, 1963–70. Technical Paper 78, U.S. Fish and Wildlife Service, Washington, D.C., USA.

MULLAN, J. W., AND R. L. APPLEGATE. 1968. Centrarchid food habits in a new and old reservoir during and following bass spawning. Proceedings of the Annual Conference of the Southeastern Association of Game and Fish Commissioners 21: 332–342.

MURDOCK, W. W., AND A. OATEN. 1975. Predation and population stability. American Naturalist 110: 351–367.

NELSON, W. R. 1974. Age, growth, and maturity of thirteen species of fish from Lake Oahe during the early years of impoundment, 1963–68. Technical Paper 77, U.S. Fish and Wildlife Service, Washington, D.C., USA.

NELSON, W. R., AND C. H. WALBURG. 1977. Population dynamics of yellow perch (*Perca flavescens*), sauger (*Stizostedion canadense*), and walleye (*S. vitreum vitreum*) in four main stem Missouri River reservoirs. Journal of the Fisheries Research Board of Canada 34:1748–1763.

NICHOLS, S. A. 1974. Mechanical and habitat manipulation for aquatic plant management: a review of techniques. Technical Bulletin 76, Wisconsin Department of Natural Resources, Madison, Wisconsin, USA.

PIERCE, B. E., AND G. R. HOOPER. 1980. Fish standing crop comparisons of tire and brush fish attractors in Barkley Lake, Kentucky. Proceedings of the Annual Conference of the Southeastern Association of Fish and Wildlife Agencies 33: 668–691.

PLOSKEY, G. R. 1982. Fluctuating water levels in reservoirs: an annotated bibliography of environmental effects and management for fisheries. Technical report E-82-5, U.S. Army Engineer Waterways Experiment Station, Vicksburg, Mississippi, USA.

PLOSKEY, G. R. 1983. A review of the effects of water-level changes on reservoir fisheries and recommendations for improved management. Technical report E-83-3, U.S. Army Engineer Waterways Experiment Station, Vicksburg, Mississippi, USA.

PRIEGEL, G. R. 1970. Reproduction and early life history of the walleye in the Lake Winnebago region. Technical Bulletin 45, Wisconsin Department of Natural Resources, Madison, Wisconsin, USA.

RAINWATER, W. C., AND A. HOUSER. 1975. Relation of physical and biological variables to black bass crops. Pages 306–309 *in* R. H. Stroud and H. Clepper, editors. Black bass biology and management. Sport Fishing Institute, Washington, D.C., USA.

ROBINS, C. R., R. M. BAILEY, C. E. BOND, J. R. BROOKER, E. A. LACHNER, R. N. LEA, AND W. B. SCOTT. 1980. A list of common and scientific names of fishes from the United States and Canada, 4th edition. Special Publication 12, American Fisheries Society, Bethesda, Maryland, USA.

RODHE, W. 1964. Effects of impoundment on water chemistry and plankton in Lake Ransaran (Swedish Lappland). Proceedings of the International Association of Theoretical and Applied Limnology 15:437–443.

SAVINO, J. F., AND R. A. STEIN. 1982. Predator-prey interaction between largemouth bass and bluegills as influenced by simulated, submerged vegetation. Transactions of the American Fisheries Society 111:255–266.

SHANE, R. M. 1981. Scheduling model development for multipurpose multiple reservoir systems: the state of the art. Section 2 *in* S. G. Hildebrand and L. B. Goss, editors. Hydroelectric operation at the river basin level: research needs to include ecological issues in basin-level hydropower planning. Electric Power Research Institute Report EPRI WS-80-155, Palo Alto, California, USA.

SHIELDS, J. T. 1958. Experimental control of carp reproduction through water drawdowns in Fort Randall Reservoir, South Dakota. Transactions of the American Fisheries Society 87:23–33.

STRANGE, R. J., W. B. KITTRELL, AND T. D. BROADBENT. 1982. Effects of seeding reservoir fluctuation zones on young-of-the-year black bass and associated species. North American Journal of Fisheries Management 2:307–315.

SUMMERFELT, R. C. 1975. Relationship between weather and year-class strength of largemouth bass. Pages 166–174 *in* R. H. Stroud and H. Clepper, editors. Black bass biology and management. Sport Fishing Institute, Washington, D.C., USA.

SUMMERFELT, R. C., AND K. E. SHIRLEY. 1978. Environmental correlates to year-class strength of largemouth bass in Lake Carl Blackwell. Proceedings of the Oklahoma Academy of Science 58:54–63.

SWANSON, G. A. 1967. Factors influencing the distribution and abundance of *Hexagenia* nymphs (Ephemeroptera) in a Missouri River reservoir. Ecology 48:216–225.

SWINGLE, H. S. 1950. Relationships and dynamics of balanced and unbalanced fish populations. Alabama Polytechnic Institute Bulletin 174, Agricultural Experiment Station, Auburn, Alabama, USA.

SWINGLE, H. S., AND W. E. SWINGLE. 1967. Problems in dynamics of fish populations in reservoirs. Pages 229–243 *in* Reservoir fishery resources symposium. Southern Divison American Fisheries Society, Bethesda, Maryland, USA.

TRIPLETT, J. R., D. A. CULVER, AND G. B. WATERFIELD. 1980. An annotated bibliography on the effects of water-level manipulation on lakes and reservoirs. Ohio Department of Natural Resources, Federal Aid Project F-57-R, Columbus, Ohio, USA.

VOGELE, L. E. 1975. Reproduction of spotted bass, *Micropterus punctulatus,* in Bull Shoals Reservoir, Arkansas. Technical Paper 84, U.S. Fish and Wildlife Service, Washington, D.C., USA.

VON GELDERN, C. E., JR. 1971. Abundance and distribution of fingerling largemouth bass, *Micropterus salmoides,* as determined by electrofishing at Lake Nacimiento, California. California Fish and Game 57:228–245.

WALBURG, C. H. 1972. Some factors associated with fluctuation in year-class strength of sauger, Lewis and Clark Lake, South Dakota. Transactions of the American Fisheries Society 101:311–316.

WALBURG, C. H. 1976. Changes in the fish population of Lewis and Clark Lake, 1956-74, and their relation to water management and the environment. Research Report 79, U.S. Fish and Wildlife Service, Washington, D.C., USA.

WEGENER, W., AND V. WILLIAMS. 1975. Fish population responses to improved lake habitat utilizing extreme drawdown. Proceedings of the Annual Conference of the Southeastern Association of Game and Fish Commissioners 28:144–161.

WEGENER, W., V. WILLIAMS, AND T. D. MCCALL. 1975. Aquatic macroinvertebrate responses to an extreme drawdown. Proceedings of the Annual Conference of the Southeastern Association of Game and Fish Commissioners 28:126–144.

WETZEL, R. G. 1975. Limnology. W. B. Saunders Company, Philadelphia, Pennsylvania, USA.

WOOD, R. 1951. The significance of managed water levels in developing the fisheries of large impoundments. Journal of the Tennessee Academy of Science 26:214–235.

WOOD, R., AND D. W. PFITZER. 1960. Some effects of water-level fluctuations on the fisheries of large reservoirs. Technical Meeting, International Union for Conservation of Nature and Natural Resources, Athens, Greece 4:118–138.

WRIGHT, J. C. 1950. The limnology of Atwood Lake, a flood control reservoir. Doctoral dissertation, Ohio State University, Columbus, Ohio, USA.

Pages 98–102 *in* G. E. Hall and M. J. Van Den Avyle, editors.
Reservoir Fisheries Management: Strategies for the 80's.
Reservoir Committee, Southern Division American Fisheries Society, Bethesda, Maryland. 1986.

Modifying Reservoir Fish Habitat with Artificial Structures

ALLEN M. BROWN

Tennessee Valley Authority
451 Evans Building
Knoxville, Tennessee 37902

ABSTRACT

Artificial structures, mainly fish attractors, have been used for many years to enhance sport fishing. The effects of attractor installation were first reported in 1931. Since that time, the art of fish attractor installation and evaluation has evolved into a variety of useful techniques, many of which remain to be evaluated or still need improvement. This paper reviews past and current uses of artificial structures. Further research to determine specific applications of habitat modification and testing of new and innovative ideas is needed to make this a more effective and predictable fishery management tool.

The ideal fish attractor is planned to be biologically effective, legal, not presenting a navigational hazard, and minimally damaging to the aesthetics of the area. It would be constructed from a selection of various materials after consideration of benefit/cost ratios, fishing pressure demands, and locations for placement (Wilbur 1974).

Biologists and fishermen have long recognized the advantages of structure to attract and hold fish. Records on known development of artificial marine reefs date back to 1804 in Japan.

C. L. Hubbs and L. Hubbs were among the first to observe the effects of fish attractors in freshwater in 1931 on Crystal Lake, Michigan (Rodeheffer 1939). Various materials have been tested through the years, but the most economical and efficient have been brush and automobile tires. Brush pile attractors were constructed in several Michigan lakes in 1933 as a project of the newly organized Civilian Conservation Corps (Rodeheffer 1945). These installations marked the beginning of brush attractors and set the basic framework after which most modern-day fish attractor programs are patterned. Officials from 28 of 45 State agencies that responded to a recent survey conducted by Prince et al. (1975) reported use of artificial attractors. Installation programs typically have three major objectives: (1) to increase harvest; (2) to increase numbers of fishing trips; and (3) to increase localized and/or overall biological production in a reservoir.

Numerous studies have been conducted evaluating the catch rate and harvest around attractors. The estimated harvest of fish by both number and weight increased after addition of artificial structures in Wewoka Lake, Oklahoma (Wright 1979). Short-term evaluations of fish attractors in seven Florida lakes indicated that areas with attractors produced signficantly higher angler catch rates than control areas (Wilbur and Crumpton 1974). Fishing over artificial stake beds produced 6.6 white crappie (*Pomoxis annularis*) per hour, as opposed to 1.8 per hour on control areas (Pettit 1973). The hook-and-line harvest in standing timbered areas on Bull Shoals Lake, Missouri, in 1958 was 3,054 pounds per acre, as opposed to 113 pounds per acre in the remainder of the lake (Burress 1961). However, not all evaluations of the effects of artificial structure on catch rates have been positive; Patriarche (1959) recorded no significant improvement in catch at Devoe Lake, Michigan, after installation of fish shelters.

Paxton and Stevenson (1979) surveyed anglers on Killdeer Reservoir, Ohio, and observed three general trends: (1) harvest of bluegill (*Lepomis macrochirus*), rock bass (*Ambloplites rupestris*), and smallmouth bass (*Micropterus dolomieui*) was significantly greater from structured reefs than from open-water stations; (2) catch of green sunfish (*Lepomis cyanellus*) was greatest from nonstructural reefs (e.g., an island is essentially a shallow water, nonstructural reef) and significantly lower from open-water areas; and (3) harvest of yellow perch (*Perca flavescens*) was significantly higher from open-water nonstructural areas than from all other stations.

Few evaluations have been published on the effects of fish attractors on biological production in large reservoirs. The addition of artificial tire reefs in the euphotic zone of Smith Mountain Lake, Virginia, resulted in a primary productivity and trophic value of the periphyton community several times greater than that of the littoral phytoplankton (Prince et al. 1977). Increases in the availability of food resulted in increased body condition and growth of sunfishes and their predators in localized areas of the reservoir (Prince et al. 1979). Detection of increased productivity on a reservoir-wide scale is difficult, but some authors believe that this is not necessary if fish production can be shown to be improved in localized areas.

Types of Attractors

Fish attractors can be categorized by the type of material used. The major ones are (1) brush, (2) automobile tires, (3) stake beds, (4) standing timber, and (5) shoreline vegetation. Many other materials have been used including gravel beds, log cribs, flue tile, and precast concrete forms, but these types have not gained wide acceptance.

Brush

Opinions concerning the effectiveness of different species of trees for concentrating fish are almost as numerous as the individuals who offer them; to date, no detailed studies comparing the use of different tree species have been reported. Hardwood brush is considered to be the most durable; on Douglas Lake, Michigan, brush shelters were intact and concentrated game fish after 30 years (Thomas et al. 1968). Configuration may also be important in prolonging the life of brush attractors. Ladder shelters that suspended brush off the bottom lasted 10 years and were more durable than circular, individual tree, and clump shelters attached to the bottom (Rodeheffer 1945).

Brush has been shown to be more effective than many other materials used to construct attractors. Herring and Anderson (1975) compared brush shelters with artificial gravel beds in four Mississippi reservoirs and found largemouth bass (*Micropterus salmoides*), bluegill, and crappie often on the brush but no fish around the gravel beds. Reefs constructed from automobile tires, brush, clay pipes, and cement blocks in Lake Tohopekaliga, Florida, were evaluated with electrofishing gear, underwater observations, creel survey, and experienced anglers; more fish were observed and caught around the brush than the other materials, but all types attracted more fish than open-water control areas (Wilbur and Crumpton 1974). Wilbur (1978) designed an experiment to obtain a more precise evaluation of the effects of various types of structure on angler success and determined that brush yielded a slightly higher success rate than attractors constructed from pipe. An evaluation comparing brush and automobile tires was conducted on Barkley Lake, Kentucky, and the combined weight of four species (channel catfish [*Ictalurus punctatus*], bluegill, largemouth bass, and white crappie) was 1,038 pounds per acre in the brush areas, 330 pounds per acre around the tires, and 37 pounds per acre in control areas (Pierce and Hooper 1980).

Brush attractors have been installed in various configurations, typically either (1) lying flat on the substrate and usually not more than 3 ft high; (2) tied on ladder-like frames that may extend several feet above the substrate; and (3) midwater reefs suspended above the bottom. Attractors lying on the bottom require less time to construct than those on frames or suspended in the water column, and they also present less of a hazard for other recreational uses. On some substrates, however, attractors on the bottom can be covered with silt, which reduces their effectiveness.

Automobile Tires

Tires are considered to be non-polluting (biologically) and will last indefinitely. Although installation is labor intensive, tires may be more cost-effective than other types in the long run because of their durability and minor maintenance requirements. Some problems include removal of air pockets and settling into soft-bottomed substrates. Tires that are visible above the water's surface detract from the aesthetic value of natural areas.

A study at Smith Mountain Lake determined the relative abundance of fish before (1973) and after (1974) a 6,750-ft long reef of scrap tires was installed. Ratios of fish counted in 1974 compared with 1973 were 12:1 in spring, 5.6:1 in summer, and 2.8:1 in fall (Prince et al. 1975).

Stake Beds

Materials other than brush and tires have been used to construct fish-concentrating areas with varying degrees of success. In 1968, hammer-driven and portable, pre-fabricated stake beds were installed in Kentucky Reservoir, Tennessee, to aid fishermen in harvesting more crappie. Over 128 hours of fishing produced 6.6 crappie per hour on stake beds (Pettit 1973). These structures are more expensive and labor-intensive than some of the other types, making it difficult to supply adequate quantities in large reservoirs. They were effective primarily for crappie and sunfish.

Standing Timber

Burress (1961) determined that angler harvest in timbered areas of Bull Shoals Lake, Missouri, was 3,054 pounds per acre, whereas in the remainder of the census area it was 113 pounds per acre. In terms of success being defined as catching at least one fish, anglers experienced higher success (94.8%) but a lower catch rate (1.25 fish per hour) in timbered areas as opposed to other areas (90.6% and 1.36 fish per hour). Davis and Hughes (1971) found similar conditions in timbered areas of Bussey Lake, Louisiana.

Shoreline Vegetation

Shoreline revegetation is particularly useful for impoundments that undergo extreme water level fluctuations. Exposed shorelines can be seeded during periods of low water to provide cover for spawning and young-of-year fishes and to promote nutrient recycling. Aggus and Elliott (1975) found that the amount and duration of flooded shoreline vegetation strongly influenced the survival of young largemouth bass in Bull Shoals Lake, Arkansas. Shoreline areas on Lake Nottely, Georgia, seeded with winter rye had a positive effect on abundance of young-of-year bass but no statistically significant effect on growth (Strange et al. 1982). According to Lagler (1956) and Bennett (1970), the most significant determinants of largemouth bass production in fluctuating reservoirs are probably the quantity and quality of available plant cover.

Other Materials

An artificial rock reef was constructed on Lake Michigan, with overall dimensions of approximately 900 ft by 80 ft, to increase the catch rate of yellow perch as well as other game species (Fogle 1979). Forty gravel beds, each consisting of 5 to 8 cubic yards of gravel, were placed in eleven lakes in Mississippi; every space of the bed was used by adult bluegills for spawning from May through August (Majure and Herring 1977).

Midwater attractors constructed from plywood panels were suspended at different depths from buoys in Lewis Smith Reservoir, Alabama. Bluegill and spotted bass (*Micropterus punctulatus*) were the species found to be most associated with these structures (Reeves et al. 1978). Further evaluation was conducted later on the same sites with structures of fiberglass panels and automobile tires. It was determined that a greater number of fish utilized structures over points of land than in coves (Smith et al. 1981), but in all areas, numbers of harvestable fish around suspended materials were low. Midwater structures might prove useful in pump-storage reservoirs, where extreme water-level fluctuations limit the use of attractors on the bottom.

Conclusions

Although some evaluations have been conducted on mechanisms and fish attractor benefits, comprehensive information needed to significantly advance the uses of artificial structures as management tools on large reservoirs has not been obtained. Questions relevant to development of adequate structure-management plans were outlined in a 1979 project proposal at the Ohio State University: (1) what types and configurations of structure do various fish species prefer?; (2) where should these structures be placed?; (3) what motivates fish to utilize structure?; (4) what effects do artificial structures have on the productivity of a lake and fish production?; and (5) how much structure does a lake need to meet specific management objectives? (David Johnson, School of Natural Resources, Ohio State University, Columbus, Ohio, USA [personal communication]). To date, the Smith Mountain Lake Project (Prince et al. 1979) has been the only study to address most of these questions. Since any reservoir can have unusual or unique biological or operational characteristics, it is likely that most fish attractor management plans would need to be individually developed following consideration of the questions posed above.

Techniques of structure installation need to be evaluated. The possibility of concentrating different desired fish species by using a combination of dissimilar attractor types and installation procedures has not been explored. Practices such as artificial baiting, attraction with underwater lights, environmentally safe dyes that provide shading in clear water lakes, sound waves, bubble curtains, pheromones, and cage culture of catfish (which often attracts wild fish) are some of the possible new areas for investigations in fish attraction and concentration.

Habitat manipulation offers many opportunities to increase the harvest of underutilized sport fishes in reservoirs. However, much work is needed to determine the impacts of many physical and biological variables on reservoir fish populations. A better understanding of these impacts along with habitat manipulation techniques may improve management decisions. For example, would spawning structures strategically placed in a reservoir with a history of low spawning success due to high wind action on shoreline areas be effective in increasing fish production? Could a reservoir with low predator abundance and a prey community consisting primarily of sunfish be improved with the addition of cover?

References

Aggus, L. R., and G. V. Elliott. 1975. Effects of food and cover on year-class strength of largemouth bass in Bull Shoals Lake. Pages 317–322 *in* R. H. Stroud and H. Clepper, editors. Black bass biology and management. Sport Fishing Institute, Washington, D.C., USA.

Bennett, G. W. 1970. Management of lakes and ponds. Van Nostrand Reinhold Company, New York, USA.

Burress, R. M. 1961. Fishing pressure and success in areas of flooded standing timber in Bull Shoals Reservoir, Missouri. Proceedings of the Annual Conference of the Southeastern Association of Game and Fish Commissioners 15:296–298.

Davis, J. T., and J. S. Hughes. 1971. Effects of standing timber on fish populations and fisherman success in Bussey Lake, Louisiana. Pages 255–264 *in* G. E. Hall, editor. Reservoir fisheries and limnology. Special Publication 8, American Fisheries Society, Bethesda, Maryland, USA.

Fogle, N. E. 1979. Barrier fishing reef, Muskogen, Michigan. Final report, Federal Aid Project F-37-D, Michigan Department of Natural Resources, Lansing, Michigan, USA.

Herring, J., and W. Anderson. 1975. Investigation of state lakes' fishing resources. Final report, Federal Aid Project F-37-1, Mississippi Game and Fish Commission, Jackson, Mississippi, USA.

Lagler, K. F. 1956. Freshwater fishery biology, second edition. W. C. Brown Company, Dubuque, Iowa, USA.

Majure, T. C., and J. Herring. 1977. Investigation of State lakes: addition of fish attractors to lakes with underwater observation. Final report, Federal Aid Project F-40-1, Mississippi Game and Fish Commission, Jackson, Mississippi, USA.

Patriarche, M. H. 1959. Progress report on the effect of the Hoad fish shelter on fishing success. Michigan Department of Conservation, University Museums Annex, Ann Arbor, Michigan, USA.

Paxton, K. O., and F. Stevenson. 1979. Influence of artificial structures on angler harvest from Killdeer Reservoir, Ohio. Pages 70–76 *in* D. L. Johnson and R. A. Stein, editors. Response of fish to habitat structure in standing water. Special Publication 6, North Central Division, American Fisheries Society, Bethesda, Maryland, USA.

Pettit, G. C., III. 1973. Stake beds as crappie concentrators. Proceedings of the Annual Conference of the Southeastern Association of Game and Fish Commissioners 26:401–406.

Pierce, B. E., and G. R. Hooper. 1980. Fish standing crop comparisons of tire and brush fish attractors in Barkley Lake, Kentucky. Proceedings of the Annual Conference of the Southeastern Association of Fish and Wildlife Agencies 33:688–691.

Prince, E. D., O. E. Maughan, D. H. Bennett, G. M. Simmons, Jr., J. Stauffer, Jr., and R. J. Strange. 1979. Trophic dynamics of a freshwater artificial tire reef. Pages 459–473 *in* R. H. Stroud and H. Clepper, editors. Predator-prey systems in fisheries management. Sport Fishing Institute, Washington, D.C., USA.

Prince, E. D., R. F. Raleigh, and R. V. Corning. 1975. Artificial reefs and centrarchid basses. Pages 498–505 *in* R. H. Stroud and H. Clepper, editors. Black bass biology and management. Sport Fishing Institute, Washington, D.C., USA.

Prince, E. D., R. J. Strange, and G. M. Simmons, Jr. 1977. Preliminary observations on the productivity of periphyton attached to a freshwater artificial tire reef. Proceedings of the Annual Conference of the Southeastern Association of Fish and Wildlife Agencies 30:207–215.

Reeves, W. C., G. R. Hooper, and B. W. Smith. 1978. Preliminary observations of fish attraction to artificial mid-water structures in freshwater. Proceedings of the Annual Conference of the Southeastern Association of Fish and Wildlife Agencies 31:471–476.

Rodeheffer, I. A. 1939. Experiments in the use of brush shelters by fish in Michigan lakes. Papers of the Michigan Academy of Science, Arts, and Letters 24:183–193.

Rodeheffer, I. A. 1945. Fish populations in and around brush shelters of different size placed at varying depths and distances apart in Douglas Lake, Michigan. Papers of the Michigan Academy of Science, Arts, and Letters 30:321–345.

Smith, B. W., G. R. Hooper, and C. S. Lawson. 1981. Observations of fish attraction to improved artificial mid-water structures in freshwater. Proceedings of the Annual Conference of the Southeastern Association of Fish and Wildlife Agencies 34:404–409.

Strange, R. J., W. B. Kittrell, and T. D. Broadbent. 1982. Effects of seeding reservoir fluctuation zones on young-of-the-year black bass and associated species. North American Journal of Fisheries Management 2:307–315.

Thomas, P. M., R. O. Legault, and G. F. Carpenter. 1968. Durability and efficiency of brush shelters installed in 1937 in Douglas Lake, Michigan. Journal of Wildlife Management 32:515–519.

Wilbur, R. L. 1974. Florida's freshwater fish attractors. Fisheries Bulletin 6, Florida Game and Fresh Water Fish Commission, Tallahassee, Florida, USA.

Wilbur, R. L. 1978. Two types of fish attractors compared in Lake Tohopekaliga, Florida. Transaction of the American Fisheries Society 107:689–695.

Wilbur, R. L. and J. E. Crumpton. 1974. Florida's fish attractor program. Pages 39–46 *in* L. Colunga and R. B. Stone, editors. Proceedings of an international conference on artificial reefs. Sea Grant Publication 74–103, Texas A&M University, College Station, Texas, USA.

Wright, G. 1979. Evaluation of fish habitat improvement in Holdenville and Wewoka lakes. Final report, Federal Aid Project F-32-R, Oklahoma Department of Conservation, Oklahoma City, Oklahoma, USA.

Pages 103–109 *in* G. E. Hall and M. J. Van Den Avyle, editors.
Reservoir Fisheries Management: Strategies for the 80's.
Reservoir Committee, Southern Division American Fisheries Society, Bethesda, Maryland. 1986.

Limnological and Ecological Changes Associated with Reservoir Aging

BRUCE L. KIMMEL AND ALAN W. GROEGER

Environmental Sciences Division
Oak Ridge National Laboratory
Oak Ridge, Tennessee 37830

ABSTRACT

Much scientific attention has been devoted to the "trophic upsurge and depression" observed in recently impounded reservoirs; however, little is known of the longer-term consequences of reservoir aging. Changes in the trophic status of water bodies are often a consequence of man-induced alterations of the watershed, rather than a result of the natural, gradual accumulation of nutrients and sediments. Because the formation of a man-made impoundment frequently promotes additional land-use change and technological development within the reservoir watershed, changes in water quality and biological productivity due to "natural" reservoir aging may be minor compared to the effects of man-induced alterations of watershed-reservoir interactions. Future trends in reservoir water quality and sport fisheries production will depend more on our ability to effectively conduct coordinated terrestrial and aquatic resources planning and management on a whole-watershed scale than on the consequences of reservoir aging per se. Coordinated research on selected impoundments and their watersheds is needed to determine the sensitivity of reservoir ecosystems to various levels of watershed perturbation, and to evaluate the effectiveness of integrated management strategies.

The construction of large impoundments in the United States has slowed due to previous utilization of the optimal reservoir sites and an increased awareness of environmental costs versus economic benefits of large reservoirs. Therefore, our United States reservoirs are aging, although the construction of new hydropower impoundments continues in other parts of the world, especially in Canada and the tropics. Because reservoirs are such recent landscape features, we know very little about the long-term ecological consequences of reservoir aging. Does our knowledge of natural lake succession and eutrophication processes allow us to anticipate ecological changes that will accompany reservoir aging? What type of management approach will be required to optimize water quality and fisheries production in aging impoundments? The purpose of this paper is to evaluate the limnological and ecological changes associated with reservoir aging within the broader perspective of the aging process in natural lakes and to assess the significance of changes attributable to "natural" reservoir aging relative to changes due to other influences.

THE AGING PROCESS IN LAKES AND RESERVOIRS

The gradual alteration of an oligotrophic lake through stages of increasing eutrophy, to dystrophy, basin extinction, and, eventually to phases of terrestrial succession, is a familiar sequence to students of ecology and limnology (Goldman 1968; Whittaker 1975). Comparisons of lakes perceived to form a successionary sequence imply a gradual, unidirectional progression from an oligotrophic (nutrient-poor, unproductive) to a eutrophic (nutrient-rich, productive) condition. However, paleolimnological evidence shows that changes in lake trophic status are not necessarily gradual, unidirectional, or natural. If undisturbed, lakes can remain in a state of trophic equilibrium for thousands of years (Fig. 1). Indeed, changes in lake nutrient status and productivity are often a consequence of altered nutrient loading due to man-induced impacts (e.g., forest clearing, road building, cultivation, sewage) rather than of the natural, gradual filling in of lake basins (NAS/NRC 1969; Likens 1972; Hutchinson 1973; Brugam 1978).

Natural and man-made lakes age by the same basic processes, nutrient loading and basin filling, but at very different rates (Ryder 1978). Most North Temperate zone natural lakes were created by Pleistocene glacial activity and are at least 10 to 12 thousand years old. In sharp contrast, most United States reservoirs were constructed within the past 50 years. However, reservoirs age much more rapidly than natural lakes for rather obvious reasons. Most reservoirs have significantly

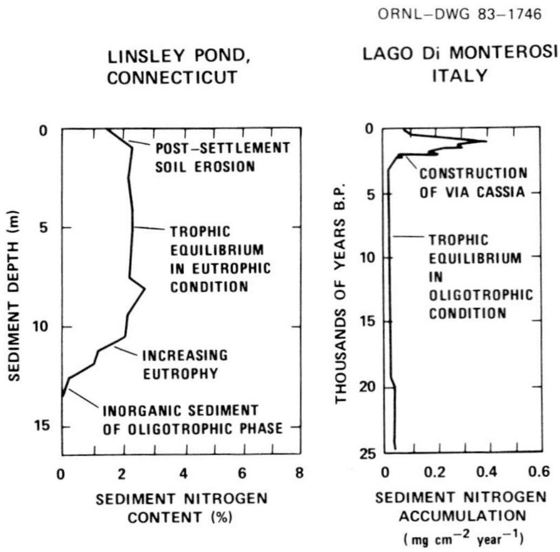

Figure 1. The trophic history of Linsley Pond, Connecticut, and Lago di Monterosi, Italy, as reflected by the organic nitrogen content of lake sediments (modified from Hutchinson 1973). Long periods of trophic equilibrium occurred in both naturally eutrophic Linsley Pond and in oligotrophic Lago di Monterosi. The effects of recent man-induced watershed disturbance (forest clearing and cultivation leading to soil erosion at Linsley Pond; road construction for Lago di Monterosi) are apparent in both systems.

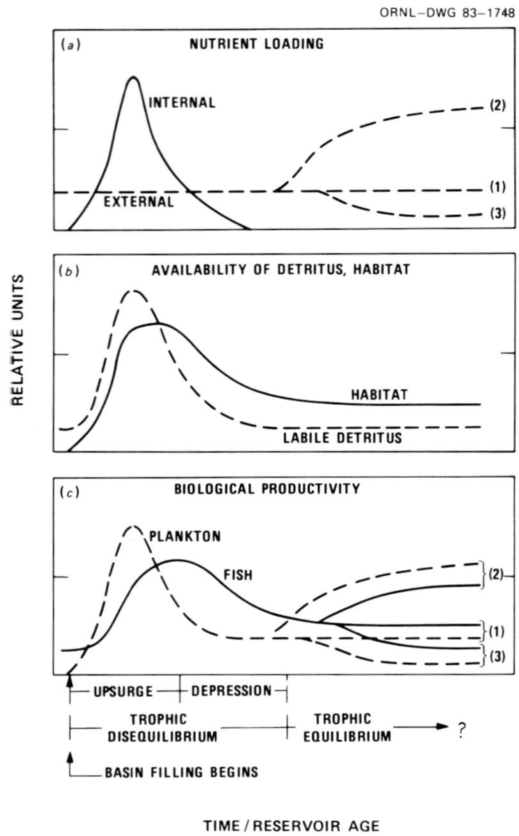

Figure 2. Changes in factors influencing reservoir water quality and biological productivity as a reservoir matures and ages: (a) internal and external nutrient loading, (b) availability of fish and invertebrate habitat and labile detritus, and (c) plankton and fish production. The brief initial period of trophic disequilibrium characteristic of new impoundments yields to a less productive but potentially more stable period of trophic equilibrium as internal nutrient loading and basin-derived detritus decrease (1). Man-induced alterations of the watershed or successional changes in terrestrial vegetation may cause increases (2) or decreases (3) in external nutrient loading and biological production.

higher drainage area-to-lake surface area ratios and receive higher sediment and nutrient loading (per unit surface area per annum) than do most natural lakes (Thornton et al. 1980; Benson 1982). Reservoirs, especially those having hypolimnetic outlets, are considerably more efficient at trapping suspended particles than at retaining nutrients (Heinemann et al. 1973) and, therefore, the filling of reservoir basins with river-borne silts and clays is the dominant aging process.

Suspended particles transported into reservoirs by tributaries settle out as the flow velocity decreases. Consequently, the highest rates of sediment deposition occur in the upper portions of reservoir basins and embayments rather than at the deepest point of the basin as usually occurs in natural lakes (Davis 1968; Lehman 1975; Kimmel 1978). Shallow deltas formed by sediment deposition at the mouths of reservoir tributaries permit the growth of rooted macrophytes and the development of reservoir-associated wetlands (Silvey and Stanford 1978), which will become more important both ecologically and economically as our relatively young "cohort" of large United States reservoirs continues to age.

Basin filling by siltation can occur rapidly in small reservoirs located in erodible watersheds. Dendy et al. (1973) summarized the results of sediment deposition surveys on 1,105 United States reservoirs and reported a strong inverse relationship between the rate of basin filling and the initial reservoir storage capacity. The mean annual rate of storage capacity loss for reservoirs of less than 1.23×10^5 m^3 (<100 acre-feet) volume was 2.7% as compared to 0.16% for reservoirs of greater than 1.23×10^9 m^3 (>1 million acre-feet) volume. The median storage capacity loss rate of 1.5% per year for the smaller impoundments (<100 acre-feet) implied that 50% of them would completely fill with sediment within about 67 years and that their usefulness would be seriously impaired within about 30 years when their basins would be 30 to 50% filled. Certainly, such rates of basin filling justify concerns regarding reservoir aging for small impoundments, especially since sport fishing is often a primary purpose. However, the storage capacity loss rates of most large reservoirs suggest useful lifetimes of several centuries.

Phases of Reservoir Aging

The Initial Trophic Disequilibrium

Natural lakes and reservoirs reach trophic equilibrium relatively rapidly after their basins fill with water, and for both, trophic status is determined initially by the nature of the drainage basin. In reservoirs, a highly productive period, termed the "trophic upsurge" (Baranov 1961), occurs prior to the establishment of a trophic equilibrium. The "trophic upsurge" is followed by a "trophic depression," which is, in fact, the initial approach of the reservoir ecosystem toward its natural equilibrium level (Fig. 2). This "trophic disequilibrium" phase has received much attention from reservoir limnologists and fishery biologists because (1) it has coincided temporally with new impoundment formation and reservoir research activities, and (2) the initial period of high biological productivity has often produced false hopes for a higher level of fisheries production than can be sustained.

The trophic upsurge in new impoundments is due to a combination of several factors: (1) a large influx of organic detritus and inorganic nutrients from the inundated reservoir basin, (2) an abundance of high quality habitat and food for benthic organisms, and (3) a rapidly expanding lacustrine environment (Baxter 1977; Ploskey 1981; Benson 1982) (Fig. 3). Similarly, the subsequent decline in biological productivity, the "trophic depression," also has multiple causes: (1) decreased internal nutrient loading, (2) a decline in biologically labile organic detritus, (3) the cessation of habitat expansion, and (4) a reduction of favorable habitat. The magnitude and duration of the trophic disequilibrium phase are quite variable among impoundments because of differing basin inundation rates, internal and external nutrient loading rates, flushing rates, the quality and quantity of new habitat, the fish assemblages present, and reservoir operations (Ploskey 1981). Recently developed empirical models (Ostrofsky and Duthie 1978, 1980; Duthie and Ostrofsky 1982; Grimard and Jones 1982), based on the phosphorus loading-trophic response approach (Vollenweider 1968, 1976) and the kinetics of nutrient leaching from inundated soils and vegetation, predict both the internal nutrient loading from the flooded reservoir basin and the external nutrient loading from the watershed and thereby permit an estimation of the probable extent of the trophic upsurge and depression periods.

Fish populations may fluctuate sharply during the trophic disequilibrium phase, but usually stabilize within 5 to 10 years (Ploskey 1981; Benson 1982; Adams et al. 1983). Jenkins (1967, 1977) and Jenkins and Morais (1971), using data from over 100 United States reservoirs, reported that sport fish harvest was negatively correlated with reservoir age; however, reservoir age had no effect on total fish standing stocks due to increases in clupeid populations. Although the decline in sport fish production is partially a function of the reduced biological productivity following the cessation of internal nutrient loading (Fig. 2), concurrent decreases in available terrestrial organic detritus (Ploskey 1981) and in habitat quality as cover disappears (Benson 1982) are also quite important (Fig. 3). Some reservoirs continue to have high nutrient levels long after fish production has declined (McCammon and von Geldern 1979; O'Brien, in press).

Reservoir Maturation and Man-Induced Changes

Following the dynamic environmental and biological fluctuations characteristic of the early years of a reservoir's existence, the magnitude and variability of biological production within

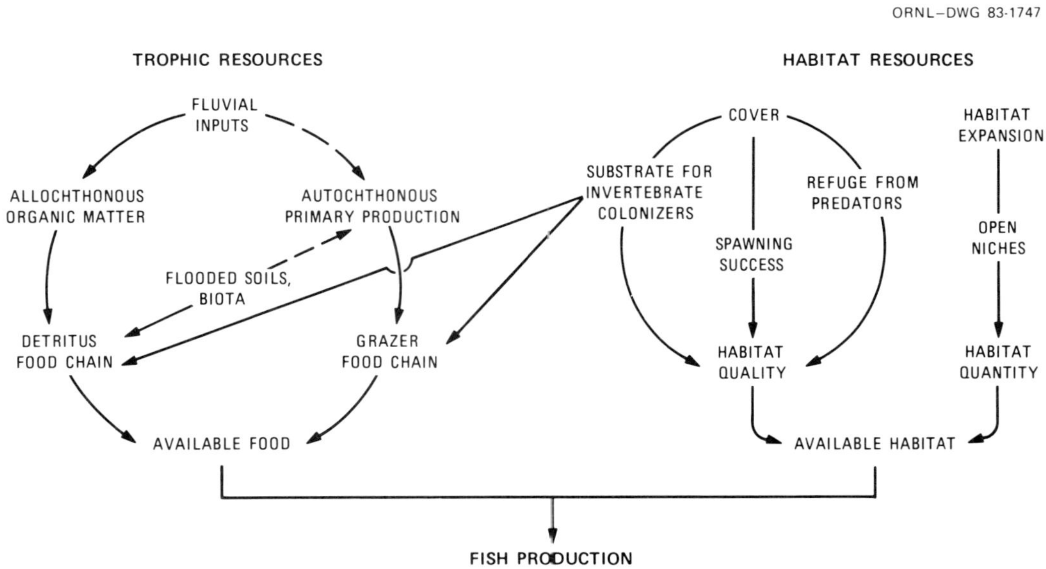

Figure 3. Reservoir fish production is controlled by a complex of factors affecting trophic and habitat resources. Relative to trophic resources, solid arrows indicate organic matter fluxes and dashed arrows represent nutrient fluxes. Flooded soils, vegetation, and terrestrial biota contribute significantly to trophic and habitat resources in new impoundments, but decline in importance as reservoirs mature. Similarly, both the quality and quantity of available habitat are reduced as reservoirs age.

the maturing reservoir become dependent on inputs of nutrients and organic matter from the watershed, as in natural lakes. Because of large drainage basins relative to reservoir surface areas and volumes, fluvial inputs are the most important sources of nutrients for most reservoir ecosystems (Gloss et al. 1980; Kimmel et al., in press). In addition to point sources of nutrients (e.g., industries, municipalities), patterns of land use within watersheds are primary determinants of the nutrient loading to aquatic systems and, thereby, of lake and reservoir productivity (Likens 1972; Hutchinson 1973). Undisturbed terrestrial ecosystems are usually characterized by runoff low in dissolved substances; however, pastures, croplands, and urban areas contribute much greater nutrient loads to aquatic systems (Likens 1975). Therefore, land-use patterns will have long-term impacts on reservoir productivity and water quality.

If reservoirs were permitted to age without being otherwise perturbed, one could project (based on our present understanding of the relationships between basin morphology, nutrient loading rates, and lacustrine productivity) that reservoir productivity should either remain relatively constant over time (for reservoirs which fill in slowly) or gradually increase as mean depth decreases (for reservoir basins undergoing rapid siltation). Additionally, one could predict that, by virtue of siltation in the upper portions of reservoir basins, aquatic macrophytes will become increasingly important as sources of organic matter production and as fish spawning and nursery habitat. As overall productivity increases, one might also expect dissolved oxygen depletion to occur more frequently, water quality to decrease, and the fish standing stock to become increasingly dominated by nongame species. However, because (1) the construction of a man-made impoundment often promotes additional land-use changes and technological development within reservoir watersheds and (2) their large watersheds focus both point and diffuse sources of nutrients into reservoir basins, we suspect that water quality and productivity changes attributable to "natural" reservoir aging will be small compared to the effects of man-induced impacts on watershed-reservoir interactions.

Summary and Conclusions

We have little direct information about the long-term ecological consequences of reservoir aging. To what extent, then, can the questions posed in the introduction to this paper be addressed? We believe that our present knowledge of natural lake succession and cultural eutrophication, along with recent improvements in our understanding of land-water interactions, provide a basis for anticipating ecological changes which will accompany reservoir aging and for suggesting a management approach for optimizing water quality and fisheries production in aging impoundments.

After the initial period of trophic disequilibrium, reservoir ecosystems will experience changes with aging similar to those occurring in natural lakes, but at a faster rate due to higher suspended sediment loading rates and, thus, more rapid basin filling by siltation. Because reservoir construction is often precedent to or concurrent with agricultural, industrial, and/or residential development of the same drainage basin, many United States reservoirs will not experience a "trophic equilibrium" in the sense of Hutchinson (1973), i.e., one unaffected by man-induced disturbances (Fig. 2). However, nutrient losses from impoundments by hypolimnetic outflows may be of particular importance in regard to ecosystem responses to watershed perturbations. Small natural lakes are very sensitive to changes in nutrient loading (Brugam 1978) because the nutrients received accumulate and recycle within their basins. Reservoirs with hypolimnetic outlets are much less retentive of nutrients and, therefore, may be considerably less sensitive to periodic watershed disturbance than natural lakes or surface-outflow impoundments. Thus, while their large drainage basins often result in high rates of nutrient and sediment loading to reservoirs, their rapid flushing rates and lower nutrient retention efficiencies may impart a relatively high degree of resilience to certain reservoir ecosystems. Unless stressed too severely or only unidirectionally, deep-outflow reservoirs may attain a dynamic trophic equilibrium consisting of episodic fluctuations about an average productivity level.

It seems apparent that the productivity level or fishery yield during the initial years of a reservoir's existence should not be used as a basis for formulating long-term management strategies. Application of the morphoedaphic index (MEI = total dissolved solids in mg per liter divided by mean depth in meters) can provide a reasonable first estimate of the potential equilibrium-level reservoir productivity (e.g., Henderson et al. 1973). However, because the ionic strength of water in a new impoundment may be elevated by basin leaching (Mitchell 1973; Duthie and Ostrofsky 1975), the total dissolved solids level of the major tributary should be used in the MEI computation to provide a more conservative value. Although the MEI has proven to be a robust predictor of fishery yield in lakes and reservoirs (Oglesby 1977, 1982; Jenkins 1982; Ryder 1982), the relationship between the MEI and fishery yield is a correlative one, and fishery yields predicted for individual impoundments may not be particularly reliable. Phosphorus loading models can also be used to estimate the approximate equilibrium-level productivity of a reservoir (Duthie and Ostrofsky 1982; Kerekes 1982); however, these models are also empirically based. The appropriate use of empirical models for management purposes requires careful consideration of their underlying assumptions, the limitations of the data sets upon which they are based, and the degree of uncertainty associated with their predictions (Reckhow 1979).

We believe that the future of reservoir water quality and fisheries production will more likely depend on our ability to implement integrated terrestrial and aquatic resources planning and management than on the consequences of "natural" reservoir aging. Reservoir water quality and fishery resources managers must learn to communicate with foresters, soil conservationists, agriculturists, range managers, and land-use and city planners, as well as with reservoir operations engineers, in order to participate effectively in developing integrated whole-watershed management strategies. Terrestrial and aquatic research scientists must also play a role; interdisciplinary research on selected impoundments and their watersheds will be required to (1) determine the sensitivity of river-reservoir systems to various levels of watershed perturbation and (2) evaluate the effectiveness of coordinated reservoir watershed, water quality, and fishery management strategies. Such an integrated, cooperative effort represents a major challenge to terrestrial and aquatic resources planners, managers and scientists, and to State and Federal agencies. However, such cooperation will almost certainly be required to circumvent the deleterious effects of both natural and accelerated reservoir aging processes.

Acknowledgments

We thank S. M. Adams, C. C. Coutant, W. Van Winkle, C. W. Gehrs, and three anonymous referees for reviewing the manuscript. Research was sponsored by the Office of Health and Environmental Research, U.S. Department of Energy, under contract W-7405-eng-26 with Union Carbide Corporation. Publication No. 2280, Environmental Sciences Division, Oak Ridge National Laboratory.

References

ADAMS, S. M., B. L. KIMMEL, AND G. R. PLOSKEY. 1983. Organic matter sources for reservoir fish production: A trophic-dynamic analysis. Canadian Journal of Fisheries and Aquatic Sciences 40: 1480–1495.

BARANOV, I. V. 1961. Biohydrochemical classification of reservoirs in the European U.S.S.R. Pages 139–183 *in* P. V. Tyurin, editor. The storage lakes of the U.S.S.R. and their importance for fishery. Israel Program for Scientific Translations, Tel Aviv, Israel.

BAXTER, R. M. 1977. Environmental effects of dams and impoundments. Annual Review of Ecology and Systematics 8:255–283.

BENSON, N. G. 1982. Some observations on the ecology and fish management of reservoirs in the United States. Canadian Water Resources Journal 7:2–25.

BRUGAM, R. B. 1978. Human disturbance and the historical development of Linsley Pond. Ecology 59:19–36.

DAVIS, M. B. 1968. Pollen grains in lake sediments: Redeposition caused by seasonal water circulation. Science 162:796–799.

DENDY, F. E., W. A. CHAMPION, AND R. B. WILSON. 1973. Reservoir sedimentation surveys in the United States. Pages 349–358 *in* W. C. Ackermann, G. F. White, and E. B. Worthington, editors. Man-made lakes: Their problems and environmental effects. Geophysical Monograph 17, American Geophysical Union, Washington, D.C., USA.

DUTHIE, H. C., AND M. L. OSTROFSKY. 1975. Environmental impacts of Churchill Falls (Labrador) hydroelectric project: A preliminary assessment. Journal of the Fisheries Research Board of Canada 32:117–125.

DUTHIE, H. C., AND M. L. OSTROFSKY. 1982. Use of phosphorus budget models in reservoir management. Canadian Water Resources Journal 7:337–348.

GLOSS, S. P., L. M. MAYER, AND D. E. KIDD. 1980. Advective control of nutrient dynamics in the epilimnion of a large reservoir. Limnology and Oceanography 25:219–228.

GOLDMAN, C. R. 1968. Aquatic primary production. American Zoologist 8:31–42.

GRIMARD, Y., AND H. G. JONES. 1982. Trophic upsurge in new reservoirs: A model for total phosphorus concentrations. Canadian Journal of Fisheries and Aquatic Sciences 39:1473–1483.

HEINEMANN, H. G., R. F. HOLT, AND D. L. RAUSCH. 1973. Sediment and nutrient research on selected corn belt reservoirs. Pages 381–386 *in* W. C. Ackermann, G. F. White, and E. B. Worthington, editors. Man-made lakes: Their problems and environmental effects. Geophysical Monograph 17, American Geophysical Union, Washington, D.C., USA.

HENDERSON, H. F., R. A. RYDER, AND A. W. KUDHONGANIA. 1973. Assessing fishery potentials of lakes and reservoirs. Journal of the Fisheries Research Board of Canada 30:2000–2009.

HUTCHINSON, G. E. 1973. Eutrophication. American Scientist 6:269–279.

JENKINS, R. M. 1967. The influence of some environmental factors on standing crop and harvest of fishes in U.S. reservoirs. Pages 298–321 *in* Reservoir fishery resources symposium. Southern Division American Fisheries Society, Bethesda, Maryland, USA.

JENKINS, R. M. 1977. Prediction of fish biomass, harvest, and prey-predator relations in reservoirs. Pages 282–293 *in* W. Van Winkle, editor. Assessing the effects of power plant-induced mortality on fish populations. Pergamon Press, New York, USA.

JENKINS, R. M. 1982. The morphoedaphic index and reservoir fish production. Transactions of the American Fisheries Society 111:133–140.

JENKINS, R. M., AND D. I. MORAIS. 1971. Reservoir sport fishing effort and harvest in relation to environmental variables. Pages 371–384 *in* G. E. Hall, editor. Reservoir fisheries and limnology. Special Publication No. 8, American Fisheries Society, Bethesda, Maryland, USA.

KEREKES, J. J. 1982. The application of phosphorus load-trophic response relationships to reservoirs. Canadian Water Resources Journal 7:349–354.

KIMMEL, B. L. 1978. An evaluation of recent sediment focusing in Castle Lake (California) using a volcanic ash layer as a stratigraphic marker. Proceedings of the International Association of Theoretical and Applied Limnology 20:393–400.

KIMMEL, B. L., O. T. LIND, AND L. J. PAULSON. (In press). Reservoir primary production. *In* K. W. Thornton, editor. Perspectives on reservoir ecosystems. John Wiley and Sons, New York, USA.

LEHMAN, J. T. 1975. Reconstructing the rate of accumulation of lake sediment: The effect of sediment focusing. Quaternary Research 5:541–550.

LIKENS, G. E. 1972. Eutrophication and aquatic ecosystems. Pages 3–13 *in* G. E. Likens, editor. Nutrients and eutrophication: The limiting-nutrient controversy. American Society of Limnology and Oceanography Special Symposium 1, Allen Press, Lawrence, Kansas, USA.

LIKENS, G. E. 1975. Nutrient flux and cycling in freshwater ecosystems. Pages 14–348 *in* F. G. Howell, J. B. Gentry, and M. H. Smith, editors. Mineral cycling in southeastern ecosystems. U.S. Energy Research and Development Agency, CONF-740513, National Technical Information Service, Springfield, Virginia, USA.

McCAMMON, G. W., AND C. E. VON GELDERN, JR. 1979. Predator-prey systems in large reservoirs. Pages 431–442 *in* R. H. Stroud and H. Clepper, editors. Predator-prey systems in fishery management. Sport Fishing Institute, Washington, D.C., USA.

MITCHELL, D. S. 1973. Supply of plant nutrient chemicals in Lake Kariba. Pages 165–169 *in* W. C. Ackermann, G. F. White, and E. B. Worthington, editors. Man-made lakes: Their problems and environmental effects. Geophysical Monograph 17, American Geophysical Union, Washington, D.C., USA.

NAS/NRC (NATIONAL ACADEMY OF SCIENCES/NATIONAL RESEARCH COUNCIL). 1969. Eutrophication: Causes, consequences, correctives. National Academy Press, Washington, D.C., USA.

O'BRIEN, W. J. (In press). Perspectives on fish in reservoir ecosystems. *In* K. W. Thornton, editor. Perspectives on reservoir ecosystems. John Wiley and Sons, New York, USA.

OGLESBY, R. T. 1977. Relationships of fish yield to lake phytoplankton standing crop, production, and morphoedaphic factors. Journal of the Fisheries Research Board of Canada 34:2271–2279.

OGLESBY, R. T. 1982. The MEI symposium—overview and observations. Transactions of the American Fisheries Society 111:171–175.

OSTROFSKY, M. L., AND H. C. DUTHIE. 1978. An approach to modelling productivity in reservoirs. Proceedings of the International Association of Theoretical and Applied Limnology 20:1562–1567.

OSTROFSKY, M. L., AND H. C. DUTHIE. 1980. Trophic upsurge and the relationship between phytoplankton biomass and productivity in Smallwood Reservoir, Canada. Canadian Journal of Botany 58: 1174–1180.

PLOSKEY, G. R. 1981. Factors affecting fish production and fishing quality in new reservoirs, with guidance on timber clearing, basin preparation, and filling. Technical Report, U.S. Army Engineer Waterways Experiment Station, Vicksburg, Mississippi, USA.

RECKHOW, K. H. 1979. Empirical lake models for phosphorus: Development, applications, limitations and uncertainty. Pages 193–221 *in* D. Scavia and A. Robertson, editors. Perspectives on lake ecosystem modeling. Ann Arbor Science, Ann Arbor, Michigan, USA.

RYDER, R. A. 1978. Ecological heterogeneity between north-temperate reservoirs and glacial lake systems due to differing succession rates and cultural uses. Proceedings of the International Association of Theoretical and Applied Limnology 20:1568–1574.

RYDER, R. A. 1982. The morphoedaphic index—use, abuse, and fundamental concepts. Transactions of the American Fisheries Society 111:154–164.

SILVEY, J. K. G., AND J. A. STANFORD. 1978. A historical overview of reservoir limnology in the southwestern United States. Pages 1–18 *in* J. Cairns, E. F. Benfield, and J. R. Webster, editors. Current perspectives of river-reservoir ecosystems. North American Benthological Society Publication 1, Blacksburg, Virginia, USA.

THORNTON, K. W., R. H. KENNEDY, A. D. MAGOUN, AND G. E. SAUL. 1980. Reservoir water quality sampling design. Pages 349–355 *in* Restoration of lakes and inland waters. EPA-440/5-81-010, U.S. Environmental Protection Agency, Washington, D.C., USA.

VOLLENWEIDER, R. A. 1968. Scientific fundamentals of the eutrophication of lakes and flowing waters, with particular reference to nitrogen and phosphorus as factors in eutrophication. Technical Report DAS/CSI/68.27, Organization for Economic Cooperation and Development, Paris, France.

VOLLENWEIDER, R. A. 1976. Advances in defining critical loading levels for phosphorus in lake eutrophication. Memorie dell Istituto Italiano di Idrobiologia 33:53–83.

WHITTAKER, R. H. 1975. Communities and ecosystems, 2nd edition. Macmillan Publishing Co., New York, USA.

Pages 110–114 *in* G. E. Hall and M. J. Van Den Avyle, editors.
Reservoir Fisheries Management: Strategies for the 80's.
Reservoir Committee, Southern Division American Fisheries Society, Bethesda, Maryland. 1986.

Review of Water Level Management on Kansas Reservoirs

DAVID W. WILLIS

Kansas Fish and Game Commission
P.O. Box 1525
Emporia, Kansas 66801

ABSTRACT

Water level management plans in Kansas typically consist of a spring rise to flood terrestrial vegetation, a summer drawdown of approximately 4 ft to allow regrowth of vegetation and concentrate predators and prey, an autumn rise of approximately 2 ft to flood some terrestrial vegetation and attract waterfowl, and a winter drawdown to again concentrate predators and prey and protect remaining vegetation from water damage. Netting data indicate that this program has increased population densities of walleye (*Stizostedion vitreum*), white crappie (*Pomoxis annularis*), and white bass (*Morone chrysops*) in impoundments where 20% or more of the basin area was exposed during drawdown. Largemouth bass (*Micropterus salmoides*) densities were negatively impacted, presumably because of the reduction of vegetative habitat resulting from mid-summer drawdown. Water transparencies typically increased with use of this water level management plan.

Water level management is considered an important tool for reservoir fisheries management (Jenkins 1970; Keith 1975). The interest in and value of water level management has resulted in voluminous literature on this topic as indicated in bibliographies by Triplett et al. (1980) and Ploskey (1982). Water resource agencies involved with Kansas reservoirs asked the Kansas Fish and Game Commission (KFG) to document the value of annual water level management plans. The purpose of this paper is to update the position of KFG relative to its water level management plan, as first reported by Groen and Schroeder (1978).

The basic approach for managing water levels of Kansas reservoirs (Fig. 1) consists of: (1) a spring rise and hold to flood terrestrial vegetation, (2) a summer drawdown of about 4 ft to allow regrowth of vegetation and concentrate predators and prey, (3) an autumn rise of about 2 ft to flood some terrestrial vegetation and attract waterfowl, and (4) a winter drawdown to once again concentrate predators and prey and protect remaining vegetation from water damage. The objectives of this general plan, as indicated by Groen and Schroeder (1978), are to increase population densities and growth rates of sport fish and to improve water quality (transparency) in reservoirs where the normal aging process often creates opposite effects. Target species of these management efforts include the walleye (*Stizostedion vitreum*), white crappie (*Pomoxis annularis*), white bass (*Morone chrysops*), and largemouth bass (*Micropterus salmoides*).

METHODS

This study was undertaken as a monitoring project—data were taken from annual sampling by management biologists. Changes in gear types and sampling regimes over the years often preclude absolute confidence in data sets. However, data presented are indicative of trends, and they represent the best source of information now available. Because of the changes in gear types, data sets will be explained as they are discussed. All mesh sizes are listed as bar measure. Statistical analyses (t-tests) were performed using the Statistical Analysis System computer package (SAS Institute, Incorporated, Cary, North Carolina). Names and physical parameters for three reservoirs included in this review are listed in Table 1.

RESULTS

Walleye

At Kanopolis Reservoir, where water level management began in 1975, walleye catch rates in gill nets were significantly higher ($P < 0.01$) during years in which water levels were managed than in pre-management samples (Table 2). This trend is best-illustrated by the data for multifilament experimental gill nets used for two years preceding water level management (1973–1974)

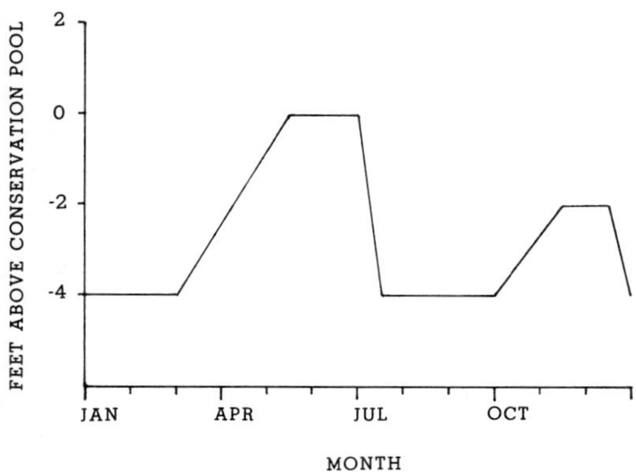

Figure 1. Typical water level management sequence used for Kansas reservoirs.

and three years during management (1975–1977). The immediate (same-year), positive response of walleye in 1975 could be detected because young-of-the-year walleye in Kansas reservoirs attain a size effectively sampled in 0.75- and 1.0-in mesh netting by autumn. The 4.5-ft drawdown used at Kanopolis Reservoir exposed 21.5% of the reservoir area.

White Crappie

Beam (1983) reported that white crappie abundance increased following initiation of water level management at Elk City Reservoir in 1977 (Table 3). Samples were collected with a nightly complement of two large fyke nets (4-ft hoop, 1.0-in mesh) and two small fyke nets (3-ft hoop, 0.5-in mesh). Interpretation of trends in white crappie abundance, however, is hampered by changes in sampling seasons during the period of record. White crappie catch per net night from 1972 to 1978 consisted of spring Age-1 fish whereas samples for 1979 to 1981 were autumn Age-0 fish. The high catch of Age-1 white crappie in spring 1978 suggests that a strong year class was produced in the first management year; Beam (1983) felt that only one strong year class of white crappie was produced in pre-management years of 1972–1976. The relatively high abundance of Age-0 white crappie during autumns of 1979–1981 may reflect production of strong year classes during those managed years as well, but pre-management data suitable for comparison were not collected. The 4-ft drawdown at Elk City Reservoir exposed 20.2% of the basin area.

White Bass

Catch rates of white bass in Kanopolis Reservoir were higher in managed years (1975–1977)

Table 1. Physical parameters of Kansas reservoirs included in assessment of water level management plans.

	Reservoir		
Parameter	Kanopolis	Elk City	John Redmond
Area (acres)	3,550	4,400	9,700
Mean depth (ft)	15	11	7
Mean annual water level fluctuation (ft)	20	10	5
Storage ratio	0.23	0.15	0.07
Shoreline development ratio	3.6	5.4	5.1

Table 2. Catch of walleye and white bass with various types of gill nets in Kanopolis Reservoir. Water level management began in 1975.

	Number collected per 1,000 ft² of gill net		
Year	Walleye	White bass	Gear[a]
1957	1.8	4.3	1, 2, 3
1958	3.0	12.3	1, 4
1959	1.5	3.8	1, 3
1961	4.9	4.2	1
1963	2.5	25.8	4
1965	1.3	13.5	1
1969	1.2	4.1	5, 6
1973	2.2	4.2	7
1974	3.5	6.7	7
1975	11.1	16.2	7
1976	16.5	22.8	7
1977	17.0	12.0	7
1978	39.8	6.7	8
1979	15.0	22.1	9
1980	37.3	31.7	9
1981	22.3	30.8	9

[a] 1 = Multifilament experimental; 125 ft × 6 ft, 0.75- to 2.0-in mesh.
2 = Multifilament flag; 210 ft × 4 ft, 2.5-in mesh.
3 = Multifilament flag; 150 ft × 4 ft, 2.5-in mesh.
4 = Multifilament experimental; 250 ft × 6 ft, 1.0- to 2.0-in mesh.
5 = Multifilament experimental; 250 ft × 6 ft, 2.0- to 4.0-in mesh.
6 = Multifilament experimental; 180 ft × 6 ft, 1.5- to 3.0-in mesh.
7 = Multifilament experimental; 180 ft × 6 ft, 0.75- to 2.5-in mesh.
8 = Monofilament flag; 150 ft × 8 ft; 0.75-, 1.25-, and 2.0-in meshes.
9 = Monofilament flag; 100 ft × 8 ft; 1.0-, 1.5-, 2.5-, and 4.0-in meshes.

Table 3. White crappie abundance at Elk City Reservoir as indicated by catch in fyke nets (from Beam 1983). Water level management began in 1977.

Year	Number collected per net-night	
	Age 0 (autumn)	Age 1 (spring)
1972	–	4.7
1973	–	0.1
1974	–	2.8
1975	–	1.2
1976	–	0.1
1977	–	0.5
1978	–	7.8
1979	6.5	–
1980	5.2	–
1981	8.1	–

Table 4. Relative year-class strength (estimated number alive at Age 1) for largemouth bass in Elk City Reservoir, using a 1980 electrofishing sample as base data and assuming a constant 50% annual survival rate. Water level management began in 1977.

Year class	Number collected in 1980	Estimated number alive at Age 1
1979	1	1
1978	5	10
1977	11	44
1976	19	152
1975	9	144
1974	9	288
1973	1	64

than in pre-management years (1973–1974) when similar gears were used (Table 2). The increase would have been statistically significant at $P = 0.07$, whereas a comparison using all available years of data (ignoring bias associated with different gears) was significant at $P < 0.02$.

Largemouth Bass

Largemouth bass populations probably are adversely affected by water level management plans that include a June or early July drawdown, but effects are difficult to ascertain because most Kansas reservoirs do not contain adequate largemouth bass habitat even when water level is not manipulated. In 1980, a sample of largemouth bass was collected by electrofishing from Elk City Reservoir and aged to determine year-class frequency. Using a technique described by Gabelhouse (1976), a constant 50% annual survival rate was assumed, and year-class frequency was adjusted to estimate relative numbers of each year class alive at Age 1 (Table 4). This approach also assumes that the electrofishing sample was representative of the age distribution of the bass population; hence, data from the 1979 year class should probably be ignored because electrofishing does not generally sample these small fish effectively. The index of relative abundance at Age 1 indicates that largemouth bass recruitment has declined since the water level management plan was implemented.

Water Transparency

Water transparency was significantly greater in managed years than in pre-management years at Kanopolis and John Redmond reservoirs ($P < 0.05$). Fourteen Secchi disk readings taken at Kanopolis Reservoir from 1957–1974 (pre-water level management) averaged 12.1 in (range of 7 to 18 in) whereas 27 readings from 1975–1981 (water levels managed) averaged 30.6 in (range of 12 to 55 in). Three Secchi disk readings at John Redmond Reservoir in 1976 ranged from 5 to 8 in with a mean of 6.3 in, whereas six readings in 1977 and 1978 (flooded vegetation from water level management first occurred in 1977) ranged from 10 to 21 in with a mean of 15.7 in. Beam (1983) attributed increases in water transparency at Elk City Reservoir to water level management.

Discussion

Kansas water level plans, when drastic enough, have apparently benefitted species which produce young-of-the-year that typically inhabit open-water (limnetic or pelagic) areas. This behavior has been shown for walleyes (Forney 1966; Priegel 1970), white crappies (Siefert 1969; Mayhew 1977), and white bass (Houser and Bryant 1967). These species probably benefit most from the increased food supply that results from flooded vegetation and are not adversely affected by a moderate July drawdown. Kansas water level plans may have a negative effect on species such as largemouth bass, in which young-of-the-year fish inhabit shoreline areas. Spring flooding of vegetation apparently allows a good spawn to occur and provides abundant food for the young largemouth bass. However, the July drawdown may move these small fish away from the structure necessary as escape cover or protection from wind and waves. Aggus and Elliott (1975) and Rainwater and Houser (1975) reported the importance of maintaining high water levels for enhancing recruitment of largemouth bass.

Ploskey (1983) recognized the need to look at basin area exposure as a function of the amplitude of fluctuation when considering water level management as a fisheries tool. Plans that have exposed 20% or more of the basin area appear to be successful for Kansas reservoirs like Kanopolis and Elk City. Many other case histories exist in which water level management has not increased populations of walleye, white bass, or white crappie in Kansas reservoirs. In essentially all of these cases, the amount of basin area exposed was less than 13%. For example, neither white crappie nor walleye populations were improved by annual drawdowns of about 12% in Pomona (4,000 acres) and Council Grove (3,285 acres) reservoirs.

Ploskey (1986) reported that increased densities of many young-of-the-year fishes after spring flooding usually are nullified if summer drawdown is equally extensive. This is apparently true for largemouth bass populations in Kansas reservoirs and may also explain why basin area exposures that were less than 13% produced no increases of walleye, white bass, or white crappie. Exposures of 20% or more may be necessary for the benefits of vegetation flooding to outweigh the negative influences of mid-summer drawdown.

Increased water transparency resulting from water level management benefits aesthetics as well as fisheries. Some authors have reported that inundated terrestrial vegetation results in lower turbidity (Keith 1968; Fox et al. 1977). In addition, compaction of mudflats during drawdown can increase water transparency. Compacted, dried mudflats often remain intact for several months after reflooding, and thus keep colloidal particles in the sediments rather than in suspension.

Management Implications

The four-step water level management technique discussed here for Kansas reservoirs appears to be a viable plan for impoundments where walleye, white bass, and white crappie are the most important sport fish. Because a majority of Kansas reservoirs are windswept and turbid (with little largemouth bass production), this plan appears to be suitable in many cases. The basin area exposure during drawdown appears to be a major factor; exposure of at least 20% seems to be necessary.

Where largemouth bass populations are important (and perhaps in situations where extensive vertical drawdowns are necessary to obtain desired exposure), more benefits might be achieved by asking for an occasional drastic drawdown over the course of an entire year. Water resource agencies, particularly those responsible for water-based recreation, may be more willing to accept an occasional year of low, though stable, water levels rather than four fluctuations within a year.

Regardless of whether the standard plan or a drastic drawdown is used, we can probably benefit walleye populations by providing a spring "buffer pool" capacity in all impoundments. This could be done by lowering reservoir levels several feet prior to ice-up each year. This storage capacity would then be available during walleye spawning in the spring, which often coincides with periods of high inflow. The extra drawdown necessary in the winter would usually be small because water levels typically decrease gradually throughout the previous summer. In addition, drawdowns just prior to ice-up, with a return to conservation pool in early spring, would probably have relatively small sociopolitical impacts because reservoir use is minimal during winter.

Acknowledgments

Many individuals, including Charles Bever, Thomas Bowman, D. Leo Dowlin, Thomas Giffin, James Goudzwaard, Calvin Groen, Richard Hunter, Leonard Jirak, Lanny Jones, Ronald Marteney, Arnold McAlexander, Kenneth McCloskey, Steven Price, Randy Schademann, Gordon Schneider, and James Stephen collected information used in this review. Contribution of data from James Beam, Troy Schroeder, and Bruce Zamrzla was especially helpful, as was input from Bruce Taggart. Robert Hartmann and Thomas Mosher assisted with manuscript review. Charlene Ritchie typed several drafts of the paper. Donald Gabelhouse, Jr., provided assistance through both technical guidance and review. Partial funding was provided by Federal Aid Project FW-9-P-2.

References

Aggus, L. R., and G. V. Elliott. 1975. Effects of cover and food on year-class strength of largemouth bass. Pages 317–322 *in* R. H. Stroud and H. Clepper, editors. Black bass biology and management. Sport Fishing Institute, Washington, D.C., USA.

BEAM, J. H. 1983. The effect of annual water level management on population trends of white crappie in Elk City Reservoir, Kansas. North American Journal of Fisheries Management 3:34–40.

FORNEY, J. L. 1966. Factors affecting first-year growth of walleyes in Oneida Lake, New York. New York Fish and Game Journal 13:146–167.

FOX, J. L., P. L. BREZONIK, AND M. A. KEIRN. 1977. Lake drawdown as a method of improving water quality. United States Environmental Protection Agency, EPA-600/3-77-005, Corvallis, Oregon, USA.

GABELHOUSE, D. W., JR. 1976. Form and function of redear sunfish populations. Master's thesis, University of Missouri, Columbia, Missouri, USA.

GROEN, C. L., AND T. A. SCHROEDER. 1978. Effects of water level management on walleye and other coolwater fishes in Kansas reservoirs. Pages 278–283 *in* R. L. Kendall, editor. Selected coolwater fishes of North America. Special publication 11, American Fisheries Society, Bethesda, Maryland, USA.

HOUSER, A., AND H. E. BRYANT. 1967. Sampling reservoir fish populations using midwater trawls. Pages 391–404 *in* Reservoir fishery resources symposium. Southern Division American Fisheries Society, Bethesda, Maryland, USA.

JENKINS, R. M. 1970. Reservoir fish management. Pages 173–182 *in* N. G. Benson, editor. A century of fisheries in North America. Special publication 7, American Fisheries Society, Bethesda, Maryland, USA.

KEITH, W. E. 1968. Turbidity control and fish population renovation on Blue Mountain Lake, Arkansas. Proceedings of the Annual Conference of the Southeastern Association of Game and Fish Commissioners 21:495–505.

KEITH, W. E. 1975. Management by water level manipulation. Pages 489–496 *in* R. H. Stroud and H. Clepper, editors. Black bass biology and management. Sport Fishing Institute, Washington, D.C., USA.

MAYHEW, J. 1977. The effects of food management regimes on larval fish and fish food organisms at Lake Rathbun. Fisheries Research Technical Series Number 77-2, Iowa Conservation Commission, Des Moines, Iowa, USA.

PLOSKEY, G. R. 1982. Fluctuating water levels in reservoirs; an annotated bibliography of environmental effects and management for fisheries. Technical report E-82-5, U.S. Army Engineer Waterways Experiment Station, Vicksburg, Mississippi, USA.

PLOSKEY, G. R. 1983. A review of the effects of water-level changes on reservoir fisheries and recommendations for improved management. Technical report E-83-3, U.S. Army Engineer Waterways Experiment Station, Vicksburg, Mississippi, USA.

PLOSKEY, G. R. 1986. Effects of water levels and hydrology on reservoir fisheries, with implications for management. Pages 86–97 *in* G. E. Hall and M. J. Van Den Avyle, editors. Reservoir fisheries management: strategies for the 80's. Reservoir Committee, Southern Division American Fisheries Society, Bethesda, Maryland, USA.

PRIEGEL, G. R. 1970. Reproduction and early life history of the walleye in the Lake Winnebago region. Technical bulletin 45, Wisconsin Department of Natural Resources, Madison, Wisconsin, USA.

RAINWATER, W. C., AND A. HOUSER. 1975. Relation of physical and biological variables to black bass crops. Pages 306–309 *in* R. H. Stroud and H. Clepper, editors. Black bass biology and management. Sport Fishing Institute, Washington, D.C., USA.

SIEFERT, R. E. 1969. Biology of the white crappie in Lewis and Clark Lake. Technical paper 22, United States Bureau of Sport Fisheries and Wildlife, Washington, D.C., USA.

TRIPLETT, J. R., D. A. CULVER, AND G. B. WATERFIELD. 1980. An annotated bibliography on the effects of water-level manipulation on lakes and reservoirs. Ohio Department of Natural Resources, Federal Aid Project F-57-R, Columbus, Ohio, USA.

Pages 115–121 *in* G. E. Hall and M. J. Van Den Avyle, editors.
Reservoir Fisheries Management: Strategies for the 80's.
Reservoir Committee, Southern Division American Fisheries Society, Bethesda, Maryland. 1986.

Effects of Environmental Factors on Growth of Largemouth Bass in Texas Reservoirs

L. ESTEBAN MIRANDA

Texas Parks and Wildlife Department
Heart of the Hills Research Station
Ingram, Texas 78025

PHILIP P. DUROCHER

Texas Parks and Wildlife Department
Austin, Texas 78744

ABSTRACT

The effects of 13 environmental factors on growth of largemouth bass (*Micropterus salmoides*) in 55 Texas reservoirs were examined through regression analyses. Growth was measured as the average age at which largemouth bass in a population reached 254-mm total length (the statewide minimum length limit). Growth was positively correlated ($P \leqq 0.05$) with $\log_{10}$ reservoir standing stock and negatively correlated with elevation above mean sea level, $\log_{10}$ total alkalinity, age and average depth of reservoirs, the prey-predator ratio, and the morphoedaphic index. No significant correlations were found between growth rate and length of growing season, average water temperature in July, total dissolved solids, surface area, shoreline development, or annual water level fluctuation. The best predictive model of growth ($R^2 = 0.45$) was a multiple regression that included the independent variables $\log_{10}$ reservoir age, average reservoir depth, $\log_{10}$ reservoir standing stock, and the prey-predator ratio.

Creel surveys conducted on 12 Texas reservoirs from March through August 1978–1981 indicated that 40% of the total fishing effort was directed towards largemouth bass (*Micropterus salmoides*). To fulfill angler demand and maintain or improve catch rates, fishery managers are confronted with optimizing production of this species. Fish production is directly related to growth (Royce 1972); hence, identification of the principal factors limiting largemouth bass growth could allow management efforts to be specifically directed toward those factors.

Growth, the increase in size of an organism, is the outcome of a series of metabolic processes that begin with food intake. Growth is determined by interactions of many abiotic and biotic environmental factors, but identification and quantification of such factors is difficult under natural conditions. Therefore, to develop predictive or descriptive models, some exactness must be sacrificed, and efforts are directed toward identifying a few variables that either affect growth directly or play important roles as moderators of growth.

In this study we have attempted to identify key environmental factors controlling growth of largemouth bass in Texas reservoirs. Factors used in the analyses are routinely recorded in Texas reservoir surveys, and they include several physical, morphometric, chemical, and fishery parameters.

METHODS

Data Description and Source

Data on 13 environmental variables (Tables 1 and 2) were available from 55 reservoirs located throughout Texas. Physical, morphometric, and chemical descriptions were obtained from published reports of the Texas Water Development Board, the Texas Agricultural Experiment Station, and the Texas Parks and Wildlife Department. Fisheries data were collected from 1975 through 1981 by biologists of the Texas Parks and Wildlife Department using electrofishing, gill netting, and cove rotenone sampling as part of standardized fish population surveys.

Growth of largemouth bass was described as the average number of years required for fish in a population to reach a given body length; thus, a positive correlation with age is equivalent to a negative correlation with growth rate. The number of years necessary for largemouth bass to reach the statewide 254-mm minimum total

Table 1. Definition of 13 independent variables used in correlation analyses with growth data for largemouth bass in Texas reservoirs.

Variables	Symbol	Definition
Growing season	GS	mean number of frost-free days in the year
Mean water temperature in July	T_w	mean of all available temperature (C) measurements taken at depth 30 cm in July
Elevation	EL	vertical distance in meters between average annual pool level and mean sea level
Age of reservoir	AGE	age in years at time of growth determination
Annual water level fluctuation	AF	mean annual vertical fluctuation from average pool level in meters
Surface area	A	hectares at average annual pool level
Shoreline development	SD	ratio of shoreline length (at average annual pool level) to circumference of a circle equal in area to that of the reservoir
Mean depth	$\bar{z}$	mean depth in meters at average annual pool level
Total alkalinity	TA	mean of all available measurements of total concentration of bases in the reservoir expressed as mg/liter equivalent calcium carbonate
Total dissolved solids	TDS	mean of all available measurements of total residue remaining after evaporation (at 180 C) of a water sample first filtered to remove suspended matter, expressed as mg/liter. When not available, TDS was approximated as 65% of specific conductance (Texas Water Development Board 1973)
Morphoedaphic index	MEI	TDS divided by $\bar{z}$ (Ryder 1965)
Standing stock	SS	mean of all available standing stock measurements (kg/hectare) on a body of water determined by cove rotenone sampling conducted from June through September. Means are unadjusted for incomplete recovery and not weighted for cove size
Prey-predator ratio	P/P	ratio of combined weight of shads, sunfishes, and cichlids to weight of gars, bowfin, pikes, catfishes ≥395 mm, temperate basses ≥165 mm, black basses ≥165 mm, and crappies ≥165 mm, as determined from SS measurements

length limit (Age_{254}) was chosen as the best measure of growth because it relates to the time of recruitment to harvestable size; it does not necessarily reflect growth rates at other ages or lengths. Age_{254} was determined for each largemouth bass population using the von Bertalanffy growth model (Everhart et al. 1975). Annual growth estimates used in the model were calculated using the Lee method of back-calculating lengths of fish from scales (Everhart et al. 1975). These estimates are approximations because (1) the time of annulus formation varies among fish and years; (2) the period between hatching and formation of first annulus is normally less than 1 year; and (3) they describe only growth rates achieved by survivors.

Analysis of Relationships

Stepwise multiple regression analysis was used to determine if significant ($P \leqq 0.05$) relationships existed between environmental variables and Age_{254}. As the first step in this analysis, simple correlation coefficients were estimated to help identify general forms of relationships. These coefficients do not account for interactions among environmental variables, and thus mainly are useful for identifying trends. The second step of the multiple regression analysis identified the best model for predicting Age_{254} from various combinations of the independent variables. Only those environmental variables that accounted for a significant amount of variability in Age_{254} at $P \leqq 0.05$ were included in the final model.

Evaluations were initially performed using logarithmic (base 10) transformations of all variables in addition to the untransformed data. Transformation of Age_{254} never significantly improved the regression equations, hence, those results are not shown. All statistical analyses were performed using programs in Dixon (1975).

Results

Age_{254} was significantly correlated ($P \leqq 0.05$) with seven of the 13 environmental variables

Table 2. Average values and ranges of 13 environmental variables (defined in Table 1) and correlation coefficients with the average age at which largemouth bass reached 254-mm total length (Age_{254}) in Texas reservoirs.

Variable	Number of reservoirs	Mean	Range	Correlation coefficients of Age_{254} with: Untransformed values of independent variable	Correlation coefficients of Age_{254} with: Log_{10}-transformed values of independent variable
GS	55	246	214–304	−0.03	−0.03
T_w	52	28	24–34	−0.24	−0.25
EL	53	238	25–688	0.27[a]	0.24
AGE	55	23	2–67	0.42[a]	0.40[a]
AF	47	1.6	0–6.1	0.27	0.27
A	55	6,691	52–73,494	0.14	0.07
SD	51	7.1	1.6–23.8	0.08	0.05
$\bar{z}$	55	6.5	1.8–18.9	0.30[a]	0.26
TA	54	119	25–268	0.26	0.27[a]
TDS	54	358	54–1,330	0.02	0.03
MEI	54	79	13–479	0.28[a]	0.05
SS	55	315.5	93.6–807.4	−0.24	−0.27[a]
P/P	55	3.9	0.4–8.8	0.48[a]	0.46[a]
Age_{254}	55	1.81	1.19–2.88	—	—

[a] Statistically significant ($P \leq 0.05$).

(Table 2). Positive correlations occurred with prey-predator ratio, reservoir age, elevation, mean depth, $\log_{10}$ of total alkalinity, and morphoedaphic index, indicating a trend for slower growth of largemouth bass at high values of these variables. The only significant negative correlation was with $\log_{10}$ of total fish biomass, suggesting more rapid growth where standing stock was greatest.

Stepwise multiple regression analysis produced a significant four-variable model that accounted for 45% of the variability of Age_{254} among reservoirs ($R^2 = 0.45$). The regression equation, with independent variables listed in order of decreasing statistical significance, was:

$$\begin{aligned} Age_{254} = {} & 1.85 + 0.054\ (\text{P/P}) \\ & + 0.342\ (\log_{10}\text{AGE}) \\ & + 0.02\ (\bar{z}) \\ & - 0.374\ (\log_{10}\text{SS}); \end{aligned}$$

abbreviations of independent variables are defined in Table 1. This equation suggests that largemouth bass growth in Texas is reduced in old and deep reservoirs having relatively high prey standing crop, but is faster where total fish standing stock is high. The logarithmic relationship between Age_{254} and reservoir age indicates that growth rates decline rapidly in the early years after impoundment but the reduction eventually becomes minor. Similarly, it appears as if growth is most affected by changes in standing stock when total biomass is relatively low.

Discussion

The inverse relationship between largemouth bass growth and the prey-predator ratio probably is not reflective of a direct cause-effect mechanism. As measured (Table 1), the prey-predator ratio included all sizes of prey species, and as such, may not indicate true forage availability for the various sizes of bass present in a given reservoir. High prey-predator ratios can occur in systems with overabundant adult prey that are too large to serve as forage for recruit-sized largemouth bass, leading to relatively slow growth. Additionally, decreased growth could occur at very low prey-predator ratios resulting from predator crowding.

Largemouth bass normally grow rapidly during the first few years after impoundment and grow more slowly as reservoirs age (Jackson 1958; Patriarche and Campbell 1958; Chance et al. 1975). This sequence has been attributed to changes in productivity and fish population density. New reservoirs provide ample space for fish populations to expand, but increased interspecific competition depresses individual and population growth rates in subsequent years (Jenkins and Hall 1953). Above-normal productivity during early years results from inputs of nutrients present in the pre-impoundment basin; these nutrients eventually are reduced through water discharge (Kimsey 1958) and fish harvest. Also contributing to high initial production is a food chain based on bacterial degradation of flooded vegetation, a process which is virtually complete in a few years (Murphy 1962).

Growth of largemouth bass was faster in shallow Texas reservoirs. This likely reflects differences in nutrient availability, stratification, and water temperature. The supply of nutrients from bottom sediments is directly proportional to area, but availability is inversely proportional to water volume, thus to mean depth (Hayes and Anthony 1964). For example, if two reservoirs of equal surface area have similar nutrient inputs and one is twice as deep as the other, the deeper lake would contain half as many nutrients per unit volume. Thermal stratification is usually limited or nonexistent in shallow reservoirs, allowing nutrients from bottom sediments to mix and recycle more readily than in deep reservoirs.

Log_{10} of total fish standing stock was the last significant variable to be added to the predictive growth model. The positive relationship cannot be attributed to any specific cause-effect mechanism because measures of total standing stock integrate the effects of numerous factors regulating fish production. Relatively fast growth by largemouth bass in reservoirs having high total fish biomass could be reflective of forage availability or could indicate that production of all fish and growth of young bass were mutually correlated with some other factors.

Three environmental variables that had significant simple correlations with Age_{254} were not retained in the multivariate model. Correlations with reservoir elevation and total alkalinity appear to be an artifact of changing climatic conditions, geological formations, and elevation in a westward direction within Texas. Growth generally declined westward as elevation and alkalinity increased (Fig. 1, Table 2). Effects of alkalinity on production are generally most pronounced at levels below those observed in Texas reservoirs. Moyle (1946) indicated that yield of fish in Minnesota increased with total alkalinity up to 40 mg/liter, but no relation existed thereafter. Boyd (1979) contended that liming Alabama ponds to increase alkalinity would not stimulate fish production when alkalinity was greater than about 20 mg/liter. Hayes and Anthony (1964) observed that fish productivity was directly proportional to alkalinity; their data included several alkalinity values <20 mg/liter.

The relationship between MEI and growth of largemouth bass in Texas was best described by a parabolic model ($R^2 = 0.22$; $P \leq 0.05$). Growth was fastest where MEI's were 100–150 and declined above and below that range. Jenkins (1982) reported that yield of largemouth bass in 294 United States reservoirs (including some data from Texas) was maximum at MEI values of 10–20. This discrepancy is probably attributable to the fact that Texas reservoirs have high total dissolved solids and are relatively shallow (giving high MEI values) in comparison to the national sample. Decreasing yields at high MEI values have been attributed to osmoregulatory complications and to inflation to MEI by high concentration of ions unrelated to fish yields (Ryder et al. 1974; Jenkins 1982; Ryder 1982; Schlesinger and Regier 1982).

Several factors shown to be important determinants of fish growth or yield in other studies were not significantly related to Age_{254} in Texas reservoirs. For some variables, such as length of growing season and water temperature, the range of the observations among reservoirs within the limited geographical area was too narrow to be of predictive value. No significant correlations were found between Age_{254} and surface area or shoreline development. In Oklahoma, growth of largemouth bass was noted to be faster in large impoundments (>200 hectares) than in small impoundments (Jenkins and Hall 1953), but the data were not tested statistically. Reports for southern reservoirs indicate no correlation between black bass (*Micropterus* spp.) yields and surface area or shoreline development (Jenkins 1967; Jenkins and Morais 1971; Campbell et al. 1978), but yield is not always equivalent to production and growth since it is partially determined by fishing effort.

Age_{254} was not significantly correlated with annual water level fluctuation. Despite reports of positive effects of water level fluctuations on growth of largemouth bass (Heman et al. 1969; Zweiacker et al. 1973), no relation between growth rates and the extent of water level fluctuations could be detected for Texas reservoirs. Past studies considered single reservoirs with levels fluctuating over several years. In this study we considered mean annual water level fluctuation in several reservoirs. Drawdowns of equal magnitude in reservoirs with unequal depths affect water volume and fish differently; therefore, vertical fluctuation may not adequately measure the effects of water level on growth when more than

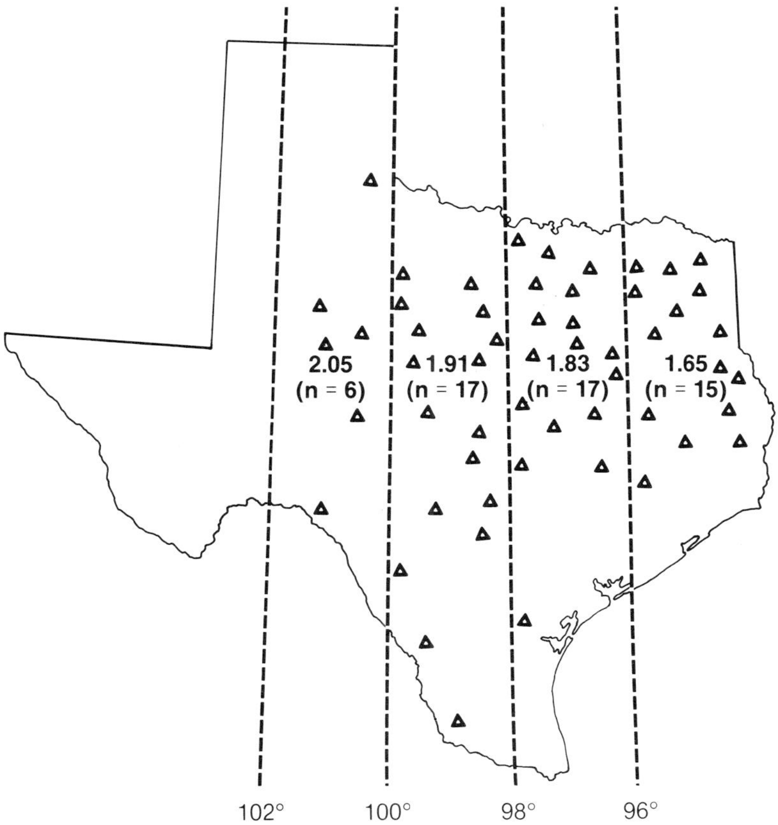

Figure 1. Average age (years) at which largemouth bass reach 254-mm total length across Texas. Dashed lines indicate longitudes, and n is the number of reservoirs evaluated in each zone. Triangles show locations of reservoirs included in the analyses.

one reservoir is being examined. Future analyses might be improved by including a descriptor of annual water level fluctuation in terms of relative water volume change and duration of changes.

Although the positive influence of total dissolved solids (TDS) on fish production has been illustrated by various authors (Ryder 1965; Jenkins 1967; Matuszek 1978), TDS was not significantly related to Age_{254}. Dissolved solids may affect growth by dictating energy expenditure required for maintenance of internal ionic balance or by determining availability of basic ions for organic synthesis. Values observed in Texas reservoirs were within the regulatory capabilities of largemouth bass (Tebo and McCoy 1964; Carver 1967; Bulkley 1975) and probably do not affect growth considerably. At low TDS, nutrients become limiting and can restrict primary productivity and fish growth (Northcote and Larkin 1956; Oglesby 1977).

Management Implications

Through these analyses we have attempted to explore the effects of several environmental factors on growth of largemouth bass. The factors examined are generally considered important to fish populations and routinely recorded in reservoir fishery surveys. Of the factors evaluated, a four-variable equation including reservoir age, mean depth, standing stock, and the prey-predator ratio provided the best model. Efforts to increase largemouth bass production should be concentrated on these four variables.

Conditions conductive to rapid growth present during early years of impoundment may be partially recreated through periodic drawdowns and subsequent inundations. This succession causes a reduction in standing stock followed by rapid expansion. Such cycles occur naturally in many west Texas reservoirs, where prolonged droughts followed by abundant precipitation produce very rapid growth rates.

Losses of nutrients from a reservoir due to hypolimnetic withdrawals during periods of thermal stratification, a common occurrence as mean depth increases, may be offset by withdrawing water from above the thermocline (Murphy 1962). This practice would thin surface layers by stirring the nutrient-rich hypolimnion and mixing it with the epilimnion, thus boosting basic productivity. Mixing could also contribute to cooler temperatures in littoral areas, often avoided by largemouth bass during part of the summer. However, Jenkins (1975) found that largemouth bass biomass was increased where deeper reservoir outlets were used. Further investigations of effects of outlet depth on fish populations is needed to understand its value as a management tool.

Fertilization of large reservoirs to increase basic productivity and standing stock is generally precluded by cost and inability to control water exchange rate. However, fertilization could be feasible in smaller reservoirs. Partial fertilization of Lake Welsh, a 550-hectare east Texas reservoir with minimum exchange rate, immediately increased production and condition of largemouth bass at a relatively low cost (Toole 1982).

Prey-predator ratio is lowered by reducing the standing stock of intermediate and large prey and by raising standing stock of largemouth bass (Noble 1981). Predator standing stock could be raised by restricting harvest, improving habitat, or redirecting biomass of intermediate and large forage into biomass of small, readily-consumed prey. Increased predator diversity could provide additional predation on intermediate and large forage, thus creating a dynamic prey community with relatively few adults and many small individuals. Under those conditions, largemouth bass production is expected to increase, compensating for a possible decrease in their biomass caused by added interspecific competition.

Many factors not considered in this study undoubtedly accounted for some of the unexplained variation in the predictive model for Age_{254}. The significance of those factors warrants further study to lead to a better understanding of factors limiting growth of largemouth bass. Such investigations should emphasize evaluation of variables that are readily measured and offer some potential for manipulation by fishery managers.

References

BOYD, C. E. 1979. Water quality in warmwater fish ponds. Auburn University Agricultural Experiment Station, Auburn, Alabama, USA.

BULKLEY, R. V. 1975. Chemical and physical effects on the centrarchid basses. Pages 286–294 *in* R. H. Stroud and H. Clepper, editors. Black bass biology and management. Sport Fishing Institute, Washington, D.C., USA.

CAMPBELL, W. J., E. J. HAYES, W. R. CHAPMAN, AND W. SEAWELL. 1978. Angling pressure and sport fish harvest in the predator-stocking evaluation reservoirs. Proceedings of the Annual Conference of the Southeastern Association of Fish and Wildlife Agencies 30:114–119.

CARVER, D. C. 1967. Distribution and abundance of the centrarchids in the recent delta of the Mississippi River. Proceedings of the Annual Conference of the Southeastern Association of Game and Fish Commissioners 20:390–404.

CHANCE, C. J., A. O. SMITH, J. A. HOLBROOK II, AND R. B. FITZ. 1975. Norris Reservoir: a case history in fish management. Pages 399–407 *in* R. H. Stroud and H. Clepper, editors. Black bass biology and management. Sport Fishing Institute, Washington, D.C., USA.

DIXON, W. J., editor. 1975. BMDP (Biomedical Computer Programs). University of California Press, Los Angeles, California, USA.

EVERHART, W. H., A. W. EIPPER, AND W. D. YOUNGS. 1975. Principles of fishery science. Cornell University Press, Ithaca, New York, USA.

HAYES, F. R., AND E. H. ANTHONY. 1964. Productive capacity of North American lakes as related to the quantity and the trophic level of fish, the lake dimensions, and the water chemistry. Transactions of the American Fisheries Society 93:53–57.

HEMAN, L. H., R. S. CAMPBELL, AND L. C. REDMOND. 1969. Manipulation of fish populations through reservoir drawdown. Transactions of the American Fisheries Society 98:293–304.

JACKSON, S. W., JR. 1958. Comparison of the age and growth of four fishes from Lower and Upper Spavinaw Lakes, Oklahoma. Proceedings of the Annual Conference of the Southeastern Association of Game and Fish Commissioners 11:232–249.

JENKINS, R. M. 1967. The influence of some environmental factors on standing crop and harvest of fishes in U.S. reservoirs. Pages 298–321 *in* Reservoir fishery resources symposium. Southern Division American Fisheries Society, Bethesda, Maryland, USA.

JENKINS, R. M. 1975. Black bass crops and species associations in reservoirs. Pages 114–124 *in* R. H. Stroud and H. Clepper, editors. Black bass biology and management. Sport Fishing Institute, Washington, D.C., USA.

JENKINS, R. M. 1982. The morphoedaphic index and reservoir fish production. Transactions of the American Fisheries Society 111:113–140.

JENKINS, R. M., AND G. E. HALL. 1953. The influence of size, age, and condition of waters upon the growth of largemouth bass in Oklahoma. Report Number 30, Oklahoma Fisheries Research Laboratory, Norman, Oklahoma, USA.

JENKINS, R. M., AND D. I. MORAIS. 1971. Reservoir sport fishing effort and harvest in relation to environmental variables. Pages 371–384 *in* G. E. Hall, editor. Reservoir fisheries and limnology. Special Publication 8, American Fisheries Society, Bethesda, Maryland, USA.

KIMSEY, J. B. 1958. Fisheries problems in impounded waters of California and the lower Colorado River. Transactions of the American Fisheries Society 87:319–332.

MATUSZEK, J. E. 1978. Empirical predictions of fish yields of large North American lakes. Transactions of the American Fisheries Society 107:385–394.

MOYLE, J. B. 1946. Some indices of lake productivity. Transactions of the American Fisheries Society 76:322–334.

MURPHY, G. I. 1962. Effect of mixing depth and turbidity on the productivity of fresh-water impoundments. Transactions of the American Fisheries Society 91:69–76.

NOBLE, R. L. 1981. Management of forage fishes in impoundments of the southern United States. Transactions of the American Fisheries Society 110:738–750.

NORTHCOTE, T. G., AND P. A. LARKIN. 1956. Indices of productivity in British Columbia lakes. Journal of the Fisheries Research Board of Canada 13: 515–540.

OGLESBY, R. T. 1977. Phytoplankton summer standing crop and annual productivity as functions of phosphorous loading and various physical factors. Journal of the Fisheries Research Board of Canada 34:2255–2270.

PATRIARCHE, M. H., AND R. S. CAMPBELL. 1958. The development of the fish population in a new flood-control reservoir in Missouri, 1948–1954. Transactions of the American Fisheries Society 87:240–258.

ROYCE, W. F. 1972. Introduction to fishery sciences. Academic Press, New York, USA.

RYDER, R. A. 1965. A method for estimating the potential fish production of north-temperate lakes. Transactions of the American Fisheries Society 94:214–218.

RYDER, R. A. 1982. The morphoedaphic index—use, abuse, and fundamental concepts. Transactions of the American Fisheries Society 111:154–164.

RYDER, R. A., S. R. KERR, K. H. LOFTUS, AND H. A. REIGER. 1974. The morphoedaphic index, a fish estimator—review and evaluation. Journal of the Fisheries Research Board of Canada 31:663–688.

SCHLESINGER, D. A., AND H. A. REGIER. 1982. Climatic and morphoedaphic indices of fish yields from natural lakes. Transactions of the American Fisheries Society 111:141–150.

TEBO, L. B., JR., AND E. G. MCCOY. 1964. Effect of sea-water concentration on the reproduction and survival of largemouth bass and bluegills. The Progressive Fish-Culturist 26:99–106.

TEXAS WATER DEVELOPMENT BOARD. 1973. Selected water-quality records for Texas surface waters, 1971 water year. Report 176, Texas Water Development Board, Austin, Texas, USA.

TOOLE, J. E. 1982. Existing reservoir and stream management recommendations: Lake Welsh. Federal Aid Project F-30-R-7, Texas Parks and Wildlife Department, Austin, Texas, USA.

ZWEIACKER, P. L., R. C. SUMMERFELT, AND J. N. JOHNSON. 1973. Largemouth bass growth in relationship to annual variations in mean pool elevations in Lake Carl Blackwell, Oklahoma. Proceedings of the Annual Conference of the Southeastern Association of Game and Fish Commissioners 26:530–540.

Pages 122–136 *in* G. E. Hall and M. J. Van Den Avyle, editors.
Reservoir Fisheries Management: Strategies for the 80's.
Reservoir Committee, Southern Division American Fisheries Society, Bethesda, Maryland. 1986.

Spatial Heterogeneity in Fish Parameters within a Reservoir

J. Robert Siler, William J. Foris, and Michael C. McInerny

Duke Power Company
Production Environmental Services
Route 4, Box 531
Huntersville, North Carolina 28078

Abstract

Various components of the Lake Norman fish community exhibited spatial heterogeneity. Harvest of sport fish exhibited a longitudinal trend within the reservoir as did total phosphorus and chlorophyll *a* concentration. Harvest of largemouth bass (*Micropterus salmoides*), crappie (*Pomoxis* spp.), striped bass (*Morone saxatilis*), and white bass (*Morone chrysops*) was lowest at the most downlake area (3 kg/hectare), increased steadily to the uplake riverine area (51 kg/hectare), and highest at the discharge of a steam-electric station (316 kg/hectare). Fishing pressure varied similarly. Threadfin shad (*Dorosoma petenense*) standing stocks in October exhibited a longitudinal gradient that was similar to the gradient observed for phytoplankton standing crop.

Interreservoir predictive models did not account for the intrareservoir variability in fish parameters, most likely because the models failed to consider the heterogeneity of habitat and physico-chemical variables within the reservoir. The existence of heterogeneity in fish parameters indicates that management decisions based on data collected from one or two stations within the reservoir could be erroneous and that managing the reservoir as a biological entity may be ineffective. An understanding of gradients within reservoirs would help managers collect representative samples with improved precision.

Spatial heterogeneity in fish standing stock, harvest, and mortality in Lake Norman indicates that the potential of the fishery is not realized because of species-specific habitat and physical constraints. Density-independent mortality appeared to regulate the biomass of striped bass and threadfin shad in the reservoir because of their relatively narrow thermal requirements. Removal of these constraints would allow the system to become regulated by density-dependent phenomena, thus improving the efficiency and stability of the system.

Variability in fishery parameters such as harvest and standing stock among reservoirs and natural lakes has received considerable attention during the past two decades (Rawson 1960; Ryder 1965; Jenkins 1967; Jenkins and Morais 1971; Oglesby 1977a; Jenkins 1982). These contributions have implied that a relationship exists between nutrient dynamics and fisheries (Ryder 1982) that is useful for understanding and predicting differences in fish yields among bodies of water. However, consideration has rarely been given to spatial variability in fishery parameters within a body of water. Intrareservoir heterogeneity in fish parameters might be expected based on nutrient gradients documented within reservoirs (Gloss et al. 1980; Paulson and Baker 1980; Kennedy et al. 1982; Thornton et al. 1982). Also, reservoirs are usually physically diverse, and spatial heterogeneity in fishery parameters is expected due to species-specific behavior and habitat preferences.

Research conducted from 1974 through 1982 on Lake Norman, North Carolina, has indicated spatial variations in fishery parameters including standing stock and angler harvest, success, and pressure as well as other biological, chemical, and physical variables. The objectives of this paper are to document the spatial heterogeneity in selected fishery parameters, assess relationships between these parameters and morphometric, physico-chemical, and biological variables, and discuss the management implications of these relationships.

Study Site

Lake Norman, located on the Catawba River in the Piedmont of North Carolina, was impounded in 1963 to provide water for hydroelectric power and cooling of steam-electric stations. It is a warm-monomictic reservoir with an oligo-mesotrophic status based on primary production and phytoplankton biomass (Rodriguez

Table 1. Morphometric, physical, and chemical characteristics of Lake Norman, North Carolina.

Characteristic	Description
Drainage area	771 km² (excludes drainage area of upstream reservoir)
Surface area	13,156 hectares (full pool)
Mean depth	10.2 m (full pool)
Maximum depth	36.6 m (full pool)
Outlet depth (midline depth)	5 m (full pool, controlled by height of skimmer weir)
Thermocline depth (1 Aug)	11 m (typical), 8 m (using definition in Jenkins [1967])
Elevation	231.6 meters mean sea level (full pool)
Mean annual fluctuation	2.6 m
Storage ratio	0.57
Shoreline development	21.5 (with islands)
Dissolved solids	30 mg/liter
Chemical type-inflow[a] (μ equivalents/liter)	HCO_3-240, Na-155, Ca-139, Mg-109, K-39
Sediment load[b]	<280 mg/liter
Growing season[b]	200 days

[a] Perkins and Whisenant (1982).
[b] Jenkins (1967).

1982a). Physical and chemical characteristics of the reservoir are summarized in Table 1.

Several power generation facilities (Fig. 1) influence hydrological, thermal, and biological components of the reservoir. Lookout Shoals Hydroelectric Station on the Catawba River discharges water from 11 to 17 m below full pool from Lookout Shoals Reservoir and provides the primary input of water into Lake Norman. Marshall Steam Station, a 1,900-megawatt coal-fired steam electric station, began operation in 1965. The intake design of Marshall is atypical of most steam stations in North Carolina, in that a skimmer wall built across the mouth of the intake cove permits only bottom water to enter the cooling system of the plant. Water temperatures in the Marshall Steam Station discharge can be as high as 11 C above ambient surface temperatures during winter. During summer, discharge temperatures are similar to ambient surface water temperatures, but the discharge has chemical characteristics of the hypolimnetic water of Lake Norman (Hogan and Adair 1982). McGuire Nuclear Station (2,360 megawatts) was under construction during most of the study period but began circulating unheated water in 1978 and heated water during periods of 1981 and 1982. Cowans Ford Hydroelectric Station (360 megawatts) is housed in the dam that forms Lake Norman and discharges epilimnetic water due to a submerged weir located in the dam forebay which retains water below 11 m (at full pool).

Methods

Lake Norman was stratified into nine sampling zones (Fig. 1). Boundaries for Zones 1 and 4 were chosen according to areas affected or projected to be affected by discharges of McGuire and Marshall Stations, respectively. Zone 7 was the intake cove for Marshall Steam Station and was not sampled in this study because it was inaccessible by boat from the reservoir. Zones 8 and 9 were the discharge canals of Marshall and McGuire Stations, respectively. Zones 2, 3, and 5 were defined according to main drainage areas of the reservoir. Zone 6 included the tailrace of Lookout Shoals Dam and the riverine portion of the reservoir. Area, volume, mean depth, shoreline length, and drainage area were determined for each zone from preimpoundment topographic surveys (Table 2).

Fishing pressure (angler-hours per hectare) and distribution of sport fishermen were estimated by counting boat and bank/pier fishermen during randomly scheduled airplane or helicopter flights (greater than 60 per year, allocated seasonally) from December 1977 through November 1978 and December 1981 through November 1982. Fishing success (kilograms and number of fish per angler-hour) was estimated by interviewing fishermen by boat (192 days annually) during the above time periods. Harvest by species was estimated as the product of success and pressure and excluded night fishing. Fishing pressure, success, and catch composition was estimated seasonally for each survey zone.

Population and standing stock estimates of threadfin shad (*Dorosoma petenense*) were made in October 1979 through 1982, December 1979 and 1980, February 1980 and 1981, and April 1979 through 1982. All samples were taken at night with a midwater Tucker trawl (Siler 1983). Population estimates, based on approximately 90 samples per sampling period, were made using the direct enumeration methodology (Regier and Robson 1967), whereby randomly selected segments of the reservoir were sampled, and catch rates were expanded to determine the population size of all ages of threadfin shad in each of the sampling zones (Siler 1985). Additionally, densities of clupeid larvae were estimated weekly

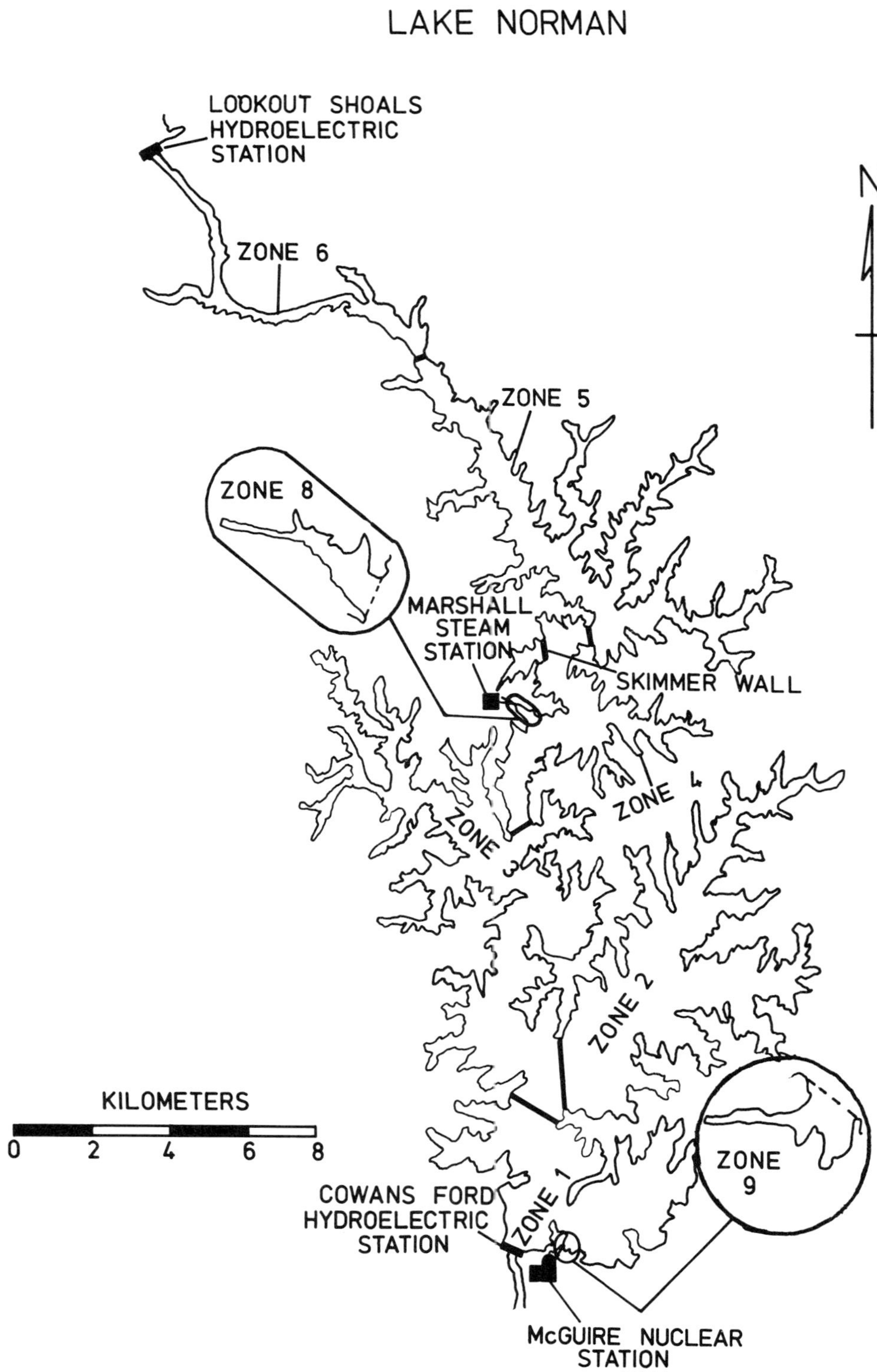

Figure 1. Lake Norman, North Carolina, with sampling zones and electric generating stations.

Table 2. Morphometric characteristics of sampling zones on Lake Norman, North Carolina.

Zone	Area (hectares)	Volume (m^3)	Mean depth (m)	Shoreline length (km)	Shoreline development	Cumulative drainage area (km^2)[a]	Immediate drainage area to volume ratio (m^2/m^3)[b]
1	2,261	2.58×10^8	11.4	86.6	5.1	765.8	0.09
2	3,300	3.33×10^8	10.1	133.4	6.1	742.7	0.21
3	3,555	4.02×10^8	11.3	243.6	11.5	674.1	0.37
4	1,216	1.42×10^8	11.7	124.5	10.1	524.1	0.14
5	2,125	1.93×10^8	8.7	197.4	12.1	503.9	1.08
6	341	8.18×10^6	2.4	51.5	7.9	304.3	37.18
8	24	2.26×10^6	9.4	3.8	2.1	—	—
9	22	2.22×10^6	10.1	3.5	2.2	—	—

[a] Excludes drainage area of an upstream reservoir.
[b] Immediate drainage excludes drainage area of upstream reservoir or zone.

during March through August 1975 from nine locations, representing five of the sampling zones.

Mark-recapture population estimates using single-census techniques (Ricker 1975) and abundance estimates at randomly selected segments of shoreline were conducted for largemouth bass (*Micropterus salmoides*) at Zones 1 through 5 during April 1982. Sampling was conducted at night with electrofishing gear. Population estimates were limited to largemouth bass longer than 200 mm. Standing stock estimates of largemouth bass and other species were obtained from cove rotenone sampling during August 1978 through 1981 at six 1.2-hectare locations in Zones 1, 3, 5, 8, and 9.

Total phosphorus, orthophosphate-phosphorus, specific conductance, and turbidity data were collected monthly in 1982 and 1983 at 19 fixed locations representing Zones 1 through 6. Additionally, 50 total phosphorus and turbidity samples were randomly collected reservoir-wide during November 1982. All samples were replicated and taken at the surface. Analytical methods were those approved by U.S. Environmental Protection Agency (1974). Chlorophyll *a* concentrations were determined (Strickland and Parsons 1972) biweekly during 1979 at nine sampling locations representing Zones 1 through 6. Water was collected at depths of 0, 2.5, and 5 m and mixed to provide one composite sample representative of the epilimnion. Samples were taken monthly in 1982 at 10 locations representing Zones 1 through 5. Water samples were collected at the surface, the bottom of the euphotic zone, and at one-half the depth of the euphotic zone. Replicate chlorophyll determinations were made for a composite of these samples. Surface chlorophyll *a* concentrations were determined at 90 randomly selected locations throughout the reservoir in parallel with the April and October 1982 threadfin shad population samples. Fishery, total phosphorus, and chlorophyll *a* data were compared among zones using analysis of variance.

RESULTS

Fishing Pressure

Fishing pressure exhibited considerable spatial heterogeneity in Lake Norman (Table 3). Relatively large increases occurred longitudinally from downlake (Zones 1 and 2) to uplake (Zones 5 and 6). This pattern was similar for boat and bank fishing pressure and varied little among seasons and years. The major exception to this trend was the high fishing pressure at the discharge canals from steam-electric stations (Zones 8 and 9), especially during winter and spring.

Sport Fish Harvest and Abundance

Harvest of largemouth bass (Table 3) gradually increased in the uplake direction, with the exception of the small discharge zones (8 and 9). The discharge zones had high catches per unit area but accounted for only a small portion of the total reservoir harvest because of their small area. Densities determined from population estimates of largemouth bass and catch rates from randomly chosen electrofishing samples (Table 4) showed the same spatial characteristics as harvest data. Electrofishing catch rates were significantly ($P < 0.0001$) higher in Zone 5 than other zones. Zones 3 and 4 were statistically similar, and Zone 4 had a higher catch rate than Zones 2 and 1. Cove rotenone data from August were variable and indicated no differences among zones except at Marshall discharge (Zone 8), where

Table 3. Creel survey estimates (95% confidence limits in parentheses) by sampling zones, Lake Norman, North Carolina, 1982.

Zone	Annual fishing pressure (angler-hours/hectare)	Fishing success		Annual harvest (kg/hectare)					
		Number/angler-hour	kg/angler-hour	Largemouth bass	Crappies	Striped bass	White bass	Ictalurids	All species
1	21.9 (±4.9)	0.47 (±0.87)	0.13 (±0.18)	1.26 (±0.52)	1.39 (±0.77)	0.12 (±0.15)	0	0.05 (±0.05)	2.93 (±1.02)
2	23.0 (±3.4)	0.76 (±1.24)	0.19 (±0.18)	1.22 (±0.63)	2.10 (±0.95)	0.69 (±0.45)	0	0.13 (±0.18)	4.46 (±1.33)
3	42.8 (±5.4)	0.88 (±0.99)	0.19 (±0.18)	2.06 (±0.89)	5.08 (±1.64)	0.19 (±0.12)	0.18 (±0.24)	0.19 (±0.21)	7.97 (±2.09)
4	69.6 (±11.3)	0.70 (±0.82)	0.13 (±0.11)	2.87 (±1.05)	4.26 (±1.99)	0.98 (±0.84)	0.34 (±0.45)	0.18 (±0.14)	9.16 (±2.87)
5	114.4 (±12.5)	1.29 (±0.85)	0.25 (±0.18)	3.09 (±1.11)	19.48 (±5.75)	1.08 (±1.69)	0.47 (±0.73)	1.85 (±1.18)	28.30 (±5.87)
6	275.1 (±30.5)	0.53 (±0.57)	0.21 (±0.65)	5.77 (±2.71)	7.48 (±3.41)	19.66 (±8.94)	17.48 (±9.97)	3.35 (±2.31)	57.97 (±14.32)
8	3,392.5 (±52.2)	0.38 (±0.30)	0.24 (±0.22)	24.17 (±18.6)	14.83 (±22.8)	192.49 (±115.9)	84.71 (±50.7)	56.8 (±33.1)	829.04 (±388.82)
9	846.1 (±29.8)	0.37 (±0.50)	0.21 (±0.24)	19.24 (±19.1)	3.84 (±5.06)	126.90 (±102.0)	7.51 (±10.9)	12.7 (±12.3)	179.34 (±103.27)

standing stocks averaged 16 kg/hectare and were almost three times the stocks in other zones (Table 5).

Average electrofishing catch rate per distance of shoreline correlated well with the annual harvest rate for each zone ($r = 0.91$, $P < 0.05$, $N = 5$) and spring harvest rates ($r = 0.90$, $P < 0.05$, $N = 5$) in 1982. Density estimates (number/hectare) in each zone (Table 4) also were correlated with harvest estimates in spring ($r = 0.97$, $P < 0.01$, $N = 5$) and with annual rates ($r = 0.94$, $P < 0.01$, $N = 5$).

Crappie (*Pomoxis* spp.) harvest varied considerably among zones, and was highest in Zone 5 in 1979 and 1982 (Table 3). Crappie harvest from downlake Zones 1 and 2 was less than 50% of the harvest in other zones. Standing stock estimates of crappie from cove rotenone samples were highest in Zone 8 (14 kg/hectare), followed by Zones 3 and 5 which averaged 6 kg/hectare (Table 5). Crappie standing stocks averaged less than 1 kg/hectare in Zone 1.

White bass (*Morone chrysops*) and striped bass (*M. saxatilis*) harvests (kg/hectare) were higher at the riverine and discharge areas than at the large downlake zones (1 through 5) (Table 3). Both species make spawning migrations to the inflows at Zones 6, 8, and 9. These areas comprised only 3% of the total area of the reservoir but accounted for 64% and 76% of the total 1982 harvest of striped bass and white bass, respectively.

Annual harvest of ictalurids was greatest in the discharge areas followed by Zones 6 and 5 (Table 3). Mean standing stocks of ictalurids were greatest in Zones 5 (15 kg/hectare) and 8 (12 kg/hectare) and ranged from 4 to 6 kg/hectare in Zones 1, 3, and 9 (Table 5). Channel catfish (*Ictalurus punctatus*), flathead catfish (*Pylodictis olivaris*), and white catfish (*I. catus*) comprised the major portion of the uplake ictalurid stock. Bullheads (*Ictalurus* spp.) were absent from the most uplake cove, but along with the white catfish dominated the downlake ictalurid stock. Flathead catfish were absent and channel catfish were uncommon downlake.

Prey Species Abundance

Threadfin shad, sunfish (*Lepomis* spp.), and yellow perch (*Perca flavescens*) are the major prey fishes in Lake Norman (Siler et al. 1982). Sunfish and yellow perch were insignificant in the sportfish harvest because of their small average size.

Table 4. Electrofishing catch rates, densities, and population estimates (95% confidence intervals in parentheses) for largemouth bass in Lake Norman, North Carolina, April 1982.

Zone	Electrofishing catch rate (number/100 m)	Population estimates	
		Number/hectare	Number in zone
1	2.2 (±0.71)	4.0 (2.0–8.6)	8,253 (4,097–18,053)
2	2.9 (±0.96)	7.7 (3.6–17.7)	23,082 (10,905–53,265)
3	4.5 (±1.65)	11.1 (5.5–24.2)	36,073 (17,909–78,910)
4	6.6 (±2.01)	13.8 (7.6–27.5)	15,826 (8,743–31,651)
5	11.3 (±2.27)	14.9 (8.4–28.7)	28,982 (16,433–55,930)

Sunfish standing stocks averaged 25 kg/hectare in Zone 5 as compared to 10 to 17 kg/hectare in Zones 1, 3, 8, and 9 (Table 5). Yellow perch standing stocks were highest in Zones 9 (17 kg/hectare) and 8 (8 kg/hectare) and ranged from 3 to 5 kg/hectare in Zones 1, 3, and 5.

Gizzard shad (*Dorosoma cepedianum*) were usually too large to be consumed by sportfish based on predator-prey equations of Jenkins and Morais (1978). Standing stocks in Zones 8 and 9 averaged 77 and 87 kg/hectare, respectively; mean stocks in Zones 1, 3, and 5 ranged from 30 to 33 kg/hectare (Table 5). Juvenile gizzard shad were not abundant in samples from Lake Norman but were always more numerous in uplake areas (Table 6).

Threadfin shad distribution in Lake Norman was heterogeneous. Larval clupeid (primarily threadfin shad) abundance was considerably higher uplake except for Zone 6 (Table 6). During October 1979, 1980, and 1981, threadfin shad standing stocks were highest in Zone 5, but not in October 1982. During December 1979 and 1980 standing stocks were lower in Zones 1 and 2 than in Zones 3 and 4, and Zone 5 had the highest standing stock (Table 6), but differences were statistically significant ($P < 0.01$) only between uplake Zones 3, 4, and 5 and downlake Zones 1 and 2 (Fig. 2). In February of 1980 and 1981, threadfin shad were most abundant in areas warmed by the thermal plume of Marshall Station (Zones 4 and 8) (Fig. 3); few threadfin shad were collected at any location with water temperatures below 7 C. Spatial distribution patterns in April were similar to those in February except the shad were more dispersed around the thermal plume.

The population size of threadfin shad in October appeared to be independent of the number of spawners in the previous April; for example, a low population size of spawners in April 1981 produced a high population size of recruits in October 1981 (Table 6). Numbers of new recruits in October per adult from the preceding April ranged from 10 to 181 recruits/spawner from 1979 through 1982. Additionally, an index of the number of clupeid larvae produced in Zone 1 did not coincide with the estimated population size of spawners in Zone 1 or the entire reservoir for 1979, 1980, and 1981.

Limnological Characteristics

Lake Norman's annual surface temperatures range from 1.6 to 33 C (excluding power plant discharge zones) with minimum temperatures of 1.6 to 8 C occurring in February and maximum temperatures of 29 to 33 C typically occurring in July and August. Thermal stratification generally begins in mid-April with a weak thermocline being established at 10 to 12 m. Usually, this thermal barrier is difficult to detect, and dissolved oxygen data more clearly define the limits of the epilimnion and hypolimnion in Lake Norman. Dissolved oxygen generally disappears below 11 m by August. Cooling and convective mixing begins in early fall (September–October) with overturn usually complete by mid-November. Discharge zones are generally 8 C warmer than the open reservoir during February and 2 C warmer in August.

Total phosphorus concentrations in Lake Norman range from undetectable (<5 mg/m^3) to 150 mg/m^3. The highest concentrations, those exceeding 30 mg/m^3, are found during the winter-spring runoff period, and the lowest concentrations are observed during the summer. Longitudinal gradients of total phosphorus are most pronounced during the spring runoff period. Spatial distribution of total phosphorus in the reservoir is closely associated with turbidity ($r = 0.90$, $P < 0.0001$, N = 59). In general, total phosphorus concentrations are two times higher in uplake than downlake regions (Table 7). Total phosphorus data from summer 1982 and spring

Table 5. Mean standing crop estimates (95% confidence intervals in parentheses) for selected species determined from cove rotenone sampling in Lake Norman, North Carolina, 1978–1981.

Zone	Mean standing crop (kg/hectare)							
	Gizzard shad	Common carp	Ictalurids	Largemouth bass	Crappie	Sunfish	Yellow perch	All species
1	33.6 (±9.0)	38.8 (±13.4)	5.5 (±2.6)	5.1 (±2.1)	<0.1 (±<0.1)	15.9 (±3.4)	3.2 (±2.0)	107.4 (±25.6)
2	33.7 (±11.3)	27.1 (±11.1)	5.8 (±3.8)	5.7 (±1.1)	6.0 (±5.8)	15.1 (±4.1)	4.6 (±2.6)	137.7 (±60.7)
3	30.2 (±15.4)	48.2 (±17.5)	15.2 (±6.7)	6.3 (±1.9)	6.4 (±5.6)	25.1 (±12.9)	4.2 (±0.8)	174.5 (±77.9)
4	76.8 (±23.8)	41.2 (±15.9)	12.0 (±6.0)	16.0 (±5.3)	13.7 (±7.7)	16.8 (±8.4)	8.3 (±4.2)	190.1 (±44.0)
5	87.4 (±43.8)	17.9 (±8.5)	4.0 (±2.9)	2.5 (±0.4)	1.4 (±1.0)	10.6 (±5.1)	17.0 (±14.3)	157.0 (±23.7)

1983 were more variable in uplake than downlake areas due to variable inputs from rainfall and runoff. Samples collected in November 1982 illustrate the total phosphorus gradient during a period of low runoff. Zone 6 had significantly ($P < 0.0001$) higher total phosphorus concentrations than other areas; Zones 4 and 5 were similar but had significantly higher phosphorus concentrations than Zones 1, 2, and 3.

Longitudinal gradients in chlorophyll *a* concentrations were also observed. Uplake regions typically had 50 to 100% higher concentrations than the downlake areas except in October 1982, when conditions were more uniform (Table 7). Concentrations were usually lower in Zone 6 than other zones. Reservoir-wide random samples collected in April and October 1982 revealed significant spatial variability during both months (Table 7). In April 1982, phytoplankton was more abundant in uplake areas. Chlorophyll *a* concentrations were highest in Zone 5, intermediate in Zones 3, 4, and 6, and lowest in Zones 1, 2, 8, and 9. Significant differences ($P < 0.001$) observed among zones in October 1982 did not reveal any gradient along the reservoir; however, these data were atypical because of an extensive *Microcystis aeruginosa* bloom in the uplake areas.

Interrelationships

Components of the Lake Norman fish community exhibited spatial heterogeneity. Possible explanations for the heterogeneity include trophic interactions, and species-specific habitat, behavioral, or physical preferences.

Distribution of fishing pressure in the reservoir might account for the spatial heterogeneity in sport fish harvest. Harvest was highly correlated with angler effort among zones in 1978 ($r = 0.97$, $P < 0.01$, $N = 6$) and 1982 ($r = 0.999$, $P < 0.0001$, $N = 8$). However, based on "fished for" success data (kg/hour) for largemouth bass, which increased slightly uplake, high correlations between fishing pressure and largemouth bass abundance estimates from electrofishing surveys and population estimates, and observations of fisherman distribution relative to striped bass concentrations, the spatial distribution of fishing pressure in Lake Norman seems to be a function of sportfish distribution patterns. Therefore, fishermen maximize their success by distributing their effort proportional to the abundance of sportfish. Campbell et al. (1978) reported a relatively stable catch rate and a high correlation between angler effort and harvest among a sample of 24 reservoirs; they suggested that anglers shifted their effort to other waters when a certain minimum success rate occurred. We assume that the spatial and temporal heterogeneity in harvest reflects actual heterogeneity in sportfish abundance in Lake Norman.

Various indices of productivity based on physical, chemical, and biological variables have been successful for predicting heterogeneity in fish harvest and standing stocks among reservoirs and lakes (Rawson 1960; Ryder 1965; Jenkins 1967; Oglesby 1977a; Jenkins 1982). The morphoedaphic index (MEI = total dissolved solids [TDS] divided by mean depth) is the most notable of these and has been widely applied. The basic concept of the MEI is that nutrients transported by energy influence fish production (Ryder 1982). Jenkins' (1982) MEI for reservoirs accurately predicted the Lake Norman harvest, but it failed to accurately reflect variation within the reservoir (Fig. 4). Possible explanations for this disparity are that intrareservoir habitat differences account for the variability of harvest, that TDS varies little within the reservoir and is not a correlate of the limiting nutrient, or that the index was intended to predict variation on a much broader scale of the MEI. As Oglesby (1977b) reported, numerous exceptions occur where TDS is unrelated to the availability of a limiting nutrient. Ryder (1982) suggested that in cases where phosphorus is the limiting ion TDS should be replaced with phosphorus.

A longitudinal gradient in chlorophyll *a* occurs in Lake Norman and is related to a total phosphorus gradient except in the riverine area. Phytoplankton standing crops are usually low in riverine systems because of continual washout (Oglesby 1977b) and high turbidity that limits light penetration (Kennedy et al. 1982). Rodriguez (1982b) reported that the phytoplankton community of Lake Norman was limited by phosphorus availability during the summer-stratified period.

The majority of the phosphorus in Lake Norman appears to be physically and/or chemically associated with clays, and total phosphorus was closely correlated with turbidity both temporally and vertically in the water column (Rodriguez 1982b). We observed that the relationship also occurs spatially within the reservoir. Internal phosphorus loading via hypolimnetic release in Lake Norman is probably not important relative to external loading; orthophosphate-phosphorus is usually undetectable in the bottom waters (Rodriguez 1982b). These characteristics are common in reservoirs (Serruya et al. 1974; Hannan and Broz 1976; Jones and Bachmann 1978; Canfield and Bachmann 1981) and emphasize the importance of external nutrient sources and the resuspension of sediments in the shallow areas to the productivity of the reservoir. Downlake areas of Lake Norman are somewhat less productive than uplake areas because of biological, physical, and chemical removal of nutrients in the uplake waters.

The importance of the phosphorus and chlorophyll *a* gradients to Lake Norman fish poulations was most apparent for threadfin shad. Threadfin shad abundance in October and December was generally correlated with the phytoplankton standing crop gradient, except in October 1982. Threadfin shad standing stocks in October and December 1979 correlated with chlorophyll *a* concentrations ($r = 0.95, P < 0.01$, N = 6; and $r = 0.93$, $P < 0.01$, N = 6, respectively). Threadfin shad larvae in Lake Norman feed on zooplankton, but juveniles and adults feed primarily on phytoplankton (W. T. Horton, Duke Power Company, Huntersville, North Carolina, USA [personal communication]), suggesting a direct association between threadfin shad abundance and algal standing crop. Furthermore, an uncharacteristic bloom of *Microcystis aeruginosa*—sometimes toxic to fish—in uplake areas in October 1982, coinciding with a similarily uncharacteristic decline in threadfin shad abundance, suggests an association between threadfin shad abundance and algal species composition.

Trophic interactions associated with the nutrient gradient also appeared related to the abundance and distribution of other fish species in Lake Norman, but the relative importance of these interactions was difficult to separate from species-specific behavior and habitat preferences. Because of the chemical and morphometric nature of dendritic reservoirs with long retention times like Lake Norman, mean depth, shoreline development, and size of the immediate drainage area can be correlates for habitat preferences for littoral species and total phosphorus. Total harvest, largemouth bass harvest and abundance, crappie harvest, and ictalurid standing stock and harvest correlated significantly with total phosphorus concentration,

Table 6. Abundance estimates from trawling for clupeids by sampling zones, Lake Norman, North Carolina, 1979-1982.

Zone	Average clupeid larval densities (number/ 1,000 m³)	Total threadfin shad standing stock (kg/hectare)					
		October				December	
		1979	1980	1981	1982	1979	1981
1	811	13.0	19.5	20.4	10.0	11.3	9.1
2	—	9.2	22.5	22.7	11.8	16.1	11.7
3	847	12.5	22.8	18.5	14.7	20.6	18.7
4	2,813	25.2	21.8	22.1	7.4	21.6	16.0
5	2,229	33.2	51.9	36.3	13.9	36.4	34.7
6	25	1.5	0.6	0.7	0.1	<0.1	0.1
8	—	2.9	14.8	15.3	0.3	13.4	0.6
9	—	0.1	0.7	23.2	2.6	12.6	0.7
Mean standing stock (area weighted)		16.5	26.3	23.3	12.0	21.2	17.6
95% Confidence interval		(±3.5)	(±5.8)	(±4.9)	(±3.2)	(±5.5)	(±3.1)
Total population size (millions)		304.4	589.6	560.9	246.6	187.2	194.8
95% Confidence interval		(±57.3)	(±165.6)	(±111.2)	(±68.5)	(±41.5)	(±50.6);

chlorophyll *a* concentration, and the physical features listed above. These correlations may represent associations between sport fish and preferred spawning habitat, as interactions were particularly pronounced during spawning periods of these species; they may also represent interactions between sport fish and available habitat for these littoral species because the uplake areas were shallower and more dendritic, and spatial distribution patterns for these species remained similar for spring, summer, and fall. The correlations also may reflect interactions between sport fish and abundant prey, which are related to the nutrient gradient.

Spatial heterogeneity of some species was influenced by water temperature. Water temperature controlled the distribution of threadfin shad during winter. High density-independent overwinter mortality and changes in distribution of threadfin shad emphasize the importance of the thermal plume from Marshall Station in sustaining this population in Lake Norman. Standing crop estimates in February 1980 and 1981 correlated with water temperature at the sample sites (respectively, $r = 0.44$, $P = 0.001$, N = 88; and $r = 0.55$, $P = 0.001$, N = 88). Similarly, the thermal characteristics of Marshall discharge in winter directly or indirectly attracted largemouth bass, crappie, white bass and striped bass, concentrating these species where they are more available to fishermen. Distribution of sportfish in winter was associated with the abundance of prey, but a direct attraction to the warm water could also be important.

Striped bass were distributed according to preferred or required water temperatures during summer and fall. Creel data indicated a downlake migration of striped bass coincident with the development of anoxic conditions in the hypolimnion which progressed downlake. When the hypolimnion of downlake areas became anoxic, striped bass apparently stopped feeding as they could no longer be caught by fishermen. Mortalities were observed in August 1983 and emaciated striped bass were observed during all previous years. Vertical distribution studies conducted in Zone 1 of Lake Norman indicate a change in distribution of striped bass from the surface waters of littoral and limnetic areas in May to the cooler limnetic waters of the metalimnion and hypolimnion in June through September (R. E. Lewis, in press). In late summer and early fall, striped bass at downlake areas were collected at some of the coolest water available and at dissolved oxygen concentrations below 1 mg/liter. Similar distributional shifts have been documented for striped bass in other waters (Waddle et al. 1980; Coutant and Carroll 1980; Schaich and Coutant 1980; Combs and Peltz 1982).

Table 6. Extended.

Total threadfin shad standing stock (kg/hectare)						Juvenile gizzard shad standing stock (kg/hectare)	
February		April				October	
1980	1981	1979	1980	1981	1982	1979	1980
6.7	0.1	<0.1	0.4	<0.1	<0.1	0.0	0.0
2.1	0.1	0.1	0.3	0.1	0.3	0.0	0.1
8.7	3.3	10.1	4.3	0.2	0.5	0.0	0.0
70.2	24.6	10.8	12.5	2.3	3.2	0.3	0.1
4.5	0.9	26.0	7.1	3.1	3.1	0.7	0.6
0.0	0.0	0.7	2.4	0.0	0.1	0.4	0.0
68.4	3.9	4.5	65.0	35.8	8.8	0.0	0.0
0.0	0.0	0.0	0.0	0.9	<0.1	0.0	0.0
12.4 (±6.1)	3.6 (±2.2)	8.2 (±2.5)	4.0 (±1.5)	0.9 (±0.5)	1.0 (±0.5)		
83.1 (±48.7)	17.5 (±9.3)	31.6 (±10.3)	15.5 (±5.7)	3.1 (±1.8)	2.8 (±1.4)		

The shift in distribution patterns of striped bass in Lake Norman occurred primarily with larger individuals. Food habits of large striped bass (>40 cm) during the stratified period suggest that they were somewhat isolated by habitat requirements from threadfin shad, whereas smaller striped bass occupied the waters of the epilimnion and fed on threadfin shad (R. E. Lewis, in

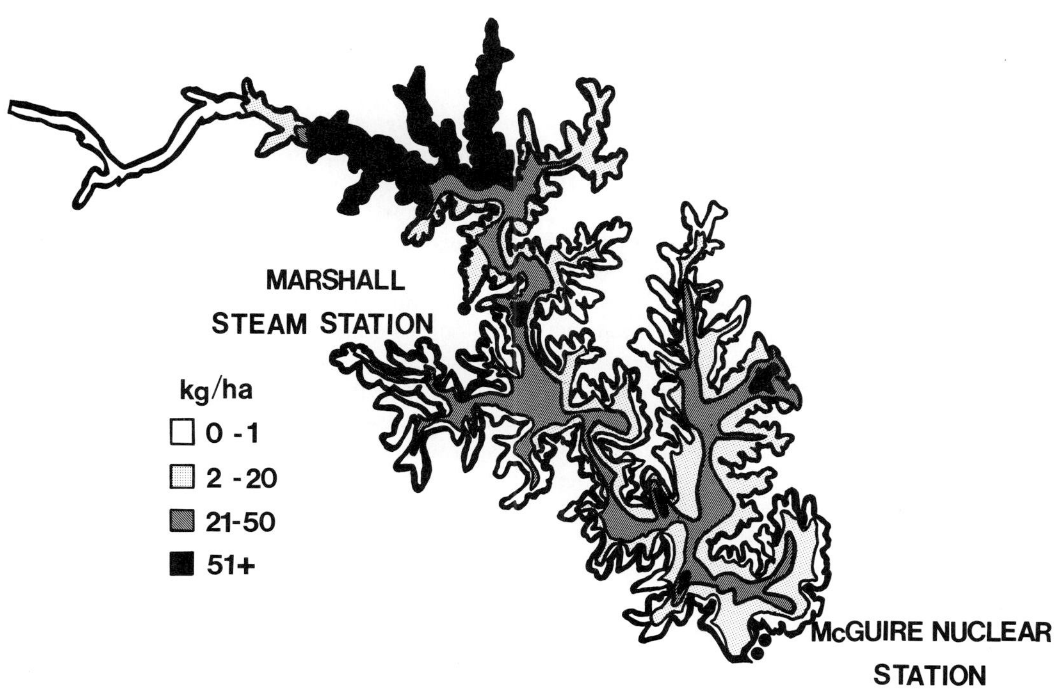

Figure 2. Distribution characteristics of threadfin shad during October 1980, in Lake Norman, North Carolina.

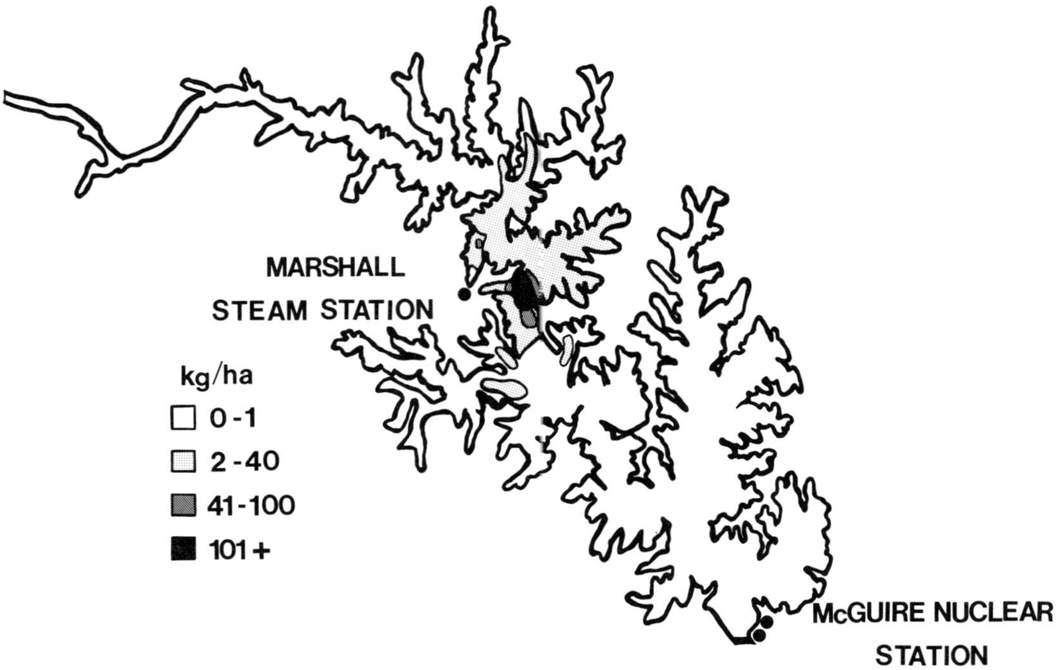

Figure 3. Distributional characteristics of threadfin shad in February 1981, in Lake Norman, North Carolina.

press). Schaich and Coutant (1980) reported that larger striped bass seemed to prefer cooler temperatures than smaller ones in Cherokee Reservoir, Tennessee. Following turnover in Lake Norman, striped bass dispersed from the metalimnion-hypolimnion and again utilized threadfin shad.

Spatial heterogeneity of some species was related to spawning requirements. For example, harvest of striped bass and white bass occurred principally in spring, when they migrated to the riverine area and the discharges of Marshall and McGuire Stations to spawn. These inflows attract white bass and striped bass, making them more accessible to boat or bank fishermen. Also, the differences between uplake and downlake species composition of ictalurids at times is likely related to spawning requirements of individual species. Although spawning requirements for some ictalurids are not well documented, others are known to have strict requirements. For example, channel catfish, a rheophilic species (Trautman 1957), requires turbid water characteristic of riverine systems to reproduce successfully (Pflieger 1975). Channel catfish were mostly restricted to upper Lake Norman, which is more dendritic, more turbid, shallower, and closer to inflows of major tributaries than lower Lake Norman.

No growth responses associated with the nutrient/productivity gradient have been observed. Growth of largemouth bass, bluegill (*Lepomis macrochirus*), and yellow perch varied little spatially, except in the discharge of Marshall where growth rates were higher (Siler et al. 1982). Increased growth in the discharge was an apparent manifestation of prolonged growing season. The observed nutrient gradients in the reservoir apparently enhanced survival of these species and had no noticeable long term effect of growth. Moore (1941) hypothesized that population size and not growth is enhanced by abundant food, and Nakashima and Leggett (1975) observed this response for yellow perch. McConnell (1965) suggested that increased growth is the principal mechanism by which fish biomass initially adjusts to higher productivity, but increased survival becomes more important with time.

Management Implications

Nutrient and productivity gradients have been documented in other reservoirs (Gloss et al. 1980; Paulson and Baker 1980; Thornton et al. 1980;

Table 7. Total phosphorus and chlorophyll *a* concentrations (95% confidence intervals in parentheses) for sampling zones on Lake Norman, North Carolina. Spring represents March, April, and May, and summer represents June, July, and August.

Zone	Total phosphorus (mg/m^3)			Chlorophyll *a* (mg/m^3)					
	Summer 1982	Spring 1983	November 1982	Summer 1979	Summer 1982	April 1982	October 1979	October 1982	December 1979
1	13 (±1.9)	18 (±3.6)	9.1 (±0.5)	4.3 (±1.7)	4.7 (±1.3)	1.7 (±0.2)	3.7	4.6 (±0.5)	5.1
2	15 (±1.6)	17 (±3.4)	9.8 (±1.0)	4.0 (±1.4)	6.2 (±2.5)	1.6 (±0.2)	4.3	6.3 (±0.5)	4.8
3	15 (±1.7)	27 (±8.9)	10.6 (±1.4)	5.0 (±1.0)	10.0 (±7.7)	2.6 (±0.3)	5.9	5.2 (±0.6)	4.8
4	19 (±1.2)	35 (±9.4)	13.0 (±1.1)	5.6 (±0.6)	13.0 (±6.0)	3.0 (±0.4)	7.4	4.7 (±0.6)	6.0
5	21 (±3.9)	34 (±6.6)	14.2 (±1.3)	7.2 (±1.3)	11.2 (±8.3)	4.8 (±0.6)	10.7	6.4 (±0.9)	6.9
6	59 (±38.2)	36 (±17.3)	20.3 (±5.6)	7.0 (±2.4)	—	3.7 (±0.5)	3.2	1.4 (±0.6)	3.5
8	65 (±83.7)	34 (±14.1)	21.0	1.7 (±0.7)	—	2.0 (±0.9)	4.0	—	5.9
9	12 (±2.0)	18 (±7.0)	9.0	—	—	1.7 (±0.2)	—	—	—

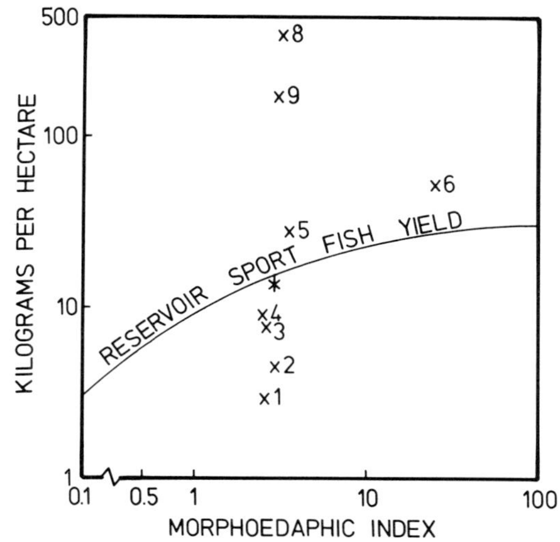

Figure 4. Sport fish harvest as a function of the morphoedaphic index (from Jenkins 1982) with respective observed values for sampling Zones 8, 9, and 1 through 6 and for reservoir-wide average harvest (*) from Lake Norman, North Carolina.

Kennedy et al. 1982; Thornton et al. 1982), as well as within natural lakes which receive substantial inflows or point source pollution (Gascon and Leggett 1977). Nutrient gradients have been linked with fish distribution patterns (Nakashima and Leggett 1975; Gascon and Leggett 1977; Rinne et al. 1981). Van Den Avyle et al. (1982) observed spatial heterogeneity in plankton and larval fish abundance in Center Hill Reservoir, Tennessee, but observed no consistent uplake-downlake gradient. These observations have strong implications to the design of limnological (Thornton et al. 1982) and fisheries sampling programs. Management decisions based on data collected from one or two stations within such bodies of water are quite apt to be erroneous. Simiarly, the approach of managing a reservoir as a biological entity may be ineffective considering the apparent heterogeneity of reservoirs; therefore, management strategies and research demonstrations might be more practical if they were designed to account for various gradients among areas within reservoirs.

The application of the concepts implied by the nutrient gradients and interreservoir models is rarely suggested. However, Prentki et al. (1981) suggested that a proposed waste water treatment facility on Lake Mead would reduce nutrient inputs into the reservoir, thereby reducing the fishery. They further stated that pumped storage facilities could be designed to enhance the productivity of reservoirs by redistributing nutrients. Fisheries managers should become active in environmental regulation, not only to prevent degradation, but to enhance fisheries where applicable. Sources for addition or redistribution of nutrients offer the potential to enhance the fisheries of particular reservoirs.

Implications of fish distribution relative to habitat requirements were also apparent. Habitat requirements were less important for the indigenous species, such as largemouth bass and crappie, than for introduced species such as striped bass and threadfin shad that have relatively narrow thermal requirements. Threadfin shad and striped bass may be regulated to some extent by density-independent factors (i.e., water temperature). These two species, despite their favorable characteristics, have some undesirable attributes; previous years' production or a major part of the adult population can be eliminated with normal annual climatic cycles. Threadfin shad biomass dropped drastically due to over-winter mortality and was lowest during optimum growing conditions of many predators; conversely, large striped bass died from high ambient summer water temperatures. The potential of striped bass may not be realized because of limited summer habitat; however, they can continue to provide a significant fishery through restocking. Increased thermal loading might add stability to threadfin shad populations in cooler reservoirs but could adversely impact striped bass populations. The yield of fisheries in Lake Norman and other reservoirs might be increased if the influences of density-independent factors, such as temperature, on survival could be reduced. This could enhance utilization of available nutrients and improve the yield of the entire system. Hybridization and selective breeding of fishes having different habitat requirements offer further solutions to this problem.

Acknowledgments

We thank Dr. J. C. Knight, M. S. Rodriguez, and S. R. Johnson for valuable suggestions during all phases of the study and manuscript preparation. B. K. Baker provided the major technical support in data analyses, data summaries, and field collections. Other personnel from Duke Power Company's Production Environmental Services Group provided data necessary for this manuscript, especially R. E. Lewis, D. J. Degan, M. S. Rodriguez, J. C. Perkins, and W. T. Horton. D. H. Davis prepared the figures; J. E. Brotherton typed the manuscript.

References

Campbell, W. J., E. Hayes, W. R. Chapman, and W. Seawell. 1978. Angling pressure and sport fish harvest in the predator-stocking-evaluation reservoirs. Proceedings of the Annual Conference of the Southeastern Association of Game and Fish Commissioners 30:114–119.

Canfield, D. E., Jr., and R. W.Bachmann. 1981. Prediction of total phosphorus concentrations, chlorophyll *a,* and Secchi depths in natural and artificial lakes. Canadian Journal of Fisheries and Aquatic Sciences 38:414–423.

Coutant, C. C., and D. S. Carroll. 1980. Temperatures occupied by ten ultrasonic-tagged striped bass in freshwater lakes. Transactions of the American Fisheries Society 109:195–202.

Combs, D. L., and L. R. Peltz. 1982. Seasonal distribution of striped bass in Keystone Reservoir, Oklahoma. North American Journal of Fisheries Management 2:66–73.

Gascon, D., and W. C. Leggett. 1977. Distribution, abundance, and resource utilization of littoral fishes in response to a nutrient/production gradient in Lake Memphremagog. Journal of the Fisheries Research Board of Canada 34:1105–1117.

Gloss, S. P., L. M. Mayer, and D. E. Kidd. 1980. Advective control of nutrient dynamics in the epilimnion of a large reservoir. Limnology and Oceanography 25:219–228.

Hannan, H. H., and L. Broz. 1976. The influence of a deep-storage and an underground reservoir on the physicochemical limnology of a permanent central Texas river. Hydrobiologia 51:43–63.

Hogan, J. E., and W. D. Adair, editors. 1982. Lake Norman summary. Duke Power Company, DUKE PWR/82-02, Charlotte, North Carolina, USA.

Jenkins, R. M. 1967. The influence of some environmental factors on the standing crop and harvest of fishes in U.S. reservoirs. Pages 298–321 *in* Reservoir fishery resources symposium. Southern Division American Fisheries Society, Bethesda, Maryland, USA.

Jenkins, R. M. 1982. The morphoedaphic index and reservoir fish production. Transactions of the American Fisheries Society 111:133–140.

Jenkins, R. M., and D. I. Morais. 1971. Reservoir sport fishing effort and harvest in relation to environmental variables. Pages 371–384 *in* G. E. Hall, editor. Reservoir fisheries and limnology. Special Publication 8, American Fisheries Society, Bethesda, Maryland, USA.

JENKINS, R. M., AND D. I. MORAIS. 1978. Prey-predator relations in the predator-stocking-evaluation reservoirs. Proceedings of the Annual Conference of the Southeastern Association of Fish and Wildlife Agencies 30:141–157.

JONES, J. R., AND R. W. BACHMANN. 1978. Phosphorus removal by sedimentation in some Iowa reservoirs. Proceedings of the International Association of Theoretical and Applied Limnology 16:1007–1017.

KENNEDY, R. H., K. W. THORNTON, AND R. C. GUNKEL, JR. 1982. The establishment of water quality gradients in reservoirs. Canadian Water Resources Journal 7:71–87.

LEWIS, R. E. (In press). Temperature selection and vertical distribution of striped bass during lake stratification. Proceedings of the 37th Annual Conference of the Southeastern Association of Fish and Wildlife Agencies.

MCCONNELL, W. J. 1965. Relationship of herbivore growth to rate of gross photosynthesis in microcosms. Limnology and Oceanography 10:593–643.

MOORE, W. G. 1941. Studies on the feeding habits of fishes. Ecology 22:91–96.

NAKASHIMA, B. S., AND W. C. LEGGETT. 1975. Yellow perch (*Perca flavescens*) biomass responses to different levels of phytoplankton and benthic biomas in Lake Memphremagog, Quebec-Vermont. Journal of the Fisheries Research Board of Canada 32: 1785–1797.

OGLESBY, R. T. 1977a. Relationships of fish to lake phytolankton standing crop, production, and morphoedaphic factors. Journal of the Fisheries Research Board of Canada 34:2271–2279.

OGLESBY, R. T. 1977b. Phytoplankton summer standing crop and annual productivity as functions of phosphorus loading and various physical factors. Journal of the Fisheries Research Board of Canada 34:2255–2270.

PAULSON, L. J., AND J. R. BAKER. 1980. Nutrient interactions among reservoirs on the Colorado River. Pages 1647–1656 *in* Symposium on surface water impoundments. American Society of Civil Engineers, Minneapolis, Minnesota, USA.

PERKINS, J. C., AND T. L. WHISENANT. 1982. Water chemistry. Pages 107–152 *in* J. E. Hogan and W. D. Adair, editors. Lake Norman summary. Duke Power Company, DUKE PWR/82-02, Charlotte, North Carolina, USA.

PFLIEGER, W. L. 1975. The fishes of Missouri. Missouri Department of Conservation, Jefferson City, Missouri, USA.

PRENTKI, R. T., L. J. PAULSON, AND J. R. BAKER 1981. Chemical and biological structure of Lake Mead sediments. Technical report number 6, Lake Mead Limnological Research Center, University of Nevada, Las Vegas, Nevada, USA.

RAWSON, D. S. 1960. A limnological comparison of twelve large lakes in northern Saskatchewan. Limnology and Oceanography 5:195–211.

REGIER, H. A., AND D. S. ROBSON. 1967. Estimating population number and mortality rates. Pages 31–66 *in* S. D. Gerking, editor. The biological basis of freshwater fish production. Blackwell Scientific Publications, Oxford, England.

RICKER, W. E. 1975. Computation and interpretation of biological statistics of fish populations. Bulletin 191, Fisheries Research Board of Canada, Ottawa, Canada.

RINNE, J. N., W. L. MINCKLEY, AND P. O. BERSELL. 1981. Factors influencing fish distribution in two desert reservoirs, Central Arizona. Hydrobiologia 80:31–42.

RODRIGUEZ, M. S. 1982a. Phytoplankton. Pages 154–260 *in* J. E. Hogan and W. D. Adair, editors. Lake Norman summary. Duke Power Company, DUKE PWR/82-02, Charlotte, North Carolina, USA.

RODRIGUEZ, M. S. 1982b. Relationships between phytoplankton growth rates and nutrient dynamics in Lake Norman, N.C. Duke Power Company, DUKE PWR/82-01, Charlotte, North Carolina, USA.

RYDER, R. A. 1965. A method for estimating the potential fish production of north-temperate lakes. Transactions of the American Fisheries Society 94:214–218.

RYDER, R. A. 1982. The morphoedaphic index—use, abuse, and fundamental concepts. Transactions of the American Fisheries Society 111:154–164.

SCHAICH, B. A., AND C. C. COUTANT. 1980. A biotelemetry study of spring and summer habitat selection of striped bass in Cherokee Reservoir, Tennessee, 1978. Oak Ridge National Laboratory, ORNL/TM-7127, Oak Ridge, Tennessee, USA.

SERRUYA, C., M. EDELSTEIN, U. POLLINGHER, AND S. SERRUYA. 1974. Lake Kinneret sediments: nutrient composition of the pore water and mud water exchanges. Limnology and Oceanography 19:489–508.

SILER, J. R. 1983. Description of a trawl handling structure for a bow-fished Tucker trawl. The Progressive Fish-Culturist 45:217–220.

SILER, J. R. 1985. A comparison of area and volumetric-based population estimates from midwater trawl catch statistics. Research report PES/85-16, Duke Power Company, Charlotte, North Carolina, USA.

SILER, J. R., R. E. LEWIS, B. K. BAKER, G. E. VAUGHAN, AND R. A. HANSEN. 1982. Fish. Pages 411-460 *in* J. E. Hogan and W. D. Adair, editors. Lake Norman summary. Duke Power Company, DUKE PWR/82-02, Charlotte, North Carolina, USA.

STRICKLAND, J. D. H., AND T. R. PARSONS. 1972. A practical handbook of seawater analysis. Fisheries Research Board of Canada, Ottawa, Canada.

THORNTON, K. W., R. H. KENNEDY, J. H. CARROLL, W. W. WALKER, R. C. GUNKEL, AND S. ASHBY. 1980. Reservoir sedimentation and water quality—a heuristic model. Pages 654–661 *in* Symposium on surface water impoundments. American Society of Civil Engineers, Minneapolis, Minnesota, USA.

Thornton, K. W., R. H. Kennedy, A. D. Magoun, and G. E. Saul. 1982. Reservoir water quality sampling design. Water Resources Bulletin 18:471–480.

Trautman, M. B. 1957. The fishes of Ohio. Ohio State University Press, Columbus, Ohio, USA.

United States Environmental Protection Agency. 1974. Methods of chemical analysis of water and wastes. Office of Technology Transfer, Washington, D.C., USA.

Van Den Avyle, M. J., R. S. Hayward, R. A. Krause, and A. J. Spells. 1982. Spatial variation in abundance of phytoplankton, zooplankton, and larval fishes in Center Hill Reservoir. Canadian Water Resources Journal 7:189–214.

Waddle, H. R., C. C. Coutant, and J. L. Wilson. 1980. Summer habitat selection by striped bass, *Morone saxatilis,* in Cherokee Reservoir, Tennessee, 1977. Oak Ridge National Laboratory, ORNL/TM-6927, Oak Ridge, Tennessee, USA.

Pages 137–143 *in* G. E. Hall and M. J. Van Den Avyle, editors.
Reservoir Fisheries Management: Strategies for the 80's.
Reservoir Committee, Southern Division American Fisheries Society, Bethesda, Maryland. 1986.

MANAGEMENT OF RESERVOIR FISH COMMUNITIES BY INFLUENCING SPECIES INTERACTIONS

Predator-Prey Interactions in Reservoir Communities

RICHARD L. NOBLE
Department of Wildlife and Fisheries Sciences
Texas A&M University
College Station, Texas 77843

ABSTRACT

Over thousands of years, natural waters have developed equilibrium between predators and prey that usually results in stable fisheries attractive to man. In contrast, predator-prey relations in reservoir communities are poorly developed. Fish reproduction is poorly synchronized with plankton, and littoral communities are frequently unstable because of water level fluctuations. Reservoir managers have tried, with relatively little success, to create predator-prey relationships similar to those so delicately developed by nature in lakes. Manipulations have been directed at piscivorous game fishes and their prey, with little consideration for the relationships between prey fishes and their prey—usually plankton or littoral invertebrates. Traditional ecological and limnological principles, developed for natural waters, have limited application to reservoir systems. Feeding ecology studies must emphasize prey availability and competitive interactions of predators. Shifts in predator-prey relationships in response to prey density and availability, particularly in relation to habitat characteristics, must be defined. Population dynamics of important predators and prey are poorly understood in reservoirs. Non-native predators and prey have potential for improving predator-prey efficiency and productivity of fisheries in reservoirs.

Predator-prey interactions should be considered as one subset of community dynamics. In that context, much of the understanding of predator-prey interactions has been developed in allied fields of science (e.g., Holling 1973; Murdoch and Oaten 1975). Predator-prey interactions as they more specifically relate to fisheries management were recently reviewed in an international symposium (Stroud and Clepper 1979); the management of forage fishes in the United States was evaluated from a regional standpoint (Ney 1981); and the use of piscivorous predator stocking in management of reservoirs of the southern United States was elaborated (Carver et al. 1978). These reviews indicate that fishery scientists are prone to consider predator-prey interactions as those between piscivorous game fishes and their prey. However, fish communities are bigger and more complex than this, with predator-prey interactions occurring among several trophic levels (O'Brien 1979).

My intent in this paper is to examine problems of game fish management in relation to overall species interactions in reservoirs and, to some extent, to abiotic influences. Since most of the theory of aquatic community ecology comes from natural systems, I try to develop an understanding of reservoir fish ecology based on what we know about natural aquatic communities. Finally, I will emphasize those areas that I perceive as vital to understand more fully if we are to successfully manage predator-prey relationships in reservoirs.

NATURAL SYSTEMS

For thousands of years, a river flows. An equilibrium develops between physical, chemical, and biological components of the river system (Cummins 1972). Relationships between biological components and their environment include adaptation to seasonal variations in physical and chemical factors. A stability, with flexibility for organisms to tolerate normal extremes of environmental variation, is reached. In addition to being reasonably stable, this system is diverse in its fauna. Diversity increases as populations establish to utilize available food and space resources. A characteristic stream fish community develops to claim its place in this system (Moyle and Li 1979). The nature of that fish community is largely determined by geographic constraints and physical characteristics of the river system. Predator-prey relationships develop among the fishes and between the fishes and other biotic components of the system (Campbell 1979). This system can be expected to be quite efficient at utilizing available resources, even though the fauna sometimes varies considerably among river systems.

Natural lakes demonstrate the results of a similar evolution, but are characterized by a less variable environment in comparison with lotic systems. Highly-evolved predator-prey relationships and delicately balanced communities frequently typify temperate lakes of glacial origin. A period of 10,000 years or so has allowed the evolution of efficient biotic relationships (MacLean and Magnuson 1977). Predator-prey relationships in natural lakes appear to be rather simple, even when the system is analyzed in a comprehensive sense (Kerr and Ryder 1977). Temperate lakes in particular are characterized by a seasonal chronology of phytoplankton blooms followed by pulses of zooplankters which graze the phytoplankton (Hutchinson 1967). Reproduction of many prey fish species is timed to exploit the zooplankton, first its larval or juvenile stages, then the adult plankton organisms. Reproduction of piscivorous fishes is timed for their young to utilize zooplankton as prey until individuals grow large enough to switch to small fish. All this synchronization may occur during just a few weeks (Mathias and Li 1982), after which the predator-prey relationship is maintained until scarcity of that prey leads the predator to switch to other prey types (Noble 1975).

Partitioning of prey among predator species seems to be relatively simple and well defined. Nevertheless, adult predators have the flexibility to switch to alternate foods when principal prey become scarce (Forney 1974). Less flexibility exists for the very young predators, which respond poorly to abrupt changes in food availability. Consequently, extreme environmental variations during early life—when synchronization with prey is vital—cause fluctuations in year-class strength of prey fishes, and, in turn, in piscivorous species with limited food alternatives. Nevertheless, compensatory mechanisms serve to buffer such impacts—e.g., a small year class of a prey species may experience more rapid growth, thereby still producing an abundance of prey biomass. The walleye (*Stizostedion vitreum*)-yellow perch (*Perca flavescens*) predator-prey relationship is a good example of the above (Forney 1980), although other systems probably differ little in function. Frequently these systems seem to be regulated by the piscivorous predator, modified by effects of environmental variation on prey abundance. Management of these systems basically involves management of the piscivore population.

Reservoir Systems

After flowing for thousands of years, a river may be impounded, creating a reservoir. It is frequently called a lake, the public may expect it to produce conventional lake fishing, and the fishery biologist begins the task of lake manager. However, a reservoir is ecologically different from a natural lake (Bayly and Williams 1973; Baxter 1977); it may vary from being nearly lentic to almost a wide spot in the river (Carline 1986). Nevertheless, fishery management objectives will likely be stated in terms of lacustrine species—almost certainly with primary emphasis on a predatory fish, probably a piscivorous one. The reservoir fishery manager usually begins to manage a predator-prey system from the "top" (e.g., Carver et al. 1978), expecting that he can use the predator to drive the system, as he might in a natural lake.

Attempts to manage and study predator-prey associations in reservoirs have usually treated predators and prey as if they exist in a vacuum (Carver et al. 1978; McCammon and von Geldern 1979; Noble 1981). Reservoir fishery biologists have assumed the existence of relatively simple predator-prey relationships of natural lakes when they categorize these waters as largemouth bass (*Micropterus salmoides*)-bluegill (*Lepomis macrochirus*), striped bass (*Morone saxatilis*)-gizzard shad (*Dorosoma cepedianum*), or walleye-yellow perch lakes and set their management goals accordingly. Obviously these systems are more complex than that; the reservoir piscivore population alone is the result of many biotic interactions, including a poorly understood set of predator-prey relationships.

Biotic relationships of this community, however, are not well-developed (Baxter 1977). Nature knows no such organism as a "reservoir species." The life expectancy of most reservoirs is minuscule in relation to the time that evolutionary processes have worked to develop the intricate systems of lakes and rivers. Numerous studies of pre- and post-impoundment fish communities indicate that, although species composition changes, species diversity remains high in reservoirs, perhaps because this habitat is an ecotone between a river and a lake. Diversity and complexity, in this case, do not seem to imply stability (May 1974), perhaps because of too little time for development of longterm homeostasis in the reservoir system. Nevertheless, basic relationships between primary producers, primary consumers, and secondary consumers will hold, although likely modified by abiotic factors.

Reservoir Fish as Predators

Asynchrony frequently exists between annual production of phytoplankton, zooplankton and planktivores. In southern reservoirs, the species most closely tied to the plankton pulses is the gizzard shad, which spawns early. Nevertheless, its year-class strength is erratic, possibly the result of quantitative and qualitative variations in plankton abundance. Young of virtually all fishes are predators upon zooplankton, yet the spawning of most species doesn't coincide with zooplankton pulses. Geographical variations in timing of plankton blooms and spawning of reservoir fishes likely account for major differences in abundance of certain species in some reservoirs.

Largemouth bass, the piscivorous predator of greatest fishery interest in southern reservoirs, spawns too late for its young to utilize the shad forage; consequently shad are not usually an important prey for young-of-year bass until late in the year when a fall spawn of shad may occur. Therefore, young bass are dependent upon the littoral community for prey. However, in many reservoirs, the littoral zone is subject to instability due to fluctuations in water level and turbidity, which can severely impact prey habitat. Macrophytes may be scarce, either because turbidity reduces the littoral zone or because its total area changes seasonally or annually with fluctuating water levels. Low water clarity and scarcity of cover therefore reduce availability of prey to young bass. At the other extreme, reservoirs which have reasonably stable water levels and clear water also are likely to have dense stands of aquatic vegetation. Abundant submerged vegetation provides substrates for invertebrate prey organisms, resulting in high biomass of prey fishes, but this same cover can prevent efficient utilization of prey and cause slow growth of young bass.

Adult bass are less affected by the timing of prey spawning, but they are affected by water level fluctuations or stability. Planned water-level reductions have been used successfully in fishery management to increase availability of littoral prey fishes (Keith 1975). Otherwise, an overabundance of submergent vegetation may force the adult bass to switch to more readily available pelagic prey.

Similar situations probably exist for other predators in these systems and for systems with other species complexes. The stress of water level fluctuations may be a selective factor in some situations which leads to greater success of species which are better adapted to limnetic existence.

Research Needs

Much remains to be learned in the area of reservoir predator-prey interactions. Biologists have tried, with varied success, to manage on the basis of principle combined with trial-and-error. Researchers have undertaken evaluations of predator-prey relationships with little regard for those elements of the community one step removed from the species of interest. Certain areas of research must be undertaken before scientifically-based progress in predator-prey management can be achieved.

Basic Ecology

Texts of general ecology or limnology typically contain little information on reservoirs. In spite of the ubiquity of reservoirs in the world, there is a scarcity of demonstrated principles of community ecology in complex reservoir systems.

Studies of biological limnology must better define reservoir plankton productivity pulses, both quantitatively and qualitatively, to the point that at least general predictions can be made for reservoirs having a particular set of nutrient and physical characteristics. Furthermore, the ecology of the littoral zone, including primary and secondary production, needs to be understood (Dykyjova' and Kvet 1978), and the relative importances of littoral, limnetic, and profundal biota need evaluation. For many reservoir fishes, the littoral zone is "where the action is," at least until its alteration leads to the plankton community playing a more dominant role. The benthic component of the system has been nearly ignored by fishery biologists in reservoirs, much as it has been in natural lakes. Benthos, obviously, is an integral part of predator-prey systems as recently demonstrated by Ploskey and Jenkins (1982). However, the extent of quantitative and qualitative variations in benthic production, their causes, and their effects on predator fishes are poorly understood.

The attention of ecologists, systems analysts and ecological modelers is needed to test applicability of the basic principles of community ecology to reservoirs. Some recognized principles and basic management concepts for natural lakes may not apply to reservoirs because of much greater variety in physical characteristics and operational patterns. On the other hand, many of the principles may actually fit, but in ways that we don't yet understand! The task is formidable, and success will depend on integrated team research.

Feeding Ecology

The literature is filled with publications on the food habits of most lake fish species, and except for larvae, is almost as well-developed on reservoir fishes. Food habits studies seldom include an analysis of the prey that was available to the predator, i.e., what its alternatives were! Food habits analyses, in themselves, don't distinguish between opportunistic feeding and food selectivity. Shifts in predator-prey relationships in response to prey density and availability must be defined, particularly in relation to habitat characteristics. Furthermore, food habits studies must focus on those fishes feeding at the same trophic levels simultaneously (e.g., Hendricks and Noble 1980; Ploskey and Jenkins 1982). Predator-prey relationships will not be completely elucidated until feeding ecology is understood in relation to prey availability and possible competitors.

Population Dynamics

Year-class fluctuations are characteristic of many fishes (Nikolsky 1961), particularly those with high reproductive potential and high susceptibility to effects of environmental variability ("steno-" species). Predators and prey fishes common to reservoirs typically have fairly high reproductive potential but most are rather flexible ("eury-") species. Nevertheless, because of the magnitude of environmental variation in reservoirs, and because the species have little evolutionary history in such waters, these "eury-" species exhibit variable "steno-" characteristics, including wide year-class variation (e.g., Nelson and Walburg 1977).

Life history characteristics are reasonably well known for most common reservoir species, and the general nature of impacts of major environmental variables on those life history characteristics are understood. In contrast, population dynamics have been rather poorly described. Especially in relation to year-class variations, compensation and depensation are integral parts of the predator-prey relationship which have been minimally explored.

As a result of recent advances in assessment (e.g., Grinstead et al. 1978), standing stocks of reservoir fishes can be estimated, but standing stock estimates must be coupled with growth and mortality data to calculate production (biomass elaboration) which is essential for determination of harvest rates. Similarly, standing stock plus production of prey is the best measure of total available forage for predators (e.g., Houser and Netsch 1971). Models of available prey biomass for piscivores, which take into account the size of prey fishes consumable by each size of predator (Jenkins and Morais 1978) have provided insight into predator-prey relationships specific to size groups and seasons (Jenkins 1979). Research must provide additional information on the various parameters of the population dynamics of predator and prey species at various trophic levels if the predator-prey models are to be refined.

Remodeling the Ecosystem

Fishery management is basically the fight against Nature, to better meet the demands and desires of humans for fish and fishing. If Nature would provide what society wants—quantitatively as well as qualitatively—fishery management would become unnecessary. Natural systems frequently produce a game fish that is highly regarded. Management then can be aimed at refinements of size and number produced. However, it seems that Nature can use some help in reservoirs, where predator-prey systems are poorly evolved.

Fishery managers have had reasonable success in improving game fish yields in reservoirs through introductions (Keith 1986). Although most introductions have focused on piscivorous predators (Carver et al. 1978), emphasis on prey fishes has increased (Noble 1981; Boxrucker 1986). Introductions of invertebrates have received little attention (e.g., Wydoski and Bennett 1981). Introductions of prey species, which are usually predators on invertebrates, have generally been made without much regard for the invertebrates in the community (Kohler et al. 1986).

Knowledge of the life history and population dynamics of the introduced species alone is insufficient for effective management through introductions. It must include an understanding of how the entire community responds. Obtaining this information is a challenge to reservoir biologists. It will likely involve a combination of sophisticated modelling coupled with more experimentation and evaluation.

Genetic engineering, a modern "buzz word," has unimaginable applicability to managing predator-prey relationships in reservoirs. Genetic manipulation, in the broad sense, can involve stocking to increase genetic variability and to improve resource use by predators, as has been the case with Florida largemouth bass (*Micropterus salmoides floridanus*) introductions (Chew 1975). In a more restrictive sense, within species, fish can be selected for higher adaptability to reservoir conditions. Other possibilities include the use of sterile hybrids, which may be more easily manipulated than species or strains. On the frontier, technologies are developing for intracellular manipulation of genetic materials. Applications to reservoir fisheries have the potential for creating new fishes with predictable characteristics.

Fisheries managers tend to be apprehensive about genetic manipulation, whether introducing non-native fishes, selecting strains, or tampering with genes. Caution is needed, not so much that we don't produce an organismic monster, but that we don't produce a community monster. Gene flow within species, from introduced to endemic, will occur in the reservoir, and additionally from the reservoir to connected natural water bodies. Irreversibility of genetic contamination in natural waters, except through thousands more years of evolution, is justification for a cautious position in this approach to improving predator-prey systems in reservoirs (Philipp et al. 1981).

Conclusion

Predator-prey interactions could have been approached from several points of view—e.g., behavioral interactions of predator and prey, life history strategies, or energetics. My approach has been to emphasize an often-forgotten aspect of predator-prey interaction—that it occurs at several trophic levels, and involves the whole community, not just the piscivorous adult game fish and its principal prey fish. Reservoir communities, because they are young and relationships are not highly evolved, offer great potential for management to produce better fish yields. However, successful management requires an understanding of the overall complex reservoir system, both biotic and abiotic components.

Acknowledgments

I particularly wish to thank P. W. Bettoli, W. J. Clark and W. E. Grant for their contributions of ideas, and Claire Armistead for her assistance in preparing the manuscript. This paper was developed through support of Texas Agricultural Experiment Station Projects S-6206 and H-6634.

References

Baxter, R. M. 1977. Environmental effects of dams and impoundments. Pages 255–283 *in* R. F. Johnston, P. W. Frank, and C. D. Michener, editors. Annual review of ecology and systematics. Annual Reviews Incorporated, Palo Alto, California, USA.

Bayly, I. A. E., and W. D. Williams. 1973. Inland waters and their ecology. Longman Australia Publishing Company, Victoria, Australia.

Boxrucker, J. 1986. Evaluation of stocking inland silversides as supplemental forage for largemouth bass and white crappie. Pages 149–155 *in* G. E. Hall and M. J. Van Den Avyle, editors. Reservoir fisheries management: strategies for the 80's. Reservoir Committee, Southern Division American Fisheries Society, Bethesda, Maryland, USA.

Campbell, K. P. 1979. Predation principles in large rivers: a review. Pages 181–191 *in* R. H. Stroud and H. Clepper, editors. Predator-prey systems in fisheries management. Sport Fishing Institute, Washington, D.C. USA.

Carline, R. F. 1986. Indices as predictors of fish community traits. Pages 46–56 *in* G. E. Hall and M. J. Van Den Avyle, editors. Reservoir Fisheries Management: Strategies for the 80's. Reservoir Committee, Southern Division American Fisheries Society, Bethesda, Maryland, USA.

Carver, D. C., G. E. Hall, and J. F. Hall. 1978. History and organization of the predator-stocking-evaluation by the Reservoir Committee, Southern Division, American Fisheries Society. Proceedings of the Annual Conference of the Southeastern Association of Fish and Wildlife Agencies 30:103–107.

Chew, R. L. 1975. The Florida largemouth bass. Pages 450–458 *in* R. H. Stroud and H. Clepper, editors. Black bass biology and management. Sport Fishing Institute, Washington, D.C., USA.

CUMMINS, K. W. 1972. What is a river?—A zoological description. Pages 33–52 *in* R. T. Oglesby, C. A. Carlson, and J. A. McCann, editors, River ecology and man. Academic Press, New York, New York, USA.

DYKYJOVA', D., AND J. KVET. 1978. Pond littoral ecosystems. Springer-Verlag, Berlin-Heidelberg, Germany.

FORNEY, J. L. 1974. Interactions between yellow perch abundance, walleye predation, and survival of alternate prey in Oneida Lake, New York. Transactions of the American Fisheries Society 103:15–24.

FORNEY, J. L. 1980. Evolution of a management strategy for the walleye in Oneida Lake, New York. New York Fish and Game Journal 27:105–141.

GRINSTEAD, B. G., R. M. GENNINGS, G. R. HOOPER, C. A. SCHULTZ, AND D. A. WHORTON. 1978. Estimation of standing crop of fishes in the predator-stocking-evaluation reservoirs. Proceedings of the Annual Conference of the Southeastern Association of Fish and Wildlife Agencies 30:120–130.

HENDRICKS, M. K., AND R. L. NOBLE. 1980. Feeding interactions of three planktivorous fishes in Trinidad Lake, Texas. Proceedings of the Annual Conference of the Southeastern Association of Fish and Wildlife Agencies 33:324–330.

HOLLING, C. S. 1973. Resilience and stability of ecological systems. Annual Reviews of Ecology and Systematics 4:1–24.

HOUSER, A., AND N. F. NETSCH. 1971. Estimates of young-of-year shad production in Beaver Reservoir. Pages 359–370 *in* G. E. Hall, editor. Reservoir fisheries and liminology. Special Publication 8, American Fisheries Society, Bethesda, Maryland, USA.

HUTCHINSON, G. E. 1967. A treatise on limnology. Introduction to lake biology and the limnoplankton, vol. 2. John Wiley and Sons, New York, New York, USA.

JENKINS, R. M. 1979. Predator-prey relations in reservoirs. Pages 123–134 *in* R. H. Stroud and H. Clepper, editors. Predator-prey systems in fisheries management. Sport Fishing Institute, Washington, D.C., USA.

JENKINS, R. M., AND D. I. MORAIS. 1978. Predator-prey relations in the predator-stocking-evaluation reservoirs. Proceedings of the Annual Conference of the Southeastern Association of Fish and Wildlife Agencies 30:141–157.

KEITH, W. E. 1975. Management by water level manipulation. Pages 489–497 *in* R. H. Stroud and H. Clepper, editors. Black bass biology and management. Sport Fishing Institute, Washington, D.C., USA.

KEITH, W. E. 1986. A review of introduction and maintenance stocking in reservoir fisheries management. Pages 144–148 *in* G. E. Hall and M. J. Van Den Avyle, editors. Reservoir fisheries management: strategies for the 80's. Reservoir Committee, Southern Division American Fisheries Society, Bethesda, Maryland, USA.

KERR, S. R., AND R. A. RYDER. 1977. Niche theory and percid community structure. Journal of the Fisheries Research Board of Canada 34:1952–1958.

KOHLER, C. C., J. J. NEY, AND W. E. KELSO. 1986. Filling the void: establishment of a pelagic fishery and its consequences to littoral fishes in a Virginia mainstream reservoir. Pages 166–177 *in* G. E. Hall and M. J. Van Den Avyle, editors. Reservoir fisheries management: strategies for the 80's. Reservoir Committee, Southern Division American Fisheries Society, Bethesda, Maryland, USA.

MACLEAN, J., AND J. J. MAGNUSON. 1977. Species interactions in percid communities. Journal of the Fisheries Research Board of Canada 34:1941–1951.

MATHIAS, J. A., AND S. LI. 1982. Feeding habits of walleye larvae and juveniles: comparative laboratory and field studies. Transactions of the American Fisheries Society 111:722–735.

MAY, R. M. 1974. Stability and complexity in model ecosystems. Princeton University Press, Princeton, New Jersey, USA.

MCCAMMON, G. W., AND C. E. VON GELDERN, JR. 1979. Predator-prey systems in large reservoirs. Pages 431–442 *in* R. H. Stroud and H. Clepper, editors. Predator-prey systems in fisheries management. Sport Fishing Institute, Washington, D.C., USA.

MOYLE, P. B., AND H. W. LI. 1979. Community ecology and predator-prey relations in warmwater streams. Pages 171–180 *in* R. H. Stroud and H. Clepper, editors. Predator-prey systems in fisheries management. Sport Fishing Institute, Washington, D.C., USA.

MURDOCH, W. W., AND A. OATEN. 1975. Predation and population stability. Pages 1–131 *in* A. MacFayden, editor. Advances in ecological research, Volume 9. Academic Press Incorporated, New York, New York, USA.

NELSON, W. R., AND C. H. WALBURG. 1977. Population dynamics of yellow perch (*Perca flavescens*), sauger (*Stizostedion canadense*), and walleye (*S. vitreum vitreum*) in four mainstem Missouri River reservoirs. Journal of the Fisheries Research Board of Canada 34:1748–1763.

NEY, J. J. 1981. Evolution of forage-fish management in lakes and reservoirs. Transactions of the American Fisheries Society 110:725–728.

NIKOLSKY, G. V. 1961. Concerning the causes of fluctuations in the abundance of fishes. Voprosy Ikhtiologii 1:659–665.

NOBLE, R. L. 1975. Growth of young yellow perch (*Perca flavescens*) in relation to zooplankton populations. Transactions of the American Fisheries Society 104:731–741.

NOBLE, R. L. 1981. Management of forage fishes in impoundments of the southern United States. Transactions of the American Fisheries Society 110:738–750.

O'BRIEN, W. J. 1979. The predator-prey interaction

of planktivorous fish and zooplankton. American Scientist 67:572–581.

PHILIPP, D. P., W. F. CHILDERS, AND G. W. WHITT. 1981. Management implications for different genetic stocks of largemouth bass (*Micropterus salmoides*) in the United States. Canadian Journal of Fisheries and Aquatic Sciences 38:1715–1723.

PLOSKEY, G. R., AND R. M. JENKINS. 1982. Biomass model for reservoir fish and fish-food interactions, with implications for management. North American Journal of Fisheries Management 2:105–121.

STROUD, R. H., AND H. CLEPPER, editors. 1979. Predator-prey systems in fisheries management. Sport Fishing Institute, Washington, D.C., USA.

WYDOSKI, R. S., AND D. H. BENNETT. 1981. Forage species in lakes and reservoirs of the western United States. Transactions of the American Fisheries Society 110:764–771.

Pages 144–148 *in* G. E. Hall and M. J. Van Den Avyle, editors.
Reservoir Fisheries Management: Strategies for the 80's.
Reservoir Committee, Southern Division American Fisheries Society, Bethesda, Maryland. 1986.

A Review of Introduction and Maintenance Stocking in Reservoir Fisheries Management

WILLIAM E. KEITH
Fisheries Division
Arkansas Game and Fish Commission
Little Rock, Arkansas 72205

ABSTRACT

Introduction stocking as a form of reservoir fisheries management has progressed at an extremely rapid rate within the last 20 years. This advancement has substantially increased the need for maintenance stocking of introduced species but has reduced maintenance and supplemental stocking of native, naturally-reproducing species. The rapid advancement of introduction stocking has not been accompanied by detailed evaluations of effects on existing fish populations, sport fishing, or fishery management activities. Current fishery management agency stocking philosophies are primarily based on informal experiences and observations of fishery managers.

Introduction of predators has provided trophy fishes, increased angler satisfaction, and redistributed fishing pressure. Impacts on prey populations have varied from unnoticeable to severe depletion, and effects on native predator fishes have been judged largely insignificant. However, displacement of ecologically similar species, predation on desirable native species, and production of intergrade populations have been documented. State fishery management administrators generally agree that there is a need for substantial increases in hatchery production, increased knowledge of reservoir conditions that control the success of fish stocking, and more definitive evaluations of the effects of stocking on existing fish populations.

Although stocking of hatchery-raised fish is one of the oldest, most visible, and in the public's mind, the most necessary form of fisheries management, only within the last 15 years has this technique become a versatile and effective management tool. Initially, the primary use of hatchery fish was for stocking waters that were newly impounded or suffered frequent fish mortalities such as winter kill. One of the early additional uses of hatchery-raised, warm-water fishes was to restore "balance" in farm ponds and small impoundments following the guidelines of Swingle (1950).

In 1967, when the first Reservoir Fishery Resources Symposium was held, only a few papers described fish stocking as a management technique. These included the stocking of rainbow trout (*Salmo gairdneri*) to provide two-story fisheries, trout stocking in cold tailwaters, and fall and winter stocking of catchable-size trout in a warm water lake to provide a short-term, high-use urban fishery. It was, however, at about this time that the great "boom" of striped bass (*Morone saxatilis*) culture, stocking, and management began to significantly change the role of fish hatcheries in managment. The late 1960's also coincided with the highly successful introductions of various salmonids into the Great Lakes.

Reports of impacts of recent developments in fish stocking activities are lacking in the literature. The purpose of this paper is to provide such an overview. Agency philosophies regarding use of fish hatcheries have developed largely through routine monitoring and informal observations rather than carefully designed experiments; therefore, much of the information in this paper was developed through personal interviews with administrators from various state fishery management agencies. The information primarily applies to agencies in the southeast and midwest areas of the United States, where warm and cool-water reservoirs are prevalent.

MANAGEMENT AGENCY PHILOSOPHY

Because of the rapid evolution of fish stocking techniques, geographic differences and the variable facilities of the different management agencies, philosophies varied considerably among the states in their use of fish stocking as a management tool. Fish stocking is virtually the only form of fisheries management in some states and is only a minor tool in others. Practices are often quite different for small and large reservoirs.

Generally, management in cold-water reservoirs involves routine stocking; however, there are increased efforts toward maintaining native populations, excluding non-native species or varieties, and managing through restrictive harvest regulations. In comparison, fisheries management activities in warm-water reservoirs are including more intensive stocking of non-native or diminishing species, in addition to habitat and population structure modifications.

Agency philosophies are often apparent in their fish stocking priorities. Normally, state-owned or controlled waters receive top priority, other public-owned waters (municipal, federal, etc.) are frequently next, and private or club lakes are of lowest priority. However, in Georgia, private waters receive the top stocking priority for pond fishes (R. M. Gennings, Georgia Department of Natural Resources, Atlanta, Georgia, USA, [personal communication]).

Fish stocking priorities are more frequently characterized by an identified need of the fishery or the fisherman. Initial stocking of a new impoundment normally ranks as a top priority. This can also include restocking waters that have been renovated through a management fish kill, depleted by pollution, or impacted by severe flushing of fishes from a reservoir. In many agencies, stocking priorities are strongly influenced by fisherman demand for stocked species. This rationale seems to be: "If fishermen can catch them and enjoy catching them, the fish should be stocked."

Introduction Stocking

Introduction stocking is the stocking of a fish species into a water body where it does not exist in notable numbers. This includes non-native species and native species that had previously been eliminated.

The hybrid produced by crossing female striped bass with male white bass (*Morone chrysops*) and pure striped bass currently receive the most interest for stocking warm water reservoirs. Other frequently introduced species include the Florida largemouth bass (*Micropterus salmoides floridanus*), walleye (*Stizostedion vitreum*), smallmouth bass (*Micropterus dolomieui*), threadfin shad (*Dorosoma petenense*), tiger muskellunge (*Esox masquinongy* female crossed with male *Esox lucius*), paddlefish (*Polyodon spathula*), and inland silversides (*Menidia beryllina*).

The expressed rationale for introduction stocking in reservoirs reflect the now-classic goals of occupying vacant niches in artificial environments, control or utilization of abundant prey fishes, providing trophy fish, and supplying additional prey species. However, experience with these introductions has led to new rationale. One of these includes the goal of providing several species of predators to provide population stability. This allows population balance through predation not to be dependent on one or very few species. Variability in survival and recruitment of a key predator species does not have as great an impact if several predator species are present. Likewise, the distribution of fishing pressure among several species rather than one or two is attractive to fishery managers as well as fishermen.

Some of these goals have been achieved. For example, the striped bass has created a late-fall and winter fishery in many southern reservoirs. Striped bass and hybrids have produced midsummer "school" fishing activities on large, open-water reservoirs as well as deep-water trolling in mountain reservoirs. Kansas has an excellent through-the-ice fishery for striped bass (G. M. Theurer, Kansas Fish and Game Commission, Pratt, Kansas, USA [personal communication]). Walleye spawning runs have advanced spring fishing much earlier in many areas.

Substantial prey reduction through introduction of large predators has not been well documented. Exceptions occurred in Texas, where excessive annual stockings of striped bass and hybrids, coupled with restricted harvest, caused prey depletion (Crandall 1979; Morris and Follis 1979). In other cases in Texas, moderate stocking rates of striped bass and liberal harvest regulations did not typically lead to prey-limited populations (R. L. Bounds, Texas Parks and Wildlife Department, Austin, Texas, USA [personal communication]). Exceptionally strong year classes of naturally reproduced striped bass in Lake Mead severely depleted the threadfin shad population and later experienced heavy post-spawning mortality due to poor body condition (B. D. Taubert, Arizona Game and Fish Department, Phoenix, Arizona, USA [personal communication]). Changes in the length distributions of prey populations in Oklahoma reservoirs are believed to be associated with introduced predators (K. E. Erickson, Oklahoma Department of Wildlife Conservation, Norman, Oklahoma, USA [personal communication]).

Competition or elimination of native fishes is the most frequently expressed concern about predator introductions. Most fishery managers suspect that some competition occurs but generally agree that the trade-offs are justified despite the absence of definitive documentation. Evidence from several reservoirs indicates that striped bass and hybrids may replace a substantial part of the white bass population, as in Texoma Reservoir (K. E. Erickson, personal communication). Hybrids characteristically switch from a clupeid diet to sunfishes (*Lepomis* spp.) when clupeid density is reduced. In some reservoirs where striped bass were in poor condition due to a shortage of clupeids, largemouth bass showed no evidence of limited prey availability (R. L. Bounds and K. E. Erickson, personal communication).

Severe reduction of native fishes from predation by introduced piscivores has not been substantiated. However, predation on stocked rainbow trout by striped bass and walleye occurs in some reservoirs in Arkansas, Oklahoma, and Georgia (Deppert 1980; R. M. Gennings, personal communication). Also, in high density crappie (*Promoxis* spp.) populations where walleye have been introduced, substantial predation on the crappie has been identified. This initially led to better growth and larger crappie in the population (R. L. Bounds, personal communication).

The introduction of the Florida largemouth bass has had varied success. Such introductions typically result in intergrade populations. Some managers insist that these introductions increase the standing crop and harvest of largemouth bass (R. L. Bounds, personal communication), but others believe that the intergrades simply replace the native population. The Florida subspecies of largemouth bass is given credit for exceptional growth potential, particularly in the third and fourth years, as far north as Oklahoma (K. E. Erickson, personal communication). Conversely, the documented higher susceptibility of Florida largemouth bass to winter-kill has caused concern for its introduction into northern waters (Cichra et al. 1982). Differences in success of Florida bass introductions are probably due in part to the different environments into which they were introduced. Although many fishery managers believe new gene pool introductions will provide results similar to new species introductions, others warn against introducing genetic compositions questionable for a specific environment (Philipp et al. 1981).

Most prey fish introductions have used threadfin shad or inland silversides. There also is interest in various species of anadromous clupeids that could be introduced into reservoirs near coastal areas. Threadfin shad introductions have increased growth by crappies, yearling bass, and other predators, but they have also led to reduced production of young largemouth bass (von Geldern and Mitchell 1975). Inland silversides introductions into reservoirs without clupeids have been credited with increased growth of young largemouth bass, adult white crappie (*Pomoxis annularis*), and white bass (Li et al. 1976; K. E. Erickson, personal communication). Silversides, however, is believed to significantly compete with young largemouth bass in some areas and to have very limited production potential (B. W. Smith, Alabama Department of Conservation and Natural Resources, Montgomery, Alabama, USA [personal communication]).

Introduction of the grass carp (*Ctenopharyngodon idella*) for aquatic vegetation control has created much controversy. The present consensus is that the species will remove aquatic vegetation of a variety of types under given circumstances with certain preferences. Techniques for selective control of vegetation and prevention of grass carp dispersal into non-target waters are major current concerns. No consistent impacts on native fish populations were identified in 31 Arkansas lakes stocked with grass carp (Bailey 1978). Determinations of appropriate stocking rates for particular situations, prevention of loss of grass carp from the local drainage, and control of natural reproduction through the use of monosex populations, sterile individuals, or hybrid grass carp are receiving the most attention. Introduction of the silver carp (*Hypophthalmichthys molitrix*), which has shown promise in intensive fish culture impoundments, is being considered in water quality management of nutrient-rich reservoirs (Henderson 1980).

Maintenance Stocking

Maintenance stocking falls into two general categories: (1) continued restocking to maintain a desired density of an introduced species that is not self-sustaining through reproduction, and (2) periodic stocking of native species to supplement natural production. A typical example of maintenance stocking is a put-and-take program such as tailwater stocking of rainbow trout below reservoirs having cold, hypolimnetic discharges. The establishment of non-reproducing populations of striped bass and hybrids has increased the need for their maintenance stocking. Maintenance stocking is often controlled by the available supply of fish rather than by the actual need. Consequently, stockings may be erratic despite plans that call for a certain schedule of number of fish. Programs have developed to overcome some of these problems. Developments include alternate-year stocking of striped bass and hybrids and various numerical combinations of the two (B. W. Smith, personal communication).

Stocking of additional numbers of native, naturally-reproducing species in established fish populations has been as controversial as some introduction of non-native fish. Loska (1982) revised published evaluations of stocking black basses (*Micropterus* spp.) and found that success typically has been poor. Notable exceptions included cases where relatively large bass were stocked and where Florida subspecies fingerlings were used. Evaluations of such stockings have been hampered by the lack of effective ways to mark stocked fish and use of imprecise creel surveys.

Several states use fish population "balance" techniques of Swingle (1950) and supplementally stock small impoundments with fingerling black bass when it is determined that sufficient prey exists. Such procedures have not been extensively used on large reservoirs. The predator-prey ratio developed by Jenkins and Morais (1978) to determine biomass of prey available to predators might be useful for identifying candidate reservoirs for stocking. Survival of stocked bass may be directly correlated with the size of fish stocked (McCammon and von Geldern 1979). Several workers have identified biomodal length distributions of Age-0 black bass in reservoirs; this apparently results from one group being able to switch to a piscivorous diet at an earlier time (Aggus and Elliott 1975; Shelton et al. 1979). This is likely related to the availability and abundance of prey at the time the predators can convert to a fish diet. From these considerations, it is apparent that the time of stocking can significantly affect success.

Certain fishes that can be economically raised to large fingerlings or catchable size are sometimes stocked to supplement poor or unsuccessful reproduction. Culture of channel catfish (*Ictalurus punctatus*) to relatively large size in ponds, raceways, or cages prior to release in reservoirs might circumvent the problem of poor survival of stocked fingerlings.

Adverse environmental conditions such as high pesticide levels, excessive turbidity, and severe water-level fluctuations may prevent successful recruitment of important native fishes in reservoirs. Annual maintenance stocking of these species may be the only means of providing successive year classes. Water temperatures in cooling water impoundments in Illinois change rapidly from winter to summer levels, and desirable temperatures are quickly by-passed for crappie spawning. This results in limited crappie reproduction, so nursery ponds are used to grow large numbers of 2–3 in crappie for stocking these waters (P. J. Paladino, Illinois Department of Conservation, Springfield, Illinois, USA [personal communication]). Some large flood control and irrigation reservoirs in the southwestern United States fluctuate 50 to 100 ft during fish spawning seasons. Subimpoundments as large as 100 acres are being designed as hatchery lakes to provide an annual black bass year class for these reservoirs (B. D. Taubert, personal communication).

Prospects for the Future

Among the fishery managers and administrators that I have contacted, there was a strong consensus that the use of hatchery-produced fish will be an increasingly important fishery management activity in the future. Considering only the need to maintain stocks of introduced species, there could be a need for at least a three-fold increase in hatchery output. Additionally, the needs for establishment of desired species have not been satisfied in many areas. The rate of construction of new impoundments has slowed, and reservoirs filled in the 1950's and 1960's now show habitat deterioration. Thus, more aggressive management activities that will include supplemental, corrective, and maintenance stocking likely will put an even greater demand on hatchery facilities. Other factors pointing toward expanding hatchery needs include recent increases in urban fishing programs that typically use stocking of catchable size fish.

Recently developed fish culture techniques have greatly expanded fish production potential. Artificial spawning techniques have been developed for most species that are frequently stocked. Successful development of a mass-produced, pellet-type fish food acceptable to predator fishes has been a major step in providing larger-sized predators for stocking. Intensive culture of fishes that readily accept inexpensive prepared feed has been advanced by the use of wire cages suspended in water bodies capable of assimilating large quantities of waste materials. The use of large pens of netting is a modification of techniques used to grow various species of salmonids in coastal waters in the northwest (Pillay 1975). Large net-pens have also been used to produce rainbow trout and channel catfish in freshwater reservoirs in Arkansas.

Despite improvements in culture techniques, there is an almost unanimous opinion that insufficient knowledge exists about the most effective times, numbers, and sizes of fish to stock as well as reservoir conditions that result in maximum survival of stocked fish. Such information is needed to optimize the production efficiency of hatchery systems. We also have insufficient knowledge of how to evaluate impacts of stocking on existing populations. Stocking programs influence fishermen, but this aspect has not been properly evaluated.

It is apparent that fish stocking will continue in reservoirs in the foreseeable future, but many questions remain for the fishery manager. Should stocking be continued at the current, rather intensive pace or should it be done more conservatively with emphasis on prestocking evaluations and post-stocking assessment? Is it more important to maximize numbers of fish cultured or optimize stocking success?

References

Aggus, L. R., and G. V. Elliott. 1975. Effects of cover and food on year-class strength of largemouth bass. Pages 317–322 *in* R. H. Stroud and H. Clepper, editors. Black bass biology and management. Sport Fishing Institute, Washington, D.C., USA.

Bailey, W. M. 1978. A comparison of fish populations before and after extensive grass carp stocking. Transactions of the American Fisheries Society 197:181–206.

Cichra, C. E., H. W. Neill, and R. L. Noble. 1982. Differential resistance of northern and Florida largemouth bass. Proceedings of the Annual Conference of the Southeastern Association of Fish and Wildlife Agencies 34:19–24.

Crandall, P. S. 1979. Evaluation of striped bass × white bass hybrids in a heated Texas reservoir. Proceedings of the Annual Conference of the Southeastern Association of Fish and Wildlife Agencies 32:588–597.

Deppert, D. L. 1980. Effects of striped bass predation on an Oklahoma trout fishery. Proceedings of the Annual Conference of the Southeastern Association of Fish and Wildlife Agencies 33:384–392.

Henderson, R. S. 1980. Production potential of catfish grow-out ponds supplementally stocked with silver and bighead carp. Proceedings of the Annual Conference of the Southeastern Association of Fish and Wildlife Agencies 33:584–590.

Jenkins, R. M., and D. I. Morais. 1978. Prey-predator relations in the predator-stocking-evaluation reservoirs. Proceedings of the Annual Conference of the Southeastern Association of Fish and Wildlife Agencies 30:141–157.

Li, H. W., P. B. Moyle, and R. L. Garrett. 1976. Effect of the introduction of the Mississippi silverside (*Menidia audens*) on the growth of black crappie (*Pomoxis nigromaculatus*) and white crappie (*Pomoxis annularis*) in Clear Lake California. Transactions of the American Fisheries Society 105:404–408.

Loska, P. M. 1982. A literature review on the stocking of black basses (*Micropterus spp.*) in reservoirs and streams. Georgia Department of Natural Resources, Atlanta, Georgia, USA.

McCammon, G. W., and C. von Geldern, Jr. 1979. Predator-prey systems in large reservoirs. Pages 431–442 *in* R. H. Stroud and H. Clepper, editors. Predator-prey systems in fisheries management. Sport Fishing Institute, Washington, D.C., USA.

Morris, J. M., and B. J. Follis. 1979. Effects of striped bass predation upon shad in Lake E. V. Spence, Texas. Proceedings of the Annual Conference of the Southeastern Association of Fish and Wildlife Agencies 32:697–702.

Philipp, D. P., W. F. Childers, and G. S. Whitt. 1981. Management implications for different genetic stocks of largemouth bass (*Micropterus salmoides*) in the United States. Canadian Journal of Fisheries and Aquatic Sciences 38:1715–1723.

Pillay, T. V. R., editor. 1975. Aquaculture development—United States. Bulletin 7. Food and Agriculture Organization of the United Nations, Rome, Italy.

Shelton, W. L., W. D. Davies, T. A. King, and T. J. Timmons. 1979. Variations in growth of initial year class of largemouth bass in West Point Reservoir, Alabama and Georgia. Transactions of the American Fisheries Society 108:142–149.

Swingle, H. S. 1950. Relationships and dynamics of balanced and unbalanced fish populations. Bulletin 274. Alabama Polytechnical Institute Agriculture Experiment Station, Auburn, Alabama, USA.

von Geldern, C. E., and D. F. Mitchell. 1975. Largemouth bass and threadfin shad in California. Pages 436–449, *in* R. H. Stroud and H. Clepper, editors. Black bass biology and management. Sport Fishing Institute, Washington, D.C., USA.

Pages 149–155 *in* G. E. Hall and M. J. Van Den Avyle, editors.
Reservoir Fisheries Management: Strategies for the 80's.
Reservoir Committee, Southern Division American Fisheries Society, Bethesda, Maryland. 1986.

Evaluation of Stocking Inland Silversides as Supplemental Forage for Largemouth Bass and White Crappie

JEFF BOXRUCKER

Oklahoma Department of Wildlife Conservation
Fishery Research Laboratory
1416 Planck
Norman, Oklahoma 73069

ABSTRACT

Approximately 5,000 adult inland silversides (*Menidia beryllina*) were introduced into Thunderbird Reservoir (2,448 hectares), Oklahoma, in March 1978 to evaluate their potential as supplemental forage for largemouth bass (*Micropterus salmoides*) and white crappie (*Pomoxis annularis*). Food habits data prior to introduction of silversides indicated that sunfish (*Lepomis* spp.) were the primary prey of small largemouth bass (≦200 mm total length, TL), and larger bass (>200 mm TL) fed on sunfish and gizzard shad (*Dorosoma cepedianum*). Small white crappie (≦200 mm TL) fed primarily on invertebrates and large white crappie (>200 mmTL) consumed gizzard shad and sunfish.

Inland silversides established a breeding population in spring 1978 and provided 95% of the littoral forage fish found in 1981 seine samples. Silversides were the primary prey of small bass during fall 1981 and were secondarily important in the summer and winter diets. Silversides were eaten by white crappies and larger bass, but never in significant amounts.

Growth rates of Age-1 largemouth bass increased following silverside introduction ($P < 0.05$). Growth of older largemouth bass and white crappie remained unchanged. Inland silversides provided supplemental forage for small bass yet did not appear to be utilized by the predator populations to the degree that their relative abundance in seine samples would suggest. It is recommended that further introductions of inland silversides be scrutinized in view of possible competitive interactions between silversides and native sunfishes and to consider stocking only in those systems where bass recruitment is limited by the availability of suitably-sized prey.

Non-native forage fishes are often introduced to prey-deficient systems by fishery managers. The ideal prey species for stocking should be prolific, stable in abundance, trophically efficient, vulnerable to predation, non-emigrating, and innocuous to other species (Ney 1981). Shad (*Dorosoma* spp.), sunfish (*Lepomis* spp.), minnows (Cyprinidae), and silversides (*Menidia* spp.) usually comprise the forage base of southern impoundments.

It has been speculated that the inland silversides (*M. beryllina*), formerly named Mississippi silversides (*M. audens*) (Chernoff et al. 1981), meets the aforementioned criteria for suitability as a forage species. Its short maximum length of about 130 mm, prolonged spawning period, and relatively high total fecundity apparently combine to make large numbers of prey available to adult and juvenile piscivorous fishes throughout the year (Mense 1967).

The inland silversides was first reported in Oklahoma from Lake Texoma in 1954 (Riggs and Dowell 1954) and has since been introduced as supplemental prey into several large reservoirs in the state. However, the effects of silverside stockings on predator fishes in Oklahoma have not been studied. This study was conducted to determine the effects of the introduction of inland silversides on the growth rates and food habits of largemouth bass (*Micropterus salmoides*) and white crappie (*Pomoxis annularis*) in Thunderbird Reservoir, which was impounded in 1965 to serve as a municipal water supply for Norman, Oklahoma. About 5,000 adult inland silversides from Boomer Lake, Oklahoma, were stocked in March 1978. Thunderbird Reservoir covers 2,448 hectares and has a shoreline development ratio of 7.9, mean depth of 6 m, and maximum depth of 21 m.

METHODS

Food habits of largemouth bass and white crappie were determined for samples collected during March–December in 1977 (pre-introduc-

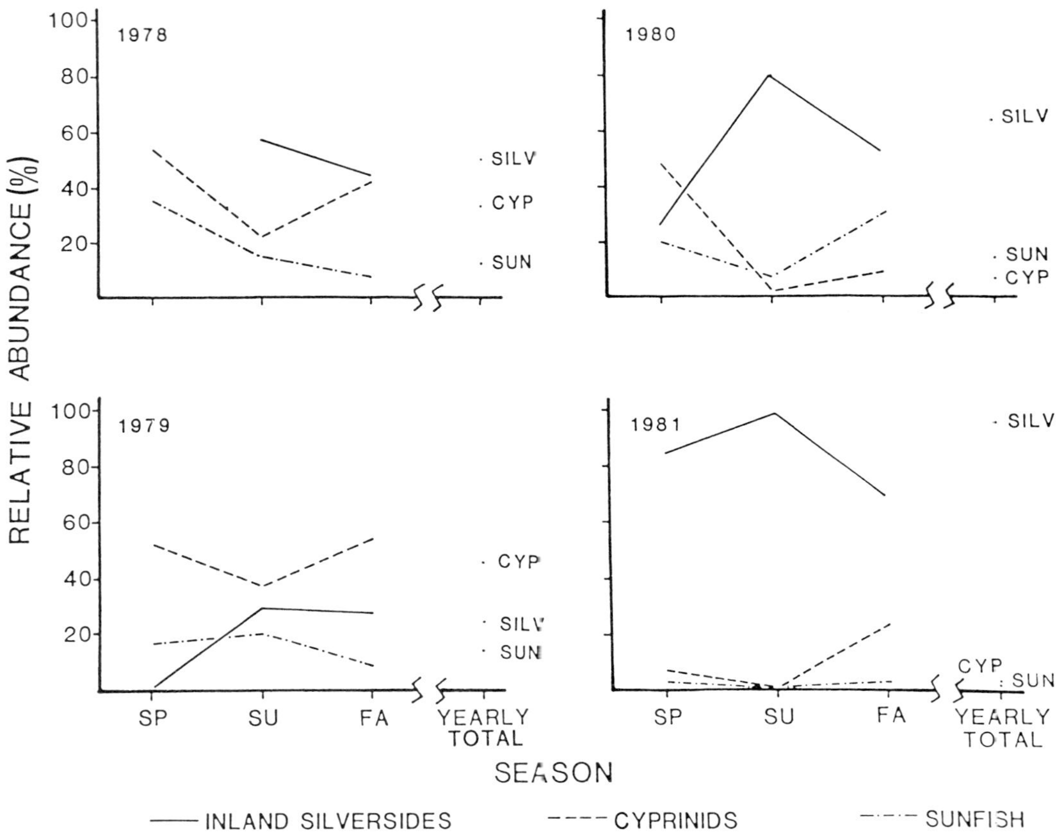

Figure 1. Relative abundance of forage fishes in spring (SP), summer (SU), and fall (FA) seine samples for 1978-1981 in Thunderbird Reservoir, Oklahoma.

tion) and in 1981 (3 years after silversides were introduced). Electrofishing was used to collect each species monthly from areas close to 16 seining stations used to sample forage fishes. In 1977, stomachs were removed from fish and the contents were preserved in 10% formalin. In 1981, glass tubes were used to evacuate the stomach contents of largemouth bass >200 mm TL and white crappie >300 mm TL (Gilliland et al. 1982). It was believed that large adult largemouth bass would not utilize silversides as forage; therefore, specimens ≧ 400 mm total length (TL) were not evaluated.

Fish found in stomachs were identified to the lowest possible taxa. Unidentifiable fish remains were classified as "unknown fish." Fish that did not comprise a significant portion of the diet were grouped as "other fish." All invertebrates were combined. Volume of the stomach contents was determined by water displacement, and percent occurrence and percent volume were calculated for all prey categories. Data were summarized by season—spring (March–May), summer (June–August), fall (September–November), and winter (December).

Total lengths (mm), weights (g), and scale samples were taken for all largemouth bass and crappies collected. Fish were aged from scales, and mean lengths for each year-class were determined from back-calculations.

Relative abundance of prey species and length frequencies of inland silversides were monitored in seine samples collected monthly April–November 1978 and March–November 1979, 1980, and 1981. Fifteen quadrant hauls (20 m² each) were taken with a 6.1 m × 1.8 m × 4.8 mm bag seine at nine permanent stations during each month except September, when 60 hauls were made at seven supplemental stations as well as at the permanent sites. Most seining sites were

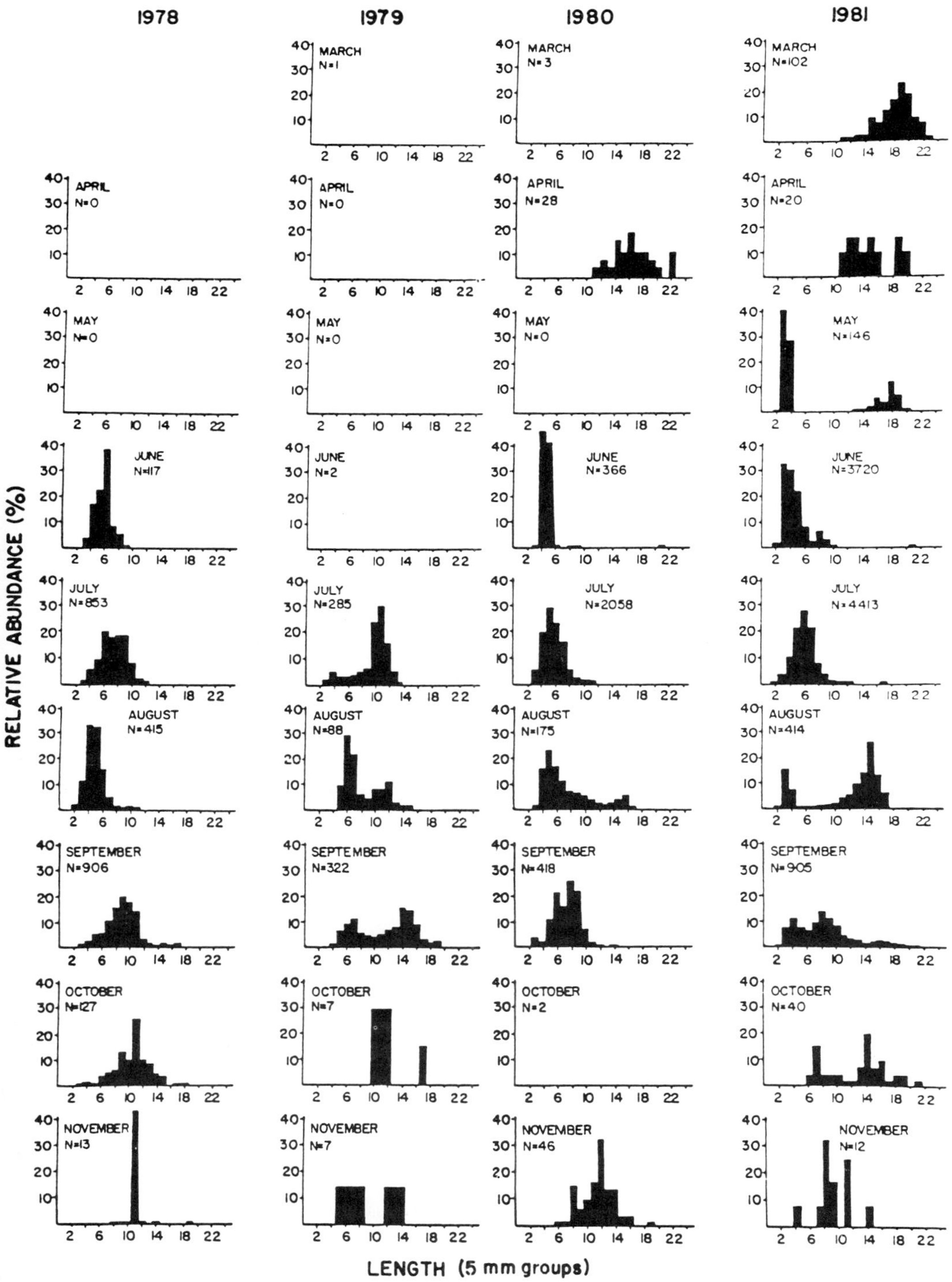

Figure 2. Length-frequencies of inland silversides from Thunderbird Reservoir, Oklahoma, 1978-1981. Values on each horizontal axis can be multiplied by 5 mm to determine total length.

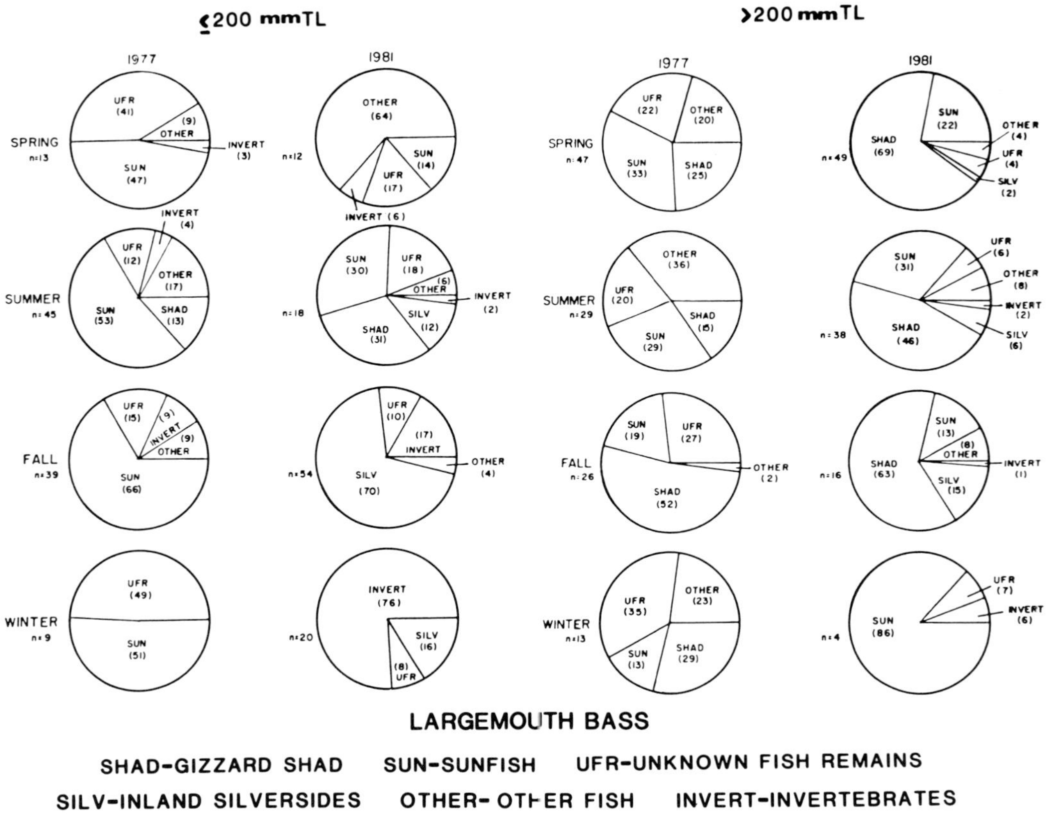

Figure 3. Percent volume of stomach contents of largemouth bass before (1977) and after (1981) the inland silversides was introduced into Thunderbird Reservoir, Oklahoma.

located near points or in small embayments in the downlake region of the reservoir. These data were summarized for the same seasons used in the food habit analyses.

Results

Inland silversides was the primary forage species collected each year except 1979, when cyprinids were most numerous (Fig. 1). Young-of-the-year silversides were first collected in June 1978. Sunfish provided 12–15% of the 1978–1980 samples, but only 0.9% of the fish collected in 1981 were sunfish, and the total number collected was only 15–25% of those in the three preceding years. Cyprinids, principally red shiners (*Notropis lutrensis*), were a large portion (33.2%) of the 1978 sample and were abundant again in 1979 (45.7%). However, relative abundance of cyprinids declined to 5.9% in 1980 and 2.7% in 1981. Gizzard shad (*Dorosoma cepedianum*) did not contribute significantly to the seine catches; however, large numbers of shad seen when electrofishing in littoral areas indicated that seining did not accurately reflect shad abundance.

Young-of-the-year silversides were found beginning in late May and early June of each year (Fig. 2). Large adults from the previous year's spawn typically were not found in samples after May. Summer and early fall were the times of peak silverside abundance, with July samples consistently yielding the highest densities. A bimodal length distribution was seen in August and September samples for 1979–1981 (Fig. 2).

Sunfish were the major identifiable prey of largemouth bass ≦200 mm TL in 1977, but during 1981, sunfish, shad, silversides, and invertebrates were seasonally important (Fig. 3). Inland silversides were consumed by small bass in summer, fall, and winter 1981, and accounted for 70% of the stomach contents in fall. Bass >200 mm TL fed primarily on gizzard shad and

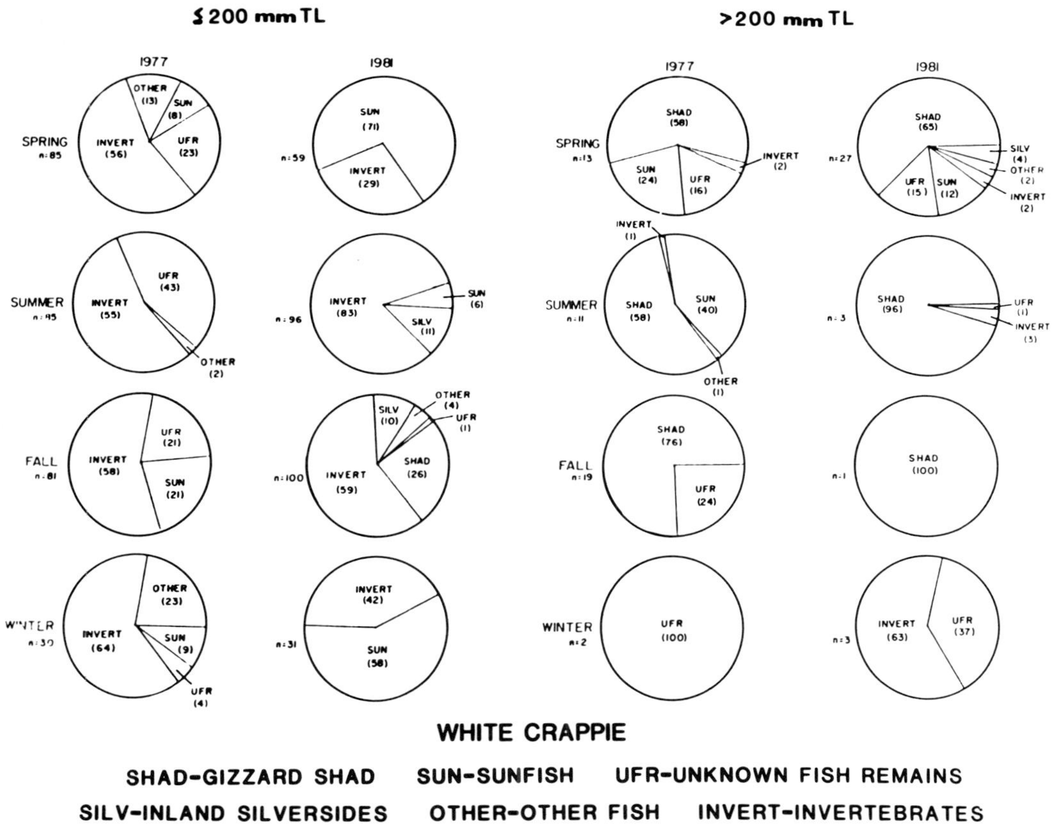

Figure 4. Percent volume of stomach contents of white crappie before (1977) and after (1981) the inland silversides was introduced into Thunderbird Reservoir, Oklahoma.

sunfish during all seasons in 1977 and 1981 (Fig. 3). Inland silversides were found in a few large bass during most seasons, but they never were a major component of the diets.

White crappie ≦200 mm TL fed primarily on invertebrates in 1977 and on invertebrates or sunfish in 1981 (Fig. 4). Inland silversides never contributed more than 11% of the food volume in small crappies and were absent from the diets in spring and winter 1981. White crappie >200 mm TL fed primarily on gizzard shad and sunfish before and after silversides introduction (Fig. 4). Too few large crappie were collected after spring 1981 to provide adequate information on the occurrence of silversides in the diet. Silversides represented only 4% of the diet of large white crappie in spring 1981.

First-year growth of largemouth bass was more rapid after silversides were introduced (Table 1). Back-calculated lengths for three year classes following the introduction (1978–1980) were significantly longer ($P \leqq 0.05$) than those calculated for the three year classes prior to the silversides introduction (1974–1976). No differences in growth rates of older largemouth bass or white crappie were detected following establishment of inland silversides.

Discussion

Despite comprising nearly 50% of the forage fish collected in 1977 (year of introduction) and about 95% of the forage fishes collected by seining in 1981, inland silversides contributed substantially only to the diet of largemouth bass ≦200 mm TL. The smallest bass (Age 0) apparently benefitted from the increased forage and showed faster growth after silversides were introduced. Largemouth bass >200 mm TL and

Table 1. Mean back-calculated lengths (standard deviation in parentheses) of largemouth bass at Age 1 for three year classes preceding and three year classes following inland silversides introduction into Thunderbird Reservoir, Oklahoma.

Year class	Number of estimates	Mean total length (mm)
Prior to introduction		
1974	49	131 (5.8)
1975	44	115 (4.7)
1976	58	109 (4.5)
After introduction		
1978	44	136 (4.8)
1979	68	140 (5.2)
1980	50	151 (5.7)

white crappies consumed minor amounts of silversides and showed no growth response to the introduction.

The abundance and wide range in lengths of silversides (Fig. 3) would appear to make them suitable forage for young-of-the-year and older game fishes throughout most of the growing season. The bimodal length distribution, seen as early as August, may have been the result of reproduction by Age-0 individuals spawned earlier in the season or an outcome of a prolonged spawning season. Inland silversides spawn from late March through July in Lake Texoma (Saunders 1959). Spawning of individual fish may occur in pulses, but spawning of the population as a whole appears to be evenly distributed over the spawning season (Mense 1967).

The lack of silversides utilization by large bass may reflect energetic inefficiency associated with foraging on small prey. A demand for larger prey items has been correlated with increases in body size (Paloheimo and Dickie 1966; Werner 1979). Also, growth may require increasingly larger prey items even if these are rare (Kerr 1971). Therefore, it may be ecologically beneficial for adult largemouth bass to select the relatively larger gizzard shad and sunfish than to prey on smaller, but more abundant, inland silversides. Results from a study at Trinidad Lake, Texas, were similar to those in Thunderbird Reservoir. Inland silversides was the major species in seine samples but provided only 5% of the diet of adult largemouth bass (Hall 1977).

Limited predation on silversides by white crappie may have been due to spatial segregation. Inland silversides generally occupy littoral habitats and when found in open water are restricted to the upper 1.5 m of the water column (Mense 1967). White crappie occur primarily at depths exceeding 1.2 m (Grinstead 1965).

The limited impact of inland silversides on white crappie in Thunderbird Reservoir contrasts other studies that have shown benefits. Inland silversides constituted 70–90% of the diet of adult black and white crappie in Clear Lake, California (Li et al. 1976). Growth rates of Age-1 and Age-2 black and white crappie declined following silverside introductions, but growth of Age-3 crappie increased. The decline in growth rate was attributed to interspecific competition for zooplankton. Richards (1977) reported increased growth of Age-2 white crappie in Carbody Reservoir, Colorado, following silverside introduction, and silversides were the only identifiable fish found in the stomachs of Age-0 white crappie from Keystone Reservoir, Oklahoma (Combs 1982). Silversides also were the most abundant food item of young-of-the-year sport fishes, including white crappie, from Lake Texoma (Harper 1978).

Changes in the littoral habitat of Thunderbird Reservoir and interactions between silversides and the native forage fishes may have masked the influences of increased silversides abundance. The density of rooted aquatic macrophytes, principally *Myriophyllum spicatum,* increased from 1977 to 1981. Sunfish were closely associated with the vegetation, and silversides occupied areas near peripheral vegetation. Since capture efficiency of a predator typically decreases as structural complexity of the environment increases (Cooper and Crowder 1979), increased availability of cover for sunfish and silversides may have negated effects of increased prey abundance. The sunfish's stronger affinity for vegetation, perhaps leading to reduced availability, may have also led to the shift in the small bass' diets between 1977 and 1981.

The potential for more direct interactions between silversides and sunfish was shown in Clear Lake, California (Moyle and Holzhauser 1978). Following the introduction of silversides, Age-0 largemouth bass showed increased consumption of silversides and decreased sunfish utilization. This shift was related to decreased sunfish abundance, which Moyle and Holzhauser (1978) attributed to competition for large zooplankters by silversides and sunfish.

The use of inland silversides as supplemental

prey in reservoirs appears to have limited applicability. Increased growth rates of Age-0 largemouth bass can result, but the overall effect on growth of sport fish in Thunderbird Reservoir was minimal. Increased prey abundance might improve survival of young largemouth bass, but this response was not evaluated. The possibility of negative impacts on existing sunfish populations, as suggested by Moyle and Holzhauser (1978), should be evaluated before further introductions of the inland silversides are made. Introductions of silversides into systems where largemouth bass recruitment is limited by low abundance of suitably-sized prey might increase predator fish densities.

References

CHERNOFF, B., J. V. CONNET, AND C. F. BRYAN. 1981. Systematics of the *Menidia beryllina* complex (Pisces: Atherinidae) from the Gulf of Mexico and its tributaries. Copeia 1981:319–336.

COMBS, D. L. 1982. Striped bass research study. Final report, Federal Aid Project F-39-R. Oklahoma Department of Wildlife Conservation, Oklahoma City, Oklahoma, USA.

COOPER, W. E., AND L. B. CROWDER. 1979. Patterns of predation in simple and complex environments. Pages 257–267 *in* R. H. Stroud and H. Clepper, editors. Predator-prey systems in fisheries management. Sport Fishing Institute, Washington, D.C., USA.

GILLILAND, E. R., C. W. KLEINHOLZ, AND M. D. CLADY. 1982. The efficiency of removing food items from fish with glass tubes. Proceedings of the Annual Meeting of the Texas Chapter of the American Fisheries Society 4:95–100.

GRINSTEAD, B. G. 1965. The vertical distribution of the white crappie, *Pomoxis annularis,* in the Buncombe Creek Arm of Lake Texoma. Master's thesis, University of Oklahoma, Norman, Oklahoma, USA.

HALL, C. R. 1977. An evaluation of Florida largemouth bass, *Micropterus salmoides floridanus,* introduction in a Texas power plant cooling reservoir. Master's thesis, Texas A&M University, College Station, Texas, USA.

HARPER, J. L. 1978. Striped bass research study. Performance report, Federal Aid Project F-29-R. Oklahoma Department of Wildlife Conservation, Oklahoma City, Oklahoma, USA.

KERR, S. R. 1971. Prediction of fish growth efficiency in nature. Journal of the Fisheries Research Board of Canada 28:809–814.

LI, H. W., P. B. MOYLE, AND R. L. GARRETT. 1976. Effect of the introduction of the Mississippi silverside, *Menidia audens,* on the growth of black crappie, *Pomoxis nigromaculatus,* and white crappie, *Pomoxis annularis,* in Clear Lake, California. Transactions of the American Fisheries Society 105:404–408.

MENSE, J. B. 1967. Ecology of the Mississippi silverside, *Menidia audens* Hay, in Lake Texoma. Fishery Research Laboratory Bulletin 6. Oklahoma Department of Wildlife Conservation, Norman, Oklahoma, USA.

MOYLE, P. B., AND N. J. HOLZHAUSER. 1978. Effects of the introduction of Mississippi silverside, *Menidia audens,* on the feeding habits of young-of-the-year largemouth bass in Clear Lake, California. Transactions of the American Fisheries Society 107:574–582.

NEY, J. J. 1981. Evolution of forage fish management in lakes and reservoirs. Transactions of the American Fisheries Society 110:725–728.

PALOHEIMO, J. E., AND L. M. DICKIE. 1966. Food and growth of fishes. III. Relations among food, body size, and growth efficiency. Journal of the Fisheries Research Board of Canada 23:1209–1248.

RICHARDS, K. R. 1977. Evaluation of the Mississippi silversides as a forage fish in Colorado. Master's thesis, Colorado State University, Fort Collins, Colorado, USA.

RIGGS, C. D., AND V. E. DOWELL. 1954. Some recent changes in the fish fauna of Lake Texoma. Proceedings of the Oklahoma Academy of Science 35: 37–39.

SAUNDERS, R. P. 1959. A study of the food of the Mississippi silversides, *Menidia audens* Hay, in Lake Texoma. Master's thesis, University of Oklahoma, Norman, Oklahoma, USA.

WERNER, E. E. 1979. Niche partitioning by food size in fish communities. Pages 311–322 *in* R. H. Stroud and H. Clepper, editors. Predator-prey systems in fisheries management. Sport Fishing Institute, Washington, D.C., USA.

Pages 156–165 *in* G. E. Hall and M. J. Van Den Avyle, editors.
Reservoir Fisheries Management: Strategies for the 80's.
Reservoir Committee, Southern Division American Fisheries Society, Bethesda, Maryland. 1986.

Fish Population Trends in Texoma Reservoir Following Establishment of Striped Bass

JACK L. HARPER AND HAROLD E. NAMMINGA
Oklahoma Department of Wildlife Conservation
P.O. Box 53465, Oklahoma City, Oklahoma 73152

ABSTRACT

Striped bass (*Morone saxatilis*) were stocked in Texoma Reservoir from 1965 through 1974 to create a sport fishery and help control abundant gizzard shad (*Dorosoma cepedianum*) and threadfin shad (*D. petenense*) populations. Most available evidence shows that the introductions were beneficial. The striped bass population expanded from 1978 to 1982 and became the most abundant sport species. Angler catch rate and harvest of striped bass nearly doubled from 1978 to 1980. White bass (*M. chrysops*) abundance declined somewhat during the study, apparently due to winterkill of threadfin shad in 1977. Threadfin shad were reintroduced in 1979 and became the most numerous species in 1980 and 1981 rotenone surveys. Predation and low water temperatures during winter led to decreased threadfin shad abundance again in 1982. Striped bass predation seemed to affect the size distribution of gizzard shad but had no apparent influence on native predator or prey species other than shad. In 1976, the length mode of gizzard shad was 254 mm, but as the striped bass population expanded, the abundance of small shad increased, and by 1982, 95% of the gizzard shad population was 102-127 mm long. Although black bass (*Micropterus* spp.) populations fluctuated, they generally increased coincident with the expansion of striped bass. Angler catch rates of black bass, white bass, and white crappie either increased or remained stable from 1978 to 1980.

The Oklahoma Department of Wildlife Conservation stocked approximately one million striped bass (*Morone saxatilis*) fingerlings in Texoma Reservoir from 1965 to 1974 to provide an additional sport fishery and a biological control on abundant gizzard shad (*Dorosoma cepedianum*) and threadfin shad (*D. petenense*). Stevens (1958) found that clupeid fish were the primary forage of striped bass in Santee-Cooper Reservoir, South Carolina, and that increased angling pressure on striped bass resulted in decreased pressure and increased catch rates for other game fish. Overall, he concluded that the striped bass had a positive influence on the reservoir and was an economic asset to the surrounding area. Bailey (1975) reported that of large reservoirs with established striped bass populations, only Percy Priest Reservoir, Tennessee, showed significantly reduced shad abundance. He also stated that fisheries managers believed striped bass introductions would not adversely affect native game fish.

The conclusions of Stevens (1958) and Bailey (1975) had not been verified for other inland reservoirs, and this investigation was initiated in March 1976 to evaluate fish population trends in Texoma Reservoir following introduction of the striped bass. This 35,600-hectare hydroelectric impoundment on the Red River, Texas and Oklahoma, was constructed in 1944 by the U.S. Army Corps of Engineers and consists of two distinct arms—the Red River arm (96 km long) and the Washita arm (72 km in length). The principal study area was the lower portion of the Washita River arm (Fig. 1). Specific objectives of this study were to determine if striped bass controlled gizzard shad, impacted native sport fishes and other forage species, and established an additional sport fishery.

METHODS

Sportfish reproduction was evaluated by shoreline seining in both arms of the reservoir at 12 locations in June–August of 1976 and 1977. From 1978 through 1981, all seining locations were in the Washita River arm. The number of seine samples ranged from 33 to 108 per year; data were expressed as numbers per hectare.

Electrofishing (alternating current) catch per unit of effort (15 minutes) was used to monitor black bass (*Micropterus* spp.) populations. Data were collected in June 1978–1981; sampling began at 0500 hours and continued approximately 4 hours each day for five consecutive days.

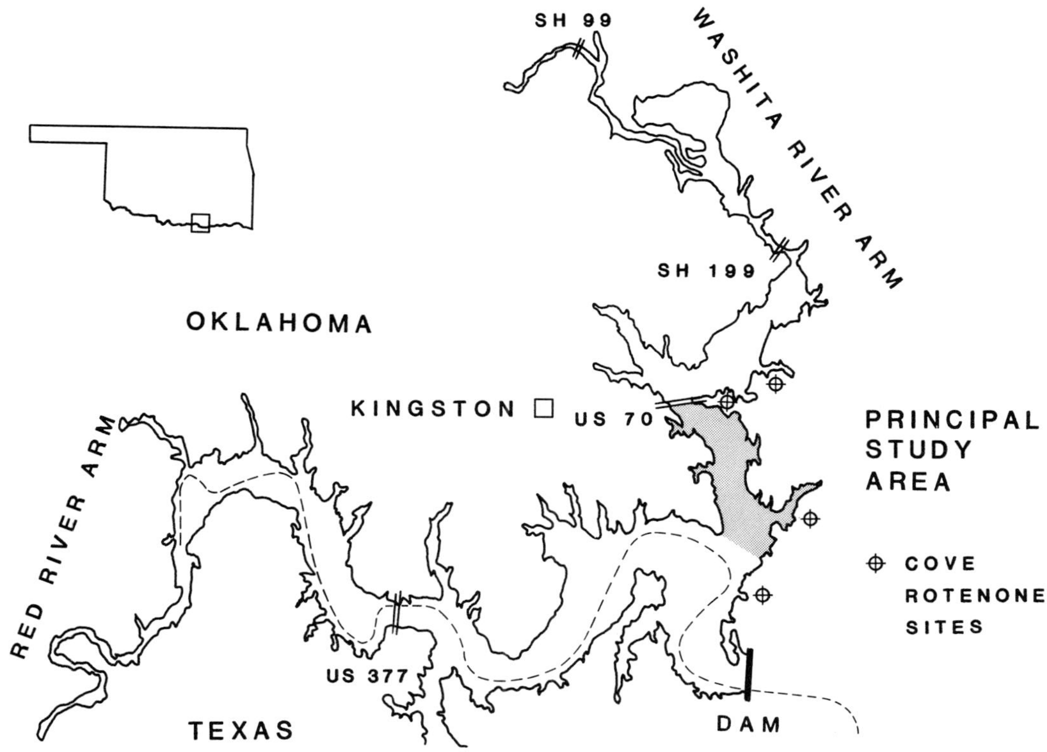

Figure 1. Map of Texoma Reservoir, showing study area and cove rotenone sites.

Gillnetting was conducted in spring (March and April) and fall (October) of 1978–1982 to determine relative abundance and mean weight of percicthyid basses (*Morone* spp.), crappie (*Pomoxis* spp.), and shad. Monofilament nets used were 60.8 m long, 1.8 m high, and had eight panels with bar meshes ranging from 19.1 mm to 101.6 mm. Two overnight sets were used per season at each of five stations in 1978. Effort was increased to three sets per station in 1979 and to four sets in 1980–1982. Catch rates were expressed as the average number collected per hour that nets were fished (about 20 hours per set).

Standing crops of fish were estimated by rotenone sampling in two coves annually (three coves in 1978) using procedures described by Grinstead et al. (1978). Four different coves were surveyed during the study (Fig. 1).

A roving creel survey was conducted March–November in 1978–1980. Pressure counts (five per day) and angler interviews were made from a boat during 60, 10-hour (daylight) days each year. The principal study area was divided into five sub-areas that were randomly selected for surveying each creel day. Twelve weekend days and eight week days were randomly selected in each 3-month quarter for interviewing anglers.

Results

Striped Bass

Trends in striped bass abundance, mean weight, and length frequencies appeared to be a function of population changes following stocking and production of a very abundant year class in 1979. Catches of Age-0 striped bass in seine samples tripled each year from 1976 to 1978 and increased sevenfold in 1979 over 1978 (Table 1). Numbers in 1980 declined to 1976 levels, but increased substantially again in 1981.

Catch rates in October gillnet samples increased about fourfold from 1978 to 1980 as Age-0 and yearling fish became susceptible to capture (Table 2). Spring gillnet catches increased throughout the study and peaked in 1982. Mean weight of striped bass in gillnets was great-

Table 1. Average number per hectare for young-of-the-year sport fishes collected by shoreline seining in June-August from Texoma Reservoir, 1976-1981.

Species	1976	1977	1978	1979	1980	1981
Striped bass	16	58	163	1,178	15	275
White bass	82	193	105	250	11	28
Largemouth bass	14	44	92	69	64	21
Spotted bass	85	13	140	50	10	41
Number of seine hauls	33	60	108	108	52	47

Table 2. Number of fish caught per hour in gillnets (C/f) and mean weight (kg) during spring (March and April) and fall (October) in Texoma Reservoir, 1978-1982.

Species	1978	1979	1980	1981	1982
			Spring samples		
Striped bass					
C/f	0.18	0.21	0.28	0.75	0.90
Mean weight	1.67	0.59	0.28	0.43	0.70
White bass					
C/f	0.25	0.21	0.11	0.01	0.07
Mean weight	0.15	0.27	0.34	0.13	0.17
Gizzard shad					
C/f	0.25	1.51	1.51	0.47	0.14
Mean weight	0.23	0.17	0.20	0.27	0.21
Threadfin shad					
C/f	0.00	0.00	1.02	0.08	0.00
Mean weight			0.03	0.03	
			Fall samples		
Striped bass					
C/f	0.24	0.99	1.03	0.64	0.47
Mean weight	1.22	0.64	0.40	0.89	0.55
White bass					
C/f	0.17	0.38	0.06	0.01	0.03
Mean weight	0.20	0.23	0.35	0.11	0.22
Gizzard shad					
C/f	1.48	0.96	0.90	0.35	0.45
Mean weight	0.16	0.17	0.17	0.18	0.17
Threadfin shad					
C/f	0.00	0.61	1.13	0.45	0.03
Mean weight		0.03	0.03	0.03	0.03

est in 1978 and lowest in 1980, when the abundant 1979 year class recruited into the samples.

The 1979 year class grew slowly and dominated the gillnet collections through 1982 (Figs. 2 and 3). Modal lengths in fall samples increased from about 150 mm in 1979 (at Age 0) to 250 mm in 1980 and 350 mm in 1981 (at Age 2). The strong 1979 year class still provided 60% of the striped bass netted in fall 1982.

Trends of striped bass abundance in cove rotenone samples were similar to those from seining. Catches were greatest in 1979 and 1981, when striped bass had the greatest density of any sport fish collected (Table 3). In both years, most of the striped bass collected were from the 1979 year class. In 1981, the high density of Age-2 striped bass resulted in the highest standing crop observed for any sport fish during the study (71 kg/hectare).

Striped bass harvest by anglers increased from 1978 through 1980, and catch rates were highest in 1980 (Table 4). Much of the increased success rate in 1980, however, was due to recruitment of the relatively small 1979 year-class fish, which led to a decline in the average size of harvested striped bass. This was most pronounced in fall 1980, when the catch rate reached 0.45 striped bass per hour.

White Bass

Trends in white bass (*Morone chrysops*) abundance and angler harvest suggest production of a strong 1979 year class as observed for striped bass. Numbers of Age-0 fish collected by seining increased from 1976 to 1979 and declined dramatically in 1980 and 1981 (Table 1). Numbers collected in rotenone surveys showed a similar trend—abundance of young increased from 1976 to 1979 and then dropped in 1980–1981 (Table 3).

The strong 1979 year class apparently led to increased angler harvest and catch per hour in 1980 (Table 4). However, the long-term influences of the 1979 year class were not clear. Catch rates of white bass in gillnets declined after 1979 (Table 2), and standing crop estimates were extremely variable in 1980–1982 (Table 3). The high density of young white bass in 1982 (Table 3) may indicate production of another strong cohort that year.

Black Basses

There were no major shifts in abundance of largemouth bass (*Micropterus salmoides*) or spotted bass (*M. punctulatus*) during the study. Numbers of young bass collected by seining (Table 1) and by cove rotenoning (Table 3) tended to be high in 1978 and 1979, whereas standing crop in coves (Table 3) and electrofishing catch rates (Table 5) were relatively high in 1981. This

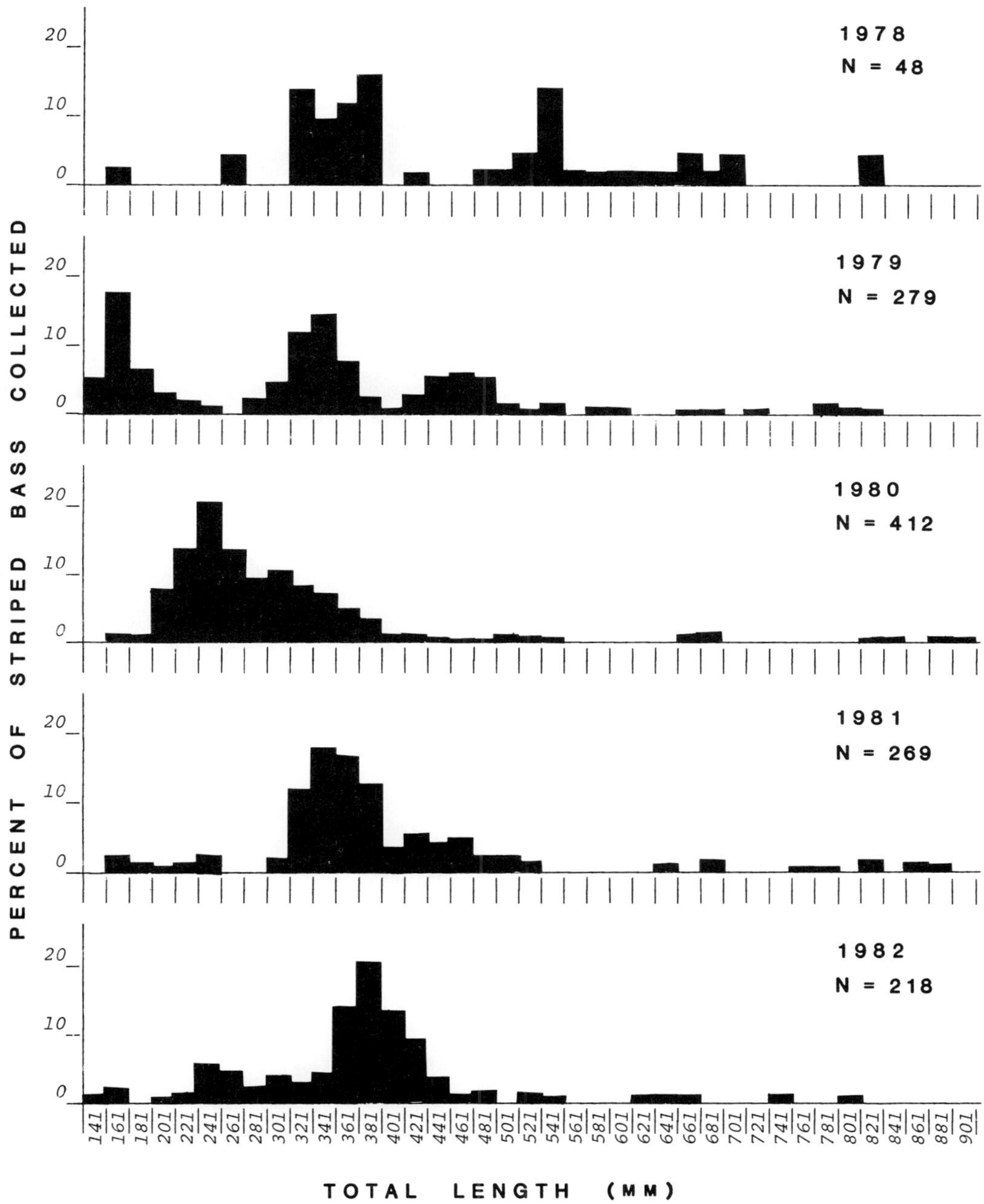

Figure 2. Length frequencies of striped bass collected by gillnetting during October in Texoma Reservoir, 1978-1982.

pattern likely reflects production of relatively abundant year classes in 1978 and, to a lesser extent, in 1979. The extremely high density of 1,128 largemouth bass per hectare in the 1982 cove rotenone sample was primarily Age-0 fish that increased the standing crop only slightly over the 1981 level (Table 3). Similarly, increased numbers of spotted bass in 1982 reflect a relatively strong year class that year.

Angler catch per hour of black bass was stable for 1978–1980 (Table 4). Therefore, the dramatic decline in numbers harvested from 16,000 in

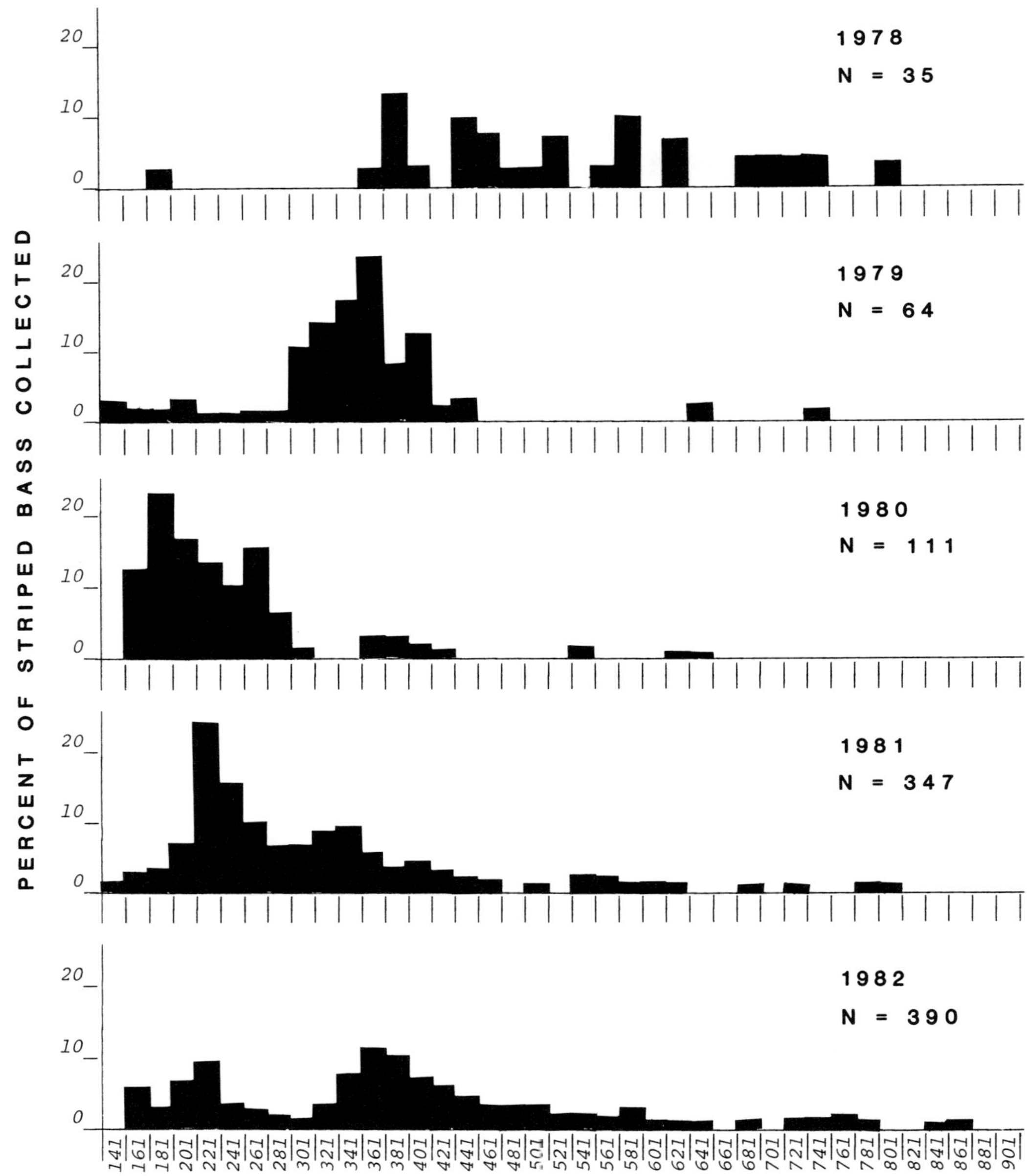

Figure 3. Length frequencies of striped bass collected by gillnetting during March and April in Texoma Reservoir, 1978-1982.

1978 to 1,900 in 1980 was apparently a function of reduced angling pressure rather than reduced black bass abundance. Average sizes of fish creeled were relatively constant over the three years (Table 4).

Crappies

Trends in white crappie (*Pomoxis annularis*) and black crappie (*P. nigromaculatus*) abundance were difficult to measure because these species were not effectively sampled by seining,

Table 3. Standing crop and density of important fish species collected by rotenone in Texoma Reservoir, 1976–1982.

Species	1976	1977	1978	1979	1980	1981	1982
			Standing crop (kg/hectare)				
Striped bass	0.3	0.9	0.2	4.7	0.6	71.5	24.1
White bass	1.1	13.7	8.7	17.0	5.6	13.7	5.9
Largemouth bass	7.2	17.8	10.5	16.0	10.0	22.7	24.3
Spotted bass	5.5	3.0	5.4	3.5	1.0	5.1	1.9
White crappie	0.9	6.0	4.3	3.7	1.1	24.5	10.8
Black crappie	0.5	0.4	0.8	1.7	1.0	0.3	1.5
Gizzard shad	535.1	269.1	371.9	320.9	265.4	339.8	1,168.6
Threadfin shad	1.9	0.0	0.0	1.8	21.8	196.1	2.4
Inland silversides[a]		2.1	0.4	7.3	1.3	0.5	2.8
Logperch	3.0	2.0	0.7	2.8	0.4	3.9	8.2
Cyprinids[b]	1.2	0.6	0.7	2.1	1.5	3.1	3.0
Bluegill	30.8	22.6	18.6	27.7	8.7	17.3	44.3
Longear sunfish	8.5	4.4	4.9	7.5	1.6	4.2	9.6
Redear sunfish	6.2	3.8	5.3	8.6	3.7	2.7	2.5
Green sunfish	0.7	0.6	0.6	1.0	0.2	0.6	2.6
			Density (number/hectare)				
Striped bass	8	90	30	925	33	991	335
White bass	15	143	100	178	88	121	350
Largemouth bass	235	308	313	430	320	258	1,128
Spotted bass	240	235	263	375	95	96	231
White crappie	13	93	40	255	25	141	668
Black crappie	3	33	23	155	10	6	203
Gizzard shad	4,148	5,398	4,003	4,088	3,263	7,837	92,508
Threadfin shad	733	0	0	253	11,275	41,012	461
Inland silversides[a]		2,628	265	6,753	1,043	436	1,874
Logperch	860	748	273	1,783	230	1,232	3,355
Cyprinids[b]	598	423	730	1,323	480	1,149	790
Bluegill	975	2,740	1,603	4,203	480	2,056	2,513
Longear sunfish	2,028	543	1,440	1,603	100	1,401	402
Redear sunfish	85	58	88	355	45	72	46
Green sunfish	35	68	80	138	13	106	101

[a] Included with cyprinids in 1976.
[b] Excluding goldfish (*Carrasius auratus*) and common carp (*Cyprinus carpio*).

gillnetting, or electrofishing. However, data from cove rotenone samples (Table 3) and creel surveys (Table 4) showed trends similar to those for white bass. Densities of young crappies in cove samples again suggested production of strong year classes in 1979 and 1982. The high angler catch per hour of crappies in 1980 (Table 4) and the high standing crop of white crappies in 1981 apparently resulted from the abundant 1979 year class. As with black bass, angler effort for crappies declined substantially from 1978 to 1980.

Forage Fish

The gizzard shad population showed a pronounced increase in abundance of young and disappearance of older fish during the study. Densities of small gizzard shad in rotenone samples were relatively consistent from 1976 through 1980, but abundance increased dramatically in 1981 and especially 1982 (Table 3). Gillnet catches of larger shad, however, declined after 1980 (Table 2) as did the catches of shad longer than 140 mm in cove samples (Fig. 4). Hence, overall gizzard shad biomass was relatively stable from 1976 through 1981 despite changes in size composition of the population. In 1982, the extremely high density of gizzard shad shorter than 140 mm provided a standing crop of about 1,200 kg/hectare (Table 3).

Trends in threadfin shad abundance were greatly influenced by winter water temperature. The species became established in Texoma Reservoir in 1957 (Riggs and Moore 1958) and was abundant until cold winters in 1976 and 1977 resulted in essentially complete mortality. Hence, none were collected in 1977 or 1978. Threadfin shad were reintroduced by Oklahoma and Texas in March and May 1979 and appeared in fall

Table 4. Effort (hours directed toward certain species), number caught per hour (C/f), total numbers caught, and mean weights of sport fish harvested by anglers in the lower Washita River arm of Texoma Reservoir, March-November in 1978-1980.

Species	1978	1979	1980
Striped bass			
Effort	15,400	28,100	20,800
C/f	0.22	0.20	0.39
Number	3,400	5,600	8,100
Mean weight (kg)	1.97	1.82	1.17
White bass			
Effort	34,400	47,900	27,100
C/f	0.77	0.44	2.28
Number	26,400	21,200	61,600
Mean weight (kg)	0.34	0.41	0.18
Black basses			
Effort	39,300	15,600	5,200
C/f	0.41	0.37	0.37
Number	16,000	5,700	1,900
Mean weight (kg)	0.62	0.58	0.52
Crappies			
Effort	5,700	3,300	2,900
C/f	0.83	0.26	1.15
Number	4,800	800	3,300
Mean weight (kg)	0.41	0.36	0.11

1979 gillnet and rotenone samples (Tables 2 and 3). They increased in abundance in 1980 and 1981, but were reduced to low levels in 1982 following another cold winter.

Many other species of fish can serve as forage for predators in Texoma Reservoir, including the inland silversides (*Menidia beryllina*), logperch (*Percina* spp.), various cyprinids, bluegill (*Lepomis macrochirus*), longear sunfish (*L. megalotus*), redear sunfish (*L. microlophus*), and green sunfish (*L. cyanellus*). Densities of nearly all of these species in cove samples were typically high in 1979 and in either 1981 or 1982 (Table 3), reflecting production of relatively large numbers of young in those years.

Discussion

The overall impact of striped bass on the Texoma Reservoir fishery appears to be beneficial. A significant year-round fishery has developed much like that described for Smith Mountain Lake, Virginia (Hart 1978). Prior to 1979, striped bass angling in Texoma Reservoir occurred mainly in spring with no concentration of anglers in any one area. Since 1979, striped bass receive greatest angling pressure during the summer when

Table 5. Number of largemouth and spotted bass caught per 15 minutes of electrofishing (C/f) and mean weight during June in Texoma Reservoir, 1978-1981.

Species	1978	1979	1980	1981
Largemouth bass				
C/f	1.33	1.48	1.36	2.79
Mean weight (kg)	0.22	0.25	0.32	0.26
Spotted bass				
C/f	0.00	0.28	0.18	1.05
Mean weight (kg)		0.13	0.18	0.11

the bigger fish gather in large schools in the main body of the lake (Summers 1982). The fish follow a daily feeding routine that beings just prior to dawn and lasts until 0900 to 1100 hours. During this time, there often are 100–300 boats fishing over the school. The apparent abundance of striped bass led to increases in the daily creel limit from three in 1977, to five in 1980, and to 15 in 1982.

The striped bass does not appear to have influenced other sport fishes or most forage species in Texoma Reservoir. Changes in abundance of several species appeared to be the result of periodic strong year classes rather than an outcome of competition with, or predation by, striped bass. The coincidence of strong 1979 year classes of white bass, crappies, and most prey species other than shad followed by additional abundant year classes in 1981 or 1982 suggests that these populations are regulated by some factors other than the striped bass.

Trends in the black bass fishery appeared to be a function of changing angler desires and production of relatively strong year classes in 1978, 1979, and 1982. The 1978 and 1979 year classes first gave high catches in seine and rotenone samples, and successful recruitment was later reflected by increased electrofishing catch rates. These increases coincided with expansion of the striped bass population. Paul Mauck (Oklahoma Department of Wildlife Conservation, Durant, Oklahoma, USA [personal communication]) also found that spring electrofishing catch rates of black bass in Texoma Reservoir during 1980 were equivalent to rates in 1977, prior to expansion of the striped bass population. The proportional stock density (PSD) of 55% in 1980 was substantially greater than the 30% level found in 1977. This improved PSD is indicative of a bass population with large numbers of quality-size bass

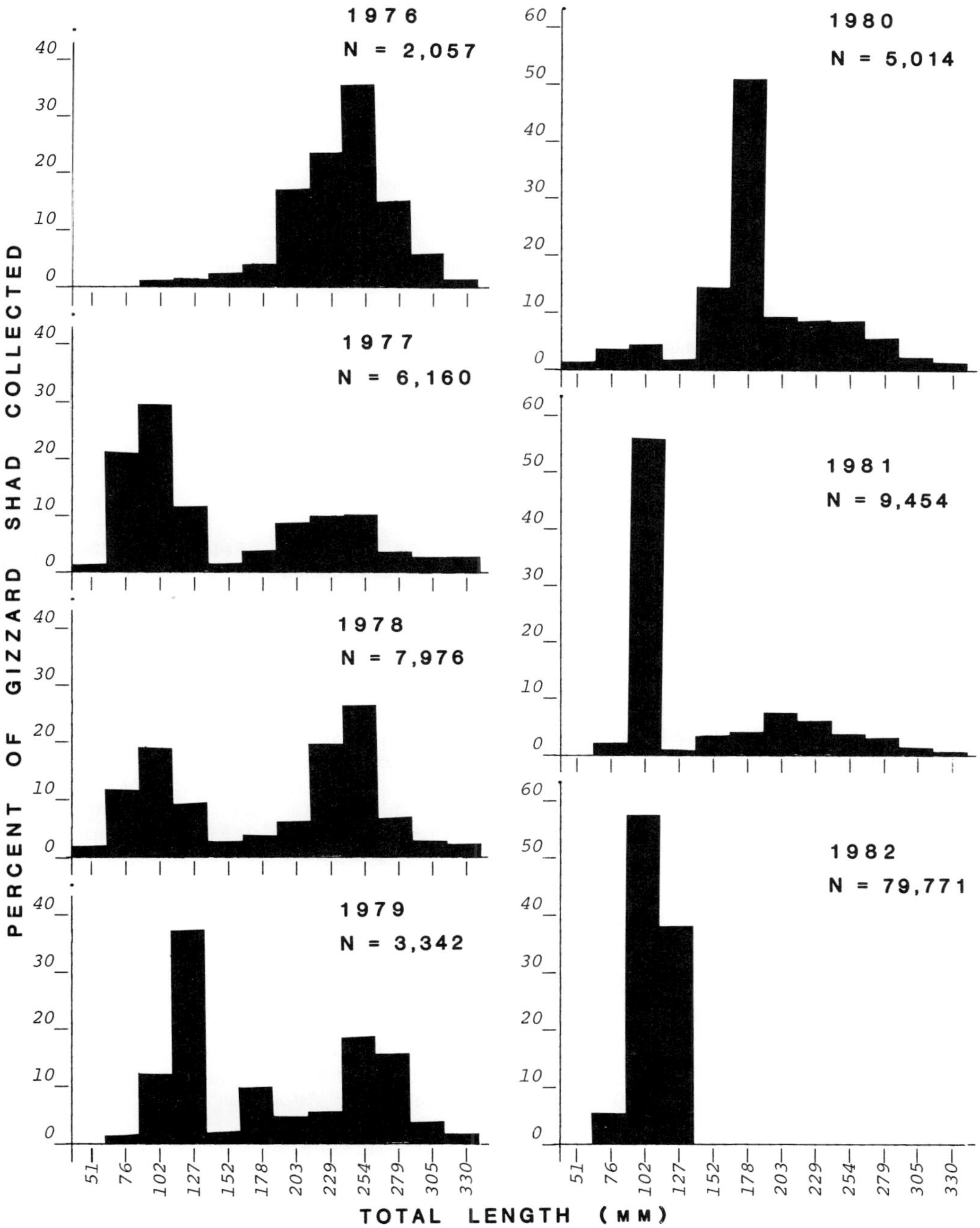

Figure 4. Length frequencies of gizzard shad collected in cove rotenone surveys in Texoma Reservoir, 1976-1982.

(304 mm or greater; Anderson and Weithman 1978). Although black bass catch per hour by anglers remained stable for 1978–1980, effort directed toward black bass decreased 88%. This suggests many bass anglers converted at least partially to striped bass fishing, and this conversion resulted in decreased pressure on largemouth bass. This phenomenon, though not to the same extent, has previously been reported in Santee-Cooper Reservoir, South Carolina (Ste-

vens 1958), and Smith Mountain Reservoir, Virginia (Hart 1978).

The striped bass has shown essentially no predation on native sport species in Texoma Reservoir (Harper 1982; William J. Matthews, University of Oklahoma Biological Station, Kingston, Oklahoma, USA [personal communication]). Rather, they primarily have a shad diet, as reported for other reservoirs (Stevens 1958; Combs 1979; Morris and Follis 1979).

There is some evidence that striped bass predation has affected gizzard shad in Texoma Reservoir whereas threadfin shad appear to have been regulated by winter weather and striped bass predation. Following elimination during extremely cold winters in 1976 and 1977 and reintroduction in the spring of 1979, threadfin shad expanded rapidly. The winter of 1981–1982 did not cause a complete winterkill, but temperatures were low enough (4 to 6 C) to make threadfin shad easier prey for striped bass.

Intense predation of threadfin shad by striped bass occurred in Texoma Reservoir during January and February 1982 (William J. Matthews, personal communication). In this case, it appears that predation may have acted in concert with temperature to nearly eliminate threadfin shad, but the relative importance of the two factors cannot be determined from available data. Rotenone and gillnet data in this study verified an extreme decline in threadfin shad numbers in 1982.

It does appear that striped bass affected the size distribution of gizzard shad. In 1976, large gizzard shad 203 to 305 mm were most abundant, but in the next three years, a second size class (102–127 mm) was also present. Factors contributing to this change probably included increased predation by striped bass, disappearance of threadfin shad, and decreased numbers of the larger gizzard shad possibly due to winter kill. In 1981 and 1982 the gizzard shad population in our study area was dominated by the 102–127 mm size group. These results are similar to those reported for Smith Mountain Reservoir (Hart 1978). Matthews (personal communication), however, found an abundance of 254–305 mm gizzard shad in the upper Red River arm of Lake Texoma in 1982 and 1983.

Structural changes of the gizzard shad population culminated in 1982 with high numbers (92,500/hectare) of small shad, which suggests that as threadfin shad declined, the gizzard shad population expanded. Under certain conditions, threadfin shad have been shown to suppress gizzard shad production (Bryant and Houser 1969), and their demise could account for the dramatic increase in gizzard shad numbers observed in 1982. Predation by striped bass also may have contributed to increased reproduction by gizzard shad in 1982. A relationship between largemouth bass predation and gizzard shad reproduction was reported by Anderson (1973) for Missouri ponds.

Acknowledgments

This study was funded by the Federal Aid Project, F-29-R, Oklahoma Department of Wildlife Conservation. Assistance by Paul Mauck and the Southcentral Region fishery crew for collection of 1982 data is appreciated.

References

Anderson, R. O. 1973. Application of theory and research to management of warmwater fish populations. Transactions of the American Fisheries Society 102:164–171.

Anderson, R. O., and A. S. Weithman. 1978. The concept of balance for coolwater fish populations. Pages 371–381 *in* R. L. Kendall, editor. Selected coolwater fishes of North America. Special Publication 11. American Fisheries Society, Bethesda, Maryland, USA.

Bailey, W. M. 1975. An evaluation of striped bass introductions in the Southeastern United States. Proceedings of the Annual Conference of the Southeastern Association of Game and Fish Commissioners 28:54–68.

Bryant, H. E., and A. Houser. 1969. Growth of threadfin shad in Bull Shoals Reservoir. Proceedings of the Annual Conference of the Southeastern Association of Game and Fish Commissioners 22: 275–283.

Combs, David L. 1979. Food habits of adult striped bass from Keystone Reservoir and its tailwaters. Proceedings of the Annual Conference of the Southeastern Association of Fish and Wildlife Agencies 32:571–575.

Grinstead, B. G., R. M. Gennings, G. R. Hooper, C. A. Schultz, and D. A. Whorton. 1978. Estimation of standing crop of fishes in the predator-stocking-evaluation reservoirs. Proceedings of the Annual Conference of the Southeastern Association of Fish and Wildlife Agencies 30:120–130.

Harper, J. L. 1982. Population trends (Texoma Reservoir). Final report, Federal Aid Project F-29-R. Oklahoma Department of Wildlife Conservation, Durant, Oklahoma, USA.

Hart, L. G. 1978. Smith Mountain Reservoir research study. Final report, Federal Aid Project F-30-R. Virginia Department of Game and Inland Fisheries, Richmond, Virginia, USA.

Morris, D. J., and B. J. Follis. 1979. Effects of striped bass predation upon shad in Lake E. V. Spence, Texas. Proceedings of the Annual Conference of the Southeastern Association of Fish and Wildlife Agencies 32:697–702.

Riggs, C. D., and G. A. Moore. 1958. The occurrence of *Signalosa petenensis* in Lake Texoma. Proceedings of the Oklahoma Academy of Science 38:64–67.

Stevens, R. E. 1958. The striped bass of Santee-Cooper Reservoir. Proceedings of the Annual Conference of the Southeastern Association of Game and Fish Commissioners 11:253–264.

Summers, G. L. 1982. Striped bass research study. Final report, Federal Aid Project F-29-R. Oklahoma Department of Wildlife Conservation, Norman, Oklahoma, USA.

Pages 166–177 *in* G. E. Hall and M. J. Van Den Avyle, editors.
Reservoir Fisheries Management: Strategies for the 80's.
Reservoir Committee, Southern Division American Fisheries Society, Bethesda, Maryland. 1986.

Filling the Void: Development of a Pelagic Fishery and its Consequences to Littoral Fishes in a Virginia Mainstream Reservoir

CHRISTOPHER C. KOHLER

Fisheries Research Laboratory
Southern Illinois University
Carbondale, Illinois 62901

JOHN J. NEY AND WILLIAM E. KELSO

Department of Fisheries and Wildlife Sciences
Virginia Polytechnic Institute and State University
Blacksburg, Virginia 24061

ABSTRACT

Riverine fishes are often unable to expand in new reservoirs, and nonnative species are routinely introduced. Between 1939 and 1980, at least 14 fish species and two hybrids were intentionally stocked in Claytor Lake, a 1,820-hectare hydroelectric impoundment of the New River in southwestern Virginia. Stockings of centrarchids predominated during the first 15 years of impoundment, whereas more recent efforts were aimed at developing a pelagic fishery for striped bass, walleye, and white bass. The alewife was introduced concurrently with striped bass and it became the dominant pelagic forage fish. Attempts to develop a two-story fishery with rainbow trout failed. Stockings of northern pike, muskellunge, and tiger musky had little or no influence on the fishery.

The pelagic fishery has not fulfilled all expectations. Striped bass exhibit slow growth and downstream emigration. White bass are abundant, but their harvest is highly seasonal. The walleye fishery, considered the best in the state until the 1960's, declined in importance following establishment of white bass and alewife. Interactions between pelagic and littoral predators are limited by trophic and habitat partitioning. Alewives may have negative effects on the young of littoral and pelagic species through predation and interspecific competition for the zooplankton resource.

The Claytor Lake experience emphasizes the need for thorough ecological analysis when considering species for introduction. Stocking strategies should be based on analysis of a species' past performance and compatibility with the resident fish assemblage to provide reasonable assessment of its risks as well as benefits.

One challenge to the reservoir fishery manager is the establishment of a species complex to utilize and provide fishing opportunity in the pelagic zone. Resident riverine species are seldom capable of colonizing the pelagic expanses which follow inundation in storage and hydroelectric reservoirs; they either disappear or are confined to near-shore areas (Timmons et al. 1979; Rainwater and Houser 1982). Consequently, managers have strived to fill the perceived vacant pelagic niches by introducing nonnative fishes with characteristics thought to be compatible with management objectives.

Reservoir introductions of pelagic fishes have generally proceeded on a trial-and-error basis, guided largely by qualitative judgements because of the absence of performance assessments in similar systems or an accepted protocol for evaluating nonnative species (Jenkins 1970; Li and Moyle 1981; Kohler and Stanley 1984). Candidate species have been selected according to a combination of three criteria: 1) availability, 2) demonstrated success in other waters, and 3) perceived attributes. Advances in fish culture have assured that almost any fish species can be made available for transplantation. Indeed, virtually every North American predator fish species has been introduced outside its original range (McCammon and von Geldern 1979). Recently, the emphasis has broadened to include prey species to expand the pelagic forage base (Jenkins 1974; Ney 1981).

Introduction of any species to enhance recreational fishing has three possible outcomes. In the most frequent scenario, the species fails either to sustain itself or to provide a fishery. Since most fishes are stocked as fry or fingerlings, the lag time between the rise in angler expectations and their ultimate deflation is usually sufficient so that public interest fades gradually, and no accounting is demanded. In the second instance, the introduction is successful, providing a new fishery or benefits to resident game fish; managers reap public approval and may stock the species in other waters. An enthused public may "help out" with illegal introductions in nontargeted systems (McCammon and von Geldern 1979). The third outcome, however, highlights the major risk. The species proves to be detrimental to the resident fish community through competition, predation, or habitat degradation so that angler harvest actually declines. If this fish is successful in establishing a self-sustaining population, the damage may be irrevocable.

As long as a trial-and-error approach prevails, both the first and third outcomes will occur with unhappy frequency. Greater assurance of a successful introduction can be obtained if the manager first obtains answers to two fundamental questions: 1) how well will the ecological requirements of the candidate species be met in this system? and 2) how will it impact the resident fish community? The answers to these questions are not easy to develop because each ecosystem is unique, and inferences regarding ecological potential and community relations can be transferred only with caution. A more serious limitation, however, may be the paucity of readily available evaluative information on which to base a decision. Long term assessment of the performance of nonnative pelagic fishes in reservoir fisheries has seldom been reported.

In this paper, we assess the record of species introductions over a 40-year period in a Virginia mainstream reservoir which has had a regionally typical stocking history. Our objectives are to consider population establishment as a function of the reservoir environment and evaluate the compatibility of introduced pelagic species and resident littoral fishes in contributing to the recreational fishery. Our broader goal is to generalize from this experience to promote a reasoned approach to manipulating the species composition of reservoirs.

Study Site

Claytor Lake was formed in 1939 as a hydroelectric impoundment of the New River in southwestern Virginia. The reservoir has a surface area of 1,820 hectares at normal pool elevation of 663 m above mean sea level and a mean depth of 15 m. Claytor Lake retains a riverine morphometry throughout its 26-km length, with widths varying from 0.29 to 0.95 km; littoral regions (<5 m depth) are limited to several sparsely-vegetated coves which grade steeply into the main channel. The average annual storage ratio is a relatively low 0.09; 80% of U.S. reservoirs surveyed by Jenkins (1967) had higher ratios. The moderately eutrophic reservoir is dimictic, and anoxic conditions occur in the hypolimnion during summer.

The fish assemblage of Claytor Lake is comprised of introduced species as well as those that were in the New River at the time of impoundment and succeeded in colonizing the reservoir. Among the latter group, the more abundant species include: smallmouth, largemouth, and spotted basses (*Micropterus dolomieui, M. salmoides,* and *M. punctulatus,* respectively); bluegill, pumpkinseed, and green sunfishes (*Lepomis macrochirus, L. gibbosus,* and *L. cyanellus,* respectively); black and white crappies (*Pomoxis nigromaculatus* and *P. annularis*); channel catfish (*Ictalurus punctatus*); flathead catfish (*Pylodictus olivaris*); yellow perch (*Perca flavescens*) and the common carp (*Cyprinus carpio*). It is uncertain whether the walleye (*Stizostedion vitreum*) population in Claytor Lake originated from an endemic river stock or through subsequent stockings (Murphy 1981).

The management history of Claytor Lake between 1939 and 1980 follows a rather typical pattern of stockings and regulations which incorporated prevailing management strategies. Initially (1939–1946), emphasis was directed toward establishing a fishery through stocking of centrarchids and walleye. The decline in angler harvest resulting from decreased habitat and fertility as the reservoir aged prompted demands for more restrictive regulations. Although size and creel limits remained modest, the reservoir was closed to all fishing from January through mid-June in the late 1940's (later reduced to April and May) primarily to protect the black basses. Restrictive regulations were gradually eased as their general ineffectiveness in warmwater reservoirs became known (Jenkins 1970).

Present regulations include only bag limits for black basses, walleye, and striped bass and minimum size limits for black basses and striped bass. Attempts to manipulate the habitat of Claytor Lake to benefit the fishery have been limited due to the demands of hydroelectric operation.

Table 1. Fish stocked in Claytor Lake, Virginia, 1939–1979 (unpublished data from Virginia Commission of Game and Inland Fisheries, Richmond, Virginia, USA).

Years	Species	Life stage stocked	Number of stockings	Number stocked
1939–1949	Walleye	fingerlings	5	650,000
	Bluegill	mixed	6	61,124
	Longear sunfish	adults	3	12,040
	Largemouth bass	mixed	8	30,448
	Smallmouth bass	fingerlings	2	3,000
	White crappie	juveniles	2	2,679
1950–1959	Largemouth bass	unknown	2	2,280
	White bass	adults	1	99
1960–1969	Muskellunge	fingerlings	6	1,410
	Northern pike	fingerlings	1	6,280
	Threadfin shad	unknown	1	unknown
	Alewife	juveniles	2	110,000
	Striped bass	fingerlings	2	6,300
	White × striped bass	fingerlings	2	700
	Rainbow trout	adults	1	22,000
	Smallmouth bass	unknown	1	5,000
1970–1979	Walleye	fingerlings	2	208,397
	Brown trout	adults	1	67,000
	Rainbow trout	adults	1	22,103
	Muskellunge	fingerlings	7	5,078
	Striped bass	fingerlings	10	645,525
	Tiger musky	fingerlings	3	1,997

Efforts have been made since the 1950's to stabilize spring water levels to promote spawning success (Hart 1981). Beginning in the 1960's, and continuing to the present, management efforts have focused on establishing and maintaining a pelagic fishery. Species have been introduced to establish self-sustaining populations and to provide increased and more diverse angler opportunity through put-grow-take stocking. Species chosen for introduction were typical of those stocked in southeastern reservoirs during the last 20 years (Jenkins 1974; Pritchard et al. 1978).

Investigations have been conducted on the Claytor Lake fishery periodically since 1948 by fisheries scientists from the Virginia Commission of Game and Inland Fisheries (VCGIF) and Virginia Polytechnic Institute and State University (VPI & SU). These studies provide the basis for our analysis of trends in the fishery and the responsible factors. This data base is limited in that it is not continuous, and quantitative information is not always comparable among studies that had different objectives and utilized different sampling techniques. When considered collectively, however, these studies provided the basis for describing the evolution of the fishery and the means to identify potential sources of change.

Species Introductions and Fates

At least 14 fish species and two hybrids have been intentionally stocked in Claytor Lake since its impoundment in 1939 (Table 1). The annual stocking density has exceeded 25 fish per hectare (Hart 1981). These introductions can be partitioned into three categories: littoral species (centrarchids); pelagic predators (percids, salmonids, esocids, percicthyids); and pelagic forage fish (clupeids). Although crappie (*Pomoxis* spp.) use both regions, the crappie fishery is primarily in open water, and we categorize them as pelagic to facilitate our comparisons.

Littoral Species

Although black basses and bluegill were stocked repeatedly during the first 15 years following impoundment, they are native to the New River, and likely became self-established. Three separate stockings of nonnative longear sunfish (*Lepomis megalotis*) failed to establish a population.

Pelagic Predators

Percids.—Walleye were stocked during the first six years following impoundment and were self-sustaining and important until the late 1960's.

Their steady decline through the mid-1970's prompted the VCGIF to initiate a fingerling maintenance-stocking program in 1975. Because of the apparent failure of walleye spawning runs upstream in the reservoir headwaters, fingerlings were obtained from lake-spawning Kansas and Nebraska stocks. Nevertheless, relatively few walleye were caught by fishermen, and the cost per harvested fish was estimated to exceed $27 (Murphy 1981).

Salmonids.—Efforts to develop a "two-story" fishery with catchable-sized rainbow and brown trout (*Salmo gairdneri* and *S. trutta*) in 1969 and 1970 failed. Claytor Lake is not suitable for trout because waters at depths with proper water temperature (21 C maximum) in August and September rarely contain sufficient oxygen (Boaze 1972).

Esocids.—Muskellunge fingerlings were stocked annually in Claytor Lake from 1963 through 1976, excluding 1965. Since 1977, tiger muskellunge (muskellunge × northern pike, *Esox lucius*) have been stocked yearly. Although musky have never contributed much to the fishery, their stocking has fulfilled the objective of providing occasional trophy fish. The single stocking of approximately 6,000 northern pike in 1969 failed to produce an established population, and no additional stockings of this species have been made.

Percicthyids.—White bass were established in 1959 from the introduction of 99 adults (Martin 1970). Harvest of white bass is seasonal, with most fish creeled in spring (spawning runs) and fall. Striped bass fingerlings have been stocked annually since 1968 at average rates of nearly 28 fish per hectare, but a significant fishery has not developed, and few fish larger than 6 kg have been creeled. Adult striped bass are frequently caught in the Claytor Lake tailrace, and their emigration from Claytor Lake has been verified. West Virginia Department of Natural Resources biologists reported that striped bass occur in fishable numbers in downstream Bluestone Reservoir and in the Ohio River (Hart 1981); these waters are, respectively, more than 100 and 250 km downstream from Claytor Lake.

Crappie.—Two white crappie stockings totaling less than 4,000 juveniles were made in the 1940s. White crappie are likely native, and were the dominant crappie species in the late 1940's (Rosebery 1950), but black crappie have become more abundant. White crappie generally prevail over black crappie when waters are turbid (Goodson 1966). The frequent occurrence of turbid conditions in the headwaters of Claytor Lake and clearer water in the remainder of the reservoir may permit both species to thrive. Crappie are second only to black basses as the species most sought-after by anglers (Hart 1981).

Pelagic Forage Fish

Threadfin shad (*Dorosoma petenense*) were introduced into Claytor Lake in 1964 and 1965 but did not become established, presumably due to intolerance of the winter temperature regime (Boaze 1972). Alewife were introduced in 1968 and 1969, and a self-sustaining population resulted. The alewife was introduced to provide forage for the walleye and white bass, as well as the pelagic predators that were stocked concurrently with alewife (i.e., striped bass, northern pike, brown trout, and rainbow trout). Alewife rapidly became a major component of the species complex of Claytor Lake (Boaze and Lackey 1974), and are now the numerically dominant pelagic species. However, precipitous declines in alewife biomass occur due to periodic and massive die-offs (Kohler and Ney 1981).

Interactions Between Pelagic Introductions and Littoral Residents

Interactions between pelagic species and littoral residents consist of predation and competition for food. As would be expected, space is effectively partitioned between openwater and shoreline areas to minimize habitat competition. Demands for spawning habitat do not generally appear to coincide in location and timing.

Predation

Pelagic on littoral species.—Prior to the introduction of alewife, the primary forage species were bluegill and crappie (Rosebery 1950; Boaze 1972). Walleye and white bass switched to alewife when they became available (Boaze 1972). Striped bass prey almost exclusively on alewife (Kohler and Ney 1982a). Consequently, alewife serve as a buffer between littoral residents and pelagic predators. However, this buffer is periodically disrupted when alewife undergo massive die-offs, such as occurred in the winter of 1977–1978 (Kohler and Ney 1981). Following that die-off, alewife utilization by piscivores was negligible, and pelagic predators switched to diets that essentially overlapped those of littoral predators, preying on bluegill, crappies, yellow perch, and crayfish. However, the alewife population recovered rapidly, and pelagic predators returned to an alewife diet within a year of the die-off.

Alewife in Claytor Lake are occasional piscivores, and were found to have consumed post-larvae (maximum 26 mm total length) of at least five fish species (largemouth bass, white bass, *Lepomis* spp., yellow perch, and golden shiner, *Notemigonus crysoleucas*), as well as their own young (Kohler and Ney 1980). Alewife piscivory appeared to be at least partially nocturnal and was considerably more prominent in littoral than in limnetic areas.

Littoral on pelagic species.—Few pelagic alewife are eaten by littoral centrarchids. Alewife have been observed to migrate into shallow coves at night, and consequently would then be available to nocturnal predators. Alewife have been observed to seasonally comprise nearly 75% volume of smallmouth bass diets, but alewife are a rare component in the diet of largemouth bass or crappie (Kohler 1980). Likewise, young pelagic predators are only occasionally consumed by littoral predators.

Competition

Piscivores.—The pattern of predation results in an effective partitioning of the prey resource between pelagic and littoral piscivores. Major breakdowns of this partitioning occur following alewife die-offs.

Planktivores.—Although alewives are primarily pelagic, they may interact with littoral communities when they move into shallow coves at night (Kelso and Ney 1985). Nocturnal movements into littoral areas had been previously reported (Hutchinson 1971; Gannon 1976), but the significance of these migrations had not been assessed. Studies of trophic overlap between Claytor Lake alewives and littoral juvenile sportfishes indicated possible competition, despite partial resource partitioning due to species differences in choice of foods and feeding periodicity. Due to the high foraging efficiencies and school densities, migrating alewives may significantly reduce the forage base available to the littoral species. Alewife may also indirectly compete with other planktivorous species or life stages by altering the zooplankton species complex. Alewife in Claytor Lake were found to be highly size-selective planktivores (Kohler and Ney 1981), as has been demonstrated in northern lakes of the United States (Brooks 1968; Hutchinson 1971; Janssen 1976). Alewife selectively prey on zooplankters >1.0 mm length with sufficient intensity to exert a major influence on species and size composition of the zooplankton community (Kohler and Ney 1982b). Alewife-induced shifts in zooplankton composition toward smaller forms represent a potentially significant adverse impact on cohabiting planktivores, including young-of-the-year sportfishes (Kohler and Ney 1981).

Trends in the Fishery

The numerous Claytor Lake fish stockings have resulted in diversification of the pelagic fishery. The net impact of these introductions on the fishery is more difficult to assess. Our interpretation is derived from time-series comparisons of species abundance, harvest, and rates of growth. A potentially confounding factor in this analysis is the accelerated eutrophication of Claytor Lake during the 1970's due to the discharge from a municipal sewage treatment plant into Peak Creek, the reservoir's major tributary (Anonymous 1975). The total standing crop of fish, as measured in summer cove rotenone collections, increased from 100 kg per hectare in the 1940's–1960's to approximately 300 kg per hectare in the late 1970's. Although the cove rotenone estimates were subject to some sampling error (Hart 1981), the trend appears real. However, differences in sampling techniques restrict our temporal comparisons to relative rather than absolute abundance and harvest. Growth rate methodology was consistent throughout the period.

Relative Abundance

The species composition of summer cove rotenone collections changed markedly between 1948 and 1978 (Fig. 1). Among littoral species, whose relative abundance should be most accurately reflected in these collections, the primary trend was the replacement of black basses by carp. Black basses have declined steadily in relative abundance, from approximately 25% of cove biomass in 1948 to <5% in the late 1970's. Conversely, carp have increased to represent one-third of cove biomass (approximately equivalent

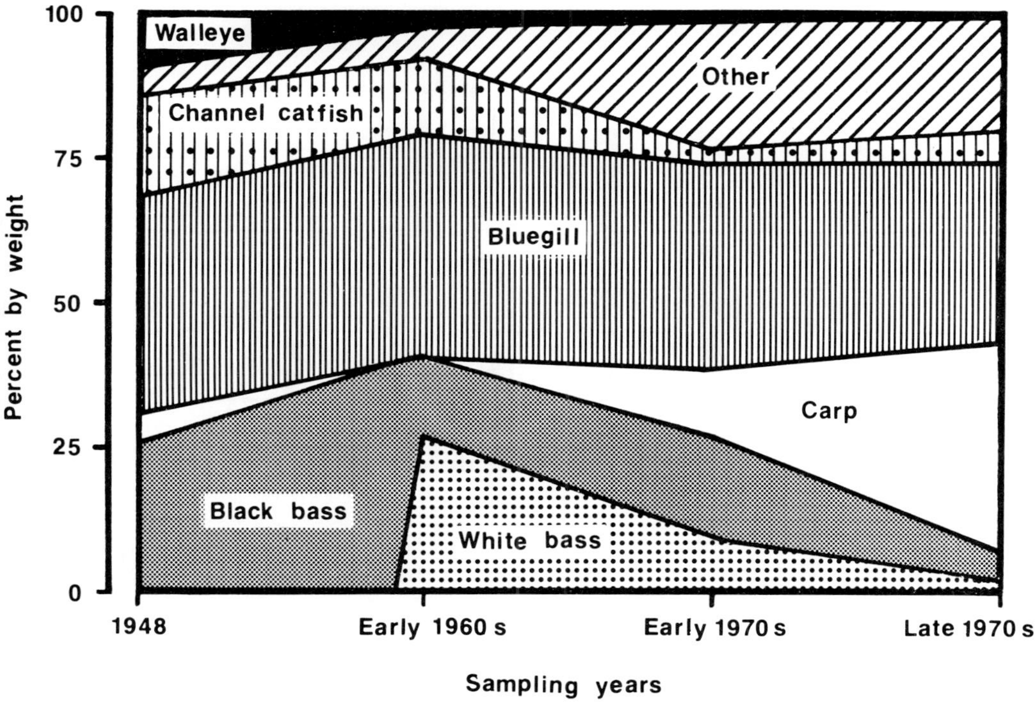

Figure 1. Percent weight of major fish species collected in cove rotenone samples at Claytor Lake, Virginia in 1948, early 1960's, early 1970's, and late 1970's (Hart 1981). "Other" includes pumpkinseed sunfish, green sunfish, crappie, and yellow perch.

to the entire cove standing crop in the first two decades of impoundment).

The composition of the pealgic fish assemblage is more difficult to describe because the species are under-represented in cove rotenone collections, and gillnets and other open-water collecting gear are size- and species-selective. However, these sources indicate major differences in relative abundance. Prior to introduction of white bass in 1959, walleye and crappie were the principal inhabitants of the pelagic zone. Following introduction, white bass expanded to the extent that even in coves they comprised one-fourth of the biomass before declining to a relatively minor component (Fig. 1). Conversely, walleye declined sharply in cove samples between 1948 and the early 1960's and constituted <1% of cove biomass in the late 1970's. Striped bass and alewife are nocturnal transients in reservoir coves and so were largely invulnerable to the cove rotenone collections during daylight hours. Fall gillnet surveys conducted between 1974 and 1979 captured striped bass, white bass, and walleye in approximately equal numbers (Hart 1981). However, striped bass numerical catch-per-unit-effort (CPUE) in gillnets averaged only 13% of the rate under equivalent sampling conditions in Smith Mountain Lake, Virginia's premier put-grow-take striped bass fishery (LaRoche 1982). Numerical CPUE of alewife in Claytor Lake gillnets exceeded that of all other species combined.

Harvest

Fishing pressure on Claytor Lake, as determined from creel surveys, has risen in recent years (Hart 1981). Daylight creel surveys conducted in 1948 and in 1976–1977 estimated angler effort at about 40 hours per hectare for the period June 20–December 31, the legal season in 1948. However, the closed season on Claytor Lake was discontinued in the 1950's; total annual angler effort in 1976–1977 doubled the 1948 level. Estimated fishing pressure rose 50% in 1978 over the previous two years, reflecting inter-annual variability in estimated angling effort. Estimated angler harvest has increased three-fold over the 30-year period, corresponding with the increase in standing crop; consequently angler CPUE is

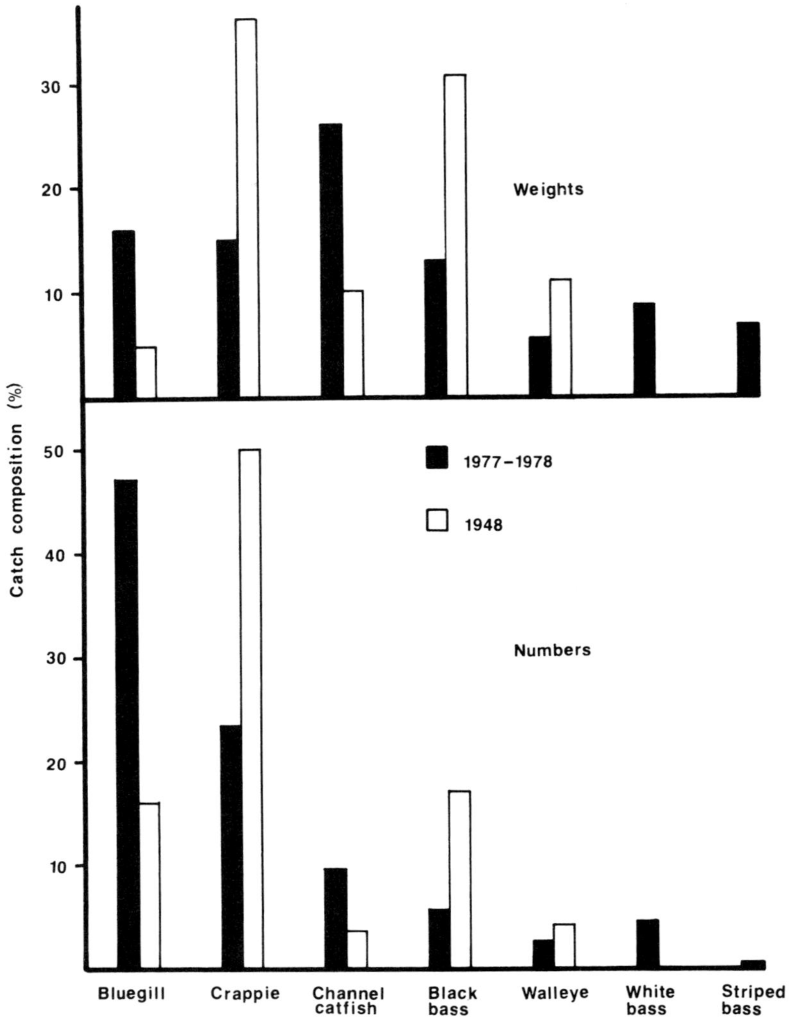

Figure 2. Comparison of angler harvest composition in 1977-1978 (Hart 1981) and 1948 (Rosebery 1950) from Claytor Lake, Virginia.

similar for both periods (approximately 0.3 fish per hour). This harvest rate is considered good for Virginia reservoirs (Whitehurst 1982).

Total harvest of black basses, crappie, bluegill, walleye, and catfish has increased since 1948, but the relative numercial and biomass contributions of these species have shifted. To compare biomass contributed by species, we assumed that mean weights of creeled fish were similar in 1948 and 1977–1978 because weights were not recorded in the earlier survey. Applying this assumption, the decline in relative contribution of black basses to the total harvest is apparent (Fig. 2). Conversely, bluegill harvest increased threefold to about half of the total number harvested and one-sixth of the biomass. The relative contribution of channel catfish to the harvest increased two and one-half times, ranking first in biomass. Carp, despite their high abundance in the reservoir, were seldom harvested by anglers in 1977–1978.

Among pelagic species, the percentages of crappies and walleye in the harvest declined by about one-half. However, introduced white and striped bass provided 8% by number and 15% by weight of the total harvest in the latter survey.

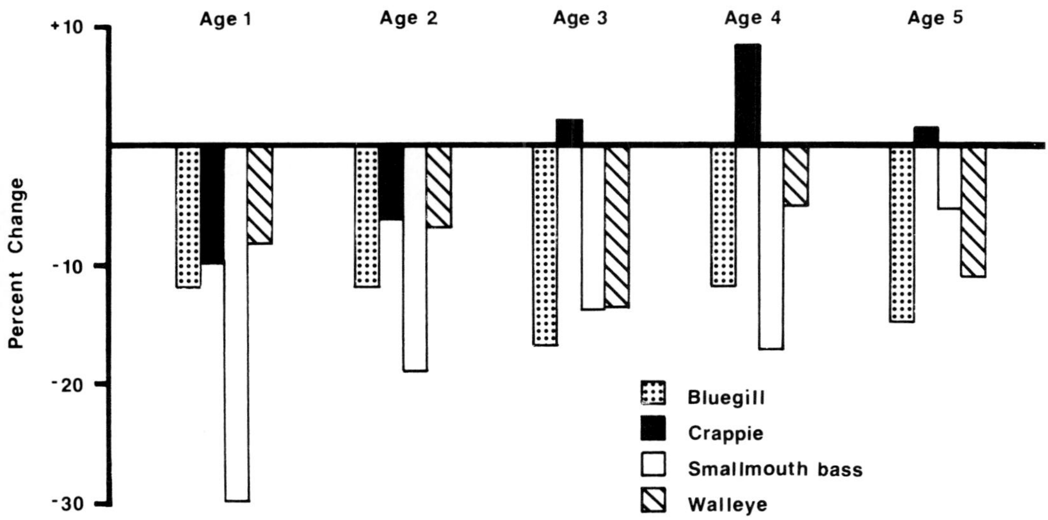

Figure 3. Changes of growth rates of bluegill sunfish, crappie, smallmouth bass, and walleye in Claytor Lake, Virginia, from the late 1940's (Rosebery 1950) to the late 1970's (Hart 1981).

The biomass contribution of the total harvest of pelagic fishes exclusive of crappie, increased from 8% in 1948 to 20% in 1977–1978.

Growth Rates

Major sport species in Claytor Lake declined in annual growth rate between 1948 and 1976–1978 (Fig. 3). Declines were most pronounced in the first two years of life in crappies and black basses, with some recovery occurring thereafter. However, bluegill total length at age was consistently depressed 12–17% throughout the lifespan, and growth rates of adult walleye also failed to show compensation. White bass growth rates were consistent through the 1970's and about average for the region (Boaze 1972). Striped bass grew relatively poorly in Claytor Lake, averaging 10–20% less in total length at each annulus than has been reported for other landlocked populations in Virginia (Hart 1981; LaRoche 1982).

Influencing Factors

Changes in species abundance, harvest, and growth rates are typical of reservoir fisheries. The influence of species manipulations on these variables is impossible to isolate from other factors such as fishing pressure, angler preference, and perhaps most importantly, the aging of the reservoir. Reservoirs characteristically experience species shifts and declines in standing crop, harvest, and growth rates as initial fertility declines. Claytor Lake differs from this pattern in that fertility has increased sharply in the last decade, promoting increases in biomass and harvest. However, species shifts have occurred and growth rates of most species have declined. These changes defy simple explanation, and we believe that interactions of introduced and resident species may be partially responsible.

The principal trend in species composition, as revealed by cove rotenone sampling, was the decline of black basses and the increase of carp. Both phenomena are probably due to habitat changes; black basses typically decline in importance in reservoirs as inundated terrestrial vegetation disappears (Aggus and Elliot 1975; Rainwater and Houser 1982), and carp are closely identified with eutrophic environments. Fewer black basses may account for the relative increase of yellow perch, pumpkinseed sunfish, and crappie residing in coves. The decline of walleye appears to be due to reproductive failure and is less easily dismissed. Although spring flood conditions in the headwaters may have impaired spawning success in some years, these were neither consistent nor unusual. Also, walleye recruitment might have been affected by predation on their larvae by introduced species. The decline in the self-sustaining walleye population coincided with the establishment of the white bass and alewife populations. White bass utilized the same headwater spawning area as did walleye and at the time when walleye fry would be present. Walleye larvae would also be vulnerable to piscivorous alewife. These hypotheses are conjectural; we know of no analysis of the compatibility of white bass or alewives and river-spawning walleye in reservoirs.

The trend in the Claytor Lake angler harvest reflects both changing angler preferences (increases in bluegill and channel catfish) and availability (decreases in black basses and walleye). A survey of angler preferences in 1976–1978 (Hart 1981) revealed that 35% of the fishermen pursued pelagic species, approximately equivalent to their contribution to the total catch. Among pelagic anglers, 40% sought crappie, white bass or striped bass, and 10% sought walleye.

The general decline in growth of juvenile fishes and bluegills despite eutrophication indicates intensifying competition for the zooplankton-benthic invertebrate food resource on which they depend. Intraspecific competition, must, by definition, be more intense than interspecific competition (MacLean and Magnuson 1977). In this respect, alewives may indirectly intensify intraspecific competition among juvenile centrarchids by buffering predation by pelagic piscivores. They also may enter into direct trophic competition through their highly efficient nocturnal foraging in coves. Cove residents might then be caught in a trophic "squeeze" between zooplanktivorous alewife and benthophagous carp and catfish for the food resource in the limited littoral area.

Black basses and crappies enjoyed relatively good growth; walleye and striped bass did not. The former fed primarily on centrarchids, while striped bass and walleye fed predominantly on alewives. Despite their high abundance, alewives were not completely available to Claytor Lake's pelagic piscivores. Alewives grew rapidly in this system (~150 mm total length at Age 1), which effectively limited their availability as suitable-sized prey to a short time early in life (Nigro and Ney 1982). However, alewife growth is extremely plastic, and could be lower in systems that are less eutrophic or have shorter growing seasons.

Implications for Management

The record of species introductions into Claytor Lake encompasses all scenarios—failures, successes, and in our opinion the establishment of at least one species with undesirable traits. The end result is a diversification of angler opportunities. Whether the fishery as a whole is more productive because of these introductions is uncertain because of other influences on the system, such as eutrophication. This record has instructional value regarding the suitability of the species involved for reservoir fisheries and the continuing evolution of strategies for species introductions. In the latter area, the challenge is two-fold: 1) establishing populations; and 2) minimizing adverse interspecific consequences.

Establishing Populations

Transplants that fail to become established occur when life cycle requirements are lacking in their new environment. Failures due to limitations in simple physical-chemical factors are usually readily discerned, e.g., the temperature-oxygen deficit for trout and low winter water temperature for threadfin shad survival. These failures are easily avoided when physical-chemical requirements of the candidate species are known and matched with the corresponding reservoir parameters. Such matching is sometimes ignored; McCammon and von Geldern (1979) refer to "a remarkable exercise in fish management futility" in which salmonids were stocked in a California lake for a decade despite a predictable summer oxygen deficit in the hypolimnion.

Striped bass did poorly in Claytor Lake, in part due to their tendency to emigrate which appears to be attributable to the high rate of flow-through. Whether striped bass are a poor candidate for mainstream reservoirs with low storage ratios remains to be determined.

Not all failures are readily attributable to physical factors. Longear sunfish (*Lepomis megalotis*) failed to establish in Claytor Lake despite apparently suitable physical habitat (Boyer and Vogele 1971). Incompatibility with resident littoral species may have been responsible. Rainwater and Houser (1982) observed that bluegill gradually replaced longear as the dominant sunfish in Beaver Lake, Arkansas. Biological incompatibility may limit even the success of pelagic species managed on a put-grow-take basis, since these are likely to be stocked as fingerlings which will initially reside in littoral areas where competition and predation are most intense. Juvenile striped bass, for example, are thought to starve when benthos is limited (Saul 1981); competition with benthophagous centrarchids, catfish, and carp may contribute to their poor performance in Claytor Lake.

Minimizing Adverse Consequences

Consequences of species incompatibility will be more severe when an introduced species establishes itself at the expense of the resident fish assemblage than if it fails because of it. In Claytor Lake, it is apparent that the alewife was suboptimal as a pelagic forage fish, buffered predation on zooplanktivorous fishes, and through its own zooplanktivory intensified trophic competition in the littoral zone. Alewife and/or white bass may be implicated in the decline of the river-spawning walleye stock. We believe that the evidence on alewife performance is sufficient to suggest that future introductions should be limited to reservoirs managed primarily for pelagic piscivores stocked on a put-grow-take basis. Whether, and under what conditions, walleye and white bass are compatible requires broader investigation of their performance in diverse situations. Assessments of this nature should precede the stocking of any species into a reservoir environment and should form the basis for a protocol of species introductions. For many species, evaluation of biological compatibility might be performed employing the data base accumulated over the long history of species manipulations in recreational fisheries. These "manipulative experiments" could be inductively analyzed for generalizable inferences (MacLean and Magnuson 1977). Comparisons of before-and-after or sympatry-allopatry performance (growth, abundance, harvest) of candidate transplant and resident species should indicate which combinations will likely be compatible. Although this approach has been criticized because of the confounding influences of extraneous temporal or ecosystem variability (Sale 1979), this unexplainable variation could be reduced by increasing the number of systems included in the analysis.

Exotic species, newly-developed hybrids, and other species contemplated for novel introductions may not be amenable to performance-based analysis. However, their habitat requirements and food habits should be sufficiently known so that they can be compared with those of resident species. Li and Moyle (1981) have incorporated loop analysis, in which species are assigned positive or negative values depending on their probable interactions (as prey, predators, or competitors) to estimate system stability. A simpler approach (MacLean and Magnuson 1977) is to determine probable overlaps along resource dimensions (time, space, food type, and size), and then to estimate their probable intensity. Kohler and Stanley (1984) have devised a protocol for evaluating proposed exotic fish introductions. These methods also elucidate critical research needs.

Conclusion

The Claytor Lake experience, which is typical of many southeastern reservoirs, underscores the need for thorough ecological analysis in screening candidates for introduction, particularly pelagic species. The ratio of successes to failures will increase if we insist on answers to questions of physical and biological compatibility before stocking, and then carefully weigh costs and benefits relative to management objectives. The answers may not always be completely satisfactory, but they represent a clear improvement over the trial-and-error strategy of the past.

References

Aggus, L. R., and G. Elliott. 1975. Effects of cover and food on year-class strength of largemouth bass. Pages 317–322 *in* R. H. Stroud and H. Clepper, editors. Black bass biology and management. Sport Fishing Institute, Washington, D.C., USA.

Anonymous. 1975. Report on Claytor Lake, Pulaski County, Virginia, EPA Region III. Working Paper No. 460. United States Environmental Protection Agency, Washington, D.C., USA.

Boaze, J. L. 1972. Effects of landlocked alewife introduction on white bass and walleye populations, Claytor Lake, Virginia. Master's thesis, Virginia Polytechnic Institute and State University, Blacksburg, Virginia, USA.

Boaze, J. L., and R. T. Lackey. 1974. Age, growth, and utilization of landlocked alewives in Claytor Lake, Virginia. The Progressive Fish-Culturist 36: 163–164.

Boyer, R. L., and L. E. Vogele. 1971. Longear sunfish behavior in two Ozark reservoirs. Pages 13–26 *in* G. E. Hall, editor. Reservoir fisheries and limnology. Special Publication 8. American Fisheries Society, Bethesda, Maryland, USA.

Brooks, J. L. 1968. The effects of prey size selection by lake planktivores. Systematic Zoology 17:272–291.

Gannon, J. E. 1976. The effects of differential digestion rates of zooplankton by alewife, *Alosa pseudoharengus,* on determinations of selective feeding. Transactions of the American Fisheries Society 105:89–95.

Goodson, L. F., Jr. 1966. Crappie. Pages 312–332

in A. Calhoun, editor. Inland fisheries management. California Department of Fish and Game, Sacramento, California, USA.

HART, L. G. 1981. Claytor Lake fish population study. Final Report, Federal Aid Project F-35-F. Virginia Commission of Game and Inland Fisheries, Richmond, Virginia, USA.

HUTCHINSON, B. P. 1971. The effect of fish predation on the zooplankton of ten Adirondack lakes, with particular reference to the alewife, *Alosa pseudoharengus*. Transactions of the American Fisheries Society 100:325–355.

JANSSEN, J. 1976. Feeding modes and prey size selection in the alewife (*Alosa pseudoharengus*) in Lake Michigan. Canadian Journal of Fisheries and Aquatic Sciences 37:177–184.

JENKINS, R. M. 1967. The influence of some environmental factors on standing crop and harvest of fishes in U.S. reservoirs. Pages 298–321 *in* Reservoir fishery resources symposium. Southern Division American Fisheries Society, Bethesda, Maryland, USA.

JENKINS, R. M. 1970. Reservoir fish management. Pages 173–182 *in* N. G. Benson, editor. A. century of fisheries in North America. Special publication 7, American Fisheries Society, Bethesda, Maryland, USA.

JENKINS, R. M. 1974. Reservoir management prognosis: migraines or miracles. Proceedings of the Annual Conference of the Southeastern Association of Game and Fish Commissioners 27:374–385.

KELSO, W. E., AND J. J. NEY. 1985. Nocturnal foraging by alewives in reservoir coves. Proceedings of the Annual Conference of the Southeastern Association of Fish and Wildlife Agencies 36:125–134.

KOHLER, C. C. 1980. Trophic ecology of an introduced landlocked alewife (*Alosa pseudoharengus*) population and assessment of alewife impact on resident sportfish and crustacean zooplankton communities in Claytor Lake, Virginia. Doctoral dissertation, Virginia Polytechnic Institute and State University, Blacksburg, Virginia, USA.

KOHLER, C. C., AND J. J. NEY. 1980. Piscivority in a landlocked alewife (*Alosa pseudoharengus*) population. Canadian Journal of Fisheries and Aquatic Sciences 37:1314–1317.

KOHLER, C. C., AND J. J. NEY. 1981. Consequences of an alewife die-off to fish and zooplankton in a reservoir. Transactions of the American Fisheries Society 110:360–369.

KOHLER, C. C., AND J. J. NEY. 1982a. Suitability of alewife as a pelagic forage fish for southeastern reservoirs. Proceedings of the Annual Conference of the Southeastern Association of Fish and Wildlife Agencies 34:137–150.

KOHLER, C. C., AND J. J. NEY. 1982b. A comparison of methods for quantitative analysis of feeding selection of fishes. Environmental Biology of Fishes 7:363–368.

KOHLER, C. C., AND J. G. STANLEY. 1984. A suggested protocol for evaluating proposed exotic fish introductions in the United States. Pages 387–406 *in* W. R. Courtenay, Jr. and J. R. Stauffer, editors. Distribution, biology and management of exotic fishes. The Johns Hopkins University Press, Baltimore, Maryland, USA.

LAROCHE, A. L. 1982. Reservoir investigations on Smith Mountain Lake. Final report, Federal Aid Project F-39, Virginia Commission of Game and Inland Fisheries, Richmond, Virginia, USA.

LI, H. W., AND P. B. MOYLE. 1981. Ecological analysis of species introductions into aquatic systems. Transactions of the American Fisheries Society 110:772–782.

MACLEAN, J. H., AND J. J. MAGNUSON. 1977. Inferences on species interactions in percid communities. Journal of the Fisheries Research Board of Canada 34:1941–1945.

MARTIN, R. G. 1970. The Claytor Lake fishery. Virginia Wildlife 37:18–19.

MCCAMMON, G. W., AND C. E. VON GELDERN, JR. 1979. Predator-prey systems in large reservoirs. Pages 431–442 *in* R. H. Stroud and H. Clepper, editors. Predator-prey systems in fisheries management. Sport Fishing Institute, Washington, D.C., USA.

MURPHY, B. R. 1981. Genetic evaluation of walleye (*Stizostedion vitreum vitreum*) stock structure and recruitment in Claytor Lake, Virginia. Doctoral dissertation, Virginia Polytechnic Institute and State University, Blacksburg, Virginia, USA.

NEY, J. J. 1981. Evolution of forage-fish management in lakes and reservoirs. Transactions of the American Fisheries Society 110:725–728.

NIGRO, A. A., AND J. J. NEY. 1982. Reproduction and early-life accommodations of landlocked alewives to a southern range extention. Transactions of the American Fisheries Society 111:559–569.

PRITCHARD, D. L., O. D. MAY, AND L. RIDER. 1978. Stocking of predators in the predator-stocking-evaluation reservoirs. Proceedings of the Annual Conference of the Southeastern Association of Fish and Wildlife Agencies 30:108–113.

RAINWATER, W. C., AND A. HOUSER. 1982. Species composition and biomass of fish in selected coves in Beaver Lake, Arkansas, during the first 18 years of impoundment (1963–1980). North American Journal of Fisheries Management 4:316–325.

ROSEBERY, D. A. 1950. Game fisheries investigations of Claytor Lake: a main stream impoundment of New River, Pulaski County, Virginia, with emphasis on *Micropterus punctulatus* (Rafinesque). Doctoral dissertation, Virginia Polytechnic Institute and State University, Blacksburg, Virginia, USA.

SALE, P. F. 1979. Habital partitioning and competition in fish communities. Pages 323–331 *in* R. H. Stroud and H. Clepper, editors. Predator-prey systems in fisheries management. Sport Fishing Institute, Washington, D.C., USA.

SAUL, B. M. 1981. Food, habits, and growth of young-of-year striped bass in Cherokee Reservoir, Tennessee. Master's thesis, University of Tennessee, Knoxville, Tennessee, USA.

TIMMONS, T. J., W. L. SHELTON, AND W. D. DAVIES. 1979. Initial fish population changes following impoundment of West Point Reservoir, Alabama-Georgia. Proceedings of the Annual Conference of the Southeastern Association of Fish and Wildlife Agencies 31:312–317.

WHITEHURST, D. K. 1982. Reservoir investigations Region II: Claytor Lake. Final report, Federal Aid Project F-35-R. Virginia Commission of Game and Inland Fisheries, Richmond, Virginia, U.S.A.

Pages 178–185 *in* G. E. Hall and M. J. Van Den Avyle, editors.
Reservoir Fisheries Management: Strategies for the 80's.
Reservoir Committee, Southern Division American Fisheries Society, Bethesda, Maryland. 1986.

Predation on Juvenile Salmonids in Columbia Basin Reservoirs

GERARD A. GRAY AND DENNIS W. RONDORF

U.S. Fish and Wildlife Service
National Fisheries Research Center
Willard Field Station, Cook, Washington 98605

ABSTRACT

Man has significantly altered the physical and biological nature of the Columbia River. Commercial exploitation of this resource began during the mid-1800's; commercial catches of chinook salmon peaked in 1883 and the mixed stocked fisheries peaked in 1895. From 1895 to about 1933, output remained fairly constant despite problems caused by overexploitation and degradation or loss of spawning habitat. By 1953, the cumulative adverse impact of dam construction became a dominant factor influencing salmon and steelhead runs above Bonneville Dam, particularly through the increased potential for predation on juvenile salmonids migrating through Columbia basin reservoirs. Factors affecting the intensity of this predation include life history characteristics of the migrating stocks, concentration of juveniles at dams, chumming effect of stunned or disoriented juveniles at turbine outfalls, limnological changes after impoundment, and changes in the predator complex. These factors have resulted in a predator-prey system in which juvenile salmonids are exposed to predation in a variety of habitats. Management of this complex system is made difficult by political, physical, and biological constraints. The most effective management program may be to reduce the susceptibility of juvenile salmonids to predation by providing maximum protection during their downstream migration. These efforts could include strategically planned hatchery releases, increased flow and spill during peak outmigration, improved screens and bypass facilities at dams, and improved transport methods.

Man has significantly altered the aquatic habitat and fish species complex in the Columbia River, and this alteration has created substantial changes in the dynamics of the predator-juvenile salmonid relationships and created serious management problems. The objectives of this paper are to outline the factors affecting predation on juvenile salmonids in Columbia basin reservoirs, discuss management measures aimed at reducing this mortality, and present possible management alternatives.

The Columbia River is 2,045 km long and drains an area of 676,810 km^2. The headwaters and about 101,000 km^2 of the drainage area are in southeastern British Columbia, and the rest of the watershed stretches over most of Oregon, Washington, Idaho, and parts of Montana, Nevada, and Wyoming. The largest tributary is the Snake River. The average daily flow at the mouth of the Columbia River is about 7,193 m^3/second, making it the fourth largest river in North America (Netboy 1980). Mean monthly flows are lowest in September and highest in June. The tremendous flows and gradient in the Columbia River offer greater hydropower potential than any other river in the United States (Netboy 1980), and numerous dams have been constructed throughout the drainage to tap this economical source of power (Table 1).

Dam construction blocked adult salmon migrations at Chief Joseph Dam (1985) on the main stem and Hells Canyon Dam (1967) on the Snake River. Of the remaining river reaches, about 88% of the main stem and 35% of the Snake River is impounded by 13 hydropower dams. Impoundments average about 62 km in length, maximum depths do not exceed 90 m, and shorelines are often steep and consist of talus or rubble substrate. Fluctuations in reservoir water levels occur daily but rarely exceed 3.0 m.

The Columbia basin's salmonid fishery was once the largest in North America, except for Alaska (Netboy 1980). Chinook salmon (*Oncorhynchus tshawytscha*) and steelhead trout (*Salmo gairdneri*) were the most abundant species, but sizable populations of chum salmon (*O. keta*), coho salmon (*O. kisutch*), and sockeye salmon (*O. nerka*) also existed. Commercial exploitation began during the mid-1800's and expanded rapidly in response to a seemingly endless supply of chinook salmon, the only commercially important west-coast species at the time. Output of

Table 1. Characteristics of impounded and free flowing sections of the Columbia and Snake rivers accessible to anadromous salmonids.

River kilometer	Feature	Date of dam completion	Reservoir length (km)	Reservoir surface area (hectares)
Columbia River				
0–234	Estuary/open river			
234	Bonneville Dam	1938	74	8,256
308	The Dalles Dam	1957	39	4,249
347	John Day Dam	1968	123	20,235
470	McNary Dam	1953	98	15,419
522	Confluence of Snake River			
568–639	Open river			
639	Priest Rapids Dam	1959	29	2,833
668	Wanapum Dam	1963	61	5,585
729	Rock Island Dam	1933	34	1,012
763	Rocky Reach Dam	1961	67	3,723
830	Wells Dam	1968	47	4,330
877	Chief Joseph Dam[a]	1955	84	3,157
Snake River				
16	Ice Harbor Dam		51	3,371
67	Lower Monumental Dam	1962	46	2,728
113	Little Goose Dam	1969	60	4,015
173	Lower Granite Dam	1970	85	3,602
258–397	Open river	1975		
397	Hells Canyon Dam[a]	1967	20	1,012

[a] No fish passage facilities.

processed chinook salmon increased from 122,400 kg in 1866 to more than 18.9 million kg in 1883 and then dipped to 8.1 million kg by 1889, largely as a result of overfishing (Craig and Hacker 1940). Poor catches of chinook salmon forced fisherman to switch to a mixed-species fishery that peaked in 1895 at about 19.4 million kg of processed fish. From 1895 to about 1933, new harvesting techniques (primarily ocean trolling) and continued exploitation of all stocks of salmon and steelhead kept output at nearly constant levels despite numerous problems caused by overexploitation and degradation of spawning and rearing habitat from mineral and timber development.

Habitat degradation and overexploitation continued to be serious problems during the hydropower development era of 1933 to 1975. However, by 1953 the cumulative adverse impact of dam construction became a dominant factor influencing salmon and steelhead runs above Bonneville Dam (Poon and Garcia 1982). The impact of dams on fish stocks such as chinook and coho salmon, which were harvested in river gillnet and ocean troll fisheries in the same year, was difficult to assess. However, catch data for up-river summer steelhead trout, a species harvested almost exclusively in the main stem fishery, did indicate a decline in catch as more dams were constructed on the Columbia and Snake rivers. This decline has been linked to a reduction in survival of hatchery and wild smolts during downstream migration (Allen 1977; Ebel 1977).

Long et al. (1968) found that direct mortality of smolts migrating through the Kaplan-type turbines used on the Columbia River ranged from 10 to 19% per dam and that unaccountable losses ranged from 8% in average flow years to 23% per dam in low-flow years. This unaccountable mor-

Table 2. Characteristics of outmigrating juvenile salmonids that pass through Columbia basin reservoirs. Lengths refer to size when migration begins.

Species	Age group	Length (mm)	Time of outmigration
Salmon			
Spring chinook	1	90–150	March–June
Summer chinook	0	75–100	July–September
Fall chinook	0	30–75	March–December
Coho	1	75–150	April–June
Sockeye	1–2	75–100	April–June
Trout			
Steelhead	1–2	150–250	March–June

tality has been linked to the creation of impoundments, which more than doubled the time required by smolts from the Snake River to pass through the system (Raymond 1979) and increased the potential for predation (Ebel 1977).

Factors Affecting Predation

Life History Factors

Columbia basin reservoirs are primarily used by juvenile salmonids as migration routes, but they may also be important rearing areas. During their migrations, juvenile salmonids are among the dominant prey fishes in littoral and pelagic habitats. The number that are consumed depends on their behavior and size during the outmigration. Juveniles produced by spring run chinook, coho, and sockeye salmon and steelhead trout move quickly through the reservoirs in pelagic areas at a size that reduces vulnerability to predators. Juveniles produced by fall and summer run chinook salmon, however, are generally smaller and follow a slow and irregular migration pattern through the reservoirs (Table 2). Sims and Miller (1982) found that average residence time for marked subyearling chinook salmon in a reservoir was 22 days, compared to 6 days for yearlings. Juvenile fall and summer chinook salmon are commonly found in littoral habitats during spring and early summer, where they may be particularly vulnerable to predation.

Concentration of Juveniles

At dams, juvenile salmonids tend to delay their downstream movement during daylight and pass as a group at night (Sims et al. 1981). The point of dam entry is apparently determined by a behavioral cue provided by high water velocity. Most of these fish, therefore, enter and leave the dam at the same general location and then are concentrated in the tailrace. Neave (1953) suggested that juvenile salmonids have adapted to predation by migrating in large concentrations; he assumed that predators consumed prey at some maximum rate and that prey migrating in large groups produced a swamping effect that reduced the proportion lost to predation. However, Peterman and Gatto (1978) found that swamping rarely occurs. The paucity of data on the relation between prey density and consumption by predators precludes a complete evaluation of the effect of concentrating juveniles at Columbia basin dams.

Turbine Effects

The percentage of migrating juvenile salmonids passing through the powerhouse at each dam depends on flow and spill; under low flow and spill nearly all migrants travel through the powerhouse (Raymond 1979). In addition to those fish killed outright in the turbines, many can be injured or stunned and left vulnerable to predators. The problem is increased by turbine flows that wash fish back to the dam face and eventually into the slack water space below the spillway (backroll) where predators can more easily reside and feed. Long et al. (1968) found survival of coho salmon released in the backroll to be 33% lower than for fish washed downstream by the current (frontroll), and he attributed this loss to predation.

Impoundment Effects

Impoundments have moderated water flows, temperature, turbidity, and depth fluctuation in the Columbia River basin, thereby exposing juvenile salmonids to a variety of habitats during their outmigration. Flows and presence or absence of impoundments appear to be major determinants affecting spring smolt migration rates, probably through their effects on water velocity. Raymond (1979) estimated that the migration of yearling juvenile salmonids from the Salmon River, Idaho, the The Dalles Dam required 26 days under low-flow, pre-dam conditions but took 65 days under low-flow, impounded conditions. The prolonged outmigration of yearling salmon under impounded conditions extends the period of exposure to predation.

Delays in migration, especially in late spring, can also result in exposure of smolts to high water temperatures, particularly in low-flow years. Exposure of smolts to temperatures higher than 13 C results in a loss of migratory behavior (Zaugg and McLain 1972) and a holdover of migrants until the next year. Holdover fish have been found during low-flow and high-flow years, but the numbers are greatly reduced during years of high flow (Raymond et al. 1973; Bentley and Stuehrenberg 1974). Reservoir residence for holdover juveniles may be increased from a few weeks to about a year, which greatly increases their time of exposure to predation in freshwater.

A decline in turbidity in the Columbia River coincided with the closure of Brownlee Dam on the Snake River and other dams in the basin during the late 1950's and early 1960's. Junge and Oakley (1966) suggested that juvenile salmon migration behavior was influenced by the reduced turbidity. Under turbid conditions at Bonneville Dam, 48–52% of all juvenile salmonids migrated during daylight. Under less turbid conditions at McNary Dam, less than 10% migrated during the day. The smaller number of juvenile salmonids migrating during the day at McNary Dam may be an adaptive response to lessen predation during the outmigration. Junge and Oakley (1966) suggested that lower turbidity was associated with higher predation mortality of juvenile sockeye salmon. Zaret (1979) concluded that decreased turbidity increased the capture efficiency of predatory fish; this conclusion is supported by the significantly higher growth rate of predators living under less turbid conditions. Thus, reduced turbidity in the Columbia River today may actually increase predation on migrating juvenile salmonids.

Major Predators

Native non-salmonid predatory fishes in the Columbia River basin are the white sturgeon (*Acipenser transmontanus*), green sturgeon (*A. medirostris*), sculpins (*Cottus* spp.), and northern squawfish (*Ptychocheilus oregonensis*). Only the northern squawfish and sculpins are still abundant in the Columbia River drainage. Introduced predatory fish in reservoirs are the walleye (*Stizostedion vitreum vitreum*), smallmouth bass (*Micropterus dolomieui*), largemouth bass (*M. salmoides*), channel catfish (*Ictalurus punctatus*), yellow perch (*Perca flavescens*), crappie (*Pomoxis* spp.), and sunfish (*Lepomis* spp.).

Of all predators that are present, only northern squawfish, walleyes, smallmouth bass, and channel catfish have the potential for significantly affecting the abundance of juvenile salmonids. These are large, abundant, and opportunistic predators that feed on a variety of prey and switch their feeding patterns when spatially or temporally segregated from a commonly consumed prey. However, there is considerable variability in the behavior of each species and habitats they occupy.

Northern Squawfish

Ricker (1941) found that northern squawfish in Cultus Lake, British Columbia, fed on redside shiners (*Richardsonius balteatus*) and threespine sticklebacks (*Gasterostus aculeatus*) in warm littoral habitats during summer and on sockeye salmon in pelagic habitats in winter. Olney (1975) noted similar distribution patterns of northern squawfish in Lake Washington, but sockeye salmon were never the principal food.

Although researchers generally agree that northern squawfish will eat juvenile salmonids, the net effect of this predation remains unclear. Ricker (1941) argued that even though the occurrence of juvenile sockeye salmon in the diets of northern squawfish from Cultus Lake was low, the impact of predation was high because northern squawfish were very abundant. This same argument has also been used to implicate northern squawfish predation as a major factor limiting survival of coho salmon in Lake Merwin, Washington (Hamilton et al. 1970), and to justify the use of numerous methods to control their abundance (Foerster and Ricker 1941; Jeppson 1957; Jeppson and Platts 1959; MacPhee and Ruelle 1969). In the Columbia and Snake rivers, the occurrence of juvenile salmonids in northern squawfish stomachs was low except at dams. Uremovich et al. (1980) estimated that northern squawfish ate 3.8 million juvenile salmonids between April 13 and August 30, 1980, in the forebay of the Bonneville Pool. It is unclear whether northern squawfish actively migrate and congregate to feed on juvenile salmonids that are delayed and funneled through dams.

Walleye

Events resulting in the introduction of walleyes into the Columbia basin are unknown, as is the impact this piscivore has on juvenile salmonids. Maule (1982) found that few juvenile salmonids were eaten by walleyes in John Day Reservoir, and suggested that the probable reason was the spatial and temporal segregation of the species when walleyes were feeding most actively. Juvenile salmonids tend to move up in the water column at night (Hoar 1958; Ali 1959), when they may be separated from walleyes feeding near the bottom. Gray et al. (1984) generally agreed with this explanation; however, they found a high frequency of occurrence (42%) of juvenile salmonids in the stomachs of walleyes collected in John Day tailrace during spring. Dams may create an atypical habitat with increased potential for spatial overlap, and therefore predation, between walleyes and juvenile salmonids. This was supported by the high occurrence of juvenile salmonids in walleye stomachs collected between 1800 and 2400 hours (Gray et al. 1984), when the greatest proportion of smolts move through the powerhouse at John Day Dam (Sims et al. 1981) and when walleyes feed most heavily (Maule 1982).

Predation by walleyes on juvenile salmonids appeared to be greatest on subyearling fish (Gray et al. 1984); about 80% of the juveniles identified in walleye stomachs were subyearlings, probably a result of their greater abundance and smaller size. Subyearling salmon less than about 100 mm long tend to migrate in shallower water than do yearlings, also increasing the likelihood of encountering walleyes (D. W. Rondorf, unpublished data).

Smallmouth Bass

The potential for smallmouth bass predation on juvenile salmonids seems high, given the behavioral characteristics of this species. In spring and early summer they inhabit rocky shoreline areas that are also inhabited by juvenile salmonids. Although the smallmouth bass was not shown to be an important predator on juvenile salmonids in either lotic or lentic habitats with resident salmonid populations (Lachner 1950; Martin and Fry 1972), recent studies in Columbia basin reservoirs and Lake Sammamish, Washington, have shown that smallmouth bass were highly predacious on outmigrating juvenile salmonids (Pflug 1981; Gray et al. 1984). However, Keating (1970) and Bennett et al. (1983) found that even though salmonids were present in Snake River reservoirs they were not important in the smallmouth bass diet. This apparent discrepancy may have occurred because of differences in salmonid stocks available as prey. In the studies where considerable predation occurred, subyearling salmonids were seasonally abundant whereas they were either scarce or absent in the Snake River impoundments studied. Subyearling juvenile salmonids may be eaten by smallmouth bass because their habitats overlap seasonally and the salmonids are ideal forage size for adult smallmouth bass.

Channel Catfish

Studies conducted in reservoirs of the Snake and Columbia rivers have shown considerable differences in the occurrence of juvenile salmonids in channel catfish stomachs. Occurrence was 1% in McNary Reservoir (D. W. Rondorf, unpublished data), 2% in John Day Reservoir (Gray et al. 1984), 58% in Little Goose Reservoir, and 76% in the tailrace of Lower Granite Dam (Bennett et al. 1983). The higher occurrence of salmonids in channel catfish diets from Little Goose Reservoir and Lower Granite Dam may have been due to sampling in tailrace habitats.

The importance of salmonid predation by channel catfish in tailrace habitats cannot be fully evaluated until the condition of juvenile salmonids eaten can be ascertained. Swingle (1950) noted a tendency by channel catfish to eat moribund or dead fish, and Bennett et al. (1983) hypothesized that channel catfish ate dead or dying juvenile salmonids below Lower Granite Dam.

Predator-prey Complex

Clearly, man's activities have reshaped the predator-juvenile salmonid interactions in the Columbia basin. The areas immediately above and below dams, where funneling and chumming have become most serious, are havens for northern squawfish. Estimates of relative abundance (Gray et al. 1984) indicated that northern squawfish were most abundant in these habitats, and population estimates at the Bonneville Dam forebay (Uremovich et al. 1980) suggested that large numbers of northern squawfish congregate in these areas. Occurrence of juvenile salmonids in northern squawfish stomachs was also high in these areas. However, near-dam habitats generally represented less than one mile of reservoir, for example, less than 2% of John Day Reservoir. Areas away from the dam but under the influence of the forebay are occupied by northern squawfish and smallmouth bass. In these areas, smallmouth bass were more abundant than northern squawfish or walleyes (Gray et al. 1984), but the occurrence of juvenile salmonids in smallmouth bass stomachs was only 1.6%. In areas affected by the high water velocities and turbulence of the tailrace, but not directly below the dam face, walleyes, channel catfish, and northern squawfish may dominate. However, walleyes congregate in these areas only in spring (Gray et al. 1984), and the abundance of channel catfish in tailrace habitats varies greatly among reservoirs.

In the remaining reservoir habitat, radical changes in flow and subtle deviations in temperature, turbidity, and community structure have created suitable habitats for all four predators. Smallmouth bass appear to prefer rocky littoral habitats (Reynolds 1965; Munther 1970); northern squawfish are more pelagic in the top 10 m of the water column (Palmer 1982), where their distribution has been positively correlated with water velocity (Patten et al. 1970; Palmer 1982); and channel catfish and walleyes are typically demersal (Scott and Crossman 1973). Relative abundance of these species can vary substantially among reservoirs. Generally, smallmouth bass and northern squawfish are the most abundant species in reservoir habitats, although channel catfish are very numerous in McNary Reservoir (D. W. Rondorf, unpublished data). The occurrence of juvenile salmonids in stomachs of predators collected away from dams was typically lower than at near-dam areas (Gray et al. 1984), but given the large area of such habitat and potential abundance of predators, it should not be overlooked. Predation in all reservoir habitats appears to be most intense on subyearling salmonids. Major determinants of this differential predation probably include size, migratory behavior, and perhaps density of subyearling fish.

The net impact of dams and the introduction of exotic piscivorous fish on the relation between predators and juvenile salmonids is not yet known, primarily because it is difficult to effectively sample the constantly changing environment of large rivers (Campbell 1979). Uremovich et al. (1980) estimated the loss of juvenile salmonids to predation by northern squawfish in a localized, near-dam habitat, and demonstrated the potential significance of predation on juvenile salmonids. The next step is to expand these estimates to areas away from the direct influence of dams.

Management Alternatives

Ricker (1952) suggested that when a prey species is temporarily massed in large numbers, predators of a given abundance take a fixed number of prey. To regulate systems in which predation is independent of prey densities, managers have generally attempted to control predator abundance. For at least two reasons this approach is generally untenable for managing Columbia basin reservoirs: (1) three of the four major predators in the system are desirable sport fishes; and (2) the large size of these reservoirs precludes the use of most management techniques. Pressed by user groups to provide fishing opportunities in areas where the fishery for salmonids has been reduced by poor returns of adults, fishery bologists are being forced to assume that enhancement of resident game fish and juvenile salmonids are not mutually exclusive. Therefore, the northern squawfish is the only predatory species in Columbia basin reservoirs for which population reduction may be acceptable.

Control of northern squawfish numbers in large reservoirs in the Columbia basin seems unlikely because available techniques have never shown long-term success in other areas (Brown and Moyle 1981). Even spot treatment with the species selective piscicide Squoxin, in such locations as dams and hatchery release sites, would be costly. The ecological ramifications of these activities are also unclear.

Rather than attempting to eliminate predators, researchers and managers should continue current efforts to reduce the susceptibility of juvenile salmonids to predation. Improved hatchery diets and rearing techniques should provide healthier fish that may better avoid predation. Strategically planned hatchery release procedures should impede the ability of northern squawfish to concentrate and feed on juvenile salmonids. Increased flow and spill during peak outmigration should speed the movement of juveniles through reservoir habitats and reduce mortality at dams. Installation of screens at turbine intakes and improved bypass facilities should reduce mortality and the chumming effect at dams. Transporting juvenile salmonids past dams with barges or trucks should reduce predation mortality within reservoirs. Development of better transport methods should reduce handling stress and decrease predation on juveniles released below Bonneville Dam. Together, these positive measures may reduce the effects of predation sufficiently to materially assist in the recovery of the salmonid fisheries.

References

ALI, M. A. 1959. The ocular structure, retinomotor and photobehavioral responses of juvenile Pacific salmon. Canadian Journal of Zoology 37:965–996.

ALLEN, R. L. 1977. Status of the upper Columbia River salmon and steelhead runs. Pages 23–30 *in* E. Schwiebert, editor. Symposium on Columbia River salmon and steelhead. Special Publication 10. American Fisheries Society, Bethesda, Maryland, USA.

BENNETT, D. H., P. BRATOVICH, W. KNOX, D. PALMER,

AND H. HANSEL. 1983. Status of the warmwater fishery and the potential of improving warmwater fish habitat in the lower Snake reservoirs. Report to the United States Army Corps of Engineers, Department of Fish and Wildlife, University of Idaho, Moscow, Idaho, USA.

BENTLEY, W. W., AND L. C. STUEHRENBERG. 1974. Effects of power peaking on juvenile salmon and trout—1974, upper Columbia and lower Snake Rivers. Report to the United States Army Corps of Engineers, National Marine Fisheries Service, Northwest and Alaska Fisheries Center, Seattle, Washington, USA.

BROWN, L. R., AND P. B. MOYLE. 1981. The impact of squawfish on salmonid populations: A review. North American Journal of Fisheries Management 1:104–111.

CAMPBELL, K. P. 1979. Predation principles in large rivers: A review. Pages 181–191 *in* R. H. Stroud and H. Clepper, editors. Predator-prey systems in fisheries management. Sport Fishing Institute, Washington, D.C., USA.

CRAIG, J. A., AND R. L. HACKER. 1940. The history and development of the fisheries of the Columbia River. United States Bureau of Fisheries Bulletin 49:133–216.

EBEL, W. L. 1977. Major passage problems. Pages 33–39 *in* E. Schwiebert, editor. Symposium on Columbia River salmon and steelhead. Special Publication 10. American Fisheries Society, Bethesda, Maryland, USA.

FOERSTER, R. E., AND W. E. RICKER. 1941. The effect of reduction of predacious fish on survival of young sockeye salmon at Cultus Lake. Journal of the Fisheries Research Board of Canada 5:315–336.

GRAY, G. A., G. M. SONNEVIL, H. C. HANSEL, C. W. HUNTINGTON, AND D. E. PALMER. 1984. Feeding activity, rate of consumption, daily ration and prey selection of major predators in John Day Reservoir. Report to the Bonneville Power Administration, United States Fish and Wildlife Service, National Fishery Research Center, Cook, Washington, USA.

HAMILTON, J. A. R., L. O. ROTHFUS, M. W. ERHO, AND J. D. REMINGTON. 1970. Use of a hydroelectric reservoir for the rearing of coho salmon, (*Oncorhynchus kisutch*). Research Bulletin 9. Washington Department of Fisheries, Olympia, Washington, USA.

HOAR, W. S. 1958. The evolution of migratory behavior among juvenile salmon of the genus *Oncorhynchus*. Journal of the Fisheries Research Board of Canada 15:391–428.

JEPPSON, P. 1957. The control of squawfish by use of dynamite, spot treatment, and reduction of lake levels. The Progressive Fish-Culturist 19:168–171.

JEPPSON, P. W., AND W. S. PLATTS. 1959. Ecology and control of the Columbia squawfish in northern Idaho lakes. Transactions of the American Fisheries Society 88:197–202.

JUNGE, C. O., AND A. L. OAKLEY. 1966. Trends in production rates for upper Columbia River runs of salmon and steelhead and possible effects of changes in turbidity. Fish Commission of Oregon Research Briefs 12:22–44.

KEATING, J. F., JR. 1970. Growth rates and food habits of smallmouth bass in the Snake, Clearwater and Salmon Rivers, Idaho, 1965–1967. Master's thesis, University of Idaho, Moscow, Idaho, USA.

LACHNER, E. A. 1950. Food, growth and habits of fingerling northern smallmouth bass, *Micropterus dolomieui* Lacepede, in trout waters of western New York. Journal of Wildlife Management 14: 50–56.

LONG, C. W., R. W. KRCMA, AND F. J. OSSIANDER. 1968. Research on fingerling mortality in Kaplan turbines. Northwest and Alaska Fisheries Center, U.S. National Marine Fisheries Service, Seattle, Washington, USA.

MACPHEE, C., AND R. RUELLE. 1969. A chemical selectively lethal to squawfish (*Ptychocheilus oregonensis* and *P. umpquae*). Transactions of the American Fisheries Society 98:678–684.

MARTIN, N. V., AND F. E. J. FRY. 1972. Lake Opeongo: effects of exploitation and introductions on the salmonid community. Journal of the Fisheries Research Board of Canada 29:795–805.

MAULE, A. G. 1982. Aspects of the life history of walleye (*Stizostedion vitreum vitreum*) in the Columbia River. Master's thesis, Oregon State University, Corvallis, Oregon, USA.

MUNTHER, G. L. 1970. Movement and distribution of smallmouth bass in the middle Snake River. Transactions of the American Fisheries Society 99:44–53.

NEAVE, F. 1953. Principles affecting the size of pink and chum salmon populations in British Columbia. Journal of the Fisheries Research Board of Canada 9:450–491.

NETBOY, A. 1980. The Columbia River salmon and steelhead trout, their fight for survival. University of Washington Press, Seattle, Washington, USA.

OLNEY, F. E. 1975. Life history and ecology of the northern squawfish *Ptychocheilus oregonensis* (Richardson) in Lake Washington. Master's thesis, University of Washington, Seattle, Washington, USA.

PALMER, D. E. 1982. Abundance, survival, distribution and movements of selected fishes in lower Snake River Reservoirs. Master's thesis, University of Idaho, Moscow, Idaho, USA.

PATTEN, B. G., R. B. THOMPSON, AND W. D. GRONLUND. 1970. Distribution and abundance of fish in the Yakima River, Washington, April 1957 to May 1958. United States Fish and Wildlife Service Special Scientific Report Fisheries 603, Washington, D.C., USA.

PETERMAN, R. M., AND M. GATTO. 1978. Estimation of functional responses of predators on juvenile salmon. Journal of the Fisheries Research Board of Canada 35:797–808.

PFLUG, D. E. 1981. Smallmouth bass (*Micropterus*

dolomieui) of Lake Sammamish: A study of their age and growth, food and feeding habits, population size, movement and homing tendencies, and comparative interactions with largemouth bass. Master's thesis, University of Washington, Seattle, Washington, USA.

POON, D., AND J. GARCIA. 1982. A comparative analysis of anadromous salmonid stocks and their population decline in selected watersheds of the eastern Pacific coast. Report to the Northwest Power Planning Council, Portland, Oregon, USA.

RAYMOND, H. L. 1979. Effects of dams and impoundments on migrations of juvenile chinook salmon and steelhead from the Snake River, 1966 to 1975. Transactions of the American Fisheries Society 108:505–541.

RAYMOND, H. L., W. W. BENTLEY, AND C. S. THOMPSON. 1973. Effects of power peaking operations on juvenile salmon and trout migrations, 1973. Report to the United States Army Corps of Engineers, United States National Marine Fisheries Service, Northwest and Alaska Fisheries Center, Seattle, Washington, USA.

REYNOLDS, J. B. 1965. Life history of smallmouth bass, *Micropterus dolomieui* Lacepede, in the Des Moines River, Boone County, Iowa. Iowa State Journal of Science 39:417–436.

RICKER, W. E. 1941. The consumption of young sockeye salmon by predacious fish. Journal of the Fisheries Research Board of Canada 5:293–313.

RICKER, W. E. 1952. Numerical relations between abundance of predators and survival of prey. The Canadian Fish Culturist 13.

SCOTT, W. B., AND E. J. CROSSMAN. 1973. Freshwater fishes of Canada. Fisheries Research Board of Canada, Bulletin 184.

SIMS, C. W., AND D. R. MILLER. 1982. Effects of flow on the migratory behavior and survival of juvenile fall and summer chinook salmon in John Day Reservoir. Report to the Bonneville Power Administration, United States National Marine Fisheries Service, Northwest and Alaska Fisheries Center, Seattle, Washington, USA.

SIMS, C. W., J. G. WILLIAMS, D. A. FAUROT, R. C. JOHNSEN, AND D. A. BREGE. 1981. Migrational characteristics of juvenile salmon and steelhead in the Columbia River basin and related passage research at John Day Dam, Volumes I and II. Report to the United States Army Corps of Engineers, United States National Marine Fisheries Service, Northwest and Alaska Fisheries Center, Seattle, Washington, USA.

SWINGLE, H. S. 1950. Relationships and dynamics of balanced and unbalanced fish populations. Agricultural Experiment Station Bulletin 274. Alabama Polytechnic Institute, Auburn, Alabama, USA.

UREMOVICH, B. L., S. P. CRAMER, C. F. WILLIS, AND C. O. JUNGE. 1980. Passage of juvenile salmonids through the ice-trash sluiceway and squawfish predation at Bonneville Dam, 1980. Report to the United States Army Corps of Engineers, Oregon Department of Fish and Wildlife, Clackamas, Oregon, USA.

ZARET, T. M. 1979. Predation in freshwater fish communities. Pages 135–143 *in* R. H. Stroud and H. Clepper, editors, Predator-prey systems in fisheries management. Sport Fishing Institute, Washington, D.C., USA.

ZAUGG, W. S., AND L. R. MCLAIN. 1972. Steelhead migration: Potential temperature effects as indicated by gill ATPase activity. Science 176:415–416.

Pages 186–195 *in* G. E. Hall and M. J. Van Den Avyle, editors.
Reservoir Fisheries Management: Strategies for the 80's.
Reservoir Committee, Southern Division American Fisheries Society, Bethesda, Maryland. 1986.

MANAGEMENT OF RESERVOIR FISH POPULATIONS BY HARVEST REGULATION

The History and Development of Warmwater Fish Harvest Regulations

LEE C. REDMOND

Division of Fisheries, Missouri Department of Conservation
Jefferson City, Missouri 65102

ABSTRACT

Earliest fish harvest regulations in the United States were directed at eliminating fish traps, nets, and seines, particularly near mouths of streams and during spawning seasons. By the early 1900's increasingly restrictive regulations progressed to minimum length limits and closed spring seasons to allow fish to spawn once before being caught. The Maximum Sustainable Yield (MSY) concept was given impetus by subsequent liberalization of regulations after the Norris Reservoir studies showed that fishing could be allowed in the spring without affecting reproduction or numbers of fish harvested in following years. Studies in the 1940's and 1950's continued to support increased sportfish harvest. Investigations in the 1960's proved that in small lakes predator fishes could be overharvested with little fishing effort, upsetting the "balance" in the fish community and seriously affecting future fishing quality for both predators and panfish. The imposing of effective size limits prevented predator overharvest and maintained balanced fish populations that provided good fishing. With use of refined fish tagging and recovery methods, biologists demonstrated that high exploitation rates of black bass and white crappie also occurred in large reservoirs. Harvest restrictions applied to large reservoirs in the 1970's showed that practical, properly applied, and well-enforced size limits could maintain sport fishing and harvest of desirable quality (Optimum Sustainable Yield [OSY]) in large impoundments as well. One challenge of the '80's is to increase public acceptance of harvest restrictions and to achieve necessary compliance. Both are required to maintain high quality fishing for a growing number of more knowledgeable and better-equipped sport fishermen.

By the time the colonists came to America, nearly all of England was in private ownership, and most of the population was denied trespass rights to fish and hunt. American settlers, therefore, jealously guarded the right to fish in this "new" land. Most of the charters of the thirteen original Colonies granted the settlers the privilege of fishing and hunting on the lands of the colony held in common. For example, a Massachusetts Bay Colony ordinance of 1641, amended in 1647, provided trespass rights through any man's property to ponds in common ownership within town boundaries for the purpose of fishing and fowling.

Fishing in those days was primarily for food. Equipped with machinery and the intent to remake the land in the image of the "old country," settlers cut and burned forests, plowed the fields, dammed and polluted the streams, and built mills. They relentlessly harvested anadromous fish runs, eliminating the Atlantic salmon (*Salmo salar*) and nearly eliminating the American shad (*Alosa sapidissima*) from the rivers south of the St. Lawrence River in Canada (Stroud 1963). It soon was apparent that fishery resources were dwindling although reasons were not known.

As new areas were settled, the citizenry imposed a few simple regulations on the taking of fish although the term "inexhaustible" was frequently applied to the fishery resources. Restraint on wasteful fishing was a first step. Recreational fishing continued to evolve and terms such as "angling for sport" and "fishing for recreation" appear frequently in diaries and newspapers predating the American Revolution. Articles on the sport of angling appeared as early as 1739. Organized angling clubs were in existence even before that; the Schuylkill Fishing Company, a social group organized complete with bylaws, was founded in 1732 in Philadelphia and has existed continuously to the present (King et al. 1962).

In this report, I examine important concepts and chronological events that have led to current strategies for regulating harvest of fish in United States reservoirs. The pendulum of change swung toward increasing restrictions between the 1600's and 1940, back toward liberalization in the mid-1900's, and is now swinging back to increased restrictions. In this report I address the events and changes chronologically under those three periods.

Liberalization → Restriction 1630–1940

Earliest restrictions to reduce excessive harvest were simple. The Massachusetts Colony, as early as 1639, passed an edict governing use of saltwater fish. Legislation passed in 1652 closed the season during the spawning period on cod, haddock, hake, pollock, and mackerel (Stroud 1963). In 1678 the court of Middlesex County, Virginia, restricted methods of taking fish by enacting a law prohibiting use of jacklights while fishing (King et al. 1962). The Common Council of New York City first defined angling by law in 1734 in an ordinance limiting "taking of fish from fresh water ponds by angling, with angle rod, hook and line only" (King et al. 1962).

Fishery resources continued to dwindle, particularly near the most populated areas, and simple restrictions proliferated. Early prototype conservation departments were born to execute and enforce these protective measures, beginning in Massachusetts in 1856 and in New Hampshire and Vermont in 1865 (Stroud 1963). In the late 1800's two conclusions were finally reached regarding declining marine fisheries: 1) fish should be allowed to spawn once before being harvested to provide sufficient eggs for future generations; and 2) small fish would continue to grow larger if allowed to live to an older age before being harvested, thereby increasing the yield in weight. Restrictions in mesh size of nets and size limits and annual quotas were enacted for some marine fishes, which restored many of the marine fisheries (Russell 1942). From these acts were developed the theories in the 1930's that fish stock abundance was linked to additions by growth and recruitment and to losses by natural and fishing mortality and from these evolved the Maximum Sustainable Yield (MSY) concept (Nielsen 1976).

Most restrictions in freshwater were directed toward outlawing nets, traps, and seines, particularly during spring spawning season, and requiring fish ladders around the mill dams so fish could move up the streams from the sea. By 1877, 27 states had enacted laws of this type (Steedman et al. 1879).

The U.S. Fish Commission and many state fish commissions came into existence in the 1870's. The U.S. Fish Commission was charged with looking into reasons for the declining fisheries, which seemed to be an assignment of an applied research nature. Instead, the commissions immediately launched into stocking fry of many species into streams and natural lakes with little consideration of habitat requirements or reasons for decline of native species.

The Federal Black Bass Act was passed in 1926 because of increasing angler concern over declining fishery resources. It outlawed interstate shipment of illegally taken black basses (*Micropterus* spp.). This was amended in 1930 and 1947 to include other species and was a milestone in conservation legislation eliminating commercial fishing for game fishes (Stroud 1966). In the 1930's the MSY concept spilled over into the inland fisheries field, and most species had small size limits (presumably to allow fish to spawn once before being caught) and low number limits. For instance, by 1940, Missouri had a 5-in size limit on bluegill (*Lepomis macrochirus*) and a daily limit of 12 bluegill (Howell 1976). Restrictions on other species were similar.

Thoughtful anglers increasingly realized that traditional regulation and indiscriminate fish stocking were often arbitrary. Much of the decline in the fisheries was due to pollution from mining, excessive timbering operations, and other environmental problems. Pressure on politically-appointed administrators prompted hiring of university fishery experts to conduct biological surveys. Another major step toward more biologically-sound regulations was the hiring of the first full time state fishery biologist in 1923 in Michigan. Many of the larger states followed in the 1930's.

A landmark in fisheries management was the first use of rotenone in 1934 by Carl Hubbs and Milton Trautman to eradicate a stunted yellow perch (*Perca flavescens*) population in a Michigan pond. This tool opened the door for biologists to observe fish population structures and thereby propose regulations based on better known conditions. A study by Carbine concerning warmwater fish spawning at Deep Lake, Michigan, indicated that fry production averaged over a half-million per acre (Stroud 1966). This was another blow to promiscuous maintenance stocking of warmwater fish and further shifted emphasis to habitat management and surveys to determine fish population conditions in lakes.

Liberalization ← Restriction 1940–1960

The second swing of the pendulum was one of increasingly liberal regulations and occurred from 1940 until the early 1960's. By 1940, 9-year-old, 60,000-acre Lake of the Ozarks, Missouri, had gained the reputation of being a "biological desert" because it contained small, slow-growing fish, had a limited food supply, and offered poor fishing (Weyer 1940). Previously-restrictive regulations—low creel limits and small size limits—were lifted and the fishing season was extended earlier in the spring. The 10-in size limit was retained on black basses, and some coves were closed to spring fishing as spawning sanctuaries. Commercial fishing was also restricted in Missouri in 1940 to the Missouri, Mississippi, and lower St. Francis rivers—about 1,100 miles of water (McKinley 1976).

The major thrust toward liberalization was the much-publicized Norris Reservoir study, where fish tagging and fish harvest studies by R. W. Eschmeyer and associates, beginning in 1940, showed that only 20% of the game fish were harvested in one of the most heavily fished parts of the lake. This was the first "concrete evidence that Norris Lake was underfished." The 59-day closed spawning season was eliminated at Norris in 1944. Fishing opportunity and harvest doubled with the open spring fishing season, and the number of fish caught in succeeding years did not noticeably decrease. This led to the generalization that most adult game fish in large reservoirs were dying of old age rather than being caught by fishermen, and liberalization of fishing seasons and limits occurred across the country. Regulations were relaxed on all but two Tennessee Valley Authority (TVA) impoundments in 1945 (Eschmeyer 1945). Many states followed this example, not only dropping closed spring seasons on reservoirs, but also dropping many other restrictions on other types of water. Stroud (1966) reported that two-thirds of the states had similarly liberalized fishing regulations after another decade.

Some states removed restrictions more carefully than others and evaluated results. A good example is Wisconsin's 10-year Five-Lakes study on 300-acre Escanaba Lake and four smaller lakes, which demonstrated high annual exploitation rates for predator species, such as a 29% average for walleye (*Stizostedion vitreum*), 46% for northern pike (*Esox lucius*), and 26% for black crappie (*Pomoxis nigromaculatus*). Even for the prolific bluegill, the annual exploitation rate averaged 42%, and reached a high of 64% in one year. Generally, conclusions from these studies were that fishing pressure varied on most lakes depending upon the quality of fishing they provided at the time (Oehmcke and Waggoner 1956). Numbers of fish harvested in years after liberalization were about the same as before although size of fish harvested decreased. Twenty-three percent of the walleye harvest for 4 years after a 13-in size limit was dropped were shorter than 13 in (Klingbiel 1953).

In the 1940's both George Bennett and Homer Swingle realized the necessity of maintaining a balance between predators and prey in small lakes and ponds but did not suggest methods of preventing overharvest of the predators (primarily largemouth bass [*Micropterus salmoides*]). Bennett (1947) suggested higher harvest of panfish (bluegill) by heavier fishing, even commercial fishing. Swingle and Smith (1950) advocated stocking more small largemouth bass to prey on overabundant small bluegill if ponds were out of balance from fishing. Bennett described the boom-bust cycle in lakes as varying between 5 and 12 years depending on the length of time required for the introduced population of fishes to expand to the point of actual competition. Competition could be delayed if "the original stock of fishes contains a high percent of carnivorous species which function with partial success as natural predators of small fish" (Bennett 1947).

Most large reservoirs went through a similar cycle, early peak fishing years followed by a substantial decline. According to G. B. Herndon (1946), Chief of Missouri's Fisheries Section, 8,500-acre Lake Wappapello

> "furnished good bass fishing during 1941 and 1942 (its first 2 years). As the water character deteriorated, good crappie (*Pomoxis* spp.) fish-

ing continued somewhat longer but all game fish populations have shown a rapid decline as the lake changed to rough fish habitat. Water level manipulation for flood control and mosquito control has created typical barren, marginal areas around the lake which cannot adequately provide foods. That Wappapello is passing now through the period of crappie dominance, with black bass fishing already at a low ebb, was revealed in a creel census made by the Conservation Commission last season (1945). A total of 2,872 anglers were interviewed on the lake, and the results showed that it took Wappapello fishermen an average of 14 hours in 1945 to land one black bass. They caught crappie at the rate of 1 for every 2.2 hours of fishing effort. To quote R. W. Eschmeyer, Chief TVA biologist, in speaking of Norris Lake, as another example: 'In the first season of the impoundment it was possible to take bass in considerable numbers .... Later it was found that fishing was only good for a few days or a few weeks after the opening of the season and then dropped off sharply.'"

Patriarche and Campbell (1957) documented the decline of the bass population and fishery in 1,650-acre Clearwater Reservoir, Missouri, even with the statewide regulation of six black bass per day. That lake was impounded early in 1948, and despite satisfactory spawnings in the reservoir, "the number of bass caught per 100 hours of fishing with artificial lures declined successively from 1949 to 1954 as follows: 19, 17, 8, 10, 5, and 1." A stocking of 10 bass per acre (8–10 in) in 1952 failed to check the downward trend. After 1951, white crappie (*Pomoxis annularis*) was the most numerous sportfish in the catch. Population estimates of 7-in and larger crappie, using tags and methods of that period, showed 80 per acre in 1951 and 72 per acre in 1954. Harvest estimates for those years were 30 per acre (38% of population estimate) and 47 per acre (65% of population estimate). Fishing pressure was high—111 hours per acre in 1951 and 88 hours per acre in 1954, much of which was for crappie. Eighty-five percent of the anglers in 1954 were "still fishing" with live bait. Only one Age-4 white crappie was taken in trap nets between 1949 and 1954, but, even with this high fishing pressure and absence of older crappie in the population, it was concluded "since less than 10% of the tagged fish have been caught by anglers each year, the apparent high mortality must be the result of natural causes."

R. W. Eschmeyer published his classic "Fish Conservation Fundamentals" in 1954, offering basic considerations for deciding upon regulations. He recognized then, among other things, that size limits on large predators were probably beneficial in some areas, that conditions change, that anglers fish to relax (and the fewer and simpler the regulations, the better), that regulations have little value unless they are enforced, and that most people obey the laws which they respect (Eschmeyer 1954). Those are key considerations today.

Martin (1958a) demonstrated an inverse relationship between fishing pressure and bass catch per unit effort in Virginia. He also showed that as few as 14 to 20 fishing trips per acre could remove the harvestable surplus of bass (up to 46% of tagged bass in a 650-acre infertile lake), and this often led to overpopulation of sunfish. At *that* time he suggested fertilization and physical population manipulation measures to increase yield rather than more restrictive regulations on public waters. Soon thereafter he recommended that owners of private ponds should not harvest bass between 1 and 2 lb so that they could remain as effective sunfish predators (Martin 1958b).

The general trend toward liberalization continued. By 1959, 34 states had removed size restrictions from bass in most waters. Missouri and Florida had established creel limits on those species classified as non-game fish to improve their image (Stroud and Jenkins 1960). *Numbers* of fish harvested were still the primary concern.

The early 1960's were a period of stabilization, consolidation, and evaluation of the effects of liberalized regulations under increasing fishing pressure, especially for bass and some of the other major predators. Fishing pressure had more than doubled at Norris Reservoir from 4 angler days per acre in 1944 to over 9 angler days per acre in 1963 while sportfish harvest increased only 11.5% (Stroud 1964). By 1967, only 14 states continued to enforce minimum size restrictions on game fish, although 46 maintained creel limits. No state imposed length limits on panfish, and only 18 states had panfish creel limits, and these were quite liberal. Alabama, Louisiana, Pennsylvania, and Tennessee had joined Missouri and Florida in imposing liberal creel limits on some non-game species in an attempt to make them more desirable (Stroud and Martin 1968).

A 6-year study of largemouth bass fishing at 396-acre Brown's Lake, Wisconsin, under liberalized regulations (5 week earlier opening and removal of 10-in size limit) demonstrated no ill effects to the bass population nor any improvement. Anglers did have a longer period to fish. The peak of bass harvest occurred the first few weeks of the season despite the earlier opening date (Mraz 1964). This is similar to results from Gladstone Lake, Minnesota, where 35% of the bass harvest occurred within two weeks of a June opening and also when the season was advanced by two weeks (Maloney et al. 1962). Bass harvest at Brown's Lake increased 40% the year after removal of the 10-in size limit and 25% the third year. Forty percent of the bass harvested were smaller than 10 in the first year and 25% were under 10 in the third year.

Liberalization → Restriction 1960–Future

In the early 1960's the pendulum began to swing back toward more restrictions and, hopefully, more biologically sound and *socially* acceptable ones. Three additional tools permitted this: development of boat-mounted electrofishing units for lake sampling in the 1940's (Larimore et al. 1950); refinement of these to more efficient direct current units in the 1960's; and development of lighter, smaller fish tags with better retention time.

Biologists in Missouri found angler exploitation of largemouth bass to be 40 to 69% in only the first four days of fishing in new lakes (22 to 205 acres) at a fishing pressure of only 9 to 38 trips per acre (Redmond 1974). Initial first-year exploitation of bass reached 72% at 205-acre Little Dixie Lake (Turner 1963) and 74% at 240-acre Pony Express Lake (Redmond 1974) at total fishing pressures of 406 and 585 hours per acre (113 and 117 trips/acre). These pressures created dramatic changes in the structure of the largemouth bass populations and within three years resulted in high populations of intermediate-size bluegill and slow-growing bass and bluegill. Fishing for bass and bluegill was poor. Analysis of creel records indicated that a daily limit of four bass, instead of the statewide limit of 10, would have saved or redistributed only 10% of the bass creeled at 205-acre Little Dixie Lake; 49% of the bass were removed as one or two fish per stringer and 59% of the fishermen had caught no bass. To be effective over a period of time, the daily limit would have had to be two or fewer; anything higher might actually have resulted in an increased harvest because of the psychological effect of a limit as a goal to strive to reach.

A number of harvest regulations were evaluated in Missouri to prevent initial overexploitation of bass in new lakes. These included a catch-and-release season prior to lake opening, leaving a lake open to fishing from time of original stocking, midwinter openings, refuge areas where fishing was not permitted, a 40% bass harvest quota, and length limits on bass. The first three techniques were intended to expose bass to fishing without significant harvest and train them to avoid lures, thus reducing their vulnerability. Anderson and Heman (1969) previously had found that adult largemouth bass from unfished ponds were caught by angling three times faster than were bass from previously-fished ponds. First-year bass harvest under each of the three learning experiences was considered excessive (Redmond 1974). Considerable movement of bass occurred out of a 40-acre refuge area in a 100-acre lake and there was some violation of the refuge area by fishermen so that bass harvest was also excessive the first year (Redmond 1974).

A 40% annual harvest rate on largemouth bass was found to be about the allowable maximum in research ponds to maintain balanced bass bluegill populations over a four year period (Graham 1974). The 40% annual harvest quota for bass was evaluated in a 9.5-acre Missouri lake opened to public fishing each spring for four seasons. This harvest quota maintained a balanced fish population that showed good growth of bass and bluegills throughout the study. The quota, however, was rapidly reached in four or five days in each year of the study at fishing pressures of only 18 to 44 trips per acre (Ming 1974)—not a very practical method to manage public fishing lakes.

The most promising technique found to manage largemouth bass populations in new lakes up to 160 acres was a minimum size limit—not the small size of early years which protected fish until they could spawn once, but a balanced size—one large enough to maintain predation on intermediate-size sunfish but not so high as to allow excessive natural mortality before fish are available for harvest. To permit a minimum size limit to function properly, new lakes were stocked with a large forage base prior to stocking a reduced number of bass fingerlings, a few adult bass, the normal number of sunfish fingerlings, and a later addition of bass fingerlings. This approach resulted in rapid bass growth, bass reproduction the following year, and the presence of a broad length range of bass when the lake was opened to fishing approximately two years later. A minimum size limit of 14 or 15 in on bass at lake opening allowed an immediate harvest of ¼ to ⅓ of the original stock and resulted in excellent fishing the first year and good bass and bluegill fishing for at least five years (Hoey and Redmond 1974; Redmond 1974). High-quality catch-and-release fishing for bass smaller than 14 in was an added benefit.

Imposition of a 12-in bass size limit on most of Missouri's relatively small, older public fishing lakes in October 1968 improved the structure of bass and bluegill populations and improved fishing for bass and bluegill for varying lengths of time (Farabee 1974; Hickman and Congdon 1974). In most of those lakes, high bass recruitment led to their stockpiling under the size limit and eventual shortages in forage (Rasmussen and Michaelson 1974). Relatively clear lakes with abundant aquatic vegetation developed crowded bass populations sooner. Introduction of gizzard shad (*Dorosoma cepedianum*) as additional forage improved bass and bluegill populations in some lakes for a time (Rasmussen and Michaelson 1974). Later management efforts included addition of northern pike and tiger muskies (*Esox lucius* × *Esox masquinongy*) (protected by a 30-in size limit) to exert more predation on the larger shad.

Use of a protected size range regulation (slot length limit) on bass to allow fishermen to harvest intermediate sizes, thus maintaining a better proportional stock density (PSD) as suggested by Anderson (1976), has improved fishing in these lakes with high bass recruitment (Eder 1984).

In smaller lakes it is necessary to maintain adequate predation on sunfish forage species to have properly-functioning fish populations that will produce good fishing. The techniques to maintain a good fish population, initially at least, have been established for smaller waters. Effective bass harvest regulations appear to be an effective tool to approach Optimum Sustainable Yield (OSY) and high quality fishing as suggested by Anderson (1975). Questions remain as to whether these techniques can be applied to effectively manage large reservoir fish populations with a shad forage base and if increased biomass and higher stocks of black bass will actually improve fish community balance to benefit yields of crappies and white bass (*Morone chrysops*).

Management techniques learned from the studies on smaller lakes were successfully applied at 25,000-acre Stockton Lake, Missouri, in the late 1960's and early 1970's to provide continuous good fishing rather than the peak and decline cycle that reservoir managers are so familiar with (Goddard and Redmond 1978, 1986). Use of minimum length limits on major predator fishes (black bass, 12 in; walleye, 18 in; northern pike and tiger muskies, 30 in) at newly developing Stockton Lake was a successful attempt to maintain high utilization, rapid turnover, and high production of gizzard shad.

A number of studies in the late 1960's and 1970's documented high exploitation of black basses in large reservoirs (Simpson 1978), similar to those rates found in Missouri's new small impoundments in the early 1960's. Among the highest rates found were in California—65% at 995-acre Merle Collins Reservoir (Rawstron and Hashagen 1972) and 58% at 20,000-acre Lake Berryessa (Rawstron and Reavis 1974).

Using Floy anchor tags and a reward system, Ming (Missouri Department of Conservation, Jefferson City, Missouri, USA, [personal communication]) found the annual exploitation rate of adult largemouth bass and spotted bass (*Micropterus punctulatus*) to be 33–42% in 43,500-acre Table Rock and 8,000-acre Pomme de Terre lakes, Missouri. Probably a more important finding by Ming was the comparison of reward tag returns with voluntary returns, which showed non-reporting of voluntary tags, even with widespread publicity, to be 30 to 62%. Considering tag loss, unknown mortality of tagged bass, and non-reporting by anglers, adult bass harvest has undoubtedly been over 50% per year for a number of years in large Missouri lakes and possibly many others. A 15-in minimum size limit was established on Lake of the Ozarks, Pomme de Terre Lake, and Table Rock Lake in 1976.

At the National Black Bass Symposium in Tulsa, Oklahoma, Fox (1975) reported that 14 states had minimum size limits on black bass, the same number as in 1967. He advised in 1975 that size limits were receiving a great deal of attention because it was recognized that they had a direct effect on population structure of bass and forage species, and the increased catch-and-release fishing provided an added value (Fox 1975). Gabelhouse (1980) reported improved crappie populations in some Kansas reservoirs managed with bass length limits. He stated, "Increased crappie growth rates, longevity, and condition factors may be a function of both increased food supply and reduced intra-specific competition. While it is felt that bass predation on small crappie reduced intra-specific competition among crappie, it is possible that increased prey allowed older crappie to survive and they simply suppressed further recruitment by subsequent year classes."

Various innovative harvest regulations are currently being evaluated. Mississippi's 16–20 in protected length range on black basses in 33,000-acre Ross Barnett Reservoir went into effect July 1, 1980 (Herring 1981). North Carolina's 14-in minimum size limit on black basses with two fish allowed below the size limit has a great deal of appeal, if bass fishing pressure on its lakes and individual fisherman harvest patterns allow the regulation to be effective. The bass harvest pattern in Missouri, where most fish are harvested as one or two bass per stringer, would likely render the North Carolina regulation ineffective in most lakes (Redmond 1974).

Using Floy anchor tags and a reward system, Missouri biologists recently found very high annual exploitation rates on crappies similar to the earlier findings in the black bass studies. Under a 30 per day creel limit restriction, anglers returned 43% of the tags from 9-in and larger crappie at 8,000-acre Pomme de Terre Lake and 56% at 25,000-acre Stockton Lake. Fishing pressure for crappie was 19 and 25 hours per acre, respectively (Colvin 1982). This high angler exploitation rate on crappie supported repeated fall trapnet findings in most large Missouri impoundments, where very few crappie Age 4 or older were captured at the end of the fishing season. Anglers have been removing a substantial segment of these crappie populations as small Age-2 and Age-3 fish. Missouri biologists are currently evaluating effects of substantially reduced creel limits of 15 crappie per day and, in Stockton and Table Rock Lakes, a 10-in minimum size limit. They already have demonstrated in one arm of Table Rock Lake that more pounds of 10-in and larger crappie can be harvested under restrictive regulations (10-in size limit and 10 per day creel limit) than were harvested for a number of years under the previous 30 fish per day limit (Colvin 1981).

Discussion

One cannot help but wonder what the biologists at Norris Lake, Tennessee, and other places in the 1940's would have found regarding angler harvest rates had they had efficient electrofishing units to readily capture fish and good lightweight *reward tags* to encourage return of tags. In those days the only methods of capturing fish for tagging were angling and hoop and gill nets; tagging was begun in the winter months when mortality may have been high (and was in some instances [Manges 1950]); and relatively heavy jaw tags were used on crappie as small as 5.5 in, white bass as small as 6 in, and spotted bass as small as 8 in (Miller 1951). Early TVA biologists relied largely on voluntary returns for their information, although there was considerable publicity and, in a few instances, prizes were given for returned tags. As crude as those early methods were, in 1940 they showed annual exploitation rates of 19% on largemouth bass, 40% on spotted bass, and 14% on walleye (Eschmeyer 1942). In the late 1940's, return rates were even higher; for instance, crappie was 29% (Manges 1950).

Other than the increased opportunity of year-round seasons, one can now wonder how many of the other liberalizations would have been justified in the 1940–1965 era. The findings of Patriarche and Campbell (1957) in the early 1950's at Clearwater Lake, Missouri, certainly would have been different. There, fishing pressure for crappie was approximately 80 to 90 hours per acre but voluntary tag returns from jaw-tagged crappie were less than 10%, and the consensus at that time was that the lack of older crappie was due to natural causes. Fishing pressure for crappie, however, was four times as high as the recent pressure at Pomme de Terre and Stockton Lakes, where a minimum of 43% and 56% of crappie were harvested annually. Netting results consistently showed lack of Age-4 and older fish in the lake as biologists are also now finding in heavily exploited populations. Would maintenance or adjustment of size limits and lower creel limits in the 1940's have produced good fishing longer in large reservoirs and eliminated the early peak and rapid decline cycle of fishing quality?

Present day biologists with more sophisticated equipment and a larger background of literature are finding fewer fishery resources underutilized. Indeed, the terms "utilize" or "underutilize" have taken on entirely different meanings than existed in the 1940's when regulations were becoming more liberalized. Sportsmen, politicians, administrators, and biologists today are recognizing the value of fishing and catching fish as recreation rather than mainly for food. Harvest of some fish and the opportunity to catch and keep a trophy fish are still valuable aspects of a fishing trip; however, there is increasing emphasis on the sport of fishing and *catching* fish, particularly for the larger predatory game fishes. The realization that quality and aesthetics both play an important role in sport fishing (McFadden 1969) opens the door to innovative fisheries management. The challenge now is to ascertain the values and needs of the public, the capabilities of individual reservoirs to meet these needs, and to develop proper regulations that are as simple and understandable as possible.

The philosophy of OSY has replaced, or is replacing, the philosophy of MSY in fisheries management. The change was made in theory in 1974 and 1975 when the International Association of Game, Fish, and Conservation Commissioners, the Sport Fishing Institute, and the American Fisheries Society resolved to adopt OSY in place of MSY. It was recognized a number of years ago with multispecies commercial fisheries that exploitation of all the stocks for MSY was not possible because of the biological interactions among species (Gulland 1968), and exploitation of a commercially-valuable prey species at MSY could reduce the yield of the predator species. It is my belief that since the mid-1940's, in managing the freshwater recreational fisheries for MSY, the exploitation of recreationally-valuable predator species has actually reduced the yield of the recreationally valuable prey species—by allowing shifts in size structures. We know this occurs in bass-sunfish combinations in relatively small waters, and evidence is accumulating that it is also true in the large bodies of water.

As the numbers of fishermen and their increasingly sophisticated knowledge and equipment continue to grow, it is becoming more apparent that desires of most fishery resource managers and most fishermen are similar. It is only in the face of growing competition for a limited amount of a desirable product that anglers want to keep 10-in walleye, 8-in bass, 7-in crappie, and 4-in bluegill. By imposing quality on the size at harvest of our major predator fishes, we should be able to properly manage our reservoirs for increased yield of sport fishes, both predator and prey. One of the greatest challenges will be properly enforcing the regulations necessary to carry this out and in informing the public of the choices they have in maximizing this potential.

The decade of the 1960's was a period of learning with respect to effective management of reservoirs, and the 1970's was a period of applying and evaluating various types of harvest regulations. An important strategy for the 1980's will be educating fishermen to the value of harvest restrictions and of demonstrating success to them. It also will be a time to evaluate the effectiveness of enforcement of these regulations. Knowing how little fishing pressure it takes to remove a high percentage of a black bass or crappie population (and possibly many others), we realize that a little non-compliance with a harvest regulation can significantly affect the tangible results. As Anderson (1975) aptly stated: "The regulation that will best approach optimum yield depends not only on the response and dynamics of bass and the fish community, but on the response of anglers." With good enforcement and with good understanding by the public of the need for sound regulations, we can both "shorten the time between bites" and achieve a high "quality of fishing that reflects a high quality of living."

Acknowledgments

I am indebted to Glenn Covington for his effort in locating much of the literature in this review.

References

Anderson, R. O. 1975. Optimum sustainable yield in inland recreational fisheries management. Pages 29–38 *in* P. M. Roedel, editor. Optimum sustainable yield as a concept in fisheries management. Special Publication 9, American Fisheries Society, Bethesda, Maryland, USA.

Anderson, R. O. 1976. Management of small water impoundments. Fisheries (Bethesda) 1:5–7, 26–28.

Anderson, R. O., and M. L. Heman. 1969. Angling as a factor influencing catchability of largemouth bass. Transactions of the American Fisheries Society 98:317–320.

Bennett, G. W. 1947. Fish management—a substitute for natural predation. Transactions of the North American Wildlife Conference 12:276–285.

Colvin, M. A. 1981. A study of the life history and

management of the crappie. Performance report, Federal Aid project F-1-R-30. Missouri Department of Conservation, Jefferson City, Missouri, USA.

COLVIN, M. A. 1982. A study of the life history and management of the crappie. Performance report, Federal Aid project F-1-R-31. Missouri Department of Conservation, Jefferson City, Missouri, USA.

EDER, S. 1984. Effectiveness of an imposed slot length limit of 12.0–14.9 inches on largemouth bass. North American Journal of Fisheries Management 4:469–478.

ESCHMEYER, R. W. 1942. The catch, abundance, and migration of game fishes in Norris Reservoir, Tennessee, 1940. Journal of the Tennessee Academy of Science 17:90–115.

ESCHMEYER, R. W. 1945. The Norris Lake fishing experiment. Division of Fisheries, Booklet, Tennessee Department of Conservation, Nashville, Tennessee, USA.

ESCHMEYER, R. W. 1954. Fish conservation fundamentals. Sport fishing Institute Bulletin 38:1–30.

FARABEE, G. B. 1974. Effects of a 12-inch length limit on largemouth bass and bluegill populations in two northeast Missouri lakes. Pages 95–99 *in* J. L. Funk, editor. Symposium on overharvest and management of largemouth bass in small impoundments. Special Publication 3, North Central Division American Fisheries Society, Bethesda, Maryland, USA.

FOX, A. C. 1975. Effects of traditional harvest regulations on bass populations and fishing. Pages 392–398 *in* R. H. Stroud and H. Clepper, editors. Black bass biology and management. Sport Fishing Institute, Washington, D.C., USA.

GABELHOUSE, D. 1980. Black bass length limit investigations. Final report, Federal Aid project F-15-R. Kansas Fish and Game Commission, Topeka, Kansas, USA.

GODDARD, J. A., AND L. C. REDMOND. 1978. Northern pike, tiger muskellunge, and walleye populations in Stockton Lake, Missouri: a management evaluation. Pages 313–319 *in* R. L. Kendall, editor. Selected coolwater fishes of North America. Special Publication 11, American Fisheries Society, Bethesda, Maryland, USA.

GODDARD, J. A., AND L. C. REDMOND. 1986. Stockton Lake: prolonging the "boom"; managing a new large reservoir with length limits. Pages 203–210 *in* G. E. Hall and M. J. Van Den Avyle, editors. Reservoir fisheries management: strategies for the 80's. Reservoir Committee, Southern Division American Fisheries Society, Bethesda, Maryland, USA.

GRAHAM, L. K. 1974. Effects of four harvest rates on pond fish populations. Pages 29–38 *in* J. L. Funk, editor. Symposium on overharvest and management of largemouth bass in small impoundments. Special Publication 3, North Central Division American Fisheries Society, Bethesda, Maryland, USA.

GULLAND, J. A. 1968. The concept of the maximum sustainable yield and fishery management. Technical Paper 70, Food and Agriculture Organization of the United Nations, Rome, Italy.

HERNDON, G. B. 1946. Can we expect good fishing in big, man-made lakes? The Missouri Conservationist 7(3):4–11.

HERRING, J. 1981. Effect of length limits and slot limits on bass populations of Mississippi waters. Fisheries Report 9, Federal Aid project F-55. Mississippi Department of Wildlife Conservation, Jackson, Mississippi, USA.

HICKMAN, G. D., AND J. C. CONGDON. 1974. Effects of length limits on the fish populations of five north Missouri lakes. Pages 84–94 *in* J. L. Funk, editor. Symposium on overharvest and management of largemouth bass in small impoundments. Special Publication 3, North Central Division American Fisheries Society, Bethesda, Maryland, USA.

HOEY, J. W., AND L. C. REDMOND. 1974. Evaluation of opening Binder Lake with a length limit for bass. Pages 100–105 *in* J. L. Funk, editor. Symposium on overharvest and management of largemouth bass in small impoundments. Special Publication 3, North Central Division American Fisheries Society, Bethesda, Maryland, USA.

HOWELL, P. 1976. Missouri's wildlife trail, part II, 1937–1976. The Missouri Conservationist 37(7): 26–61.

KING, W., J. E. HEMPHILL, A. W. SWARTZ, AND K. F. STUTZMAN. 1962. Sport fishing—today and tomorrow. Study Report 7. Outdoor Recreation Resources Review Commission, U.S. Government Printing Office, Washington, D.C., USA.

KLINGBIEL, J. 1953. Are fishing restrictions necessary? Wisconsin Conservation Bulletin 18(11):1–3.

LARIMORE, R. W., L. DURHAM, AND G. W. BENNETT. 1950. A modification of the electric fish shocker for lake work. Journal of Wildlife Management 14:320–323.

MALONEY, J. E., D. R. SCHUPP, AND W. J. SCIDMORE. 1962. Largemouth bass population and harvest, Gladstone Lake, Crow Wing County, Minnesota. Transactions of the American Fisheries Society 91:42–52.

MANGES, D. E. 1950. Fish tagging studies in TVA storage reservoirs, 1947–1949. Journal of the Tennessee Academy of Science 25:126–140.

MARTIN, R. G. 1958a. Influence of fishing pressure on bass fishing success. Proceedings of the Annual Conference of the Southeastern Association of Game and Fish Commissioners 11:76–81.

MARTIN, R. G. 1958b. More fish from farm ponds. Virginia Wildlife 19(11):10–12.

MCFADDEN, J. T. 1969. Trends in freshwater sport fisheries of North America. Transactions of the American Fisheries Society 98:136–150.

MCKINLEY, D. 1976. Missouri's wildlife trail, part I, 1700–1936. The Missouri Conservationist 37(7): 1–25.

MILLER, L. F. 1951. Fish harvesting on two TVA

mainstream reservoirs. Transactions of the American Fisheries Society 80:2–10.

MING, A. 1974. Regulation of largemouth bass harvest with a quota. Pages 39–53 *in* J. L. Funk, editor. Symposium on overharvest of largemouth bass in small impoundments. Special Publication 3, North Central Division American Fisheries Society, Bethesda, Maryland, USA.

MRAZ, D. 1964. Evaluation of liberalized regulations on largemouth bass, Brown's Lake, Wisconsin. Technical Bulletin 31, Wisconsin Department of Natural Resources, Madison, Wisconsin, USA.

NIELSEN, L. A. 1976. The evolution of fisheries management philosophy. Marine Fisheries Review 38(12):15–23.

OEHMCKE, A. A., AND D. W. WAGGONER. 1956. How liberal can you get? Wisconsin Conservation Bulletin 21(5):12–15.

PATRIARCHE, M. H., AND R. S. CAMPBELL. 1957. The development of the fish population in a new flood-control reservoir in Missouri, 1948 to 1954. Transactions of the American Fisheries Society 87:240–258.

RASMUSSEN, J. L., AND S. M. MICHAELSON. 1974. Attempts to prevent largemouth bass overharvest in three northwest Missouri lakes. Pages 69–83 *in* J. L. Funk, editor. Symposium on overharvest of largemouth bass in small impoundments. Special Publication 3, North Central Division American Fisheries Society, Bethesda, Maryland, USA.

RAWSTRON, R. R., AND K. A. HASHAGEN, JR. 1972. Mortality and survival rates of tagged largemouth bass (*Micropterus salmoides*) at Merle Collins Reservoir. California Fish and Game 58:221–230.

RAWSTRON, R. R., AND R. A. REAVIS. 1974. First year harvest rates of largemouth bass at Folsom Lake and Lake Berryessa, California. California Fish and Game 60:52–53.

REDMOND, L. C. 1974. Prevention of overharvest of largemouth bass in Missouri impoundments. Pages 54–68 *in* J. L. Funk, editor. Symposium on overharvest of largemouth bass in small impoundments. Special Publication 3, North Central Division American Fisheries Society, Bethesda, Maryland, USA.

RUSSELL, E. S. 1942. The overfishing problem. Cambridge University Press, Cambridge, England.

SIMPSON, D. 1978. A review of the literature on black bass populations in large reservoirs. Final report, Federal Aid project F-1-R-26. Missouri Department of Conservation, Jefferson City, Missouri, USA.

STEEDMAN, I. G. W., J. REID, AND S. WOODSON. 1879. Annual Report, Missouri Fish Commission, Jefferson City, Missouri, USA.

STROUD, R. H. 1963. Fish conservation. Pages 329–413 *in* I. N. Gabrielson, editor. The fisherman's encyclopedia. The Stackpole Co., Harrisburg, Pennsylvania, USA.

STROUD, R. H. 1964. Liberalized fishing at Norris. Sport Fishing Institute Bulletin 157:4–5.

STROUD, R. H. 1966. Fisheries and aquatic resources. Pages 57–89 *in* H. Clepper, editor. Origins of American conservation. Ronald Press, New York, USA.

STROUD, R. H., AND R. M. JENKINS. 1960. Fish conservation highlights 1957–1959. Sport Fishing Institute, Washington, D.C., USA.

STROUD, R. H., AND R. G. MARTIN. 1968. Fish conservation highlights 1963–1967. Sport Fishing Institute, Washington, D.C., USA.

SWINGLE, H. S., AND E. V. SMITH. 1950. Factors affecting the reproduction of bluegill bream and largemouth black bass in ponds. Circular No. 87, Alabama Polytechnic Institute, Agricultural Experiment Station, Auburn, Alabama, USA.

TURNER, J. L. 1963. A study of a fish population of an artificial lake in central Missouri with emphasis on *Micropterus salmoides* (Lacépède). Master's thesis, University of Missouri, Columbia, Missouri, USA.

WEYER, A. E. 1940. The Lake of the Ozarks: a problem in fishery management. The Progressive Fish Culturist 51:1–10.

Pages 196–202 *in* G. E. Hall and M. J. Van Den Avyle, editors.
Reservoir Fisheries Management: Strategies for the 80's.
Reservoir Committee, Southern Division American Fisheries Society, Bethesda, Maryland. 1986.

Commercial Fishing as a Reservoir Management Tool

ARNOLD W. FRITZ AND HARRY L. WIGHT
Illinois Department of Conservation, Carlyle, Illinois 62231

ABSTRACT

The use of commercial fishing to reduce populations of selected species for the benefit of sport fish populations has met with limited success. Benefits attributable to intensive commercial removal programs include increased reproductive success of some sport and forage species, improved growth rates and body condition, increased angling success, and improvement of environmental conditions favoring sport species. Lack of knowledge concerning predator-prey relationships, interspecific competition, effects of fish removal programs, and life history requirements of some reservoir fishes has limited the formation of dependable guidelines to achieve specific management goals. Conflicts with recreational interests, restrictive laws and regulations, and inability to conduct intensive removal programs due to design purposes of reservoirs have been identified as impediments restricting the utilization of commercial fishing as a management tool. Specific areas where research and technological developments are needed to quantify the use and value of commercial fishing are suggested. Whether commercial fishing will ever become a widely-utilized management tool will be determined by future fishery management needs, attitudes associated with the operation and uses of reservoirs, and availability of new information describing effects on sport and commercial fish populations.

Prior to the 1930's, fishery management consisted principally of stocking and regulating harvest of numerous species. As interest in sport fishing grew and the quality and quantity of the angler's catch declined, fishery managers sought new methods to maintain or enhance angling success. Commercial fishing was proposed and utilized as a promising tool to accomplish desired management goals. This was based on the commonly held belief that many commercial and forage species (commonly referred to as "rough" or "trash" fish) compete with game fish for available food and space; thus their removal would favor the game fishes and result in better fishing.

The value of commercial fishing was first explored on natural lakes and ponds of the Midwest. Some sport fishery benefits were realized, which prompted resource agencies to employ rough fish removal crews or to open other lakes to commercial fishing to accomplish the same purpose. This management technique was later used on many of the large reservoirs throughout the nation where angling success declined following initial years of good fishing.

Commercial fishing is currently allowed on many lakes and reservoirs, but typically it is authorized for the utilization of an available food resource rather than for the expressed purpose of fishery management. In this report, we review earlier commercial fishing efforts, identify gaps in the current use of this technique, and describe potential research needs to more effectively utilize this management tool.

LITERATURE REVIEW

Fish Population and Angling Responses

One of the earliest efforts to manipulate fish populations by commercial fishing was made by Ricker and Gottshalk (1941). They attributed increases of game fish and angling success to an intensive seine removal of 88 lb/acre of carp (*Cyprinus carpio*), carpsuckers (*Carpiodes* spp.), and buffalo (*Ictiobus* spp.) from Bass Lake, Indiana. Rose and Moen (1953) reported similar benefits in East Okoboji Lake, Iowa, following an 11-year seining program that removed 40 to 681 lb/acre of buffalo, carp, and freshwater drum (*Aplodinotus grunniens*). An intensive 12-year trawl, hoop net, and trap net harvest of freshwater drum from Lake Winnebago, Wisconsin, resulted in sport fish population benefits according to Priegel (1971). Populations of white bass (*Morone chrysops*), black crappie (*Pomoxis nigromaculatus*), yellow perch (*Perca flavescens*), sauger (*Stizostedion canadense*), and walleye (*Stizostedion vitreum*) increased significantly following several years of drum removal.

White (1959) indicated that crappie (*Pomoxis* spp.) fishing improved following legalization of commercial netting in Tennessee Valley Authority (TVA) reservoirs. However, Lambou and Stern (1959) reported inconclusive results when 90% of the gizzard shad (*Dorosoma cepedianum*) were seined from a small Louisiana lake; fishing for largemouth bass (*Micropterus salmoides*) improved, but poor panfish angling resulted from overcrowding and slow growth created by two strong year classes of sunfish (*Lepomis* spp.) and crappie. Grinstead (1975) expressed the opinion that an intensive rough fish removal program could contribute to more desirable largemouth bass populations. Limited sport fishing benefits were also attributed to commercial removal operations investigated by Cahoon (1953), Hulsey (1957), and Moyle and Clothier (1959).

Ricker and Gottshalk (1941), Rose and Moen (1953), and Cahoon (1953) attributed an additional sport fish population benefit, which they termed "improved environmental conditions," to the removal of bottom-feeding or "rooting" species such as carp and bullheads (*Ictalurus* spp.). They reported that extensive removal of these species resulted in increased water clarity favoring the growth of sight-feeding species and expansion of aquatic vegetation, which also provided better spawning sites and protective cover for young fishes. Their observations are consistent with the findings of Buck (1956), who indicated that turbid water had a detrimental effect on largemouth bass reproduction, growth, and angling success in Oklahoma waters. Carter (1954), Priegel (1971), Ager et al. (1981), and others have noted that commercial fishing had little deleterious effect on sport fish populations or on angling.

Growth and Body Condition Responses

In Massachusetts, Grice (1958) used fyke nets to intensively thin panfish and some non-game species in four ponds ranging in size from 20 to 269 acres. The species being thinned, bluegill (*Lepomis macrochirus*), yellow perch, black crappie, and white perch (*Morone americana*), exhibited increased growth rates, but there was no increase in the proportion of other game fish. Bennett (1962) attributed increased growth and improved body condition of largemouth bass and bluegills to intensive thinning of these species.

Starrett and Fritz (1965) correlated improvement in the growth and body condition of bluegill, crappie, freshwater drum, and bigmouth buffalo (*Ictiobus cyprinellus*) with the removal of commercial fish from Lake Chautauqua, Illinois, during the first half of a 10-year seining program. However, successful reproduction, recruitment of fish from the inflowing river, and water level fluctuations tended to offset any long-term measurable benefits.

Ager et al. (1981) reported that crappie and sunfish exhibited improved growth rates and body conditions following initiation of a commercial fishing program on Lake Okeechobee, Florida, where large tonnages of shad, gar (*Lepisosteus* spp.), catfish, sunfish, and black crappie were removed. Fishing for largemouth bass remained good, indicating no harm occurred to the bass population due to commercial fish removal.

Wenger (1968) reviewed studies concerning the effects of commercial netting. He noted that one investigator found slower growth and smaller size of white crappie following cessation of commercial netting at Reelfoot Lake, Tennessee.

Stephen (1986) reported that an intensive 3-year commercial fish removal program on Lovewell Reservoir, Kansas, resulted in increased growth of white bass and white crappie (*Pomoxis annularis*) in response to a significant increase in biomass of available prey.

Reproduction Responses

Hulsey (1957) investigated a combined winter water-level draw-down and commercial fish removal operation on Nimrod Lake, Arkansas. The removal of 58 lb/acre of buffalo and gizzard shad resulted in a three-fold increase in the poundage of minnows (Cyprinidae) and sunfish fry, which contributed to a higher survival of young largemouth bass and white bass.

Moyle and Clothier (1959) reported a marked reduction of the carp population and increased spawning success of crappie and black bullheads (*Ictalurus melas*) following a 5-year winter seining program in Traverse Lake, Minnesota. However, an earlier study by Moyle et al. (1950) detected no permanent reproductive benefits where an average of 97 lb/acre of carp, buffalo, and freshwater drum were removed by haul seines over a 25-year period from 14 Minnesota lakes. Scidmore and Woods (1961) studied the effects of a 3-year commercial removal program on four other Minnesota lakes and concluded that no relationship could be found between the amount of fish removed and improvement in spawning success or growth of panfish.

Crawford (1982) evaluated the effects of combined commercial fishing and drawdown programs on Nimrod and Blue Mountain lakes in Arkansas. He noted an increase in the number of young-of-the-year bluegill in Nimrod Lake and increased spawning success of sunfish, crappie, and channel catfish (*Ictalurus punctatus*) in Blue Mountain Lake. Although sport fish populations increased in both lakes, Crawford concluded that it was impossible to attribute all benefits to the removal program.

Stephen (1986) found that the available prey for 6-in predators increased from 22 to about 1,000 lb/acre from 1978 to 1981 because of a commercial fish removal program in 2,900-acre Lovewell Reservoir, Kansas. Simmons (1981) believed that removal of fish from a system caused a biological void into which forage species could expand and provide a larger forage base for game species.

Assessment of Previous Investigations

After reviewing available literature on the use of commercial fishing as a reservoir management tool, two facts are most apparent. The first is the paucity of literature evaluating commercial fish removal programs; second, studies have been conducted on such a variety of waters with different fish populations, duration of removal, types of commercial gear used, and poundages of fishes removed, that no specific implementation guidelines are apparent.

Previous investigations showed that limited sport fish population and angling benefits were produced by commercial fish removal operations. Where large populations of commercial or forage fish were present, the intensive reduction of those species frequently resulted in increased spawning success of both forage and sport fishes, higher survival rates, better growth rates, and improved angler success. Several studies showed that significant reductions of large populations of sunfish and crappies resulted in increased growth of the remaining species. Other investigations noted positive environmental and sport fish benefits from the removal of fishes responsible for inducing turbidity of the water. The size structure and abundance of forage species have also been enhanced by some removal programs. There were also indications that commercial fishing programs conducted on smaller lakes and reservoirs with more stable water levels and uniform bottoms tend to yield more positive sport fishery benefits than those on large reservoirs; this is probably because removal of targeted species is more efficient under these conditions.

Current Commercial Fishing Use

A survey of the continental states was made in early 1983 to obtain a better perspective of the permitted use of commercial fishing on inland waters and how this practice is viewed by management agencies. Responses from 46 states provided information summarized below.

Forty states permit some type of commercial fishing in their inland waters, principally on large rivers but also on reservoirs and a few glacial lakes. Only 22 states allow commercial fishing on one or more reservoirs, and of these only six indicate it was employed as a management tool to improve sport fishing. The other 16 states permit commercial fishing primarily to utilize an existing food resource. Several states felt that some positive sport fishery benefits were achieved when commercial fishing was employed. However, most agencies did not monitor commercial operations for possible benefits to sport fish populations; rather, they confined their activities to gathering catch statistics and other related data.

Only five agencies indicated that utilization of commercial fishing as a management tool might increase in their states. Most indicated that its use would probably remain the same or decline despite the presence of numerous under-utilized species such as carp, suckers, buffalo, carpsuckers, and catfish. This opinion may reflect the fact that 27 states felt their citizens were not receptive to commercial fishing. Twenty-three of the 40 agencies permitting commercial fishing indicated "mixed" to "bad" experiences working with commercial fishermen, especially in terms of compliance with laws and accurate reporting of catch statistics. Only 22 agencies indicated any efforts on their part to promote commercial harvest or public consumption of commercial and under-utilized fish species. Several agencies indicated they had been subject to increased political pressure to stop or limit commercial fishing in some areas. Those states permitting commercial fishing were in agreement that the lack of good markets is a major obstacle to future expansion of commercial fishing opportunities.

Other factors have impacted the use of commercial fishing as an acceptable management tool by states. Fish from many rivers and some reservoirs have been found to have levels of chemical contamination exceeding the U.S. Food and Drug Administration's action levels. Thus, many consumers are reluctant to purchase any freshwater fish. Also, outdoor and conservation magazines often refer to the commercial species in terms of "coarse," "rough," and "trash," and dwell on the problems they create; this has done little to enhance the public's perception or consumption of these species. The angling fraternity's disdain of commercial fishing has intensified in areas where large trophy species such as the striped bass (*Morone saxatilis*) and their hybrids have been stocked. These sport fishermen and some regulatory agencies are convinced that the use of any commercial gear will destroy the sport fishery for this species.

Investigative Needs

Research

The acceptance of commercial fishing as a useful management tool will not be attained until implementation guidelines can be established, which, when applied, will produce more specific and predictable management benefits. These guidelines can only be established by additional research to better understand relationships between species and their life histories. Such information is limited for buffalofishes, carpsuckers, gars, and gizzard shad. A reduction in competition for food and space has been the common justification for many commercial removal programs, but the extent to which various species compete and the effects of competition on production of sport species is not adequately understood. The effects of removing a substantial portion of the lake biomass and how the various populations might respond also warrant further study.

The potential benefits of removing sport species such as bullheads, catfish, crappies, sunfish, yellow perch, and white and yellow bass that tend to over-populate in some waters should be explored in greater detail. More attention should be devoted to an understanding of predator-prey relationships to improve the management of reservoir species. Commercial fishing techniques have the potential of influencing the size structure, abundance, and production of predators and prey. Little is known concerning the effects of egg predation and how such predation may influence the production of sport species. Studies on the levels of competition, predator-prey relationships, and effects of biomass removal are needed to manage fishery resources more efficiently.

Studies done under the auspices of the Reservoir Committee of the Southern Division, American Fisheries Society, have included extensive standing crop assessments on Douglas Reservoir, Tennessee (Hayne et al. 1967), and Barkley Lake, Kentucky (Aggus et al. 1980). These studies provided insight into the relative population densities in various water areas within a reservoir and indicated the dominance of commercial and forage species in reservoir fish communities. More investigations of this nature should be performed to develop better sampling tools and techniques for studying the dynamics of reservoir fish populations and assessing the impacts of reservoir fishery management programs.

Two other research topics worth pursuing are the possible presence of pheromones in reservoirs and the development of chemical attractants. The identification and development of pheromones or baits could conceivably enhance a manager's ability to manipulate fish populations. Swingle (1956) and Chew (1973) have indicated that dense populations of some species such as bluegill, carp, buffalo, and gizzard shad produce a hormone-like repressive factor that limits reproduction of largemouth bass in small lakes or in eutrophic environments. Whether a level of repressive substances is ever produced by overpopulated species within large lakes or reservoirs is not known. Knowledge of such and targeting the responsible species for intensive removal could produce significant management benefits.

Technology

Much can be done to improve or develop better commercial fishing gear and techniques. Methods currently used are essentially the same as those described by Carter (1954), Starrett and Barnickol (1955), and White (1959). The development of selective harvest gear should be pursued to increase the efficiency of removal operations for management purposes and food production.

Greenwood and Boussu (1967) suggested that some of the vessels and fishing gear used by the marine fishing industry may be adaptable to removing large tonnages of reservoir fishes. Trawls have been used successfully to harvest freshwater drum (Priegel 1971) in a large glacial lake, but Grinstead and Wrenn (1970) found its application for harvesting commercial species in TVA reservoirs to be limited. Houser and Bryant (1967) found, however, that a midwater trawl could be used for making population estimates of clupeids. Great Lakes type trap nets were effective in harvesting commercial species from TVA reservoirs (Grinstead 1969), and pound nets may have similar application.

Research should be pursued to fully explore the value of sonar and radio tracking to aid in locating concentrations of reservoir fishes. Sonar systems, which have been used extensively by the marine commercial fishing industry, are now being used by some inland commercial fishermen to locate and harvest many species. Further refinements in these detection systems could aid in the desired commercial removal of many reservoir species. Radio tracking could also become another valuable tool to increase commercial harvest. Johnson and Hasler (1977) reported the successful use of ultrasonic transmitters to locate and subsequently harvest large concentrations of carp in Lake Mendota, Wisconsin.

Marketing

Perhaps the most important factor and critical need governing the success of any fish removal program is the existence of good markets. If the species being harvested cannot be sold at a profit, the likelihood of commercial fishermen removing the desired poundages of the targeted species is remote. Subsidies or harvest incentives must be provided to insure the removal of those species having little or no monetary value (Carroll et al. 1963; Grinstead 1975). The need to develop markets for food and non-food species was recognized by Riggs (1957), Shields (1958), and Carroll et al. (1963). Greenwood and Boussu (1967) suggested means of creating new markets to utilize low-value species, but the demand for and economics of an industrial approach to utilize these species has failed to materialize.

The freshwater fish industry, which lacks the organization, direction, and working capital of the marine and aquaculture interests, has not been successful in promoting the consumption of freshwater fishes or in developing new products and markets. To remedy this situation, it may require the technological and economic assistance of State and Federal agencies. Universities can also play a role in helping meet the needs of the freshwater commercial fishing industry; personnel trained in food technology, marketing, and fishery economics are essential to the development of new markets and stability of the industry.

Conclusions and Recommendations

For reservoir managers to even consider use of intensive fish removal programs, they must have support and assistance of state fishery resource agencies. Some states do not recognize commercial fishing as a legitimate use of the resource, and in others, laws and regulations surrounding commercial fishing are so restrictive that the possibility of achieving a substantial harvest of target species is minimal. In such cases, legislative action is needed to permit its use without unnecessary limitations.

State resource agencies have often neglected to promote legitimate roles for commercial fishing through their information and education programs. Most sport fishermen remain convinced that their fishing will be ruined by commercial fishermen. Commercial fishing has often been prohibited because fishery managers and administrators desire to protect their put-and-take stockings of striped bass in reservoirs. Waterfowl hunters and power- and sail-boating interests often oppose commercial fishing on the basis that it interferes with their activities. All of this has frequently resulted in public and political pressures being applied to regulatory agencies to reduce existing commercial removal programs or banning their use entirely.

To plan an effective and acceptable removal program, careful attention should be paid to several aspects. The type of gear, mesh size, and fishing techniques must be selected to insure adequate harvest of the targeted species. The gear must have a low potential to adversely affect the sport species. The time selected for a removal operation must correspond to periods of low recreational use, especially sport fishing, yet also be allowed when commercial fishing conditions are good. Regulations must be established to define the limits of what the fishermen can or cannot do during the program. The type and number of fishing permits or contracts to be issued, and how they are allocated, must be determined. The status of existing fish markets or the development of new ones must be investigated.

When harvest of under-utilized species is a management objective, the major task is to decide when and how commercial fishing is to be permitted. Minimum supervision is required, and data collection may be limited to catch statistics and length-frequency information. Fishing may be authorized either by permit or simply opening the reservoir to harvest by all individuals who possess the required licenses and authorized gear. A non-permit system allows many fishermen to participate and share in the economic benefits of the fishery resource, but it is not conducive to the gathering of reliable harvest data. Unrestricted removal programs are also frequently characterized by numerous fishing violations, which result in a lack of public acceptance of such programs. Issuance of permits that clearly define regulations and require a performance bond is recommended.

When commercial fishing is being employed as a fishery management tool, the involvement required is more complex. To achieve sport fish population and angling goals, it is essential that removal programs have adequate funding, time, manpower, and agency commitment to assure completion and thorough evaluation. Extensive fish population data, including a good creel survey, must be collected prior to, during, and following the removal operation to monitor biological changes. Removal quotas must be established and economic incentives provided if fishermen are to remove large poundages of low-value species. Limiting the number of permits or contracts to be issued is also recommended to insure control of such removal-research programs. This can be done by holding a lottery, requiring a high entry fee, or a bid system. The advantage of the bid system is that funds are received that could support the removal of low-value or non-saleable species. Requiring the posting of a sizeable performance bond helps to insure compliance with state fishing laws, regulations, and contract provisions related to removal quotas and the reporting of harvest and gear information. It also tends to restrict the number of contract applicants to those who are financially sound, more skilled, better equipped, and more conscientious in helping with meeting program goals.

At this time, it appears we are at the crossroads of either rejecting or further exploring the roles of commercial fishing in reservoir fishery management. Based on our review of previous fish removal programs that document sport fishery benefits but provide few guidelines for effective employment, it is obvious that the management potential of removing a significant portion of a water's fish biomass has not been thoroughly assessed.

It is recommended that an intensive and comprehensive commercial fish removal program be initiated on reservoirs that are typical of specific geographic regions to examine management potentials. Since most state agencies do not have the resources to undertake such an extensive research venture, it is further recommended that state and federal agencies join forces to undertake comprehensive evaluation projects. Committees established within regional Divisions of the American Fisheries Society could coordinate such interagency efforts. Reservoirs have commercial as well as sport fishery resources that should be managed to produce the greatest overall benefits. In light of limited management tools and techniques available to fishery managers, particularly for large reservoirs, it is only logical that the use of commercial fishing as a management tool be explored in greater detail.

References

Ager, L., D. Scovell, D. McCall, D. Powell, and K. Kerce. 1981. Lake Okeechobee–Kissimmee River project. Annual report for 1980, Florida Game and Fresh Water Fish Commission, Tallahassee, Florida, USA.

Aggus, L. R., D. C. Carver, L. L. Olmsted, L. L. Rider, and G. L. Summers. 1980. Evaluation of standing crops of fishes in Crooked Bay, Lake Barkley, Kentucky. Proceedings of the Annual Conference of the Southeastern Association of Fish and Wildlife Agencies 33:710–722.

Bennett, G. W. 1962. Management of artificial lakes and ponds. Reinhold Publishing Corporation, New York, USA.

Buck, D. H. 1956. Effects of turbidity on fish and fishing. Report 56, Fishery Research Laboratory, Oklahoma Department of Wildlife Conservation, Norman, Oklahoma, USA.

Cahoon, W. G. 1953. Commercial carp removal at Lake Mattamuskett, North Carolina. Journal of Wildlife Management 17:312–317.

Carroll, B. B., G. E. Hall, and R. D. Bishop. 1963. Three seasons of rough fish removal at Norris Reservoir, Tennessee. Transactions of the American Fisheries Society 92:356–364.

Carter, E. R. 1954. An evaluation of new types of commercial fishing gear in Kentucky lakes. Kentucky Academy of Science 15(3):58–80.

Chew, R. L. 1973. The failure of largemouth bass, *Micropterus salmoides floridanus* (Le Sueur), to spawn in eutrophic, over-crowded environments. Proceedings of the Annual Conference of the

Southeastern Association of Game and Fish Commissioners 26:306–319.

CRAWFORD, T. 1982. The effect of commercial fish removal by netting on sport fish populations in two Arkansas reservoirs. Final report. Arkansas Game and Fish Commission, Little Rock, Arkansas, USA.

GREENWOOD, M. R., AND M. F. BOUSSU. 1967. Efficient fishing systems—potential for increasing effectiveness of commercial fisheries in reservoir management and utilization. Pages 467–476 *in* Reservoir fishery resources symposium. Southern Division American Fisheries Society, Bethesda, Maryland, USA.

GRICE, F. 1958. Effect of removal of panfish and trashfish by fyke nets upon fish population of some Massachusetts ponds. Transactions of the American Fisheries Society 87:108–115.

GRINSTEAD, B. G. 1969. Comparison of various designs of Wisconsin-type trap nets in TVA reservoirs. Proceedings of the Annual Conference of the Southeastern Association of Game and Fish Commissioners 22:444–457.

GRINSTEAD, B. G. 1975. Response of bass to removal of competing species by commercial fishing. Pages 475–479 *in* R. H. Stroud and H. Clepper, editors. Black bass biology and management. Sport Fishing Institute, Washington, D.C., USA.

GRINSTEAD, B. G., AND W. B. WRENN. 1970. The influence of various factors on the catch of otter trawls in a warmwater reservoir. Proceedings of the Annual Conference of the Southeastern Association of Game and Fish Commissioners 23: 462–473.

HAYNE, D. W., G. E. HALL, AND H. M. NICHOLS. 1967. An evaluation of cove sampling of fish populations in Douglas Reservoir, Tennessee. Pages 244–297 *in* Reservoir fishery resources symposium. Southern Division American Fisheries Society, Bethesda, Maryland, USA.

HOUSER, A., AND H. E. BRYANT. 1967. Sampling reservoir fish populations using mid-water trawls. Pages 391–404 *in* Reservoir fishery resources symposium. Southern Division American Fisheries Society, Bethesda, Maryland, USA.

HULSEY, A. H. 1957. Effects of a fall and winter drawdown on a flood control lake. Proceedings of the Annual Conference of the Southeastern Association of Game and Fish Commissioners 10:285–289.

JOHNSON, P. B., AND A. HASLER. 1977. Winter aggregations of carp *Cyprinus carpio*) as revealed by ultrasonic tracking. Transactions of the American Fisheries Society 106:556–559.

LAMBOU, V. W., AND H. STERN, JR. 1959. Preliminary report on the effect of the removal of rough fishes on the Clear Lake sport fishery. Proceedings of the Annual Conference of the Southeastern Association of Game and Fish Commissioners 12:36–56.

MOYLE, J. B., J. H. KUEHN, AND C. R. BURROWS. 1950. Fish population and catch data from Minnesota Lakes. Transactions of the American Fisheries Society 78:163–175.

MOYLE, J. B., AND W. D. CLOTHIER. 1959. Effects of management and winter oxygen levels on the fish population of Prairie Lake. Transactions of the American Fisheries Society 88:178–185.

PRIEGEL, G. 1971. Evaluation of intensive freshwater drum removal in Lake Winnebago, Wisconsin, 1955–1966. Technical Bulletin 47. Wisconsin Department of Natural Resources, Madison, Wisconsin, USA.

RICKER, W. E., AND J. S. GOTTSHALK. 1941. An experiment in removing coarse fish from a lake. Transactions of the American Fisheries Society 70:382–390.

RIGGS, C. D. 1957. Unexploited potential of our freshwater commercial fisheries. Transactions of the American Fisheries Society 87:299–308.

ROSE, E. T., AND T. MOEN. 1953. The increase in game fish populations in East Okoboji Lake, Iowa, following intensive removal of rough fish. Transactions of the American Fisheries Society 82:104–114.

SCIDMORE, W. J., AND D. E. WOODS. 1961. Changes in the fish populations of four southern Minnesota lakes subjected to rough fish removal. Minnesota Department of Conservation, Fish and Game Investigations, Fisheries Service 3:1–19.

SHIELDS, J. T. 1958. Fish management problems of large impoundments on the Missouri River. Transactions of the American Fisheries Society 87:356–362.

SIMMONS, E. G. 1981. Effects of commercial netting on rough fish populations. Final report. Texas Parks and Wildlife Department, Austin, Texas, USA.

STARRETT, W. C., AND P. G. BARNICKOL. 1955. Efficiency and selectivity of commercial fishing devices. Illinois Natural History Survey Bulletin 26(4):325–366.

STARRETT, W. C., AND A. W. FRITZ. 1965. A biological investigation of the fishes in Lake Chautauqua, Illinois. Illinois Natural History Survey Bulletin 29(1):1–104.

STEPHEN J. 1986. Effects of commercial harvest on the fish community of Lovewell Reservoir, Kansas. Pages 211–217 *in* G. E. Hall and M. J. Van Den Avyle, editors. Reservoir fisheries management: strategies for the 80's. Reservoir Committee, Southern Division American Fisheries Society, Bethesda, Maryland, USA.

SWINGLE, H. S. 1956. Appraisal of methods of fish population study—Part IV—Determination of balance in farm fish ponds. Transactions of the North American Wildlife Conference 21:298–322.

WENGER, A. G. 1968. Commercial netting effects. Progress report, Federal Aid project T-12-R-13. Texas Parks and Wildlife Department, Austin, Texas, USA.

WHITE, C. E., JR. 1959. Selectivity and effectiveness of certain types of commercial nets in the TVA lakes of Alabama. Transactions of the American Fisheries Society 88:81–87.

Pages 203–210 *in* G. E. Hall and M. J. Van Den Avyle, editors.
Reservoir Fisheries Management: Strategies for the 80's.
Reservoir Committee, Southern Division American Fisheries Society, Bethesda, Maryland. 1986.

Stockton Lake: Prolonging the "Boom," Managing a New Large Reservoir with Minimum Length Limits

JOHN A. GODDARD AND LEE C. REDMOND
Division of Fisheries, Missouri Department of Conservation
Jefferson City, Missouri 65102

ABSTRACT

Three-stage filling of 24,900-acre Stockton Lake, Missouri, facilitated development of an excellent predator fish population and an adequate prey base to support it. Largemouth bass (*Micropterus salmoides*) and several minnow species were established by stocking ponds in the reservoir basin prior to dam closure. Walleye (*Stizostedion vitreum*) and northern pike (*Esox lucius*) fry and fingerlings were stocked in the newly flooded terrestrial vegetation during the first 3 years. Inland silverside (*Menidia beryllina*) and threadfin shad (*Dorosoma petenense*) were released to supplement the native prey species. Implementation of minimum length limits of 12 in for black bass and 18 in for walleye within two years of impoundment permitted development and maintenance of abundant predator fish populations. A sport fishing "boom" followed by a collapse in fish population structure and reduced fishing success typical of most new reservoirs did not occur within the first eight years after impoundment of Stockton Lake. Sport fish harvest was initially good and increased for the first six years in Big Sac Arm and stabilized (±10%) at that level for at least three more years (to end of study). In Little Sac Arm, the fishery increased steadily and was highest at the end of the study (year eight of the reservoir). Gizzard shad consistently produced abundant year classes, and shad 3 in and less were available throughout each growing season. Growth rates of sport fishes were rapid throughout the study. Minimum length limits on the major predator fishes appeared to be the major factor in developing and maintaining good sport fishing, without the usual collapse, for at least eight years at Stockton Lake.

Reservoir managers have long been familiar with the "boom-bust" cycle in fishing. Bennett (1947) described this cycle in lakes as varying between 5 and 12 years depending upon the time required for the introduced population of fishes to expand to the point of competition. Competition could be delayed if "the original stock of fishes contains a high percent of carnivorous species which function with partial success as natural predators of small fish."

Very high initial harvest of largemouth bass (*Micropterus salmoides*) in small impoundments has been previously documented (Turner 1963) and found to be excessive (Redmond 1974). Overharvest of bass was the primary factor in causing overpopulation of sunfish and continued poor fishing for bass and sunfish in Little Dixie Lake, Missouri (Heman et al. 1969), and other small lakes (Redmond 1974). Early stocking with various sizes of predator and prey fishes and the prevention of overharvest of the predators by appropriate size limits provided long-term, good fishing in small lakes (Redmond 1974; Hoey and Redmond 1974). This approach, maintaining a high percent of carnivorous species (a step beyond that advocated by Bennett), is the basis of Missouri's fisheries management program for new small impoundments (Redmond 1974). These concepts were also applied to planning the development of both the warmwater and coolwater sport fisheries at Stockton Lake, a new large reservoir (Goddard and Redmond 1978).

The primary objective at Stockton Lake was to develop and maintain maximum predation to achieve high reproduction and rapid turnover of the prey species. Expected benefits were a continuing respectable harvest of sport fishes including many of quality size, a trophy fishery for large northern pike and walleye, a good catch-and-release fishery for sublegal-sized fish, and maintenance of rapid growth rates of major sport species.

Management efforts to achieve these benefits consisted of pre-impoundment stocking of ponds in the basin with largemouth bass and several minnow species; stage filling over a 2-year period to enhance production and survival of at least two year classes of predator fishes; initial releases of large predator species and prey species not native to the impounded stream; and early establishment of size limits on the five major predator fish species. Success of these efforts was evaluated on the two major arms of the lake with periodic fish population sampling, growth rate determinations, and an intensive creel census implemented in year two of the reservoir. This paper presents results from 1968 (before dam closure) through the first eight years after impoundment (ending December 1977).

Study Site

Stockton Lake is a 24,900-acre impoundment in west-central Missouri that was completed by the U.S. Army Corps of Engineers in 1969 for flood control and hydropower generation. The reservoir has a drainage area of 1,160 mi^2, a shoreline of approximately 250 mi, and a mean depth of 35.7 ft at conservation pool elevation (867 ft above mean sea level). Chemical analysis of the lake waters revealed average alkalinity readings of 116 mg/liter, total hardness of 126 mg/liter, calcium hardness of 98 mg/liter, chloride of 5.4 mg/liter, and specific conductance of 266 micromhos per cm (Jones 1977). The lake stratifies annually, and during summer, the top of the thermocline normally is 30 ft deep.

A two-stage filling plan was recommended to the Corps of Engineers by the Missouri Department of Conservation. The intent was to provide at least a second year of flooding terrestrial vegetation to enhance survival of stocked fishes and promote reproduction, survival, and growth of fishes already in the impounded stream. However, three years of springtime vegetation flooding actually occurred. The reservoir began filling in December 1969, and by April 1, 1970, and May 1, 1970, it contained 6,890 and 9,452 surface acres, respectively. The lake level reached 20,713 surface acres by April 1971, 23,166 acres by September 1971, and full pool in January 1972. The basin was completely cleared of timber except for the upper ends of the arms, but there was some regrowth of trees and shrubs during the 3 years of filling.

Methods

In spring 1968, 48 renovated farm ponds in the lake basin were stocked with varying numbers of adult fathead minnows (*Pimephales promelas*), golden shiner (*Notemigonus crysoleucas*), emerald shiner (*Notropis atherinoides*), and mosquitofish (*Gambusia affinis*). Fifteen hundred, 4–7 in largemouth bass were distributed among 13 ponds to provide additional spawners in the filling lake.

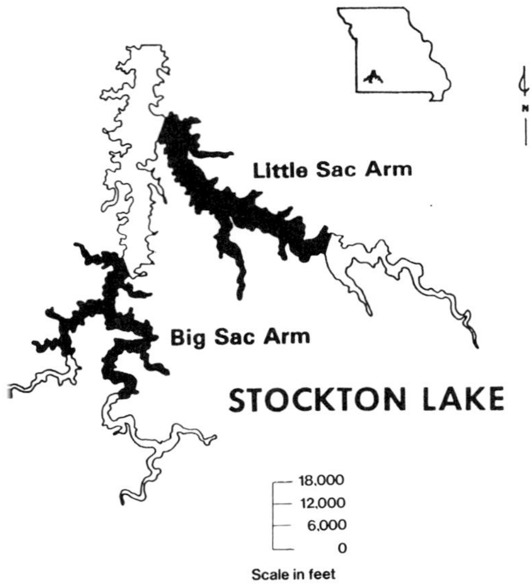

Figure 1. Map of Stockton Lake, Missouri, showing creel survey areas (shaded).

Initial reservoir stockings included 9.6 million walleye (*Stizostedion vitreum*) and 2.7 million northern pike (*Esox lucius*) fry and fingerlings during 1970 and 1971. Additional releases during April 1971 of 11,000 adult inland silverside (*Menidia beryllina*) and 6,000 adult threadfin shad (*Dorosoma petenense*), plus distribution of mats with threadfin shad eggs, completed the fish stocking program. White bass (*Morone chrysops*) were not added because of unpredictability of year-class strength and possible competition with other predators (Hanson 1965). White bass occurred in the lake by 1975 but did not appear in the fishery until 1982; they were probably introduced by anglers. A total of 15,000, 10-in tiger muskellunge (*Esox lucius* × *Esox masquinongy*) was added in 1975 and 1976 to replace a declining northern pike population.

Beginning in April 1971 and continuing through the duration of this study, a 9-month, daytime roving creel survey described by Kathrein (1953) and modified by Fry (1965) and Hanson (1969) was conducted on two areas of the lake to evaluate the success of fisheries management efforts. These areas, Big Sac Arm (4,186 acres) and Little Sac Arm (6,383 acres), represent about 38% of the reservoir's total area and shoreline length (Fig. 1).

Boat-mounted, 230-volt direct current electrofishing gear was used each spring to collect fish for age, growth, abundance, and population structure determinations.

Results

Development of the Fishery

Efforts to establish threadfin shad and emerald shiner were unsuccessful; however, mosquitofish, fathead minnows, and golden shiners provided useful forage during the early years of the lake. Inland silverside numbers expanded to approximately 56% of the total silverside population, which also included the native brook silverside (*Labidesthes sicculus*), but inland silversides disappeared from the reservoir during the winter of 1977. Small gizzard shad (≤3 in) were abundant throughout 1970–1977, even during fall and early winter months.

Walleye and northern pike introductions were successful; as quality-sized fish, they contributed significantly to the sport fishery and grew rapidly, averaging 19.1 and 23.0 in at Age 3, respectively, in 1974 (Goddard and Redmond 1978). Walleye became abundant in the lake. In Big Sac Arm, they were captured by electrofishing at the rate of 12.7 per hour in the springs of 1971–1972, with capture rates of over 100 per hour in some areas. They spawned successfully in tributary streams many years and provided a trophy fishery there, several miles above the creel census areas. During most springs, fish weighing 7–11 lb were caught regularly for 3–4 weeks. This stream fishery declined toward the end of the study period, when many walleye began spawning along the dam and on gravel points in the lake. Walleyes provided good spring fishing there and were abundant enough to be the sole source of broodstock for Missouri's hatcheries. Northern pike did not spawn sufficiently to maintain their numbers, and die-offs occurred from 1971 through 1973 (Goddard and Redmond 1978). By 1977, northern pike disappeared from the fishery. Angling interest for walleye and northern pike was high, particularly in the early years; anglers specifically seeking them accounted for 11–20% of the fishing trips in the creel census areas in 1971–1973. The tiger muskellunge grew well, but there were only four verified catches of legal-sized fish (30-in) during the study period, in May 1977 (Goddard and Redmond 1978).

Successful black bass spawning occurred during the spring of 1970, and additional year classes were produced annually throughout the study period. Largemouth bass made up more than 99% of the black bass sampled during 1970; this declined to about 90% by 1976 as numbers of native spotted bass (*Micropterus punctulatus*) and smallmouth bass (*Micropterus dolomieui*) increased. Growth of the 1970 year class of largemouth bass was rapid; total lengths of 12.0 and 14.0 in were reached at Age 2 and Age 3. Growth rates were consistently rapid throughout the study and averaged 7.0, 11.0, 13.7, 16.0, and 18.1 in at Ages 1–5, respectively.

Largemouth bass exhibited annual improvements in size structure that peaked in 1975, when 20% of the stock size bass (those >8 in) were larger than 15 in (Fig. 2). Proportional Stock Density (PSD, percent of stock size bass exceeding 12 in) values (Anderson and Weithman 1978) for 1971–1977 were 3, 41, 72, 66, 48, 37, and 32%, respectively. Fishery managers in Missouri attempt to maintain a PSD value for bass in the 40–60% range.

Sport Fishing

Angler harvest of Stockton Lake fishes began in earnest during the early fall months of 1970. Northern pike were protected under statewide regulations that included a minimum harvestable length of 30 in and a daily creel limit of two fish. Harvest of walleye, crappie (*Pomoxis* spp.), and black bass was initially restricted only by daily creel limits of 4, 30, and 10 fish, respectively.

Interviews of Stockton Lake anglers during the fall of 1970, prior to initiation of the creel census, revealed that many small black bass were being caught and kept. Problems caused by largemouth bass overharvest in small impoundments (Redmond 1974) indicated that it might be desirable to impose additional harvest restrictions on black bass here. A minimum harvestable length of 12 in was established on black bass in Stockton Lake in April 1971 to develop and maintain a sizable population of larger bass that would provide quality fishing and exert heavy predation on gizzard shad. An 18-in minimum length limit for walleye was also implemented for the same reasons (Goddard and Redmond 1978).

Total angling effort and black bass fishing effort and harvest in Stockton Lake were relatively stable throughout the first eight years of impoundment (Table 1). The Big Sac Arm supported a more intensive fishery than the Little Sac Arm

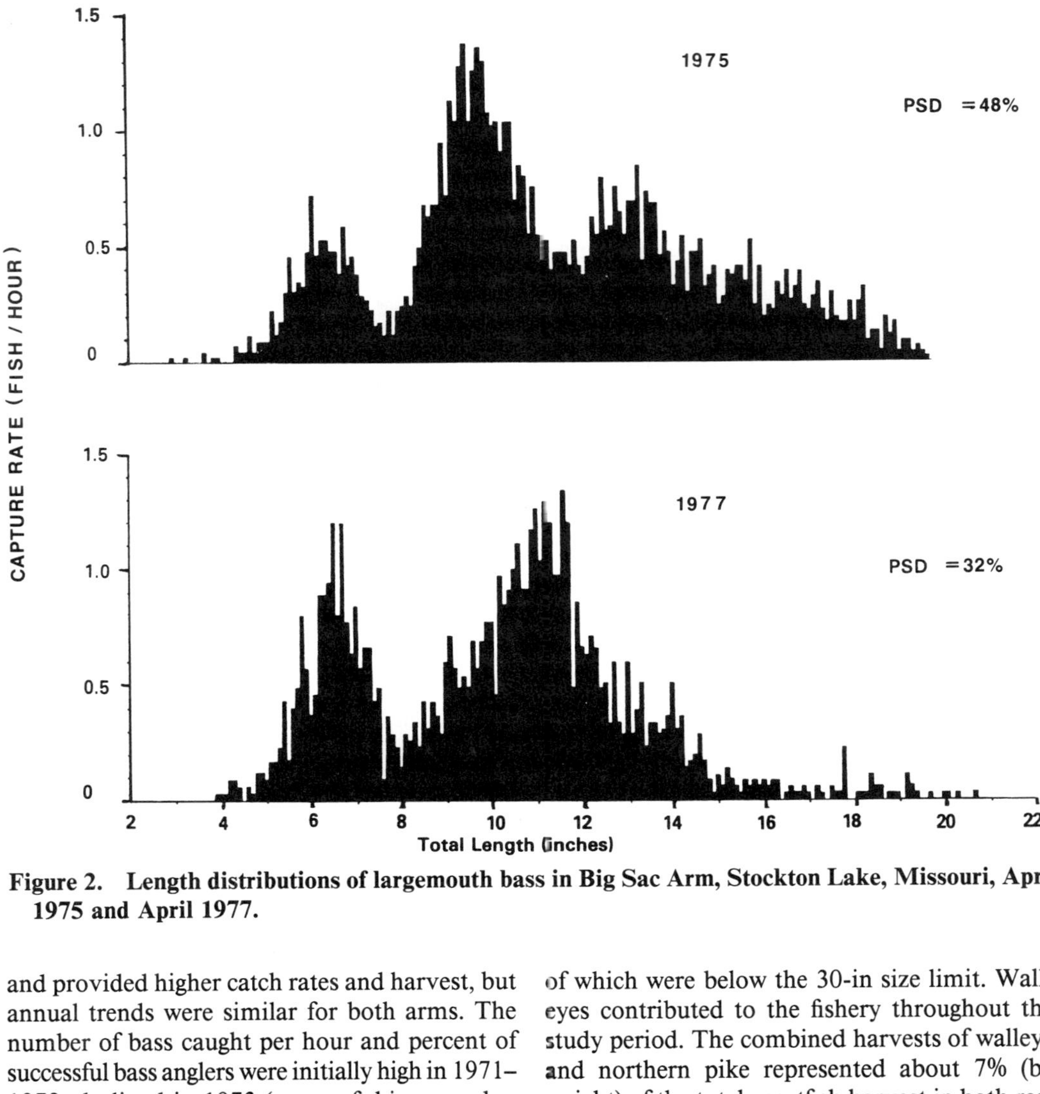

Figure 2. Length distributions of largemouth bass in Big Sac Arm, Stockton Lake, Missouri, April 1975 and April 1977.

and provided higher catch rates and harvest, but annual trends were similar for both arms. The number of bass caught per hour and percent of successful bass anglers were initially high in 1971–1972, declined in 1973 (a poor fishing year because of spring flooding), and steadily increased from 1974 through 1977. Much of the increase in total fishing pressure after 1973 was due to development of an excellent crappie fishery. Virtually all measures of quality of the black bass fishery at the end of the study were equal to or better than initial values.

Trends in total sportfish harvest in Stockton Lake were affected by changes in northern pike, walleye, and crappie populations. Northern pike disappeared from the fishery by 1977, but about 2,850 legal-sized pike weighing an average of 8.4 lb were harvested from the two creel survey areas in 1971–1976. An additional 22,000 pike were estimated to have been caught and released, most of which were below the 30-in size limit. Walleyes contributed to the fishery throughout the study period. The combined harvests of walleye and northern pike represented about 7% (by weight) of the total sportfish harvest in both reservoir arms.

Although some white crappie (*Pomoxis annularis*) and black crappie (*Pomoxis nigromaculatus*) were harvested immediately after impoundment, they first contributed substantially to the creel in 1974, when harvest was about 6.3 lb per acre in Big Sac Arm and 3.0 lb per acre in Little Sac Arm. Thereafter, crappie harvest (over 90% was white crappie) was slightly greater than 50% of total sportfish harvest.

Discussion

Early management efforts included stocking selected prey and predator species and maintaining harvest control through imposition of

Table 1. Estimated daytime black bass fishery parameters and total daytime fishing effort in Stockton Lake creel survey areas, March-November, 1971-1977.

	1971[a]	1972	1973	1974	1975	1976	1977
Big Sac Arm							
Total fishing hours/acre	32	35	27	34	42	42	39
Total pounds of sport fish harvested/acre[b]	6	11	9	13	19	18	18
Bass fishing hours/acre	15	20	16	15	17	17	15
Number bass caught/hour[c]	1.0	0.4	0.2	0.3	0.5	0.6	0.6
Percent bass anglers successful[d]	54	43	34	42	50	60	62
Number of bass ≧12 in harvested/acre	2.7	5.0	2.7	2.8	2.8	2.7	3.3
Pounds of bass ≧12 in harvested/acre	5.0	9.5	5.7	6.3	5.7	5.7	7.6
Little Sac Arm							
Total fishing hours/acre	18	25	15	19	21	20	22
Total pounds of sport fish harvested/acre[b]	4	8	4	7	7	8	10
Bass fishing hours/acre	8	13	7	10	12	9	8
Number bass caught/hour[c]	0.8	0.3	0.2	0.2	0.3	0.4	0.5
Percent bass anglers successful[d]	54	37	32	34	40	54	54
Number of bass ≧12 in harvested/acre	1.8	2.3	1.0	1.3	1.1	1.4	1.7
Pounds of bass ≧12 in harvested/acre	3.0	4.5	3.0	3.0	3.0	3.0	3.7

[a] Not surveyed during March.
[b] Includes black basses, northern pike, walleye, and crappies.
[c] Includes all sizes of bass.
[d] Catching at least one bass of any size.

minimum size limits on northern pike, walleye, and black basses. These, in conjunction with the slow, progressive filling of the lake, permitted development and maintenance of a large sport fish population that was further enhanced by an excellent forage base. During the first eight years following impoundment, the fishery yielded an average annual harvest of 20 sport fish (13 lb) per acre from the Big Sac Arm. Black basses, primarily largemouth bass, accounted for an average of 16% by number and 47% by weight of this harvest.

Sport fish harvest climbed steadily to a peak on the Big Sac Arm during year six, when 36 fish (19 lb) per acre were removed, and harvest remained near this peak in 1976 and 1977. Little Sac Arm harvest increased each year to a level of 18 fish (10 lb) per acre in the final year studied (year eight).

The relative stability of harvest at Stockton Lake, particularly the good black bass fishing, differs from most other large reservoir fisheries. By 1940, 9-year-old, 60,000-acre Lake of the Ozarks, Missouri, had the reputation of being a "biological desert" because of its poor fishing (Weyer 1940). Lake Wappapello, Missouri, went from good bass fishing its first two years, in 1941 and 1942, to poor in 1945 when it took 14 hours (the equivalent of four fishing trips in most Missouri waters) to catch one black bass (Herndon 1946). Patriarche and Campbell (1957) documented the decline of the bass population and fishery in Clearwater Reservoir, Missouri, under the fairly restrictive statewide regulation of six bass per day. Despite satisfactory bass reproduction, the number of bass caught per hour of fishing with artificial lures declined progressively from 0.19 in 1949 (first year of full impoundment) to 0.01 in 1954. This was the equivalent of one bass per 25 angling trips.

The fishery at Beaver Lake, Arkansas, showed similar trends. Eschmeyer and Scott (1982) reported:

> "After closure of Beaver Dam (Arkansas) in late 1963, sport fishing was exceptionally good for about four years (anglers caught more than 30 pounds of sport fish per acre annually). The harvest of more important sport fishes then declined through 1980, the 17th year of impoundment. Although there were substantial year-to-year fluctuations during the period, annual harvest of all sport fishes decreased about 80%. The catch of black basses and crappies showed similar declines. Fishing pressure peaked at about 45 hours/acre during the 4th year of impoundment and then slowly declined to about 10 hours/acre in 1979–1980. This decline paralleled the decrease in harvest. The pattern of declining harvest on Beaver Lake is characteristic of that in other man-made impoundments as they age."

Table 2. Pounds per acre of sport fish (including black basses, northern pike, walleye, white bass, and crappie) harvested from Beaver, Bull Shoals, and Table Rock lakes following impoundment.

Location	Year after impoundment						
	2	3	4	5	6	7	8
Beaver Lake[a]	41.4	30.7	43.6	11.0	23.8	19.4	27.7
Bull Shoals Reservoir							
Little North Fork	41.0	30.5	17.5	29.5	81.7		
White River		20.7	34.1	30.0	83.0	67.3	39.2
Table Rock Lake							
Indian Point		84.7	20.1	12.1	12.1	7.3	8.7
Long Creek	41.5	64.9	16.0	10.6	12.1	11.4	13.3
James River		14.3	18.7	24.0	113.9	15.8	7.7

[a] Includes harvest of all fishes.

Black bass harvest at Beaver Lake in its best years, years two through four, averaged 16 lb per acre (Table 2). This was higher than the seven-year average of 6.3 lb per acre at Big Sac Arm of Stockton Lake, but harvest at Beaver Lake declined thereafter and averaged 7 lb per acre for the next 10 years. Total sport fish harvest at Beaver Lake fell from an average of 39 lb per acre to only 16 lb per acre (USFWS 1980).

Burress (1965) reported outstanding fishing in two arms of Bull Shoals Reservoir, Missouri, during the first six years of impoundment (Table 2). Sport fish harvest in Little North Fork arm averaged 30 fish (35 lb) per acre and black bass harvest averaged 24 lb per acre. Burress also reported, "It was unfortunate that the creel census could not have been extended for another year on the Little North Fork arm to evaluate reports by dock owners and anglers that bass fishing declined markedly in 1959 (year seven). However, there is good reason to believe that increasing populations of white bass and crappie offered strong competition as they did on the White River arm with much the same result."

Sport fish harvest in the White River arm of Bull Shoals Reservoir during years three through eight averaged 58 fish (46 lb) per acre, including 9 lb per acre of black bass (Table 2). Total sport fish harvest peaked in year six at 83 lb per acre and declined to 39 lb per acre in year eight, when the catch was comprised mostly of crappies. Black bass harvest also peaked in year six (16 lb per acre) and declined dramatically by year eight (3 lb per acre). Hanson (1965) estimated the abundance of largemouth bass in the White River arm, and found that numbers present in years seven and eight were only about 20% of the year-six abundance. The decline was attributed to low numbers of gizzard shad and competition with increasing numbers of white bass.

Creel surveys on three arms of Table Rock Lake, Missouri, revealed somewhat different trends in early sportfish harvest (Hanson 1971; Missouri Department of Conservation, Jefferson City, Missouri, USA [unpublished data]). Harvest (total as well as black bass) in the Long Creek and Indian Point arms peaked in year three and then fell to about 15–25% of the peak levels by year seven. The fishery in the James River arm peaked in year six but declined to 25% (black bass) and 7% (all species) in just two years (year eight). Subsequent data for all three arms of Table Rock Lake show continued declines in harvest (G. Novinger, Missouri Department of Conservation, Columbia, Missouri, USA [personal communication]).

The fisheries described above provided highest sport fish harvest within the first six years of impoundment, which is typical of the boom-bust cycle described by Bennett (1947). These trends occurred under statewide harvest regulations with no size limits. The Stockton Lake fishery reached its peak harvest in year six in Big Sac Arm and maintained itself near this maximum for the additional years of the study. The Little Sac Arm fishery increased steadily through year eight.

We feel that the benefits from the program implemented at Stockton Lake far outweighed any disadvantages. A good sport fishery developed and was maintained without severe peaks and valleys typical of fishing in many less-managed reservoirs. The stocking of walleye and northern pike and their protection to quality size with length limits stimulated additional angler interest, particularly in the early years of the fishery, and provided the additional dimension of a trophy fishery (Goddard and Redmond 1978). Walleye continue to be an important, highly sought sport species at Stockton Lake. This is atypical of Missouri reservoirs. The size limits on black bass, we believe, were largely responsible for the maintenance of a stable, high-quality bass fishery as well as a good catch-and-release fishery. The catch of sublegal and legal-sized black bass in Big Sac Arm increased annually after 1973 and tripled by 1977. The catch-and-release fishery was an important benefit of the management program. The poundage of bass released was about equal to harvest in Stockton Lake.

Other components of the fish population in Stockton Lake, primarily white crappie, contributed substantially to the fishery as expected. Growth rates of white crappie were excellent throughout the period and averaged over 10 in at Age 3 (M. Colvin, Missouri Department of Conservation, Columbia, Missouri, USA [personal communication]).

Between 1973 and 1978, large numbers of small shad were always available as prey throughout the growing season. Shad in the 1.6–3.2 in length range were taken in September mid-water trawl samples at an average rate of 657 per 10-minute haul (range 309–1,115), one of the highest capture rates for this size of shad in any large reservoir in Missouri (F. Vasey, Missouri Department of Conservation, Columbia, Missouri, USA [personal communication]). We believe the maintenance of a large population of predator fishes, primarily through use of appropriate minimum size limits, was largely responsible for this excellent forage base.

By 1977 the harvest of legal-sized black bass (12 in and larger) had reduced the number of bigger bass in the population as indicated by declines in PSD and RSD values (Fig. 2). Estimated exploitation rates of legal-sized largemouth bass were 30–40% annually. Because of the continuing excellent growth and the cropping down to the length limit, the population consisted of only two age classes of largemouth bass. In this situation, the failure to successfully recruit a strong year class of largemouth bass could have seriously impacted the fishery; therefore, it appeared desirable to protect at least one more year class. In January 1978, the minimum harvestable length for black basses was increased to 15 in. By spring 1979, largemouth bass abundance and size structure began to show improvement, and evaluation continues.

References

Anderson, R. O., and S. A. Weithman. 1978. The concept of balance for coolwater fish populations. Pages 371–381 *in* R. L. Kendall, editor. Selected coolwater fishes of North America. Special Publication 11, American Fisheries Society, Bethesda, Maryland, USA.

Bennett, G. W. 1947. Fish management—a substitute for natural predation. Transactions of the North American Wildlife Conference 12:276–285.

Burress, R. M. 1965. A quantitative creel census on two arms of Bull Shoals Reservoir, Missouri. Proceedings of the Annual Conference of the Southeastern Association of Game and Fish Commissioners 16:387–398.

Eschmeyer, R. W., and Thomas G. Scott. 1982. Activities in the Divisions of Research for Fiscal Year 1982. U.S. Fish and Wildlife Service, Washington, D.C., USA.

Fry, J. P. 1965. Harvest of fish from tailwaters of three large impoundments in Missouri. Proceedings of the Annual Conference of the Southeastern Association of Game and Fish Commissioners 16: 405–411.

Goddard, J. A., and L. C. Redmond. 1978. Northern pike, tiger muskellunge, and walleye populations in Stockton Lake, Missouri; a management evaluation. Pages 313–319, *in* R. L. Kendall, editor. Selected coolwater fishes of North America. Special Publication 11, American Fisheries Society, Bethesda, Maryland, USA.

Hanson, W. D. 1965. Dynamics of the largemouth bass population in Bull Shoals Reservoir, Missouri. Proceedings of the Annual Conference of the Southeastern Association of Game and Fish Commissioners 16:398–404.

Hanson, W. D. 1969. Harvest of fish in Pomme de Terre Reservoir and its tailwater. Progress report, Federal Aid project F-1-R-18. Missouri Department of Conservation, Jefferson City, Missouri, USA.

Hanson, W. D. 1971. Harvest of fish in selected areas of Table Rock Reservoir. Final report, Federal Aid project F-1-R-19. Missouri Department of Conservation, Jefferson City, Missouri, USA.

Heman, M. L., R. S. Campbell, and L. C. Redmond. 1969. Manipulation of fish populations through reservoir drawdown. Transactions of the American Fisheries Society 98:293–304.

Herndon, G. B. 1946. Can we expect good fishing in big, man-made lakes? The Missouri Conservationist 7(3):4–11.

Hoey, J. W., and L. C. Redmond. 1974. Evaluation of opening Binder Lake with a length limit for bass. Pages 100–105 *in* J. L. Funk, editor. Symposium on overharvest and management of largemouth bass in small impoundments. Special Publication 3, North Central Division American Fisheries Society, Bethesda, Maryland, USA.

Jones, J. R. 1977. Chemical characteristics of some

Missouri reservoirs. Transactions Missouri Academy Science 11:58–71.

KATHREIN, J. W. 1953. An intensive creel census on Clearwater Lake, Missouri, during its first four years of impoundment, 1949–1952. Transactions North American Wildlife Conference 18:282–295.

PATRIARCHE, M. H., AND R. S. CAMPBELL. 1957. The development of the fish population in a new flood-control reservoir in Missouri, 1948 to 1954. Transactions of the American Fisheries Society 87:240–258.

REDMOND, L. C. 1974. Prevention of overharvest of largemouth bass in Missouri impoundments. Pages 54–58 *in* J. L. Funk, editor. Symposium on overharvest and management of largemouth bass in small impoundments. Special Publication 3, North Central Division American Fisheries Society, Bethesda, Maryland, USA.

TURNER, J. L. 1963. A study of a fish population of an artificial lake in central Missouri with emphasis on *Micropterus salmoides* (Lacépède). Master's thesis, University of Missouri, Columbia, Missouri, USA.

USFWS (UNITED STATES FISH AND WILDLIFE SERVICE). 1980. Reservoir creel census computer printout. Unpublished manuscript, U.S. Fish and Wildlife Service, National Reservoir Research Program, Fayetteville, Arkansas, USA.

WEYER, A. E. 1940. The Lake of the Ozarks: a problem in fishery management. The Progressive Fish Culturist 51: 1–10.

Pages 211–217 *in* G. E. Hall and M. J. Van Den Avyle, editors.
Reservoir Fisheries Management: Strategies for the 80's.
Reservoir Committee, Southern Division American Fisheries Society, Bethesda, Maryland. 1986.

Effects of Commercial Harvest on the Fish Community of Lovewell Reservoir, Kansas

JAMES L. STEPHEN

Kansas Fish and Game Commission, 511 Cedar, Concordia, Kansas 66901

ABSTRACT

The Kansas Fish and Game Commission authorized commercial harvest of bigmouth buffalo (*Ictiobus cyprinellus*), carp (*Cyprinus carpio*), river carpsucker (*Carpiodes carpio*), and freshwater drum (*Aplodinotus grunniens*) from Lovewell Reservoir in October 1978. This was a limited-entry commercial fishery awarded to one successful applicant on a bid system. In a 10-month period, 150 lb per acre of nonsport fish were removed, and 99% of this was bigmouth buffalo. Annual cove samples demonstrated substantial changes in fish populations following this removal. Total fish biomass increased from 66 to 1,775 lb per acre during a 4-year period. During the same period, length distributions for white crappie (*Pomoxis annularis*), white bass (*Morone chrysops*), and gizzard shad (*Dorosoma cepedianum*) improved, and growth of white bass and white crappie increased. Prey available for 6-in white bass and crappie increased from 22 to 829 lb per acre.

After a 3-year investigation that was primarily an inventory of nonsport fish populations in Federal reservoirs, the Kansas Fish and Game Commission authorized commercial fishing with limited entry on Lovewell Reservoir in October 1978. Following competitive bids, one commercial fishing contractor was awarded a contract to harvest certain species. This impoundment was selected primarily because it contained a sizable population of bigmouth buffalo (*Ictiobus cyprinellus*) dominated by four strong year classes (Stephen 1978). Although the reservoir maintained a sport fishery that varied from fair to good for walleye (*Stizostedion vitreum*) and catfishes (*Ictalurus* spp.), white crappie (*Pomoxis annularis*) and white bass (*Morone chrysops*) grew slowly, as is typical of Kansas reservoirs having abundant large (>5 in) gizzard shad (*Dorosoma cepedianum*).

Managing harvest of sportfish populations by removing nonsport fishes is not a new concept. State agency crews have attempted control of rough fish (Miller et al. 1959), but the most effective and long-lasting removal programs have utilized private commercial fishermen under various contract arrangements. However, cost efficiency analysis has resulted in the curtailment of many large removal programs (Neess et al. 1957).

Even though many studies have been inconclusive, positive results to sportfish populations have been frequently reported, and investigators have continued to attempt to define cause and effect. Moyle and Clothier (1959) stated, "It is of great practical importance to know whether the increase in crappie and bullheads was made possible by the decline in carp from seining operations. Cause and effect cannot be proven but it seems likely that these population trends are related." Rose and Moen (1953) reported increased populations of game fish in warmwater lakes after nonsport fish populations were reduced with seines. On the other hand, Scidmore and Woods (1961) indicated no direct relationship between the amount of removal accomplished and improved growth or spawning success of panfish. After studying lakes where nonsport fish had been removed for 25 years, Moyle et al. (1948) concluded that rough fish had little effect on the poundage of game fish in southern Minnesota lakes.

Kansas deliberated these implications, and chose to broaden the approach by introducing and supplementally stocking predator species in an attempt to rapidly replace the biomass resulting from removal of nonsport fish. It was thought that a broader array of predator species would better utilize the young produced by the remaining nonsport fishes. Study objectives for Lovewell Reservoir were to monitor commercial harvest and describe resultant changes in sport and nonsport fish populations.

STUDY AREA

Lovewell Reservoir, in extreme north central Kansas, was completed by the Bureau of Reclamation in 1957 as an irrigation and flood control project. At irrigation pool, the reservoir has a surface area of 2,986 acres, storage capacity of 44,100 acre-ft, average depth of 14.7 ft, and a shoreline of 44 mi. The reservoir covers an area of 5,025 acres at flood control pool. The surrounding terrain varies from bottomland with rich, silty loam soil to rolling tall- and short-grass prairie. Average rainfall is only 26 in. Water quality is generally good, but heavy silt loads are carried into the reservoir at times of large runoff.

Since its construction, Lovewell Reservoir has undergone many basic configuration changes. The most obvious are recession of the shoreline due to wave action and siltation of cove areas, which have resulted in the destruction of game fish habitat and extension of the littoral zone. Extensive areas having water depths of 1.5–6.0 ft now exist. These areas are subject to wind-induced wave action, and turbidity has increased to a point that the impoundment is classified as semi-turbid to turbid.

Inflow to the reservoir from the White Rock River drainage (380 mi^2) is supplemented May to October by water released from Harlan County Reservoir, Nebraska. This water travels 30 mi through Courtland Canal before entering the lower end of Lovewell Reservoir. During summer, however, outflows for irrigating 27,371 acres of cropland below Lovewell Reservoir normally exceed inflows. The annual vertical water level fluctuation is 8.8 ft, with drawdowns during July and August ranging from 3.3 to 7.9 ft.

Fishes inhabiting Lovewell Reservoir are typical of Kansas reservoirs (Table 1). The primary predator species are walleye and white bass; the primary prey is gizzard shad. Non-native species introductions initiated in 1978 included the muskellunge (*Esox masquinongy*), the muskellunge × northern pike (*E. lucius*) hybrid, and the white bass × striped bass (*Morone saxatilis*) hybrid. The striped bass, which first occurred in samples in 1976, was not stocked directly into the reservoir. It apparently entered via the inlet canal from the Republican River system.

Table 1. Fishes known to inhabit and stocked in Lovewell Reservoir, Kansas.

Species	Date first captured	Year stocked
Gizzard shad (*Dorosoma cepedianum*)	1957	
Northern pike (*Esox lucius*)	1974	
Northern pike × muskellunge hybrid	1979	1978
Muskellunge (*Esox masquinongy*)		1978
River carpsucker (*Carpiodes carpio*)	1957	
Quillback (*Carpiodes cyprinus*)	1961	
Bigmouth buffalo (*Ictiobus cyprinellus*)	1957	
Smallmouth buffalo (*Ictiobus bubalus*)	1976	
Shorthead redhorse (*Moxostoma hubbsi*)	1960	
Carp (*Cyprinus carpio*)	1957	
Red shiner (*Notropis lutrensis*)	1960	
Emerald shiner (*Notropis atherimoides*)	1974	
Sand shiner (*Notropis stramineus*)	1976	
Fathead minnow (*Pimephales promelas*)	1960	
Golden shiner (*Notemigonus crysoleucas*)	1976	
Suckermouth minnow (*Phenacobius mirabilis*)	1976	
Black bullhead (*Ictalurus melas*)	1957	
Channel catfish (*Ictalurus punctatus*)	1957	1957
Flathead catfish (*Pylodictus olivaris*)	1968	
Orangespotted sunfish (*Lepomis humilis*)	1959	
Green sunfish (*Lepomis cyanellus*)	1969	
Bluegill (*Lepomis macrochirus*)	1957	
Largemouth bass (*Micropterus salmoides*)	1957	1957
Black crappie (*Pomoxis nigromaculatus*)	1957	1966
White crappie (*Pomoxis annularis*)	1957	
Walleye (*Stizostedion vitreum*)	1958	1958
Longnose gar (*Lepisosteus osseus*)	1966	
Goldeye (*Hiodon alosoides*)	1959	
Freshwater drum (*Aplodinotus grunniens*)	1957	
White bass (*Morone chrysops*)	1958	
Striped bass (*Morone saxatilis*)	1976	
White bass × striped bass hybrid	1978	1978

Methods

Commercial harvest data were compiled from daily catch reports submitted to the Kansas Fish and Game Commission by contract fishermen. The harvest resulted primarily from the use of over-night sets of trammel nets, but seines and traps were also permitted. Species eligible for harvest included only bigmouth buffalo, carp (*Cyprinus carpio*), river carpsucker (*Carpiodes carpio*), and freshwater drum (*Aplodinotus grunniens*).

Commercial fishing commenced in October 1978 and continued through May 1979. Additional harvest was permitted in October–November 1979 to improve the effectiveness of the initial removal. Weekends and holidays were usually not fished, and commercial fishing was permitted during daylight only. Minimum mesh size on trammel and gill nets was set at 4 in (bar measure) to protect the walleye population. This restriction was no apparent hardship because of the large average size of the bigmouth buffalo.

Pre- and post-harvest cove rotenone samples were collected in August 1978–1981 to measure changes in biomass and densities of fish, espe-

Table 2. Commercial harvest, effort (using trammel and gill nets), and catch per unit of effort (C/f) at Lovewell Reservoir, 1978-1979.

	Harvest (lb)						
	Bigmouth buffalo	Carp	Drum	River carpsucker	Total	Effort (net days)	C/f (lb/net-day)
1978							
October	9,020	0	0	0	9,020	60	150
November	91,615	4	14	47	91,680	396	232
December	40,005	0	0	2	40,007	137	292
1979							
January	82,424	2	162	0	82,588	388	213
February	52,800	0	0	0	52,800	284	186
March	70,325	63	0	85	70,473	350	201
April	60,450	375	0	705	61,530	413	149
May	14,400	255	0	944	15,599	320	49
October	5,000	0	0	0	5,000	40	125
November	7,080	10	15	0	7,105	76	94
Total	433,119	709	191	1,783	435,802	2,464	

cially young-of-the-year. Prey-predator relations were also assessed using these samples. Cove acreages sampled in 1978–1981 were 0.9, 0.9, 1.8, and 1.5, respectively. Recovered fish were measured, and results were recorded as numbers and weights for 1-in groups per acre. The final estimates used adjustment factors to compensate for different distribution between cove and open water areas (Hayne et al. 1967) and to correct for incomplete recovery of fishes (Grinstead et al. 1978). Analyses of available prey for specific sizes of predators in Lovewell Reservoir were conducted by the National Reservoir Research Program (U.S. Fish and Wildlife Service, Fayetteville, Arkansas, USA) using the annual cove rotenone data and equations in Jenkins and Morais (1978). Additional data were collected with gillnets and fyke traps to assist with length-frequency analysis and estimate growth rates of white bass and white crappie.

Results

Commercial Harvest

Harvest of permitted species totaled 435,802 lb and was 99.4% bigmouth buffalo (Table 2). The minor catches of carp, drum, and river carpsuckers reflected the use of the 4-in mesh nets specified in the contract. Based on a surface area of 2,900 acres (conservation pool), 149 lb per acre of bigmouth buffalo, 0.24 lb of carp, 0.07 lb of drum, and 0.61 lb of river carpsucker were removed, giving a total harvest of 150 lb per acre. The reported catch of sport fish in commercial nets was negligible (0.003% by weight) and was predominately flathead catfish (*Pylodictus olivaris*).

Harvest per net-day was highest in November–December 1978 (Table 2). During this period, ice cover permitted harvest on days that otherwise could not have been fished due to high winds, and the aggregating behavior of bigmouth buffalo was conducive to fishing. Harvest, effort, and catch-per-effort were low during October–November 1979 despite knowledge of wintering habits gained in 1978.

Trends in Biomass

Standing crop in cove samples increased from 66 lb per acre in 1978 to 1,775 lb per acre in 1981 (Table 3). Gizzard shad, carp, river carpsucker, and freshwater drum accounted for most of this change, but abundance of white bass and white crappie also increased substantially in 1981. It appears as if the most rapid effect of commercial removal of bigmouth buffalo was production of abundant year classes of shad, carp, and drum in 1979. Biomass and density of these species increased dramatically through 1981, when abundant year classes were produced again. Strong year classes of river carpsucker, white bass, and white crappie were produced in 1980; collection of these fish in addition to young-of-the-year led to substantial increases in biomass in 1981.

Abundance of most other species of fish was either relatively stable following commercial harvest (Table 3) or was not reliably measured in the cove samples. Sportfishes that appeared

Table 3. Density (number/acre) and biomass (lb/acre) for important prey, commercial, and sport fishes collected in August cove samples from Lovewell Reservoir, 1978-1981.

Species	1978 Number/acre	1978 lb/acre	1979 Number/acre	1979 lb/acre	1980 Number/acre	1980 lb/acre	1981 Number/acre	1981 lb/acre
Gizzard shad	817	15	5,184	116	2,890	58	113,231	800
Carp	65	6	151	1	347	140	537	645
Freshwater drum	129	3	327	1	223	3	652	65
River carpsucker	5	5	0		16	5	111	135
Bigmouth buffalo	0		0		5	<1	10	6
White bass	21	6	0		27	2	153	43
White crappie	11	1	50	1	54	1	249	14
All other species	1,441	30	1,076	24	948	23	829	35[a]
Total	2,489	66	6,788	143	4,510	233	115,773	1,775

[a] Excludes one flathead catfish that weighed 32 lb.

to be unaffected included largemouth bass (*Micropterus salmoides*), channel catfish (*Ictalurus punctatus*), and bluegill (*Lepomis macrochirus*). Although walleyes and striped bass × white bass hybrids were important components of the sport fishery, they were not consistently sampled in coves, preventing interpretation of population trends.

Introductions of non-native predators other than the white bass × striped bass hybrid did not appear to be effective. Fingerling northern pike, muskellunge, and their hybrid were stocked annually in 1978–1980 at rates ranging from 24 to 440 fish per acre. These fish did not recruit into the fishery, and stocking was ceased after 1980.

Changes in Population Size Structure

Production of abundant gizzard shad year classes after 1978 led to dramatic shifts in population length distribution. In 1978, all gizzard shad collected were 4–5 in (Fig. 1), but in subsequent years, the distribution broadened. By 1981, shad ranged 1–8 in and about 99% of the sample was <3 in long.

Length distributions of white bass and white crappie also changed following commercial harvest. Results for white bass reflect production of a strong 1980 year class and either increased survival or growth, resulting in greater numbers of fish >10 in by 1981 (Fig. 2). Similar trends occurred for white crappie (Fig. 3). Production of abundant young in 1979–1980 led to a relatively high abundance of crappie >7 in during 1981.

Growth

Growth of white bass and white crappie was more rapid after commercial harvest. Estimated average annual increments for the first three years of life were greater after harvest except for first-year growth of white bass (Table 4). For white bass, year-2 and year-3 increments prior to harvest were about 56–64% of the post-harvest rates. Equivalent percentages for white crappie were 80, 70, and 97 for years 1–3, respectively.

Discussion

Fish community responses to commercial harvest in Lovewell Reservoir included increased reproduction and standing crop of some species and stimulated growth of others. One prey species (gizzard shad), three nonsport species (carp, freshwater drum, and river carpsucker), and two predator species (white bass and white crappie) showed higher numbers of young and standing crop almost immediately after removal of 150 lb per acre of commercially important species.

Increases of drum, carp, and river carpsucker were not likely a response to the reduction of adult fish in those species because their total commercial harvest was low. Rather, this more likely was a response to the reduction of bigmouth buffalo, as observed elsewhere by Jester (1971). Strong recruitment of these species, however, has not occurred. Intraspecific competition for food between young-of-year and older fish may be a major factor in poor recruitment of certain nonsport fishes. Adult carp probably are less-selective feeders than young carp (Sigler 1958), but if the younger fish are forced from favorable habitat into marginal habitat, recruitment may be insignificant (Johnson 1972). The decline in summer water level at Lovewell Reservoir because of irrigation could force young

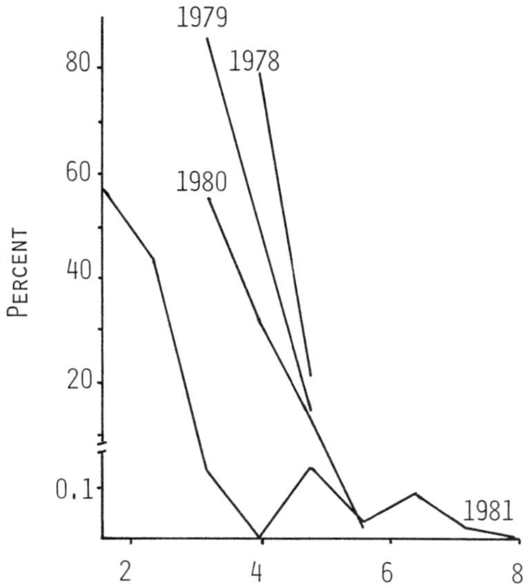

Figure 1. Length distribution of gizzard shad from cove samples at Lovewell Reservoir, 1978-1981.

nonsport fishes from favorable habitat and expose them to predation or a reduced food supply.

The occurrence of young-of-year bigmouth buffalo in the 1980 and 1981 samples was the only evidence of reproduction of this species in recent years. Density-dependent factors may have been

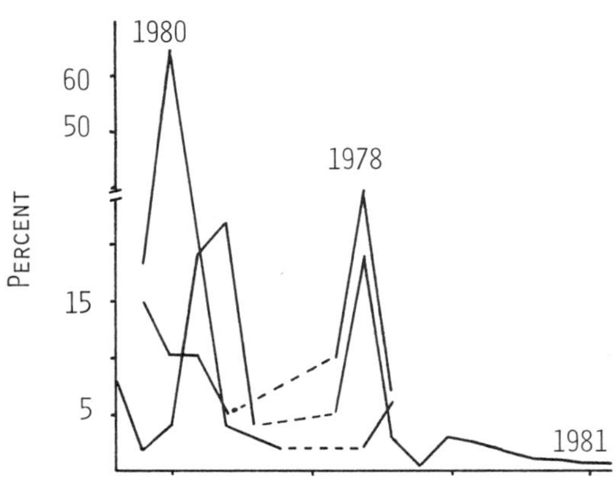

Figure 2. Length-frequency distribution of white bass from cove samples at Lovewell Reservoir, 1978-1981.

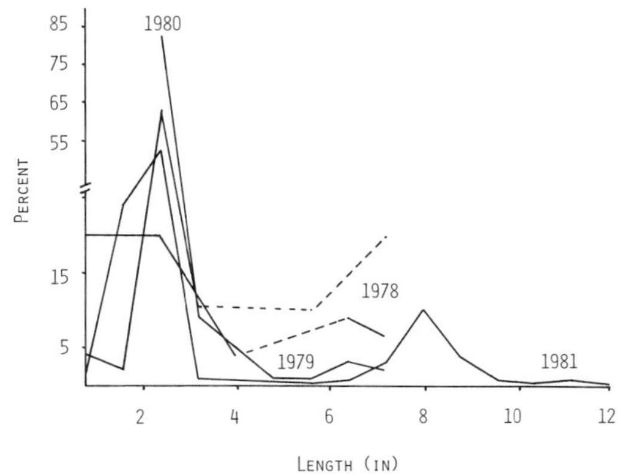

Figure 3. Length-frequency distribution of white crappie from cove samples at Lovewell Reservoir, 1978-1981.

responsible for the lack of recruitment prior to commercial harvest. Strong year classes, followed by years of extremely low recruitment, have been reported for bigmouth buffalo (Mitzner 1972) and carp (Wickers 1976).

Stimulated growth and reproduction of white bass and white crappie were judged to be beneficial responses to removal of nonsport fish. Increased growth rates were apparently related to the increased numbers of gizzard shad <5 in long that could be eaten by small predators. An evaluation of the relationship between biomass of suitably sized prey for 6-in white bass and white crappie (yearlings), using equations in Jenkins and Morais (1978), showed that the amount of available prey increased from 22 lb per acre in 1978 to 829 lb per acre in 1981 (Fig. 4). Similar increases occurred for larger predators as well.

The contract commercial fishing experiment at Lovewell Reservoir resulted in removal of a

Table 4. Average annual growth increments (in) during the first three years of life for white bass and white crappie (sample size in parentheses) after commercial harvest of fish in Lovewell Reservoir.

Species	Year of life 1	2	3
White bass			
Before harvest	4.4 (32)	3.3 (31)	2.5 (5)
After harvest	4.4 (64)	5.8 (22)	3.9 (5)
White crappie			
Before harvest	2.5 (78)	3.1 (52)	2.9 (33)
After harvest	3.1 (92)	4.4 (10)	3.0 (5)

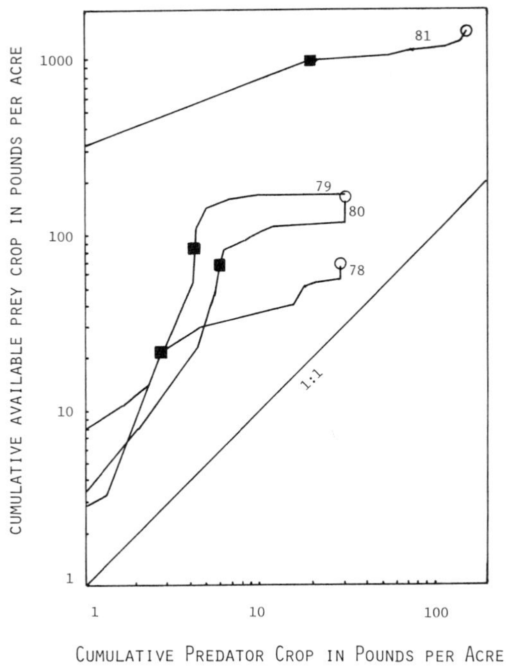

Figure 4. Logarithmic plot of cumulative AP/P ratio in Lovewell Reservoir, 1978-1981, based on adjusted August cove samples. Solid squares represent predators in the 6-in length group; open circles indicate the total AP/P ratio.

substantial poundage of nonsport fish in a short time period. This was a practical means of reducing an abundant nonsport fish and utilizing it as an economic resource. This project also demonstrated that limited-entry commercial fishing can co-exist with sportfishing. By restricting commercial fishing to weekdays, multiple-use of the reservoir can be accommodated. The use of trammel net apparently is compatible with existing sportfisheries provided that mesh sizes are large enough to protect important species. However, as in Lovewell Reservoir, use of a large mesh size (≥4 in) can preclude efficient harvest of some target species such as carp and river carpsucker.

Despite the apparent benefits of nonsport fish removal in Lovewell Reservoir, much additional evaluation needs to be done to determine if such techniques are helpful in other reservoirs having high abundance of commercially useful species. It is not apparent for Lovewell Reservoir why predator species other than white bass and white crappie did not respond. Furthermore, it is possible that the biomass of bigmouth buffalo that was removed will eventually be replaced by other nonsport fishes such as carp and river carpsucker, leading to conditions like those during pre-removal years. If conditions revert to those of previous years, knowledge of the duration of the cycle would help with evaluations of cost-effectiveness of commercial removal as a management tool.

References

GRINSTEAD, B. G., R. M. GENNINGS, G. R. HOOPER, AND D. A. WHARTON. 1978. Estimation of standing crop of fishes in the predator-stocking-evaluation reservoirs. Proceedings of the Annual Conference of the Southeastern Association of Fish and Wildlife Agencies 30:120–130.

HAYNE, D. W., G. E. HALL, AND H. M. NICHOLS. 1967. An evaluation of cove sampling of fish populations in Douglas Reservoir, Tennessee. Pages 244–297 *in* Reservoir fishery resources symposium. Southern Division American Fisheries Society, Bethesda, Maryland, USA.

JENKINS, R. M., AND D. I. MORAIS. 1978. Prey-predator relations in the predator-stocking-evaluation reservoirs. Proceedings of the Annual Conference of the Southeastern Association of Fish and Wildlife Agencies 30:140–157.

JESTER, D. B. 1971. Effects of commercial fishing, species introduction, and drawdown control on fish populations in Elephant Butte Lake, New Mexico. Pages 265–285 *in* G. E. Hall, editor. Reservoir fisheries and limnology. Special Publication 8, American Fisheries Society, Bethesda, Maryland, USA.

JOHNSON, L. 1972. Keller Lake: characteristics of a culturally unstressed salmonid community. Journal of the Fisheries Research Board of Canada 29: 731–740.

MILLER, N. J., C. L. BRYNILDSON, AND C. W. THREINEN. 1959. Rough fish control. Publication 229, Wisconsin Conservation Department, Madison, Wisconsin, USA.

MITZNER, L. 1972. Population studies of bigmouth buffalo in Coralville Reservoir with special reference to commercial harvest. Iowa Fishery Research Technical Series 72-3, Iowa Conservation Commission, Des Moines, Iowa, USA.

MOYLE, J. B., J. H. KUEHN, AND C. BURROWS. 1948. Fish populations and catch data from Minnesota lakes. Transactions of the American Fisheries Society 78:163–175.

MOYLE, J. B., AND W. D. CLOTHIER. 1959. Effects of management and winter oxygen levels on the fish populations of a prairie lake. Transactions of the American Fisheries Society 88:178–185.

NEESS, J. C., W. T. HELM, AND C. W. THREINEN. 1957. Some vital statistics in a heavily exploited population of carp. Journal of Wildlife Management 21:279–292.

ROSE, E. T., AND T. MOEN. 1953. The increase in game fish populations in East Okoboji Lake, Iowa, following intense removal of rough fish. Transactions of the American Fisheries Society 82:104–114.

SCIDMORE, W. J., AND D. E. WOODS. 1961. Changes in the fish populations of four southern Minnesota lakes subjected to rough fish removal. Fisheries Service Publication 3, Minnesota Department of Conservation, St. Paul, Minnesota, USA.

SIGLER, W. F. 1958. The ecology and use of carp in Utah. Agricultural Experimental Station Bulletin 405, Utah State University, Logan, Utah, USA.

STEPHEN, J. L. 1978. Kansas marketable fisheries investigation. Final report, Federal Aid (NMFS) project 2-272-F. Kansas Fish and Game Commission, Pratt, Kansas, USA.

WICKERS, F. 1976. Age and growth of carp (*Cyrinus carpio*) from Pathfinder Reservoir, Wyoming 1974–1975. Wyoming Game and Fish Department, Cheyenne, Wyoming, USA.

Pages 218–226 *in* G. E. Hall and M. J. Van Den Avyle, editors.
Reservoir Fisheries Management: Strategies for the 80's.
Reservoir Committee, Southern Division American Fisheries Society, Bethesda, Maryland. 1986.

Evaluation of Harvest Regulations for Largemouth Bass Populations in Reservoirs: A Computer Simulation

ANTHONY J. ZAGAR AND DONALD J. ORTH

Department of Fisheries and Wildlife Sciences, Virginia Polytechnic Institute and State University Blacksburg, Virginia 24061

ABSTRACT

Computer simulations of hypothetical largemouth bass (*Micropterus salmoides*) fisheries were used to evaluate the management implications of different harvest regulations. Data on growth, mortality, and reproduction for reservoir populations were collected for use in the simulations. Simulations were conducted with five length limit regulations and three growth rates. Effects of harvest regulations were evaluated in terms of total numbers, weights, and the number of quality-size bass harvested. Small minimum-length limits maximized numbers harvested whereas large minimum-length limits and large slot limits produced greater harvests of quality-size fish. Large slot limits and large minimum-length limits also resulted in a balanced recreational-trophy fishery and are recommended for use only when compensatory growth responses are not expected to occur.

Over one million man days were spent bass fishing in 1980, and much of this took place on man-made lakes and reservoirs across the country (USFWS 1982). Yet, even with this level of interest, reliable management programs are still in development and reservoir managers have few guidelines to follow. The purpose of this paper is to provide some management guidelines through computer simulation of several hypothetical reservoir largemouth bass (*Micropterus salmoides*) populations. Specifically, we evaluate the effectiveness of length limits in terms of numbers, weights, and sizes of fish harvested.

Length limits as management tools have produced a variety of angling results and population responses in largemouth bass fisheries (Redmond 1986). Farabee (1974), Hickman and Congdon (1974), Hoey and Redmond (1974), Johnson and Anderson (1974), and Rasmussen and Michaelson (1974) have reported that 305-mm length limits in midwestern lakes resulted in population shifts to increased numbers of sublegal bass. Furthermore, Hickman and Congdon (1974), Hoey and Redmond (1974), and Rasmussen and Michaelson (1974) found reduced growth rates due to overcrowding. Ming and McDannold (1975) found overcrowding by young fish and reduced growth rates, but the numbers and average weights of harvested fish increased substantially. Pelzman (1979) reported similar harvest results but did not observe major changes in the numbers of small fish. These varied responses may be attributed to different survivorships, relative growth rates, or food bases existing in different largemouth bass populations. Regardless of the cause, fisheries managers need decision aids to help them recommend harvest regulations.

Computer simulation is a useful tool for describing and analyzing unstable fisheries such as those existing in many large lakes and reservoirs where water levels, year-class strengths, stocking programs, and angling pressures exhibit large yearly or seasonal fluctuations. Computer simulation models can be designed with stochastic components which mimic these types of processes and then used to examine how a system may respond to changes in various factors. The simulation model used in this study represents our current knowledge of largemouth bass population dynamics in reservoirs. Therefore, through the development of the model we also identify gaps in our data. The simulation model used was an adaptation of Taylor's (1981) general inland fishery simulator (GIFSIM).

METHODS

Model Description

GIFSIM, as described by Taylor (1981), simulates a fishery by dividing a fish population into two sexes and several age classes, each of which can have distinct mortality and exploitation rates for several time intervals within a year. GIFSIM uses difference equations (Taylor 1981) to describe the dynamics of a fish stock:

$$F_{i+1} = F_i - F_i(u_i + v_i), \quad (1)$$

where: F_{i+1} = population size at the beginning of period i + 1,
F_i = population size at the beginning of period i,
u_i = fishing mortality (fraction) within period i,
v_i = natural mortality (fraction) within period i; and

$$F_{t+1} = F_t + r, \quad (2)$$

where: F_{t+1} = population size at the beginning of year t + 1,
F_t = population size at the beginning of year t, and,
r = recruits to the population in the year between t and t + 1.

GIFSIM applies equations (1) and (2) to each sex and age class for each period within the year. Thus, the total population size at any time is found by summation.

Each of the processes involved in simulating a fish population (mortality, exploitation, reproduction, and egg-survival) has either a constant value assigned for each period of the year or a computed value based on the functional models described by Taylor (1981). In this study, some of these models were modified to accommodate the available data or to overcome the model's shortcomings in describing reservoir largemouth bass population dynamics. The density-dependent growth model and the stochastic component of the natural mortality model were deleted because of insufficient data. The functional form of the egg-survival model was used, but the stochastic component was modified by using a Poisson distribution.

GIFSIM requires that population parameters (natural mortality, fishing mortality, fecundity) either be supplied or computed within the program for each period within a year. Also needed are lengths for each age class in the population.

Two functional models were used within GIFSIM. The egg-production model is a linear function of the weight of a female:

$$f = a + bW, \quad (3)$$

Table 1. Assumed lengths (mm) at age for simulated slow-, medium-, and fast-growing largemouth bass populations.

Age (years)	Length		
	Slow	Medium	Fast
1	84	118	155
2	185	215	264
3	267	287	346
4	318	341	409
5	356	389	454
6	384	434	480
7	414	463	512
8	442	495	550
9	460	510	
10	475	531	
11	495		
12	505		
13	513		
14	523		
15	533		

where: f = number of eggs produced by a female of a given weight,
W = weight of a female in grams, and
a, b = constants.

The survival rate of eggs to Age 1 was modeled as a negative exponential function of population biomass:

$$S = ae^{bW}, \quad (4)$$

where: S = the mean survival rate of the eggs to Age 1,
W = the population's biomass in kilograms, and
a, b = constants.

This model allows a density-dependent mechanism to partially control the number of recruits in a population. In the absence of data on largemouth bass reproduction under variable water stages and environmental conditions, variation in recruitment can be simulated by introducing random variability. Thus the survival rates of young-of-the-year were modeled using a Poisson distribution.

Parameter Estimation

Growth rates.—Mean length-at-age data taken from Bennett (1937), Carlander (1977), and Orth (1977) were used to define baseline values for slow-, medium-, and fast-growing populations, respectively (Table 1).

Length-weight relationship.—Length-weight

relations were derived from data on mean weights per length class from Carlander (1977). A length-weight relationship was then estimated by linear regression on the log-transformed data. The resulting equation ($r = 0.9998$; $P < 0.01$) was:

$$W = 8.939 \times 10^{-6} L^{3.0748}, \quad (5)$$

where: W = estimated weight in grams, and
L = total length in millimeters.

Fecundity.—Estimates of the number of eggs per female of a given weight were obtained from Clady (1970), Coomer (1976), Olmsted (1974), and Kelley (1962). Data were pooled and linear regression was used to estimate parameters of equation (3), giving $f = 6701.0 + 19.3W$ ($r = 0.6865$; $P < 0.01$).

Sexual maturity of largemouth bass is size-related; males typically mature before females (Heidinger 1975). The size at which females mature and spawn has been reported to be 250 mm (Heidinger 1975) and 300 mm (Bryant and Houser 1971). For our study, a value of 250 mm was used for all largemouth bass. An even sex ratio was assumed.

Mortality.—Mortality rates reported in the literature for unexploited populations, which should approximate natural mortality rates, ranged from 0.20 (Regier 1963) to 0.23 (Clady 1975), whereas natural mortality rates for exploited populations ranged from 0.11 to 0.56 (Rawstron and Hashagen 1972). In Ridge Lake, Illinois, Bennett et al. (1969) found that natural mortality rates were size-related. Multiple linear regression was used on the midpoints of the reported weight ranges to fit a second-order polynomial equation to the Ridge Lake natural mortality rates. The resulting equation ($r^2 = 0.9999$; $P < 0.01$) was:

$$v = 0.45436 - 0.66529 \times 10^{-3} W + 0.300998 \times 10^{-6} W^2, \quad (6)$$

where: v = natural mortality, and
W = weight of fish in grams.

This equation was used to estimate age-specific natural mortality rates for each hypothetical population, based on assumed lengths-at-age (Table 1) and corresponding weights from equation (5). There is some evidence for higher mortality of older male largemouth bass (Padfield 1951; Bryant and Houser 1971; Hubert 1976) which shifts the sex ratio in favor of females. However, due to a lack of suitable data, equal mortality rates were assumed for both sexes.

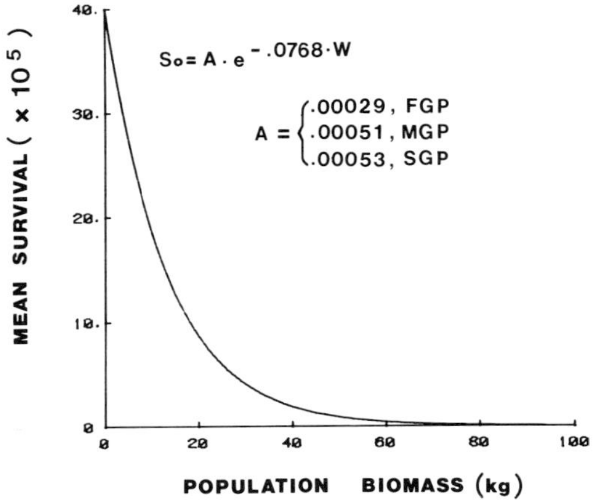

Figure 1. Relationship between mean survival rate (S_0) from egg stage to Age 1 and population biomass (W) for reservoir largemouth bass populations. Values shown for the parameter A are those used in simulations of fast-growing (FGP), medium-growing (MGP), and slow-growing populations (SGP).

We expect that fishing mortality varies among sex and age classes, but no specific data were available for largemouth bass. Most of the fishing mortality rates of largemouth bass reported in the literature fall between 30% and 50%. Based on this information, a fishing mortality rate of 40% was used for all fish greater than 250 mm in all simulation experiments. We assumed that no direct fishing mortality occurred for fish under 250 mm in length.

Hooking mortality is a potentially important mortality source, especially when angling pressure is heavy. Few studies have been done on hooking mortality; those that exist indicate that hooking mortality is quite variable and depends largely on where the fish was hooked and water temperature (Pelzman 1978). Initial mortality for tournament-caught bass ranged 11–25%, and delayed mortality was 3–17% (Table 2). Based on these data, a conservative estimate of 10% hooking mortality was used in all simulation experiments.

Young-of-the-year survival determines the population size of Age-1 fish. To obtain accurate estimates of young-of-the-year survival at average population biomass levels of 12–15 kg/hectare (Bennett et al. 1969; Bennett 1971; Jenkins 1975, 1976; Orth 1980; Paragamian 1982;

Table 2. Rates of hooking mortality for tournament-caught largemouth bass.

Reservoir	State	Mortality (%) Initial	Delayed	Days	Reference
Ross R. Barnett	Miss.	11	5	16	Wellborn and Barkley 1974
Keowee	SC.	16	8	14	Archer and Loyacano 1975
Seminole	Ga.	25	3	19	Plumb et al. 1975
Kissimmee	Fla.	16	5	6	May 1973
Sam Rayburn	Tex.	24	11		Seidensticker 1975
Toledo Bend	Tex.	23	12		Seidensticker 1975
St. Johns R.	Fla.	17	17	6	Moody 1975

Rainwater and Houser 1982), a method based on the Leslie model was used (Vaughan and Saila 1976). The equilibrium survival rate (S_0) from egg-stage to Age 1 is computed as:

$$S_0 = \frac{1}{\sum_{i=1}^{k-1}\left[m_{i+1}\left(\prod_{j=1}^{i} S_j\right)\right]},$$

where: m = the fecundity per individual of age i,
S_j = the survival rate (fraction) for age j, and
k = the maximum age obtained by any individual in the population.

For each of the three populations (slow-, medium-, and fast-growing), the equilibrium egg-survival rate was adjusted slightly to produce stable populations when used in GIFSIM. This adjusted value was used as the true value for average biomass levels (12 kg/hectare). The true value for very low (near-zero) biomass levels was assumed to be 10 times greater than that at average biomass levels. These values were used to compute the egg-survival model parameters (Fig. 1) corresponding to the slope and intercept values of equation (4).

Orth (1977, 1978) found considerable variation in egg-survival rates. Therefore, allowing egg-survival rates to vary by a factor of 10 from near-zero biomass levels to average biomass levels was deemed to be a reasonable approximation. Furthermore, when the average survival rates were being adjusted, we found that rates outside of these bounds produced unstable density fluctuations.

Initial population structure.—The population structure used to initiate simulation experiments was a stable age distribution computed from a Leslie matrix model. This distribution was used along with individual weights at age to calculate the actual number of individuals in each age class at average biomass levels.

Simulation Experiments

For each of the three populations, simulations were conducted under five harvest regulations: three minimum-length limits (254 mm, 305 mm, 356 mm) and two slot limits (254 to 356 mm and 305 to 406 mm). Output from the simulations included annual predictions of population size, numbers and weights of bass harvested, and numbers of quality-size bass (>381 mm) harvested.

Deterministic experiments showed that when simulations were initiated with the stable-age distributions, a 10-year period of population instability followed. After this period, the population assumed a new stable-age distribution. Similar results were obtained with stochastic simulation experiments that were carried out for 100-year periods. The initial period of instability was due to the stable-age distribution's response to exploitation and was not representative of previously fished systems that have had time to respond to harvest. Therefore, the results presented are averages for annual predictions made after the initial 10-year transient phase.

Results and Discussion

For all three populations, total numbers harvested were greatest for the 254-mm and 305-mm minimum-length limits (Fig. 2). The 356-mm minimum and 254–356 mm slot limits typically produced the least numbers caught.

Total yield predictions were highest for the more-restrictive length limits. Mean harvested biomass in all populations was greatest for the 356-mm minimum and 254–356 mm slot limits and least for the 254-mm minimum limit (Fig.

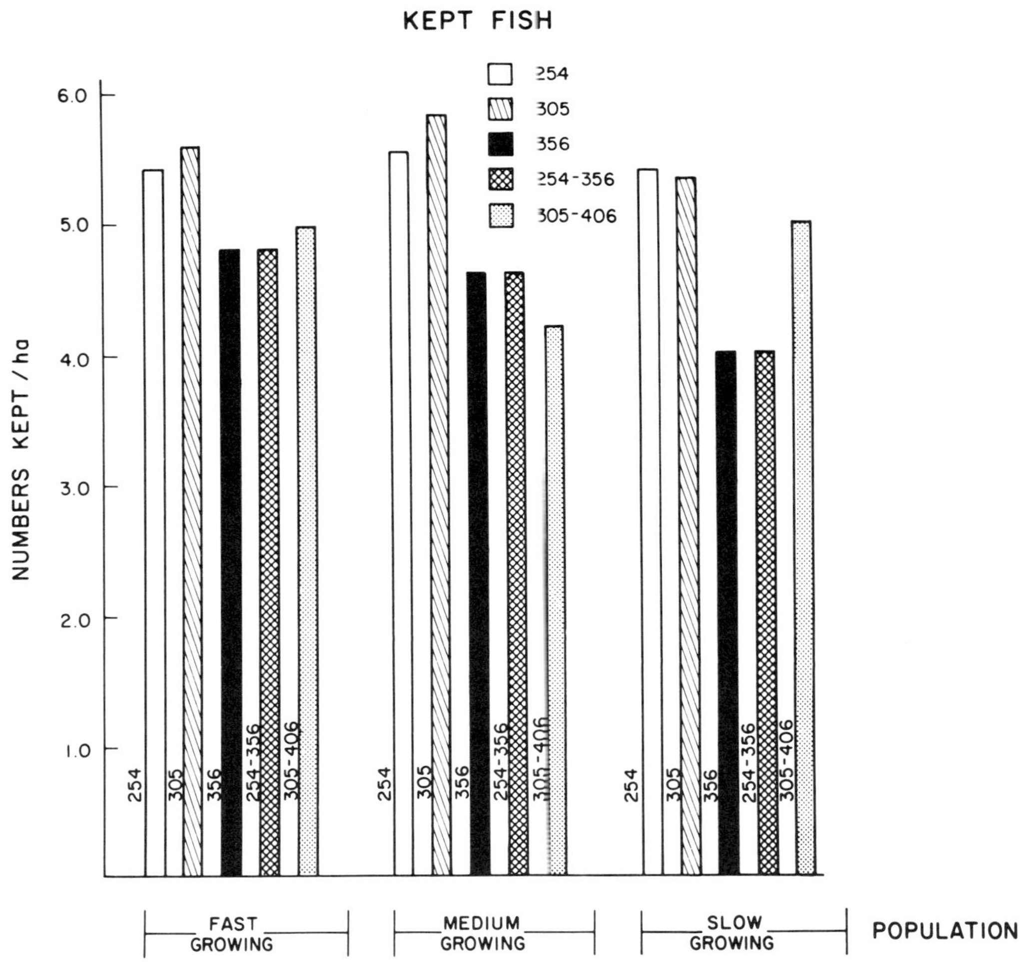

Figure 2. Simulated numbers of largemouth bass harvested per hectare for slow-, medium-, and fast-growing populations subjected to five hypothetical harvest regulations. Length limits are in millimeters.

3). For the medium- and fast-growing populations, the two slot limits and the highest minimum-length produced similar yields, with substantial reductions at lower minimum limits. However, impacts of different regulations were less-pronounced in the slow-growing population. In that case, only the 254-mm minimum-length results differed substantially from the others.

Impacts of different regulations were most pronounced for harvest of quality-size fish (Fig. 4). For all three populations, predicted harvest of bass >381 mm was three to six times greater when the highest minimum-length and slot limits were in effect than when the shortest minimum-length was used. There was a strong positive relationship between minimum-length and predicted harvest of quality bass.

These simulation results may partially explain the variety of largemouth bass population responses to harvest regulations. In the course of our experiments, we observed that over relatively short time periods, a given harvest regulation seemed to have little influence on the quality of fishing. Extrapolation to field situations leads us to believe that divergent population responses to harvest regulations may not necessarily mean a harvest regulation is not working as designed. Rather, short-term field evaluations of length limit regulations are insufficient for determining relative differences. Paragamian (1982)

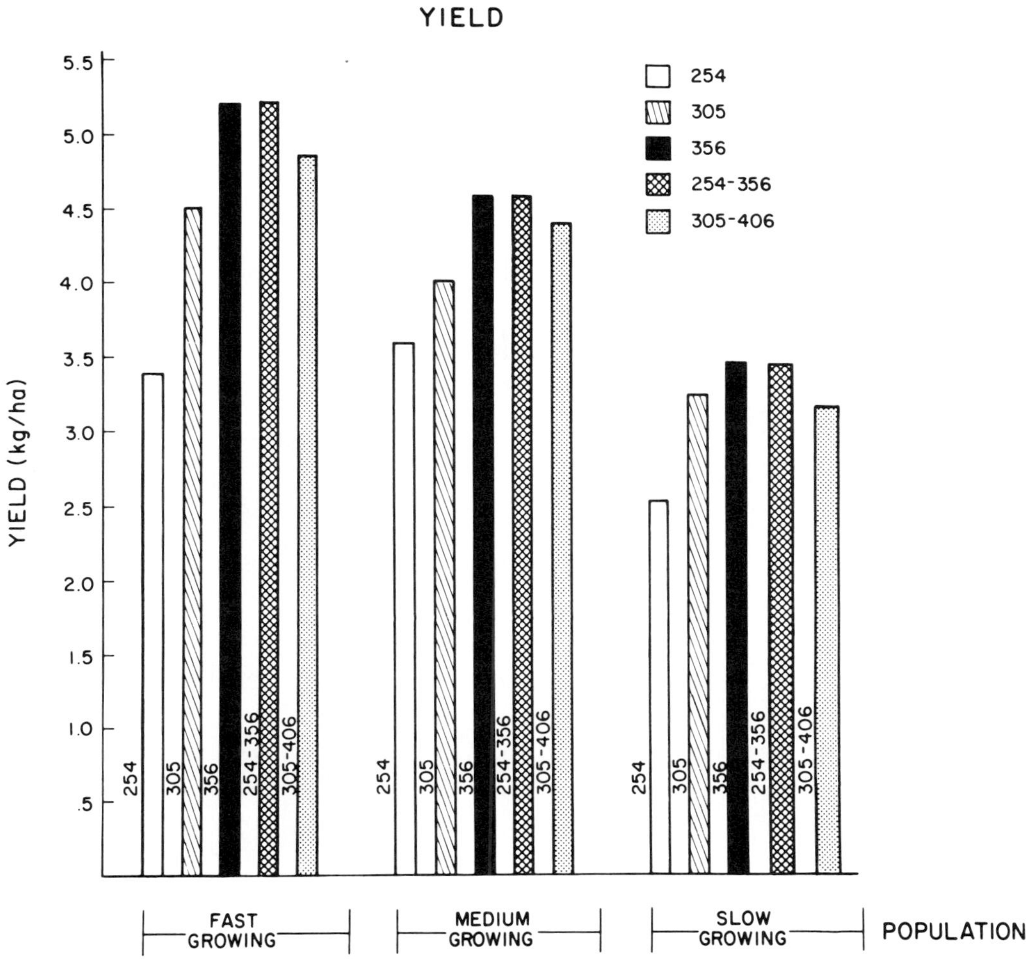

Figure 3. Simulated biomass of largemouth bass harvested per hectare for slow-, medium-, and fast-growing populations subjected to five hypothetical harvest regulations. Length limits are in millimeters.

reported an instance where a 356-mm length limit was changed to a 305-mm limit because of poor quality fishing. Later, the agency thought it should have maintained the original regulation. Gabelhouse (1980) found that in mid-western lakes, relatively large minimum-length limits were appropriate management strategies for most waters.

One important system aspect not evaluated in our simulations is the effect of density-dependent growth. Data relating growth rates and population levels were not available to incorporate into this model. However, many authors have reported reduced growth rates following implementation of length limit regulations (Hickman and Congdon 1974; Hoey and Redmond 1974; Rasmussen and Michaelson 1974; Ming and McDannold 1975; Pelzman 1979).Our simulations showed a slight stockpiling of younger age classes even with constant growth. Future simulations should be carried out after data are obtained to relate growth rates to population density. Our results, therefore, should be applied with caution only to those situations in which growth rates are not expected to change significantly following the implementation of length limit regulations.

These results can be used, however, to provide guidelines for selecting the most promising regulations for individual reservoirs. The choice de-

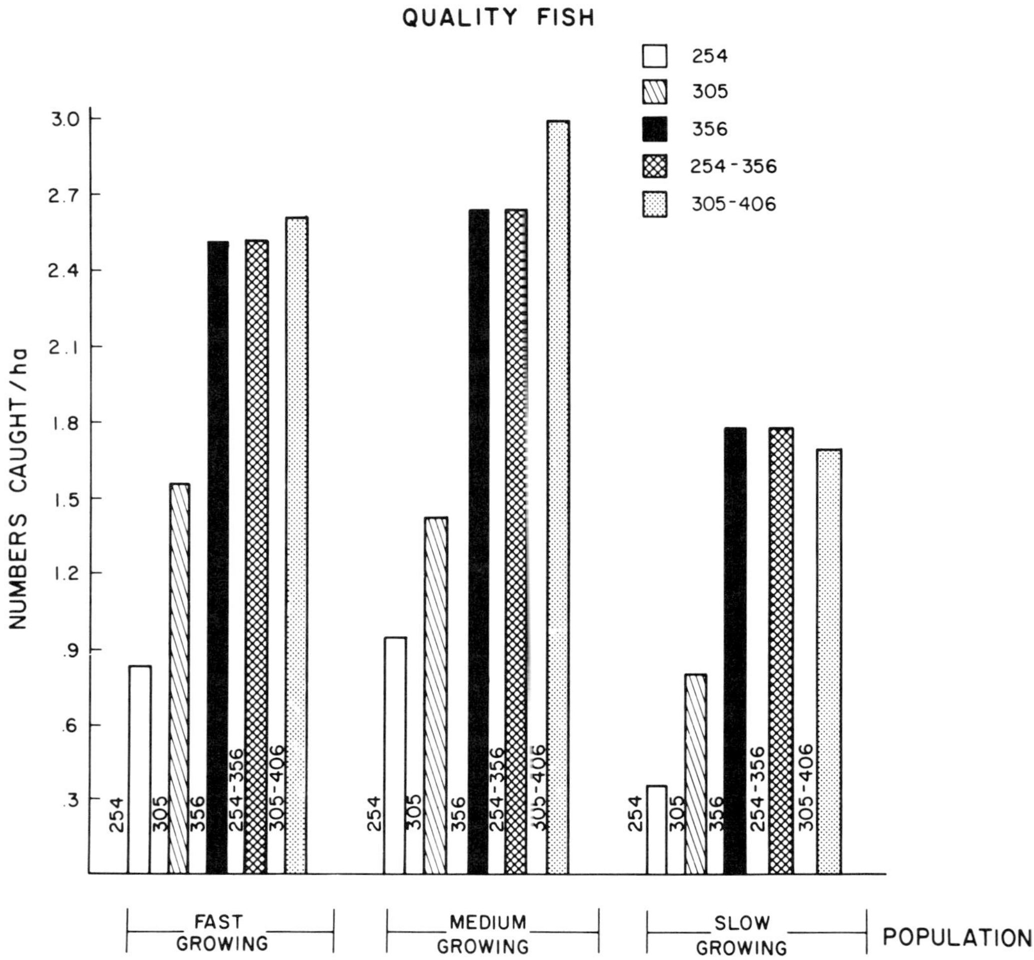

Figure 4. Simulated numbers of quality-size largemouth bass (>381 mm) harvested for slow-, medium-, and fast-growing populations subjected to five hypothetical harvest regulations. Length limits are in millimeters.

pends on the management objective as well as growth rates. If the objective is to develop a trophy fishery, we recommend a high slot limit or a high minimum-length limit. These restrictions protect potential quality-size fish at a time when they would otherwise be subjected to high fishing mortality. Therefore, a large proportion of sublegal fish will eventually enter the fishery, assuming no compensatory growth. A relatively low minimum-length limit works well if the objective is to maximize the total number of fish harvested, while higher minimum-length limits or slot limits work better for maximizing the biomass harvested. If one wants a balance between a trophy fishery and a recreational fishery, a high slot limit or a high minimum-length limit is a good choice. The minimum-length limit will provide the most balanced harvest, and the slot limit will produce a higher proportion of quality-size fish. Both, however, can provide for sustained yields.

References

Archer, D. L., and H. A. Loyacano. 1975. Initial and delayed mortalities of largemouth bass captured in the 1973 National Keowee B.A.S.S. tournament. Proceedings of the Annual Conference of the Southeastern Association of Game and Fish Commissioners 28:90–96.

BENNETT, G. W. 1937. The growth of the largemouth bass, *Huro salmoides* (Lacépède), in the waters of Wisconsin. Copeia 1937:104–118.

BENNETT, G. W. 1971. Management of lakes and ponds, Second Edition. Van Nostrand Reinhold Company, New York, USA.

BENNETT, G. W., H. W. ADKINS, AND W. F. CHILDERS. 1969. Largemouth bass and other fishes in Ridge Lake, Illinois, 1941–1963. Illinois Natural History Survey Bulletin 30(1):1–67.

BRYANT, H. E., AND A. HOUSER. 1971. Population estimates and growth of largemouth bass in Beaver and Bull Shoals reservoirs. Pages 349–357 *in* G. E. Hall, editor. Reservoir fisheries and limnology. Special Publication 8, American Fisheries Society, Bethesda, Maryland, USA.

CARLANDER, K. D. 1977. Handbook of freshwater fishery biology, Volume 2. Iowa State University Press, Ames, Iowa, USA.

CLADY, M. D. 1970. Regulation of fish populations in three lightly exploited lakes in northern Michigan. Doctoral dissertation, University of Michigan, Ann Arbor, Michigan, USA.

CLADY, M. D. 1975. The effects of a simulated angler harvest on biomass and production in lightly exploited populations of smallmouth and largemouth bass. Transactions of the American Fisheries Society 104:270–276.

COOMER, C. E., JR. 1976. Population dynamics of black bass in Center Hill Reservoir, Tennessee. Master's thesis, Tennessee Technological University, Cookeville, Tennessee, USA.

FARABEE, G. B. 1974. Effects of a 12-inch length limit on largemouth bass and bluegill populations in two northeast Missouri lakes. Pages 95–99 *in* J. L. Funk, editor. Symposium on overharvest and management of largemouth bass in small impoundments. Special Publication 3, North Central Division American Fisheries Society, Bethesda, Maryland, USA.

GABELHOUSE, D., JR. 1980. Black bass length limit investigations. Final report, Kansas Fish and Game Commission, Topeka, Kansas, USA.

HEIDINGER, R. C. 1975. Life history and biology of the largemouth bass. Pages 11–20 *in* R. H. Stroud and H. Clepper, editors. Black bass biology and management. Sport Fishing Institute, Washington, D.C., USA.

HICKMAN, G. D., AND J. C. CONGDON. 1974. Effects of length limits on fish populations in five north Missouri lakes. Pages 84–94 *in* J. L. Funk, editor. Symposium on overharvest and management of largemouth bass in small impoundments. Special Publication 3, North Central Division American Fisheries Society, Bethesda, Maryland, USA.

HOEY, J. W., AND L. C. REDMOND. 1974. Evaluation of opening Binder Lake with a length limit for bass. Pages 100–105 *in* J. L. Funk, editor. Symposium on overharvest and management of largemouth bass in small impoundments. Special Publication 3, North Central Division American Fisheries Society, Bethesda, Maryland, USA.

HUBERT, W. A. 1976. Age and growth of three black bass species in Pickwick Reservoir. Proceedings of the Annual Conference of the Southeastern Association of Game and Fish Commissioners 29: 126–134.

JENKINS, R. M. 1975. Black bass crops and species associations in reservoirs. Pages 114–124 *in* R. H. Stroud and H. Clepper, editors. Black bass biology and management. Sport Fishing Institute, Washington, D.C., USA.

JENKINS, R. M. 1976. Prediction of fish production in Oklahoma reservoirs on the basis of environmental variables. Annals of the Oklahoma Academy of Science Publication 5:11–20.

JOHNSON, D. L., AND R. O. ANDERSON. 1974. Evaluation of a 12-inch length limit on largemouth bass in Phillips Lake, 1966–1973. Pages 106–113 *in* J. L. Funk, editor. Symposium on overharvest and management of largemouth bass in small impoundments. Special Publication 3, North Central Division American Fisheries Society, Bethesda, Maryland, USA.

KELLEY, J. W. 1962. Sexual maturity and fecundity of the largemouth bass, *Micropterus salmoides* (Lacépède), in Maine. Transactions of the American Fisheries Society 91:23–28.

MAY, B. E. 1973. Evaluation of large scale release programs with special reference to bass fishing tournaments. Proceedings of the Annual Conference of the Southeastern Association of Game and Fish Commissioners 26:325–334.

MING, A., AND W. E. MCDANNOLD. 1975. Effect of length limit on an overharvested largemouth bass population. Pages 416–424 *in* R. H. Stroud and H. Clepper, editors. Black bass biology and management. Sport Fishing Institute, Washington, D.C., USA.

MOODY, H. L. 1975. Tournament catch of largemouth bass from St. Johns River, Florida. Proceedings of the Annual Conference of the Southeastern Association of Game and Fish Commissioners 28:73–82.

OLMSTED, L. L. 1974. The ecology of largemouth bass (*Micropterus salmoides*) and spotted bass (*Micropterus punctulatus*) in Lake Fort Smith, Arkansas. Doctoral dissertation, University of Arkansas, Fayetteville, Arkansas, USA.

ORTH, D. J. 1977. Development of a computer simulation model of largemouth bass population dynamics. Master's thesis, Oklahoma State University, Stillwater, Oklahoma, USA.

ORTH, D. J. 1978. Computer simulation models for predicting population trends of largemouth bass in large reservoirs. Proceedings of the Oklahoma Academy of Science 58:35–43.

ORTH, D. J. 1980. Changes in the fish community of Lake Carl Blackwell, Oklahoma (1967–77) and a test of the reproductive guild concept. Proceedings of the Oklahoma Academy of Science 60:10–17.

PADFIELD, J. H., JR. 1951. Age and growth differentiation between the sexes of the largemouth black bass, *Micropterus salmoides* (Lacépède). Journal of the Tennessee Academy of Science 26:42–54.

PARAGAMIAN, V. L. 1982. Catch rates and harvest results under a 14-inch minimum length limit for largemouth bass in a new Iowa impoundment. North American Journal of Fisheries Management 2:224–231.

PELZMAN, R. J. 1978. Hooking mortality of largemouth bass (*Micropterus salmoides*) less than 305 mm long. California Fish and Game 64:185–188.

PELZMAN, R. J. 1979. Effects of a 305 mm (12.0 inch) minimum size limit on largemouth bass (*Micropterus salmoides*) at Merle Collins Reservoir. California Fish and Game 65:141–150.

PLUMB, J. A., J. L. GAINES, AND M. GENNINGS. 1975. Experimental use of antibiotics in preventing delayed mortality in a bass tournament on Lake Seminole, Georgia. Proceedings of the Annual Conference of the Southeastern Association of Game and Fish Commissioners 28:87–90.

RAINWATER, W. C., AND A. HOUSER. 1982. Species composition and biomass of fish in selected coves in Beaver Lake, Arkansas, during the first 18 years of impoundment (1963–1980). North American Journal of Fisheries Management 4:316–325.

RASMUSSEN, J. L., AND S. M. MICHAELSON. 1974. Attempts to prevent largemouth bass overharvest in three northwest Missouri lakes. Pages 69–83 *in* J. L. Funk, editor. Symposium on overharvest and management of largemouth bass in small impoundments. Special Publication 3, North Central Division American Fisheries Society, Bethesda, Maryland, USA.

RAWSTRON, R. R., AND K. A. HASHAGEN, JR. 1972. Mortality and survival rates of tagged largemouth bass (*Micropterus salmoides*) at Merle Collins Reservoir. California Fish and Game 58:221–230.

REDMOND, L. C. 1986. The history and development of warmwater fish harvest regulations. Pages 186–195 *in* G. E. Hall and M. J. Van Den Avyle, editors. Reservoir fisheries management: Strategies for the 80's. Reservoir Committee, Southern Division American Fisheries Society, Bethesda, Maryland, USA.

REGIER, H. A. 1963. Ecology and management of largemouth bass and bluegills in farm ponds in New York. New York Fish and Game Journal 10: 1–89.

SEIDENSTICKER, E. P. 1975. Mortality of largemouth bass for two tournaments utilizing a "don't kill your catch" program. Proceedings of the Annual Conference of the Southeastern Association of Fish and Game Commissioners 28:83–86.

TAYLOR, M. W. 1981. A generalized inland fishery simulator for management biologists. North American Journal of Fisheries Management 1:60–72.

USFWS (U.S. DEPARTMENT OF THE INTERIOR, FISH AND WILDLIFE SERVICE, AND U.S. DEPARTMENT OF COMMERCE, BUREAU OF THE CENSUS). 1982. 1980 National survey of fishing, hunting, and wildlife-associated recreation. U.S. Government Printing Office, Washington, D.C., USA.

VAUGHAN, D. S., AND S. B. SAILA. 1976. A method for determining mortality rates using the Leslie matrix. Transactions of the American Fisheries Society 105:380–383.

WELLBORN, T. L., JR., AND J. H. BARKLEY. 1974. Study on the survival of tournament released bass on Ross R. Barnett Reservoir, April 1973. Proceedings of the Annual Conference of the Southeastern Association of Game and Fish Commissioners 27: 512–519.

Pages 227–237 *in* G. E. Hall and M. J. Van Den Avyle, editors.
Reservoir Fisheries Management: Strategies for the 80's.
Reservoir Committee, Southern Division American Fisheries Society, Bethesda, Maryland. 1986.

MANAGEMENT IMPLICATIONS OF ENERGY DEVELOPMENT

Fishery Management in Cooling Impoundments

LARRY L. OLMSTED
Duke Power Company
Production Environmental Services
Huntersville, North Carolina 28078

JAMES P. CLUGSTON
U.S. Fish and Wildlife Service
National Fisheries Research Laboratory
Gainesville, Florida 32606

ABSTRACT

Impoundments constructed primarily for the dissipation of waste heat represent a unique aquatic habitat. Although cooling impoundments typically support substantial fisheries, increasing fishing demand dictates that these impoundments be managed productively. Virtually all fishery management techniques used in noncooling lakes and reservoirs can be applied to cooling impoundments with only minor modifications. Numerous additional management options unique to cooling impoundments are available as a result of the pumping and heating of large quantities of water. These options range from altering stocking strategies to capitalize on the altered thermal regime to attempting to influence water quality and nutrient availability by controlling mixing patterns. We describe various management options that warrant consideration during the siting, design, construction, and operation of cooling impoundments.

Steam electric plants presently account for 89% of the electricity produced annually in the United States (Edison Electric Institute 1982). All steam electric plants release large quantities of heat to the environment. Nuclear plants, with their lower efficiency and no stack losses, produce water heat loads 60–70% higher than those of fossil-fueled plants. The most efficient coal-fired and nuclear facilities in the United States have thermal efficiencies of approximately 38 and 32%, respectively. A typical 2,000-megawatt nuclear plant produces waste heat at the condenser at a rate of 13×10^9 BTU/hour (Ryan and Harleman 1973). Design considerations dictate that cooling water flows at such a plant are usually greater than 85 m^3/second and that the rise in temperature (ΔT) of water flowing through the condenser is approximately 11 C. The challenge facing utilities is to dispose of this heat in an economical and environmentally compatible manner.

Cooling towers have been proposed as the ultimate solution to the "thermal pollution problem." However, several factors detract from the desirability of cooling towers: high cost, potential fogging problems, chemical additions required for biofouling, high evaporative losses, and aesthetic considerations. A common alternative is to discharge the condenser cooling water directly to a cooling impoundment. This paper considers a cooling impoundment to be any impoundment that serves to dissipate heat from a steam electric plant.

The United States has 135 existing or planned surface-water cooling impoundments, representing approximately 0.5 million hectares (Clugston et al. 1982). Although many of these (about 47) are either in or scheduled to be in Texas, appreciable numbers are in the Midwest and Southeast (Fig. 1). Tetra Tech, Inc. (1981) listed the physical characteristics of most cooling impoundments in the United States and the operational characteristics of the associated power plants.

Impoundments constructed primarily for the dissipation of waste heat represent a unique aquatic habitat. The use and discharge of power

Figure 1. Location of 135 planned and existing cooling impoundments in the United States (modified from Tetra Tech, Inc. 1981).

plant cooling water provides for three phenomena of fundamental limnological importance.

1) *Elevated water temperature.*—Thermal discharges from a power plant can have a profound influence on water temperature distribution in the receiving impoundment. The importance of water temperature in regulating life processes of aquatic organisms in reservoirs has been extensively documented (Miller and DeMont 1974; Duke Power Company 1977; Larimore and Tranquilli 1981). In discharge areas of cooling impoundments, the thermal influence may be sufficient to alter seasonal processes such as spawning (Miller and DeMont 1974; Siler and Clugston 1975) and growth (Siler and Clugston 1975; Sule et al. 1981; Smithson et al. 1986). Thermal loading has also been related to physiological phenomena such as gas bubble disease (Adair and Hains 1973) and parasitism (Eure and Esch 1973). Controlling the quantity of heat and the manner in which it is discharged into the impoundment represent major challenges and opportunities in managing the fishery of a cooling impoundment.

2) *Movement of large volumes of water.*—Pumping of cooling water by a power plant results in the horizontal transport of large volumes of water, which provides some control over movements of water masses. Horizontal water movements not only influence the distribution of heat, dissolved gases, nutrients, and other physical and chemical variables, but also the distribution of microorganisms and fish.

3) *Control of vertical structure.*—Design of intake and discharge structures, outlet depth of associated dams, and varying operations based on the relation between cooling water flow and ΔT all have the potential of altering vertical, physical, and chemical gradients. This alteration may in turn influence the distribution of plankton (Duke Power Co. 1977; Wilde et al. 1977), benthos (Brinkhurst 1974), larval fish (Ruelle et al. 1977), and adult fish (Siler and Foris 1985).

Statement of Problem

Cooling impoundments have provided an expanding fishery resource in the last several decades due to construction programs by utilities. However, several factors signal an end to this era. First, the rate of growth of electrical demand has slowed. Whereas most utilities were experiencing annual growth rates of approximately 7.5% in 1974, annual growth in electrical demand in 1982 was approximately 3.3% (North

American Electric Reliability Council 1982). This reduced demand for electricity has resulted in the postponement and cancellation of construction of numerous plants and cooling impoundments. A second factor is that regulatory agencies do not consider cooling impoundments as the "best available technology" as a heat sink for power plants. Finally, depending on size and location, the capital costs of developing a cooling impoundment may be high.

Given the unusual limnological characteristics of cooling impoundments, the question then is "How can cooling waters best be managed to ensure a viable sport fishery?" Virtually all conventional management techniques have been applied to cooling impoundments with various degrees of success. The focus of this paper is to recognize how these techniques must be modified to be more successful and to identify effective management strategies for cooling impoundments.

Working Hypothesis

The regulation of harvest by catch and size limits to maintain an acceptable yield is the most common approach to fishery management. Although regulations are sometimes the only feasible management technique, we believe that initial emphasis should include an attempt to increase yield by increasing production. Jenkins (1967, 1982), Matuszek (1978), Oglesby (1977), Ryder (1965, 1978), and others have intensively investigated the use of abiotic and biotic variables as predictors of fish yield or biomass; however, the ability to link these variables with fish yield has seldom been incorporated into management plans concerned with increasing yield. This slow transition from statistical classification and prediction to management application reflects the limited ability of managers to control the driving variables in most natural lakes and impoundments. However, the mechanism of infusing energy into the system through pumping and heating large volumes of water provides the fishery manager of a cooling impoundment an opportunity to influence some potentially "limiting" variables and alter yield.

Siting, Design, and Construction Considerations

The siting, design, and construction of cooling impoundments associated with power plants establish the ultimate constraints on available management options. Each of these activities offers opportunities to enhance the fisheries.

Siting

Proximity to electrical load centers, existing transmission lines and rights of way, and transportation arteries are of primary importance in siting power plants and their cooling impoundments. Within these constraints, selection of a site on the basis of environmental factors may significantly affect the fishery. The morphology of a lake's basin has a major influence on its nutrient dynamics, oxygen regime, heat budget, and productivity (Ryder 1982). In areas where nutrients commonly limit the productivity of reservoirs, optimal reservoir characteristics would be small surface area, shallow mean depth, large drainage area, and high shoreline development. Shallow mean depth increases nutrient availability by reducing the development of, and loss of nutrients to, the tropholytic zone (Ryder 1978). Attempts to limit mean depth for fish production must, however, be weighed against considerations to establish multihabitat fisheries, maintain diverse recreational uses of the lake, minimize land inundated to provide the necessary heat sink, and maximize head if hydropower generation is planned. High shoreline development increases nutrient flux by increasing the land-water interface where nutrient exchange is high (Ryder 1978). If the manager is attempting to manage a littoral-oriented community, maximizing shoreline development should be a prime objective. In addition to enhancing productivity, increasing shoreline development augments preferred areas for spawning and feeding.

Retention time also influences productivity in cooling lakes. Jenkins (1967) noted the utility of storage ratio (a correlate of retention time) in predicting standing crop and commercial harvest of fishes in the United States. Retention time should be adequate for high plankton production. Brook and Woodward (1956) stated that the water exchange rate must be greater than 18 days for significant development of zooplankton. Vollenweider and Dillon (1974), studying nutrient budgets, noted the importance of retention time on biological productivity. On the other hand, extremely long retention times in cooling impoundments can result in the concentration of potentially toxic compounds (Olmsted et al. 1986) and lead to eutrophication and generally poor water quality.

Design

The withdrawal and discharge depths of power plants and associated dams are of primary importance in determining thermal gradients in reservoirs. In high nutrient systems, optimal design would call for the power plant to withdraw from and discharge into the surface layers. This scheme would intensify stratification, and nutrients would be effectively isolated in the hypolimnion (Wright 1968). Conversely, where nutrient limitation is anticipated, withdrawal by the power plant from the hypolimnion will bring nutrient-rich water to the surface and will also tend to disrupt thermal stratification of the impoundment. Release depth of the dam should be in the upper portion of the water column if the manager wishes to retain nutrients within the cooling impoundment (Wright 1968). Variable-depth releases in dams are becoming increasingly common and allow flexibility to manage physical-chemical conditions of impoundments and tailwaters.

Passage of organisms through the cooling system of a power plant (entrainment) and the impingement of larger organisms on the intake screens are possible environmental consequences that have received substantial investigation (Jensen 1976). Although these losses may have only negligible effects on fisheries (Porak and Tranquilli 1981), poor siting and design may allow impingement and entrainment to become important in population dynamics. Recent symposia have focused on designs of intakes that control these factors (Sharma and Palmer 1978; Dorn and Johnson 1981), and the Federal Government has issued guidance documents regarding "best available technology" for intakes (U.S. Environmental Protection Agency 1975, 1976).

The physical design of discharge structures should reflect ecological requirements of the fish community being managed. Yield of rheophilic species may be increased by providing a preferred habitat in a long discharge canal. However, long discharge canals may be undesirable in situations where gas bubble disease, cold shock, or chemical discharges are a concern because the canal would tend to confine and exacerbate the effect.

Consideration must be given during design to minimize the potential for toxic effects due to plant operation. Historically, these concerns have revolved primarily around chlorination for control of biofouling, leaching of metals from condenser tubing, and proper handling of ash. Recently the utility industry has made significant strides in reducing chlorine use, primarily through changes in operational procedures (Garey et al. 1980). Included in the design of new plants has been a method of circulating abrasive rubber balls through the condenser to reduce or eliminate the need to use a biocide in this system.

Leaching of metals from condenser tubes is reduced by using highly inert materials such as zirconium and stainless steel. Use of leachable metals, especially copper and brass, has sometimes led to environmental problems and the eventual need for replacement of tubing. For example, poor reproduction and deformities in bluegills (*Lepomis macrochirus*) in Lake Robinson, South Carolina, were believed to be a result of copper and zinc in the acidic water. Biologists with Carolina Power and Light speculated that metals were being introduced into Lake Robinson as products of erosion and corrosion of brass condenser tubes in the generating plant (Carolina Power and Light Company and Lawler, Matusky, and Skelly, Engineers 1981; Carolina Power and Light Company 1982). When the brass tubes at the plant were replaced with stainless steel tubes, the bluegills in Lake Robinson recovered.

The importance of proper ash disposal has recently received much attention. Selenium and arsenic leachates from flyash discharged into Belews Lake, North Carolina, have been identified as the primary causes of a substantial decline in the fisheries (Cumbie and Van Horn 1979). Discharges from ash basins also have been implicated in declining fisheries in several other cooling impoundments. Proper design of ash disposal facilities usually involves discharging from ash basins to bodies of water with low retention times, chemical treatment of ash basin discharges, or inclusion of dry ash handling capabilities (Olmsted et al. 1986).

Construction

Management options during construction are limited. Ploskey (1981) reviewed site preparation procedures for impoundments and noted that concerns for reservoir clearing depend on trade-offs among factors such as mosquito control, water quality, recreation, and fish production. In areas of inherently low fertility it appears that retention of herbaceous vegetation in the reservoir basin is conducive to the establishment of high fish production during the early years of impoundment. Vegetation left in highly fertile basins where reservoirs will be created could contribute to water quality problems such as algal blooms and low dissolved oxygen.

Consideration should be given to habitat enhancement during construction. Excavation and disposal of material from the construction site often occurs within the impoundment basin. Properly conducted, this may increase the structural heterogeneity and provide topographic relief, increased spawning habitat, and potential thermal refugia. Prior geological testing should be performed to ensure that this practice will not cause problems by exposing strata of rocks that contain potentially toxic minerals.

The large number of variables to be considered during siting, design, and construction that can affect standing crops of fishes in a planned cooling impoundment suggests that a statistical model would be useful in assessing the anticipated fishery potential. Two general approaches exist for predicting fish standing crops in reservoirs. Jenkins (1967) and Jenkins and Morais (1971) used multiple linear regression to relate the standing crop of a single species to a set of abiotic variables. To predict species abundance in a cooling impoundment, one measures the appropriate set of abiotic variables and substitutes them into the appropriate regression equation. The output predicts standing crops of individual species as a function of the abiotic variables.

Tetra Tech, Inc. (1982) published a methodology in which cluster analysis, linear and multiple regression, and discriminant analysis are used to predict characteristics of the fish community in a new impoundment. The equations combine information from nine limnological variables into a single index of classification. Based on the value of this index, the lake is classified into one of four distinct groups. Fishery characteristics of the indicated group are then used to predict fish community structure and potential recreational fishery use.

Both methods are useful as a first approximation, and application of either technique during siting, design, and construction should provide a reasonable estimate of species biomass and community type at a particular site. Selection of the design predicting highest projected yield of desired species should enhance the fishery resources of the impoundment. An obvious extension of the methodology would allow decisions on siting, design, and construction to be made in a framework of cost/benefit analyses relating to the fisheries expected under various options. To date, this methodology has not been applied in the site selection process for cooling impoundments.

Operational Considerations and Management Implications

The presence of a thermal discharge substantially increases fishery management options available. Within the operating constraints of the plant, it is sometimes possible to vary the amount of cooling water used, with an inverse relation to ΔT. By capitalizing on this relation, vertical thermal gradients can be influenced within broad bounds. Thus, in impoundments where nutrients are abundant, withdrawal of smaller volumes of surface water with a higher ΔT enhances stratification and concentrates nutrients in the hypolimnion. In nutrient-poor impoundments, withdrawing large volumes of water from the hypolimnion and discharging them near the surface at slightly above-surface water temperatures tends to increase the availability of nutrients by relocating them to the trophogenic zone. However, this arrangement may result in lower dissolved oxygen and higher turbidity in the epilimnion.

Within limits, the added heat increases the rate of autotrophic production in nonnutrient-limited impoundments. Increased temperatures also increase respiration, and in nutrient-limited systems this added respiration could decrease net production. Heat loading also has the effect of artificially extending the growing season. Numerous researchers have shown that an increased growing season has a positive influence on fish yield (Jenkins 1967; Ryder et al. 1974; Ryder 1978).

Each fish species demonstrates a maximum growth rate within a specific temperature regime. Maximum production of desired species under existing trophic conditions could probably be achieved by following this regime as closely as possible within efficient operating confines of the power plant. However, this is difficult in many situations due to thermal limits imposed by regulatory agencies, which are usually based on statewide water quality standards and may have little relation to fish production.

Fish attractors are a commonly used technique for enhancing fisheries in impoundments. Alterations in thermal profiles resulting from power plant operations impose some restrictions on the placement of fish attractors in cooling impoundments. Since seasonal water temperature and dissolved oxygen profiles may differ from nearby noncooling impoundments, these factors must be considered to ensure that structures will be placed in usable habitat of target species. Number and location of attractors must also be considered so that water flow and cooling efficiencies are not impeded (Clugston et al. 1982). Additionally, attractors must be built of highly resistant materials and securely anchored to prevent operational problems from components being drawn into the power plant intakes.

An additional management option arises from the substantial and rapid thermal changes in the near-field area of the power plant discharge. Since numerous species use discharge areas for spawning and overwintering, utilities could help reduce certain nongame species by providing "thermal shock" at strategic times. Such a practice would require substantial coordination and planning to avoid affecting nontarget species; it is also inconsistent with regulatory pressures to minimize plant-induced mortalities. We are unaware of any attempts to apply this technique.

Altered thermal conditions in cooling impoundments provide unique opportunities for fish stocking. Since the waters are artificially warmed, thermal regimes in northern cooling impoundments may be similar to those in southern impoundments, and thus stocking strategies used in the South may be appropriate (Clugston et al. 1982). Likewise, tropical species and strategies may be appropriate for southern impoundments. Elevated temperatures have made it possible to introduce exotic species, where state regulations allow such introductions, to provide additional forage, more efficient predators, and weed control. If tropical species are stocked in cooling impoundments, concerns regarding proliferation in the event of escape are minimized since these species are usually unable to overwinter in unheated temperate waters.

Blue tilapias (*Tilapia aurea*) were introduced into a Texas cooling lake where they became important prey for largemouth bass (*Micropterus salmoides*). They have also been introduced and are flourishing in Lake Julian, North Carolina (Bobby J. Ward, Carolina Power and Light Company, New Hill, North Carolina, USA [personal communication]). The threadfin shad (*Dorosoma petenense*) is the most commonly introduced forage species and has been stocked in most cooling impoundments. The operation of a power plant on a reservoir is a requisite for overwintering of this species in northern latitude impoundments, since heavy mortality usually occurs at temperatures below 9 C (Strawn 1965). With this overwintering capability, cooling impoundments have substantially extended the range of threadfin shad in the United States. Many states now use these impoundments as the prime source of adult threadfin shad for annual restocking in other waters. Fishery managers also should be aware that a complete shutdown for an extended time during winter could eliminate all temperature-sensitive nonnative fish.

The stocking of predators in cooling impoundments is generally well received by sportsmen. Predators are usually introduced to help control populations of nongame fish, particularly gizzard shad (*Dorosoma cepedianum*), and to provide trophy fisheries. Attention has recently been focused on stocking of sterile hybrids on a "put-grown-and-take" basis. The hybrid of the muskellunge (*Esox masquinongy*) and northern pike (*E. lucius*) appears to have management potential in northern cooling lakes, and the white bass (*Morone chrysops*) and striped bass (*M. saxatilis*) hybrid has performed well in northern and southern cooling lakes. The Texas Parks and Wildlife Department has been particularly active in the stocking of nonnative predators. In addition to assessing freshwater predators from Africa and South America (Rutledge and Lyons 1978), the Texas agency has introduced some estuarine and marine species (Lasswell et al. 1979) into relatively saline cooling reservoirs.

Conversely, the altered thermal regimes of cooling impoundments can represent suboptimum conditions for predators that may be successful in nearby unheated reservoirs. Coutant (1980) determined that striped bass have a thermal niche that changes with age of the fish. He found that waters with summer temperatures near 24–26 C provide optimum thermal conditions for growth of juveniles. However, temperature requirements change as striped bass grow, and reservoirs without well-oxygenated water below about 22 C may be unsuitable for large fish (>5 kg). Stocking of predators has met with various degrees of success, but where adequate consideration has been given to matching the environmental requirements of fish with prevailing conditions in cooling impoundments, impressive fisheries have often resulted.

Clugston et al. (1982) noted that the potential for aquatic weed problems is higher in cooling impoundments than in noncooling lakes because of the longer growing season that results from the heat load and buildup of nutrients through evaporation and recycling of the cooling water. Exotic fish species have been a valuable weed control option in cooling impoundments. The Carolina Power and Light Company (1983) has successfully used the redbelly tilapia (*Tilapia zilli*) to control bladderwort (*Utricularia* sp.) in Lake Sutton, North Carolina, and is experimenting with it in Lake Hyco, North Carolina. The Santee-Cooper Power Authority uses thermally altered ponds to culture tilapias that are sold to numerous agencies for use in weed control.

Cooling impoundments have horizontal thermal gradients that may result in segregation of species on the basis of thermal preferenda (Tranquilli et al. 1981). Segregation of species enhances the potential of several management strategies. Selective removal by toxicants is a management technique used to increase production of desirable species by reducing the numbers of undesirable ones. Spatial segregation allows more nearly complete removal of the target species without killing the nontarget species. Likewise, segregation could also allow commercial fishing, with reduced risk of incidental capture of sport species. Regulations restricting gear, season, and areas based on distribution, as influenced by plant operation, could control the harvest of both sport and commercial species. This strategy also has the potential of reducing animosity between sport fishermen and commercial fishermen.

The movement of large volumes of water may foster spatial segregation of species based on rheotactic responses, thus offering similar management opportunities as segregation by thermal selection. Such segregation has been shown at McGuire Nuclear Station on Lake Norman, North Carolina, by Lewis (in press). Before the start of operations and the discharge of heat, pumps were run for several years to supply station service water; the resultant current attracted rheophilic species such as white bass, striped bass, and yellow perch, *Perca flavescens* (Siler et al. 1981).

Current also offers an inexpensive and dependable means of distributing chemical compounds. These compounds could be a flocculant such as alum, or lime if the alkalinity is low, or fertilizer if the impoundment is characterized by low nutrients. Although fertilizer is infrequently applied to impoundments because of high cost, the potential for fertilization of cooling ponds is greater due to the smaller size and extremely small loss of nutrients through downstream discharge. Trickling of appropriate nutrients into the power plant discharge has possibilities for increasing productivity in nutrient-limited impoundments. Caution must be exercised, however, because the addition of nutrients to a cooling impoundment with high heat load could promote undesirable algal growth, water quality problems, and biofouling of condenser tubes.

Commercial culture of fish, shrimp, and other organisms in heated discharges is an accepted practice, and certain variations of the process have sport fish management potential. Proposals have been advanced to culture sport fish in cages in the heated discharge of power plants, and to release a certain proportion to the lake at the time of harvest. This idea appears to have appeal to cooperating agencies. The aquaculturist could realize a profit from the operation, the managing agency could gain through lease fees, and the resource could benefit through stocking of adult fish. This approach would be most useful where natural recruitment is low. Concerns surrounding this technique involve liability, security, and impacts on water quality. In actuality, uneaten food could increase production of fishes in the vicinity of rearing cages in nutrient-poor impoundments.

Commonwealth Edison, in conjunction with researchers at Southern Illinois University, has taken a novel approach to the cooling impoundment-aquaculture relation. Lewis and Heidinger (1981) are investigating the use of cooling ponds as nursery areas for receiving waters. The researchers stocked fry and fingerlings of selected species in Dresden Pond, Illinois, and released the fingerlings to receiving waters after they had grown rapidly in the heated water. Preliminary results suggest that this technique has potential for some species, especially striped bass and walleyes (*Stizostedion vitreum*). As Clugston et al. (1982) pointed out, however, the use of a cooling pond as a nursery area could be hindered by the lack of a means for passing fish to another body of water.

Size, creel, and season restrictions are applied

to cooling impoundments for the same reason that they are applied to noncooling impoundments—primarily to prevent overharvest. Clugston et al. (1982) noted several peculiarities of cooling impoundments that made them more subject than noncooling impoundments to overharvest. Fish in cooling lakes often congregate near the power plant discharge, and fishing pressure and harvest in this area may be intensive (Siler et al. 1986). The altered thermal regime often allows predators to feed actively during extended portions of the year which, in addition to increasing production, increases their chances of being harvested. Additionally, lack of ice cover on northern cooling ponds has provided a winter fishery, further increasing the chances of over-exploitation. One approach toward reducing harvest is that of closing the discharge area to fishing during seasons when fish are congregated there. However, such closures should be considered only in extreme situations since the discharge fishery is usually a heavily advocated aspect of cooling impoundment fisheries.

Primary problems of managing fisheries in cooling impoundments tend to vary on a regional basis. In the Midwest, major problems are usually a paucity of spawning areas and excessively abundant gizzard shad. Habitat enhancement, primarily by means of increasing shallow areas and installing riprap, is the preferred method of compensating for lack of spawning areas. Where this is impossible or inadequate, consideration should be given to using nursery areas as practiced by Illinois Power Company. This company made nursery areas of two farm ponds that drain directly into Lake Clinton, Illinois. The ponds are left dry through the winter and seeded with rye. In spring the ponds are filled and stocked with mature largemouth bass and threadfin shad. In autumn the ponds are drained into Lake Clinton and the process is repeated. This technique has been successful in supplementing largemouth bass production in the cooling pond (James A. Smithson, Illinois Power Company, Clinton, Illinois, USA [personal communication]). Some situations may require intensive annual stocking to ensure adequate recruitment. Commonwealth Edison, for example, has built a hatchery to provide fish for its cooling impoundments (Richard Monzingo, Commonwealth Edison, Chicago, Illinois, USA [personal communication]).

Control of gizzard shad in cooling impoundments is important not only for fishery management, but also for plant operation due to impingement concerns. Ideally, a predator should be stocked that would control gizzard shad, but such efforts have been largely ineffective. Control of gizzard shad with toxicants, as now practiced by most state agencies and many utilities, represents the best technique.

Except in Florida, the principal management problem in southeastern cooling impoundments is generally low production. Although production may be increased by manipulating physical processes to keep nutrients in the euphotic zone, fertilization often results in immediate and substantial increases in productivity (Larsen 1983). High cost is usually the reason fertilization is not more widely applied. Experience in Texas has shown that inorganic agricultural fertilizers are not required; suitable results can be gained by using organic wastes from nearby industries (Jay A. Pruett, Southwest Electric Power Company, Shreveport, Louisiana, USA [personal communication]). A major management concern in most Florida cooling impoundments is invasion by undesirable exotic fish species and vegetation. The state of Florida has enacted restrictions on exotic fish aquaculturists and established penalties for violators.

Future Needs

Although many cooling impoundments provide excellent fisheries, significant improvements can be made.

1) Fisheries should be an integral consideration in siting and design phases of project development. Siting and design impose ultimate constraints on fisheries, and the fishery manager should be prepared to provide timely, effective input. Existing fishery models may be useful at this stage.
2) Formal fishery management plans need to be developed for each cooling impoundment. Each plan must contain a concise statement of management objectives and, on the basis of existing or proposed conditions, should identify problems and propose actions to achieve the objectives. The results should be monitored, and management plans reviewed and updated, at regular intervals.
3) Although some utilities and state fishery agencies cooperate closely, others do not, and there is a general need for improvement. Although most states maintain responsibility for

management, the responsibility for monitoring can often be shared or assumed by the utility.

4) The primary research need is the evaluation of various management techniques for increasing fish yield. Too often, management strategies are applied with inadequate attention given to evaluation. As a result, little conclusive information is available as to what works and what does not and why. Accurate evaluation is paramount in the transition of fishery management from "hit-or-miss magic" to productive, applied science.
5) It is extremely difficult to maximize the usefulness of a resource when the public is excluded from it. Although many utilities are reexamining their decision to deny public access to cooling impoundments, several have no plans to open them. Primary reasons for restricting access are safety and liability considerations. Several utilities have skirted the liability concern by leasing the facilities, along with liability, to government or quasi-government agencies. Safety concerns sometimes can be answered by instituting relatively inexpensive physical modifications or restricting access to certain portions of cooling impoundments. Closure of the impoundment to the public is sometimes appropriate, but suitable alternatives can usually be found.
6) Regulatory constraints represent an obstacle in the management of many cooling impoundments. The primary interest of regulatory agencies has been to reduce adverse effects, rather than to capitalize on positive consequences such as managing the heat load to maximize fish production. End-of-pipe thermal discharge limits, or the provision of small mixing zones are the most common means of regulating thermal discharges. These means have often led to arbitrarily defined discharge limits and the imposition of onerous monitoring programs on the utilities to ensure compliance. Similar restrictions aimed at protecting "water quality" hinder the development of aquaculture enterprises and other projects that would enhance the productivity of most systems. If proper attention were given to public health effects, most systems could benefit by having secondary treated sewage discharged into cooling impoundments.
7) Fishery management is often characterized by a willingness to study what is there and either "regulate it" or "stock it." There is an acute need for more open and professional discussion and understanding among fishery biologists, limnologists, engineers, and regulators. The anticipated end product of such interaction is the development of innovative, but technically and economically sound, proposals for increasing the fish yield of cooling impoundments.

References

ADAIR, W. D., AND J. J. HAINS. 1973. Saturation values of dissolved gases associated with the occurrence of gas-bubble disease in a heated effluent. Pages 59–78 *in* J. W. Gibbons and R. R. Sharitz, editors. Thermal ecology. U.S. Atomic Energy Commission, Washington, D.C., USA.

BRINKHURST, R. O. 1974. The benthos of lakes. St. Martin's Press, New York, New York, USA.

BROOK, A. J., AND W. B. WOODWARD. 1956. Some observations on the effects of water inflow and outflow on the plankton of small lakes. Journal of Animal Ecology 25:22–35.

CAROLINA POWER AND LIGHT COMPANY, AND LAWLER, MATUSKY, AND SKELLY, ENGINEERS, 1981. Investigation of deformities and lowered recruitment of bluegill (*Lepomis macrochirus*) in Robinson Impoundment: H. B. Robinson Steam Electric Plant. Carolina Power and Light Company, New Hill, North Carolina, USA.

CAROLINA POWER AND LIGHT COMPANY. 1982. 1981 studies of bluegill deformities and recruitment in Robinson Impoundment: final report. Carolina Power and Light Company, New Hill, North Carolina, USA.

CAROLINA POWER AND LIGHT COMPANY. 1983. An evaluation of three methods of controlling *Egeria densa* in Hyco Reservoir. Carolina Power and Light Company, New Hill, North Carolina, USA.

CLUGSTON, J. P., D. H. BARWICK, AND T. T. FENDLEY. 1982. Management of cooling ponds and lakes for fish and wildlife—a guidance manual. U.S. Fish and Wildlife Service, FWS/OBS-82/36, Kearneysville, West Virginia, USA.

COUTANT, C. C. 1980. Environmental quality for striped bass. Pages 179–187 *in* H. Clepper, editor. Proceedings of the fifth annual marine recreational symposium: striped bass. Sport Fishing Institute, Washington, D.C., USA.

CUMBIE, P. M., AND S. L. VAN HORN. 1979. Selenium accumulation associated with fish mortality and reproductive failure. Proceedings of the Annual Conference of the Southeastern Association of Game and Fish Agencies 32:612–624.

DORN, P. B., AND J. T. JOHNSON, editors. 1981. Proceedings of the workshop of advanced intake technology for power plant cooling water systems.

Electric Power Research Institute, Palo Alto, California, USA.

DUKE POWER COMPANY. 1977. Oconee Nuclear Station Environmental Summary Report 1971–1976. Duke Power Company, Charlotte, North Carolina, USA.

EDISON ELECTRIC INSTITUTE. 1982. Statistical year book on the electric utility industry. Washington, D.C., USA.

EURE, H. E., AND G. W. ESCH. 1973. Effects of thermal effluent on the population dynamics of helminth parasites in largemouth bass. Pages 207–215 *in* J. W. Gibbons and R. R. Sharitz, editors. Thermal ecology. U.S. Atomic Energy Commission, Washington, D.C., USA.

GAREY, J. F., R. M. JORDEN, A. H. AITKEN, D. T. BURTON, AND R. H. GRAY. 1980. Condenser biofouling control-symposium proceedings. Ann Arbor Science Publishers, Inc., Ann Arbor, Michigan, USA.

JENKINS, R. M. 1967. The influence of some environmental factors on standing crop and harvest of fishes in U.S. reservoirs. Pages 298–321 *in* Reservoir fisheries resources symposium. Southern Division American Fisheries Society, Bethesda, Maryland, USA.

JENKINS, R. M. 1982. The morphoedaphic index and reservoir fish production. Transactions of the American Fisheries Society 111:133–140.

JENKINS, R. M., AND D. I. MORAIS. 1971. Reservoir sport fishing effort and harvest in relation to environmental variables. Transactions of the American Fisheries Society 107:385–394.

JENSEN, L. D., editor. 1976. Third national workshop on entrainment and impingement, section 316(b)—research and compliance. Ecological Analysts, Inc., Melville, New York, USA.

LARIMORE, R. W., AND J. A. TRANQUILLI, editors. 1981. The Lake Sangchris study: case history of an Illinois cooling lake. Illinois National History Survey Bulletin 32, Article 4, Champaign, Illinois, USA.

LARSEN, L. 1983. Bass in hot water. Outdoor Life 171(1):58–59.

LASSWELL, J. F., G. GARZA, AND W. H. BAILEY. 1979. Status of marine fish introductions into the fresh waters of Texas. Proceedings of the Annual Conference of the Southeastern Association of Fish and Wildlife Agencies 31:399–403.

LEWIS, R. E. In Press. Temperature selection and vertical distribution of striped bass during lake stratification. Proceedings of the 37th Annual Conference of the Southeastern Association of Fish and Wildlife Agencies.

LEWIS, W. M., AND R. C. HEIDINGER. 1981. Sport fishing potential of power plant cooling reservoirs. Report RP1743, Electric Power Research Institute, Palo Alto, California, USA.

MATUSZEK, J. E. 1978. Empirical predictions of fish yields of large North American lakes. Transactions of the American Fisheries Society 107:385–394.

MILLER, R. W., AND D. J. DEMONT. 1974. Fisheries research. Pages 187–216 *in* L. D. Jensen, editor. Environmental responses to thermal discharges from Marshall Steam Station, Lake Norman, North Carolina. Report 11. Electric Power Research Institute and Johns Hopkins University, Baltimore, Maryland, USA.

NORTH AMERICAN ELECTRIC RELIABILITY COUNCIL. 1982. Electric power supply and demand 1982–1991. Princeton, New Jersey, USA.

OGLESBY, R. T. 1977. Relationship of fish yield to lake phytoplankton standing crop, production and morphoedaphic factors. Journal of the Fisheries Research Board of Canada 34:2271–2279.

OLMSTED, L. L., D. J. DEGAN, J. S. CARTER, AND P. M. CUMBIE. 1986. Ash basin effluent as a concern of fisheries managers: a case history and perspective. Pages 261–269 *in* G. E. Hall and M. J. Van Den Avyle, editors. Reservoir fisheries management: strategies for the 80's. Southern Division American Fisheries Society, Bethesda, Maryland, USA.

PLOSKEY, G. R. 1981. Factors affecting fish production and fishing quality in new reservoirs, with guidance on timber clearing, basin preparation, and filling. Technical report E-81-11. U.S. Army Engineer Waterways Experiment Station, Vicksburg, Mississippi, USA.

PORAK, W., AND J. A. TRANQUILLI. 1981. Impingement and entrainment of fishes at Kincaid Generating Station. Pages 631–655 *in* R. W. Larimore and J. A. Tranquilli, editors. The Lake Sangchris study: case history of an Illinois cooling lake. Illinois Natural History Survey Bulletin 32, Article 4, Champaign, Illinois, USA.

RUELLE, R., W. LORENZEN, AND J. OLIVER. 1977. Population dynamics of young-of-the-year fish in a reservoir receiving heated effluent. Pages 46–67 *in* W. Van Winkle, editor. Proceedings of the conference on assessing the effects of power plant-induced mortality on fish populations. Pergamon Press, New York, New York, USA.

RUTLEDGE, W. P., AND B. W. LYONS. 1978. Texas peacock bass and nile perch: status report. Proceedings of the Annual Conference of the Southeastern Association of Fish and Wildlife Agencies 30:18–23.

RYAN, P. J., AND D. R. F. HARLEMAN. 1973. An analytical and experimental study of transient cooling pond behavior. Report 161. Ralph M. Parsons Laboratory for Water Resources and Hydrodynamics, Massachusetts Institute of Technology, Cambridge, Massachusetts, USA.

RYDER, R. A. 1965. A method for estimating the potential fish production of north-temperate lakes. Transactions of the American Fisheries Society 94:214–218.

RYDER, R. A. 1978. Fish yield assessment of large lakes and reservoirs—a prelude to management. Pages 403–423 *in* S. D. Gerking, editor. Ecology of freshwater fish production. Blackwell Scientific Publications, Oxford, England.

RYDER, R. A. 1982. The morphoedaphic index—use, abuse, and fundamental concepts. Transactions of the American Fisheries Society 111:154–164.

RYDER, R. A., S. R. KERR, K. H. LOFTUS, AND H. A. REGIER. 1974. The morphoedaphic index, a fish yield estimator—review and evaluation. Journal of the Fisheries Research Board of Canada 31: 663–688.

SHARMA, R. K., AND J. B. PALMER, editors. 1978. Proceedings of a workshop on larval fish exclusion systems for power plant cooling intakes. Argonne National Laboratory, NUREG/CP-002ANL/ES-66, Argonne, Illinois, USA.

SILER, J. R., AND J. P. CLUGSTON. 1975. Largemouth bass under conditions of extreme thermal stress. Pages 333–341 *in* R. H. Stroud and H. Clepper, editors. Black bass biology and management. Sport Fishing Institute, Washington, D.C., USA.

SILER, J. R., R. E. LEWIS, B. K. BAKER, G. E. VAUGHAN, AND R. A. HANSEN. 1981. Fish. Chapter 10 *in* J. E. Hogan and W. D. Adair, editors. Lake Norman Summary. Technical report Duke PWR/82-02. Duke Power Company, Charlotte, North Carolina, USA.

SILER, J. R., W. J. FORIS, AND M. C. MCINERNY. 1986. Heterogeneity in fish parameters within a reservoir. Pages 122–136 *in* G. E. Hall and M. J. Van Den Avyle, editors. Reservoir fisheries management: strategies for the 80's. Southern Division American Fisheries Society, Bethesda, Maryland, USA.

SMITHSON, J. A., K. F. KURZAWSKI, AND T. V. CLEVENGER. 1986. Management implications of thermal discharges on a largemouth bass population in a perched cooling pond. Pages 255–260 *in* G. E. Hall and M. J. Van Den Avyle, editors. Reservoir fisheries management: strategies for the 80's. Southern Division American Fisheries Society, Bethesda, Maryland, USA.

STRAWN, K. 1965. Resistance of threadfin shad to low temperatures. Proceedings of the Annual Conference of the Southeastern Association of Game and Fish Commissioners 17:290–293.

SULE, M. J., J. M. MCNURNEY, AND D. R. HALFFIELD, JR. 1981. Food habits of some common fishes from heated and unheated areas of Lake Sangchris. Pages 500–519 *in* R. W. Larimore and J. A. Tranquilli, editors. The Lake Sangchris study: case history of an Illinois cooling lake. Illinois Natural History Survey Bulletin 32, Article 4, Champaign, Illinois, USA.

TETRA TECH, INC. 1981. Assessment methodology for new cooling lakes. Vol. III. Limnological and fisheries data and bibliography. Electric Power Research Institute EPRI RP1488-1, Palo Alto, California, USA.

TETRA TECH, INC. 1982. Assessment methodology for new cooling lakes. Vol. III. Development of empirical multivariate relationships for evaluating fish communities in new cooling lakes. Electric Power Research Institute, EPRI RP1488-1, Palo Alto, California, USA.

TRANQUILLI, J. A., R. KOCHER, AND J. M. MCNURNEY. 1981. Population dynamics of the Lake Sangchris fishery. Pages 413–499 *in* R. W. Larimore and J. A. Tranquilli, editors. The Lake Sangchris Study: case history of an Illinois cooling lake. Illinois Natural History Survey Bulletin 32, Article 4, Champaign, Illinois, USA.

U.S. ENVIRONMENTAL PROTECTION AGENCY. 1975. Guidelines to determine best available technology for the location, construction, and capacity of cooling water intake structures for minimizing adverse environmental impact. Section 316(a), P.L. 92-500, Washington, D.C., USA.

U.S. ENVIRONMENTAL PROTECTION AGENCY. 1976. Development document for best technology available for the location, design, construction and capacity of cooling water intake structure for minimizing adverse environmental impact. EPA 400/1-76/015-a, Washington, D.C., USA.

VOLLENWEIDER, R. A., AND P. J. DILLON. 1974. The application of the phosphorus loading concept to eutrophication research. NRCC No. 13690. National Research Council of Canada, Burlington, Ontario, Canada.

WILDE, E. W., L. L. OLMSTED, AND A. GNILKA. 1977. Some observations concerning the effects of a power station's thermal effluent on phytoplankton dynamics. Journal of the Tennessee Academy of Science 52:10–14.

WRIGHT, J. C. 1968. Effect of impoundments on productivity, water chemistry, and heat budgets of rivers. Pages 188–199 *in* Reservoir fisheries resources symposium. Southern Division American Fisheries Society, Bethesda, Maryland, USA.

Pages 238–246 *in* G. E. Hall and M. J. Van Den Avyle, editors.
Reservoir Fisheries Management: Strategies for the 80's.
Reservoir Committee, Southern Division American Fisheries Society, Bethesda, Maryland. 1986.

An Overview of Reservoir Fisheries Problems and Opportunities Resulting from Hydropower

NORVILLE S. PROSSER

Sport Fishing Institute
608-13th Street, N.W.
Washington, D.C. 20005

ABSTRACT

Interest in hydropower in the United States is undergoing a resurgence, and some 1,900 sites have been identified for new hydropower development. Hydropower facilities affect fisheries by creating impoundments, imposing migration barriers, and by altering physio-chemical characteristics and biota of impounded water. Opportunities for recreational fisheries development are commonly associated with hydropower impoundments. Fisheries biologists must prepare management recommendations with an understanding of operational impacts on fish passage, reproduction, and survival.

In the United States, water managers have harnessed natural river flows for the production of electricity for over 100 years. Recognizing that stored water represents stored energy, the producers of hydropower have impounded streams and rivers and created many ponds and reservoirs. Today, there are 77,175 megawatts of installed hydropower, which constitutes approximately 11.4% of the United States' total installed electric power capacity (INTASA, Inc. 1981). Several advantages are associated with hydropower compared to other power production technologies. Hauck and Edson (1976) listed rapid start-up and loading, long life, low operating and maintenance costs, and low outage rates.

Capital construction costs, rising fuel costs, and perceived safety and environmental considerations associated with other electrical generation alternatives have stimulated renewed interest in hydropower construction. A recent congressionally mandated study identified 1,900 sites feasible for hydropower development and concluded that the nation has developed less than one-third of its hydroelectric capability (INTASA, Inc. 1981). Most of the sites (1,350) were located at existing dams, but most of the potential energy production was related to undeveloped sites. Between 300 and 400 of the undeveloped new sites were judged likely to be operated as storage projects. Loar and Hildebrand (1983) reported that although interests are presently focused on retrofitting existing dams for hydropower, the greater energy potentially available at undeveloped sites may attract increased attention as the United States strives for energy independence.

Hydropower impoundments can be subdivided into two types: mainstream, where the storage ratio (reservoir volume divided by annual discharge volume) is less than 0.165; and higher ratio storage reservoirs (Jenkins 1982). The operational modes include conventional hydropower, which may provide either base-load or peaking power, and pumped-storage operations, which provide peaking power production.

Operations of hydro facilities affect fishery resources in several ways. Several of the impacts noted by Hildebrand (1979) are reviewed in this paper: creation of impoundments, imposition of barriers to instream movement (physical presence of dam), and effects of operations after impoundment. Although alteration of the physical and chemical characteristics of downstream flow is a major consequence of hydropower generation, this paper addresses constraints and opportunities associated with creation and operation of the dam, appurtenant structures, and the resultant reservoir.

CREATION OF IMPOUNDMENTS

Potential power output of a hydropower plant is a function of flow and hydraulic head as well as turbine and generator efficiencies. Impoundments that are constructed to extend the period during which adequate flow is available and to increase hydraulic head alter important habitat parameters. Modification of depth, velocity, and residence time of water alters thermal characteristics, which may produce density stratification within the impoundment. Mixing within the water column is restricted during stratification, and the dissolved oxygen content of hypolimnetic waters can be depleted while the concentration of nutrients such as nitrogen and phosphorous and other elements (e.g., manganese and iron) may increase (Loar and Hildebrand 1983). Reservoirs having longer hydraulic residence and greater depth are more likely to stratify. Jenkins (1982) reported that in a sample of 294 reservoirs (representing 18% of United States reservoirs), hydropower impoundments were typically larger and deeper but had shorter average water retention times than nonhydropower reservoirs.

Changes in fish species composition and levels of production can occur as a result of physical and chemical changes induced by impounding free-flowing streams. Hildebrand (1979) noted that some riverine species may disappear from the reservoir environment whereas others might survive at substantially different levels of abundance. Hydroelectric projects located on warmwater streams usually result in an increase in total fish standing crop over that of the free-flowing streams, but on coldwater streams, the populations (primarily salmonids) generally decline (Ebel 1979).

Hydropower reservoirs may exhibit high initial sport fish standing crops immediately after impoundment followed by diminished game fish populations in subsequent years (Ploskey 1981). Initially high reservoir productivity has been attributed to retention of allochthonous and autochthonous materials during initial impoundment, release of nutrients from inundated soils, and availability of organic material from inundated substrates and vegetation.

Many hydropower projects provide an opportunity to diversify fish stocks available to recreational anglers. Introductions of striped bass (*Morone saxatilis*) in southern and southwestern reservoirs (Bailey 1975) and *Morone* hybrids (Ware 1975) are notable examples. Predator stocking in hydropower impoundments often has been stimulated by real or perceived availability of excessive limnetic prey species, especially gizzard shad (*Dorosoma cepedianum*). Jenkins (1979) sounded a cautionary note for introduction of predators in impoundments of the southern United States by noting that available food is the principal factor limiting predator production in reservoirs. Upon analysis of available prey to predator biomass ratios in 26 southern reservoirs, Jenkins and Morais (1978) concluded that half of the reservoirs were deficient in available prey and that stocking of additional predators should be discontinued in those cases. A recent review of forage fish management included applications in hydropower reservoirs (Ney 1981).

The availability of reservoirs having areas with low temperature and adequate dissolved oxygen has permitted the establishment of put-grow-and-take, two-story fisheries for salmonids including the rainbow trout (*Salmo gairdneri*) and brown trout (*Salmo trutta*) (Oliver et al. 1979). At Lake Taneycomo, a 700-hectare hydroelectric impoundment in Missouri, an excellent rainbow trout fishery was established and supported by the hypolimnetic discharge from upstream Table Rock Lake. The Taneycomo trout fishery would not exist without Table Rock hydropower releases.

Reservoirs, including those constructed for purposes of generating electricity, annually support millions of man-days of recreational angling. In fact, over 40% of all anglers fished in reservoirs at some time during 1980 (USFWS 1982). Expanded angling opportunities have obviously resulted from the construction of impoundments to operate hydropower plants. As noted by Eicher (1983), in the preponderance of cases, the greater water area of an impoundment can provide significantly greater fishing opportunities than provided by the unimpounded stream.

To enhance the recreational use of hydropower impoundments, fishery managers, in cooperation with the operating agency, can implement a host of management activities including protection of areas of standing timber, timing of filling, artificial shelters, water level manipulation, access, and stocking. Several literature reviews and guidance manuals discuss such practices (Nelson et al. 1978; Leedy et al. 1980; Ploskey 1981; and Clugston et al. 1982).

Barriers to Migration

Upstream Migrants

Construction and operation of hydropower facilities pose difficulties for the movement of migratory and resident fish stocks. For migratory species requiring access to upstream spawning habitat, physical obstructions created by dams and turbulence from power plant discharges can have significant consequences. Fishery bioengineering has developed technology for passing upstream migrants around dams. Three basic types of bypass systems are currently in use: fishways that allow fish to swim through a series of progressively higher pools, locks where fish are crowded into a chamber which is then filled with water to raise the fish over the dam, and water-filled lifts that are mechanically raised to release fish above the dam (Loar and Hildebrand 1983). Trap-and-haul methods have also been developed which are usually low cost and particularly desirable where distance is a factor (Eicher 1983). Most of the research and engineering on fish passage has taken place in the Pacific Northwest. Much of the available information on fishways has been described by Clay (1961) and Bell (1973).

The entrance is the most important feature of a fishway. It must have sufficient flow to attract fish and cannot be too far downstream from the barrier or too far from the main stream flow (Nelson et al. 1978). Inadequate flow in and near the fishway at John Day Dam on the Columbia River delayed upstream migration of adult salmonids, increased exposure to waters supersaturated with atmospheric gases, and resulted in substantial mortality (Beiningen and Ebel 1970). Gas supersaturation is caused by entrainment of air by water as it falls into a plunge basin. The supersaturation problem has been severe in the Pacific Northwest where there were 58 major hydroelectric dams in 1976 (INTASA, Inc. 1981). Raymond (1979) observed that in the Columbia River system during years of high flow, most juvenile mortality resulted from prolonged exposure to high concentrations of dissolved gases.

Actions which may be taken to resolve gas supersaturation problems at hydropower dams include reducing spillage during critical periods of fish migration, adjusting generation schedules to pass water through turbines rather than spill over the dam, and modifying spillways to reduce the plunge depth (Hildebrand 1979). Weitkamp and Katz (1980) reviewed the extensive literature on dissolved gas supersaturation and presented a summary of steps that have been employed to overcome the problem.

Hauck and Edson (1976) suggested that pumped-storage projects can create added problems by attracting upstream migrants and resident species in reservoirs to the powerhouse during the generating mode. This attraction could delay upstream migration.

Downstream Migrants

Safe passage of downstream migrant juveniles, eggs, and larvae has proven even more problematical than passage of adults at hydropower dams. Raymond (1979) reported that survival of juvenile chinook salmon (*Oncorhynchus tshawytscha*) from the Salmon River to Ice Harbor Dam on the lower Snake River declined from about 85% to 33% following installation of two additional mainstream hydropower reservoirs having inadequate juvenile fish passage facilities. Losses of young fish during out-migration may result from predation and other losses within a reservoir. At Brownlee Reservoir in Idaho, juvenile salmon and steelhead did not move successfully through the 92-km-long reservoir, and the $3.5 million downstream migrant collection facility located directly above the dam was abandoned (Leedy et al. 1980).

Mortalities from passage of fish through hydropower turbines have been documented to range from a few percent per dam to as high as 30% per dam (Loar 1983). Mortality generally is greater during low water years, when the amount of spill is minimal and all river flow is passed through turbines. Two alternatives have been developed to improve downstream passage of fish through hydroelectric facilities: 1) provide turbines that are relatively safe for passage of fish; and 2) install screens to bypass fish around turbines (Eicher 1983).

Francis and Kaplan turbines were commonly installed in hydropower projects of the Northwest. More recently, technological advancements have led to design and utilization of the bulb, tube, and Ossberger crossflow turbines, which can provide improved passage (Gloss and Wahl 1983). Knapp (1981) suggested that out-migration might be facilitated at hydropower projects by installing multiple turbines of different designs and by operating turbines that result in minimal fish loss, either by design or screening facilities, during critical out-migration periods. Gloss and Wahl (1983) demonstrated a size-related mortality caused by Ossberger turbines and recommended that smolts be released at the smallest size possible to minimize turbine losses.

The operation of pumped-storage facilities can attract downstream migrants and resident species to the pumping plant if waterflows are of sufficient velocity and volume to compete with downstream waterflows. As a consequence, the attracted fish can be pumped into the upper reservoir (Hauck and Edson 1976). Some mortality has been noted as a consequence of passing fish through such pumping units (Robbins and Mathur 1976).

Operational Impacts of Hydroelectric Facilities

Operations of hydroelectric facilities affect the physiochemical properties of impounded water and biological communities. The impacts are related to water level changes, the location and operation of withdrawal facilities, and the shape of the reservoir basin. Run-of-river projects with minimal storage have less impact on the original stream's biological community than do storage projects that alter the extent and frequency of natural fluctuations (Loar and Hildebrand 1981). Use of impounded water to generate electricity can affect reproduction of fish and benthos, energy transfer and recruitment, and entrainment of nutrients, fish, and other biota through the dam. Planned manipulation of water storage and release can be an effective management tool, and the practice is receiving increasing attention by fishery biologists.

Fish Reproduction

High water levels during the spawning period produce environmental conditions favoring reproductive success for many species of fish. Abundant food and refuge from predation due to the presence of submerged terrestrial vegetation during spring and summer have been identified as contributing to greater abundance of young-of-year black basses (*Micropterus* spp.) (von Geldern 1971; Aggus and Elliott 1975; Houser and Rainwater 1975; Martin et al. 1981; Clugston et al., 1982; Strange et al. 1982). Keith (1975) recommended that terrestrial vegetation be inundated immediately prior to spawning to obtain maximum results. Other investigators have reported that inundated vegetation plays a major role in reproductive success of northern pike (*Esox niger*) (Hassler 1970; Groen and Schroeder 1978) and white crappie (*Pomoxis annularis*) (Beam 1983).

Other studies have demonstrated that increases in the volume of stored water during spawning periods produce increased numbers of young bass even in the absence of submerged terrestrial vegetation (Lambou 1959; Pierce et al. 1965). Bennett (1970) suggested that the reduction of sunfish during a period of drawdown may result in high survival of bass at the next bass spawning season. Smith (1976) reported that inhibition of largemouth bass spawning was linked to interactions with overly abundant forage species. Increasing the area of standing water habitat during spawning periods benefits reproductive success as the aquatic community expands to fill the newly available and sparsely populated habitat. This increase in young organisms is typical of newly formed impoundments and older impoundments where mortality or wash-out during water level reductions have reduced standing crops.

Receding water levels during spawning may result in desertion of nests, disruption of spawning, and direct egg mortality (Ploskey 1981). Species differ with regard to the frequency and magnitude of water level fluctuation that can be tolerated before production of young is affected. These differences depend upon the life cycles of individual species, including such factors as length of spawning season, number of spawning attempts per season, duration of incubation, and spawning depth.

Largemouth bass (*Micropterus salmoides*) and some other species are capable of adapting to substantial water level variations (Schoumacher 1976), and spawning success adequate to sustain populations within fluctuating reservoir systems has been documented in several cases (Kimsey 1957; Jester 1971; Wegener and Williams 1975; Robbins and Mathur 1976). In one case, no direct relationship was found between water-level fluctuations of 2 to 3 m per week and spawning success of bass and panfish in Leesville Reservoir, a pumped-storage impoundment in Virginia (Estes 1971). Water withdrawals at pumped-storage projects may be of sufficient magnitude to limit the reproduction of nest-building fishes, as was reported for the Muddy Run Pumped Storage Project, where the upper impoundment is operated with a maximum drawdown of 15.2 m per week (Robbins and Mathur 1976).

Declining water levels and natural reproduction are incompatible for littoral spawning species with long incubation and emergence times such as kokanee (*Oncorhynchus nerka*), lake whitefish (*Coregonus clupeaformis*), and burbot (*Lota lota*) (Stober 1983). Declining water levels have also been correlated with reduced spawning success for many important prey species including gizzard shad, minnows (Cyprinidae), white bass (*Morone chrysops*), white crappie, and yellow perch (*Perca flavescens*) (Keith 1975; Groen and Schroeder 1978).

Survival and Recruitment

Operation of hydropower impoundments to foster successful spawning is no guarantee of subsequent survival to yearlings the following year. For instance, neither total bass biomass nor density was correlated with water levels during the spawning season at Beaver Lake, Arkansas, despite a positive relationship between spring water level and number of young-of-year bass (Rainwater and Houser 1982). Jenkins (1979) concluded that the numbers of young predators successfully spawned is not as important to ultimate year class strength as is availability of suitable food and resulting growth and condition.

As noted by Ney (1981), production of fishes at higher trophic levels depends upon efficient energy transfer from lower trophic levels. Elemental plant nutrients and detrital materials resident within the lake basin and flowing into a reservoir from the watershed are the base of lower trophic level production. Jenkins (1974) reported that under circumstances of similar total dissolved solids, reservoirs with higher exchange rates (shorter hydraulic residence), such as mainstream hydropower reservoirs, supported larger standing crops of fish. Aggus and Lewis (1978) suggested that the total quantity of nutrients passing through a reservoir may be of greater importance to production than nutrient concentration.

Transfer of energy through algal production or detrital pathways to higher trophic levels may depend on substrate conditions and water quality within the littoral zones of reservoirs. Planktonic algae utilize nutrients supplied by inflowing water and nutrients released by microbial decomposition of submerged organic material. Reservoir drawdown, which exposes unconsolidated substrate materials to oxidation, and subsequent reflooding can increase the availability of nutrients (Wegener and Williams 1975; Noble 1981). However, hydropower operations that fluctuate reservoir water levels may cause wave action to erode unconsolidated shoreline deposits and cause increased turbidity, which can reduce sunlight penetration and impair biological production (Bulkley 1975; Ploskey 1981; Jenkins 1982). A positive relationship between phytoplankton standing crop (measured as chlorophyll *a*) and yield of sport fishes was demonstrated by Jones and Hoyer (1982). Gross photosynthesis and fish yield were highly correlated in studies by McConnell et al. (1977).

Water clarity can be improved if reservoir littoral zones are stabilized and protected from wave action erosion by rooted vegetation (Groen and Schroeder 1978; Beam 1983). Inundated herbaceous vegetation provides an important source of nutrients for algae and detritivores, which increases the food supply to higher trophic levels (Ploskey 1981). Strange et al. (1982) reported greater quantities of zooplankton in early summer in inundated terrestrial vegetation. Allan and Romero (1975) observed greater numbers of zooplankton in littoral areas and in the diets of largemouth fingerlings in areas of inundated terrestrial vegetation in Lake Mead. Aggus (1971) reported greater densities of benthos in association with flooded herbaceous plants.

The benefits of submerged shoreline vegetation to water clarity, autochthonous production, and detrital food pathways should be recognized and employed by fishery biologists to recommend seeding of exposed shorelines at hydropower reservoirs during spring and/or summer growing seasons. This practice has resulted in higher levels of recruitment and improved sport fisheries (Groen and Schroeder 1978; Beam 1983). Ploskey and Jenkins (1982) suggested use of water-level manipulation to inundate terrestrial vegetation to increase production of food organisms in DeGray Lake, Arkansas.

Growth of young piscivores prior to their first winter of life is important in establishing subsequent year class strength (Aggus and Elliot 1975; Oliver and Holeton 1979; Timmons et al. 1981). Kimsey (1957), Shelton et al. (1979), and Ploskey and Jenkins (1982) emphasized the favorable influence of an abundance of available prey to predators in late summer and fall to overwinter survival. Noble (1981) and Wegener and Williams (1975) suggested fall drawdowns to concentrate prey, thereby making them more vulnerable to predation as a means to improve growth and survival of piscivorous fish. Large drawdowns at the Muddy Run Pumped Storage Project made gizzard shad more vulnerable to predation, leading to accelerated growth of smallmouth bass (Heisey et al. 1980). Dewatering may force prey organisms out of submerged littoral vegetation. Submerged vegetation has been demonstrated to reduce predation mortality of juvenile bluegills by largemouth bass (Savino and Stein 1982). Ploskey (1981) cautioned that use of fall drawdown should be based upon estimates of total standing crop and available prey.

Entrainment Losses

The depth and configuration of the intakes that control flow into penstocks and turbines can influence the physical and chemical characteristics of the impoundment as well as the loss of reservoir biomass through the dam. Epilimnetic discharge may remove plankton and reduce reservoir productivity (Hildebrand 1979). Temperature and dissolved oxygen concentrations in impounded water may lead to concentration of prey and/or predator species in refugia at certain times of the year (Coutant and Carroll 1980). Losses would be quite severe if water for hydropower generation were withdrawn from refugia.

Aspects of outlet bioengineering are complex and site-specific. However, in general, the ability to selectively withdraw water from various depths is important to reservoir management because this allows operational strategies that minimize losses of nutrients, biota, or water having desirable chemical and physical characteristics.

Conclusion

Development of water resources for hydropower generation has beneficial and damaging aspects with regard to reservoir fisheries. Artificial impoundments, including those created for purposes of hydropower generation, have contributed significantly to the United States' recreational fishing resource base. However, hydropower generation may entail competing uses of water which pose difficult challenges for fishery managers. To minimize such conflicts, which may intensify as greater demands are placed on water supplies for power generation, fisheries biologists must improve their understanding of fundamental reservoir limnology and ecology. Only then can they place priorities on their management objectives and negotiate for construction and operational strategies that result in predictable fishery compensation. Hydropower proponents should adopt a willingness to provide the greatest possible flexibility in project design and operation so as to accommodate fishery needs.

The Pacific Northwest Electric Power Planning and Conservation Act (P.L. 96-501), which passed in December 1980, represented a significant forward step in hydropower-fisheries interactions. The Act requires the conservation and restoration of the Columbia River Basin's fish and wildlife resources that have been adversely affected by hydroelectric development and operations. This congressionally mandated guidance for responsible natural resources management, including fisheries, should serve as a protocol leading to similar commitments by the hydropower industry throughout the country.

References

Aggus, L. R. 1971. Summer benthos in newly flooded areas of Beaver Reservoir during the second and third years of filling 1965–1966. Pages 139–152 *in* G. E. Hall, editor. Reservoir fisheries and limnology. Special Publication 8, American Fisheries Society, Bethesda, Maryland, USA.

Aggus, L. R., and G. V. Elliott. 1975. Effects of cover and food on year-class strength of largemouth bass. Pages 317–322 *in* R. H. Stroud and H. Clepper, editors. Black bass biology and management. Sport Fishing Institute, Washington, D.C., USA.

Aggus, L. R., and S. A. Lewis. 1978. Environmental conditions and standing crops of fishes in predator-stocking-evaluation reservoirs. Proceedings of the Annual Conference of the Southeastern Association of Fish and Wildlife Agencies 30:131–140.

Allan, R. C., and J. Romero. 1975. Underwater observations of largemouth bass spawning and survival in Lake Mead. Pages 104–112 *in* R. H. Stroud and H. Clepper, editors. Black bass biology and management. Sport Fishing Institute, Washington, D.C., USA.

Bailey, W. M. 1975. An evaluation of striped bass introduction in the southeastern United States. Proceedings of the Annual Conference of the Southeastern Association of Game and Fish Commissioners 28:54–68.

Beam, J. H. 1983. The effect of annual water level management on population trends of white crappie in Elk City Reservoir, Kansas. North American Journal of Fisheries Management 3:34–40.

Beiningen, K. T., and W. J. Ebel. 1970. Effect of John Day Dam on dissolved nitrogen concentrations and salmon in the Columbia River, 1968. Transactions of the American Fisheries Society 99:664–761.

Bell, M. C. 1973. Fisheries handbook of engineering requirements and biological criteria. Contract report DACW75-68-C-0086. U.S. Army Corps of Engineers, North Pacific Division, Portland, Oregon, USA.

Bennett, G. W. 1970. Management of lakes and ponds. Van Nostrand Reinhold, New York, New York, USA.

Bulkley, R. V. 1975. Chemical and physical effects on centrarchid basses. Pages 286–294 *in* R. H. Stroud and H. Clepper, editors. Black bass biology and management. Sport Fishing Institute, Washington, D.C., USA.

Clay, C. H. 1961. Design of fishways and other fish facilities. Queen's Printer, Ottawa, Canada.

Clugston, J. P., D. H. Barwick, and T. T. Fendley. 1982. Management of cooling ponds and lakes for fish and wildlife. U.S. Fish and Wildlife Service FWS/OBS-82/36, Kearneysville, West Virginia, USA.

Coutant, C. C., and D. S. Carroll. 1980. Temperature selection by ten ultrasonic-tagged striped bass in freshwater lakes. Transactions of the American Fisheries Society 109:195–202.

Ebel, W. 1979. Effects of hydroelectric projects on fish populations. Pages 170–176 *in* Hydropower: a national energy resource. U.S. Government Printing Office, Washington, D.C., USA.

Eicher, G. J. 1983. Fisheries enhancement from hydroelectric development. Pages 72–77 *in* G. A. Barnhart, editor. Hydropower and fisheries: impacts and opportunities. Northeastern Division American Fisheries Society and New York State Department of Environmental Conservation, Albany, New York, USA.

Estes, R. D. 1971. The effects of the Smith Mountain Pump Storage Project on the fisheries of the lower reservoir, Leesville, Virginia. Doctoral dissertation, Virginia Polytechnic Institute and State University, Blacksburg, Virginia, USA.

Gloss, S. P., and J. R. Wahl. 1983. Mortality of juvenile salmonids passing through Ossberger Crossflow turbines at small-scale hydroelectric sites. Transactions of the American Fisheries Society 112:194–200.

Groen, C. L., and T. A. Schroeder. 1978. Effects of water level management on walleye and other coolwater fish in Kansas reservoirs. Pages 278–283 *in* R. L. Kendall, editor. Selected coolwater fishes of North America. Special publication 11, American Fisheries Society, Bethesda, Maryland, USA.

Hassler, T. J. 1970. Environmental influences on early development and year-class strength of northern pike in Lakes Oahe and Sharpe, South Dakota. Transactions of the American Fisheries Society 99:369–375.

Hauck, F. R., and Q. A. Edson. 1976. Pumped-storage: its significance as an energy source and some biological ramifications. Transactions of the American Fisheries Society 105:158–164.

Heisey, P. G., D. Mathur, and N. C. Magnusson. 1980. Accelerated growth of smallmouth bass in a pumped-storage system. Transactions of the American Fisheries Society 109:371–377.

Hildebrand, S. G. 1979. Potential environmental impacts of hydroelectric development: an overview. Pages 322–329 *in* Hydropower: a national energy resource. U.S. Government Printing Office, Washington, D.C., USA.

Houser, A., and W. C. Rainwater. 1975. Production of largemouth bass in Beaver and Bull Shoals Lakes. Pages 310–316 *in* R. H. Stroud and H. Clepper, editors. Black bass biology and management. Sport Fishing Institute, Washington, D.C., USA.

INTASA, Inc. 1981. National hydroelectric power resources study. Report DACW72-80-C-0002. U.S. Army Corps of Engineers Institute for Water Resources, Ft. Belvoir, Virginia, USA.

Jenkins, R. M. 1974. Reservoir management prognosis: migraines or miracles. Proceedings of the Annual Conference of the Southeastern Association of Game and Fish Commissioners 27:374–385.

Jenkins, R. M. 1979. Predator-prey relations in reservoirs. Pages 123–134 *in* R. H. Stroud and H. Clepper, editors. Predator-prey systems in fisheries management. Sport Fishing Institute, Washington, D.C., USA.

Jenkins, R. M. 1982. The morphoedaphic index and reservoir fish production. Transactions of the American Fisheries Society 111:133–140.

Jenkins, R. M., and D. J. Morais. 1978. Prey-predator relations in the predator-stocking-evaluation reservoirs. Proceedings of the Annual Conference of the Southeastern Association of Fish and Wildlife Agencies 30:141–157.

Jester, D. B. 1971. Effects of commercial fishing, species introduction, and drawdown control on fish populations in Elephant Butte Reservoir, New Mexico. Pages 265–285 *in* G. E. Hall, editor. Reservoir fisheries and limnology. Special publication 8, American Fisheries Society, Bethesda, Maryland, USA.

Jones, J. R., and M. V. Hoyer. 1982. Sportfish harvest predicted by summer chlorophyll *a* concentration in midwestern lakes and reservoirs. Transactions of the American Fisheries Society 111:176–179.

Keith, W. E. 1975. Management by water level manipulation. Pages 489–497 *in* R. H. Stroud and H. Clepper, editors. Black bass biology and management. Sport Fishing Institute, Washington, D.C., USA.

Kimsey, J. B. 1957. Fisheries problems in impounded waters of California and the lower Colorado River. Transactions of the American Fisheries Society 87:319–332.

Knapp, W. E. 1981. Hydropower 1980: implications for fishery science and management. Paper presented at the 37th Northeast Fish and Wildlife Conference, Virginia Beach, Virginia, USA.

Lambou, V. 1959. Fish populations in backwater lakes in Louisiana. Transactions of the American Fisheries Society 88:7–15.

Leedy, D. L., L. E. Dove, and T. M. Franklin. 1980.

Compatibility of fish, wildlife, and floral resources with electric power facilities. Edison Electric Institute, Washington, D.C., USA.

LOAR, J. M. 1983. Impacts of hydropower development on downstream fish passage. Pages 25–33 *in* G. A. Barnhart, editor. Hydropower and fisheries: impacts and opportunities. Northeastern Division American Fisheries Society and New York State Department of Environmental Conservation, Albany, New York, USA.

LOAR, J. M., AND S. G. HILDEBRAND. 1983. A comparison of environmental issues related to development of small hydropower resources at new versus existing sites. Pages 279–297 *in* T. N. Veziroglu, editor, Alternative Energy Sources III, Volume 9, Policy/Environment. Hemisphere Publishing Corporation, Washington, D.C., USA.

MARTIN, D. B., L. J. MENGEL, J. F. NOVOTNY, AND C. H. WALBURG. 1981. Spring and summer water levels in a Missouri River reservoir: effects on age-0 fish and zooplankton. Transactions of the American Fisheries Society 110:370–381.

MCCONNELL, W. J., S. LEWIS, AND J. E. OLSON. 1977. Gross photosynthesis as an estimator of potential fish production. Transactions of the American Fisheries Society 106:417–423.

NELSON, R. W., G. C. HORAK, AND J. E. OLSON. 1978. Western reservoir and stream habitat improvements handbook. U.S. Fish and Wildlife Service FWS/OBS-78/56, Ft. Collins, Colorado, USA.

NEY, J. J. 1981. Forage-fish management in the United States. Transactions of the American Fisheries Society 110:725–781.

NOBLE, R. L. 1981. Management of forage fishes in impoundments of the southern United States. Transactions of the American Fisheries Society 110:738–750.

OLIVER, J. D., AND F. G. HOLETON. 1979. Overwinter mortality of fingerling smallmouth bass in relation to size, relative energy stores, and environmental temperatures. Transactions of the American Fisheries Society 108:130–136.

OLIVER, J. L., P. L. HUDSON, AND J. P. CLUGSTON. 1979. Effects of a pumped-storage hydroelectric plant on reservoir trout habitat. Proceedings of the Annual Conference of the Southeastern Association of Fish and Wildlife Agencies 31:449–457.

PIERCE, P. C., J. E. FREY, AND H. M. YAWN. 1965. An evaluation of fishery management techniques utilizing winter drawdown. Proceedings of the Annual Conference of the Southeastern Association of Game and Fish Commissioners 17:347–363.

PLOSKEY, G. R. 1981. Factors affecting fish production and fishing quality in new reservoirs, with guidance on timber clearing, basin preparation, and filling. Technical report E-81-11. U.S. Army Corps of Engineers, Waterways Experiment Station, Vicksburg, Mississippi, USA.

PLOSKEY, G. R., AND R. M. JENKINS. 1982. Biomass model of reservoir fish and fish-food interactions, with implications for management. North American Journal of Fisheries Management 2:105–121.

RAINWATER, W. C., AND A. HOUSER. 1982. Species composition and biomass of fish in selected coves in Beaver Lake, Arkansas, during the first 18 years of impoundment (1963–1980). North American Journal of Fisheries Management 4:316–325.

RAYMOND, H. L. 1979. Effects of dams and impoundments on migrations of juvenile chinook salmon and steelhead from the Snake River, 1966 to 1975. Transactions of the American Fisheries Society 108:505–541.

ROBBINS, T. W., AND D. MATHUR. 1976. The Muddy Run Pumped Storage Project: a case history. Transactions of the American Fisheries Society 105:165–172.

SAVINO, J. F., AND R. A. STEIN. 1982. Predator-prey interactions between largemouth bass and bluegills as influenced by simulated submersed vegetation. Transactions of the American Fisheries Society 111:255–266.

SCHOUMACHER, R. 1976. Biological considerations of pumped-storage development. Transactions of the American Fisheries Society 105:155–157.

SHELTON, W. L., W. D. DAVIES, T. A. KING, AND T. J. TIMMONS. 1979. Variations in the growth of the initial year class of largemouth bass in West Point Reservoir, Alabama and Georgia. Transactions of the American Fisheries Society 108:142–149.

SMITH, S. L. 1976. Behavioral suppression of spawning in largemouth bass by interspecific competition for space within spawning areas. Transactions of the American Fisheries Society 105:682–685.

STOBER, Q. J. 1983. Overview of fisheries management in fluctuating reservoirs in the northwestern United States. Pages 26–27 *in* H. H. Allen and L. R. Aggus, editors. Effects of fluctuating reservoir water levels on fisheries, wildlife, and vegetation. Miscellaneous paper E-83-2, U.S. Army Corps of Engineers, Waterways Experiment Station, Vicksburg, Mississippi, USA.

STRANGE, R. J., W. B. KITTRELL, AND T. D. BROADBENT. 1982. Effects of seeding reservoir fluctuation zones on young-of-the-year black bass and associated species. North American Journal of Fisheries Management 2:307–315.

TIMMONS, T. J., W. L. SHELTON, AND W. D. DAVIES. 1981. Early growth and mortality of largemouth bass in West Point Reservoir, Alabama-Georgia. Transactions of the American Fisheries Society 110:489–494.

USFWS (U.S. DEPARTMENT OF INTERIOR, FISH AND WILDLIFE SERVICE, AND U.S. DEPARTMENT OF COMMERCE, BUREAU OF THE CENSUS). 1982. 1980 National survey of fishing, hunting and wildlife-associated recreation. U.S. Government Printing Office, Washington, D.C., USA.

VON GELDERN, C. E., JR. 1971. Abundance and distribution of fingerling largemouth bass, *Micropterus salmoides,* as determined by electrofishing at

Lake Nacimiento, California. California Fish and Game 57:228–245.

WALBURG, C. H. 1976. Changes in the fish population of Lewis and Clark Lake, 1956–1974, and their relation to water management and the environment. Research report 79. U.S. Fish and Wildlife Service, Washington, D.C., USA.

WARE, F. J. 1975. Progress with *Morone* hybrids in freshwater. Proceedings of the Annual Conference of the Southeastern Association of Game and Fish Commissioners 28:48–54.

WEGENER, W., AND V. WILLIAMS. 1975. Fish population responses to improved lake habitat utilizing an extreme drawdown. Proceedings of the Annual Conference of the Southeastern Association of Game and Fish Commissioners 28:144–160.

WEITKAMP, D. E., AND M. KATZ. 1980. A review of dissolved gas supersaturation literature. Transactions of the American Fisheries Society 109:659–702.

Pages 247–254 *in* G. E. Hall and M. J. Van Den Avyle, editors.
Reservoir Fisheries Management: Strategies for the 80's.
Reservoir Committee, Southern Division American Fisheries Society, Bethesda, Maryland. 1986.

Fisheries Problems Associated with the Truman Dam Pumped Storage Hydroelectric Project in West Central Missouri

KEVIN R. RICHARDS, RONALD J. DENT, JR., AND WILLIAM H. DIEFFENBACH
Missouri Department of Conservation
P.O. Box 180
Jefferson City, Missouri 65102

ABSTRACT

Harry S. Truman Dam was completed in 1977 to provide flood control, recreation, and pumped storage hydropower production. An excellent tailwater fishery developed immediately below Truman Dam; however, problems associated with the project have adversely impacted many aspects of the fishery of Lake of the Ozarks. Spawning migrations of several species have been blocked, and reproductive success of species able to spawn downstream from the dam has been affected by water level fluctuations and variable flows. Fish kills have occurred due to gas supersaturation, low dissolved oxygen concentrations, and periodic dewatering of the turbine draft tubes for maintenance purposes. Mortality caused by pumpback also is a severe problem, and is being investigated in detail. Less obvious, long-term impacts of the operation of Truman Dam are also being studied. Placement of this type of pumped storage hydroelectric facility between two large warm-water reservoirs is unique. Hopefully, fishery problems documented at the Truman Dam project can be avoided in future energy development projects.

Harry S. Truman Reservoir is a U.S. Army Corps of Engineers (USACE) multi-purpose project on the Osage River in west central Missouri (Fig. 1). It covers 55,600 surface acres at normal conservation pool and 209,300 acres at maximum flood pool, making it the largest reservoir in the state. At conservation pool Truman Reservoir has a maximum depth of 74 ft and a mean depth of 22 ft. It is located immediately upstream from the second largest reservoir in Missouri, Lake of the Ozarks, which covers 55,000 acres at normal conservation pool. Lake of the Ozarks was created in 1931 by the completion of Bagnell Dam across the Osage River, approximately 93 mi downstream from Truman Dam.

Lake of the Ozarks and Bagnell Dam are owned by Union Electric Company; the adjacent land is privately owned, with extensive residential and recreational development. Lake of the Ozarks has developed into one of Missouri's most important tourist and recreational areas. From 1976 through 1980 the lake provided over 2 million hours of fishing annually for visitors (Richards 1982).

Truman Reservoir was authorized as a single-purpose, flood control project in 1954; it became a multiple-purpose project in 1962 when Congress added hydropower and recreation as purposes. The 1962 recommendations for operation specified a total project output of 100,000 kw using three generators. The maximum discharge through each turbine would have been 11,000 cubic feet per second (cfs). Studies to determine the final scope of hydropower continued after Congress provided initial funding for construction.

In 1966, the project was modified to a pumped storage hydropower plant that utilized six reversible pump-turbines with a firm capacity of 160,000 kw with releases of 55,000 cfs to 65,000 cfs. The power plant is unique in the United States and Canada (IEEE 1976) as it is the only one to utilize an inclined-axis design for the turbine/generator assembly (Fig. 2). Other unusual features include its location on the headwater area of Lake of the Ozarks, and use of Lake of the Ozarks as an afterbay.

Construction of Truman Dam started in 1964. The Osage River channel was closed in 1977 upon completion of the dam. The spillway gates were closed in fall 1979 and the lake was allowed to fill to conservation pool; the first generating unit came on line in August 1981. The pumpback

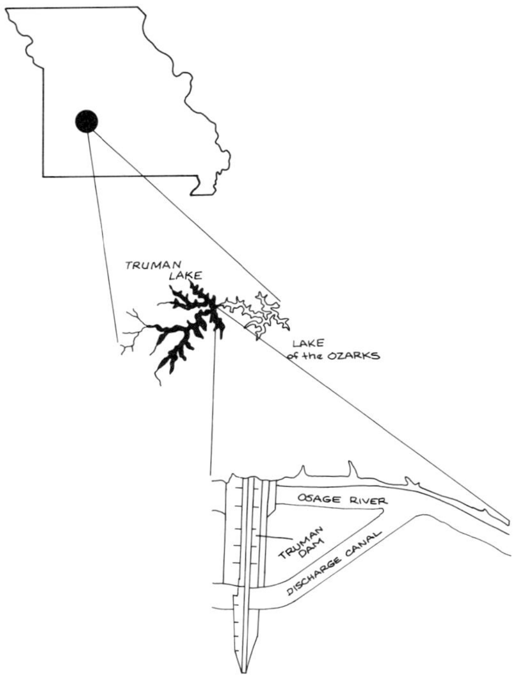

Figure 1. Location of Harry S. Truman Reservoir, Lake of the Ozarks, and Harry S. Truman tailwater area, Missouri.

capability of the project was tested in April 1982, but had not been operated again prior to the preparation of this report.

Concerns regarding the effects of construction and operation of the Truman project on the fishery of Lake of the Ozarks were well documented prior to project completion (Hanson 1975; Sparrowe 1977; Russell et al. 1980). The fisheries and environmental problems that have been observed at this project are directly associated with the location, design, and planned operational pattern of Truman Dam. If the project had been constructed and operated at the level approved in 1962, some fisheries problems relating to the physical barrier of the dam and fluctuations of water levels, velocities, and volumes would likely have occurred, since such problems are characteristic of many flood control and hydroelectric projects (Hildebrand 1980; Turbak et al. 1980; Walburg et al. 1981). However, the 1966 decision to increase the generating capacity of the project and to incorporate pumped storage capability increased the scope and magnitude of the problems. The purpose of this paper is to discuss the proposed operation and many of the

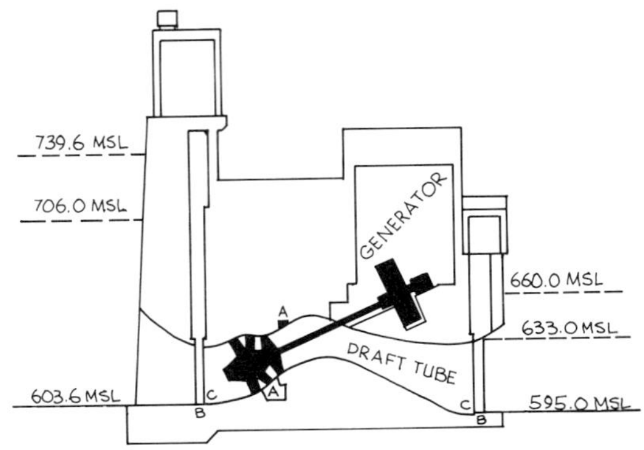

Figure 2. Cross-sectional diagram of Harry S. Truman Dam, showing locations of access hatches (A), bulkhead slots (B), and fish rescue areas during dewatering (C).

problems that have developed at Truman Dam, with the hope that similar problems can be avoided at future hydroelectric energy development projects.

Proposed Operation

The proposed operation of the Truman project is described in detail by USACE (1980). The following is a summary of that proposed operation and its predicted hydrological effects.

Flood control operation.—The spillway has four tainter gates used to release water from the reservoir. The spillway is not used except during periods of high inflow for flood control.

Hydropower generation.—The power plant will generate an average of 309 million kw hours of electrical energy annually. A large portion of the energy would be produced as peaking power from June through September. During this period generation would average 7.5 hours per day, 5 days per week. Hydropower releases of 55,000–65,000 cfs would produce 160,000–184,000 kw.

Pumpback.—Turbines would be reversed and water would be drawn from Lake of the Ozarks when power releases exceed inflow. Peak pumpback would occur at a rate of 27,500 cfs for 8.5 hours per week night and 40 hours on weekends from June through September.

Hydrological effects.—During a typical day of maximum power generation the water level immediately below the dam will rise 5 to 6 ft.

The rise is expected to be less than 1 ft 30 mi downstream. When Lake of the Ozarks is at full pool and Truman Dam is at capacity generation, water velocities in the upper 4 mi of Lake of the Ozarks will range from 3.8 to 5.1 ft/second. This is a significant increase from non-flood velocities prior to Truman project construction which averaged less than 1.0 ft/second, with a mean daily flow of 2,500 cfs. For comparison, the flood of July 1967 produced a flow of 67,000 cfs with a velocity of 4.4 ft/second. During a typical night of maximum pumpback operation the water level immediately below the dam will decrease approximately 2 ft; the decrease will be negligible 30 mi downstream. Velocities during pumpback will range from 2.0 to 2.6 ft/second with Lake of the Ozarks at full pool. Velocities during generation and pumpback will be greater when water levels at Lake of the Ozarks are lower.

Some expected downstream effects of operation described in the USACE (1980) report include erosion of banks subjected to increased flows, erosion of banks and sediment bars subjected to inundation, changes in aquatic and terrestrial vegetation, and potential safety hazards to unskilled boaters. Each of these effects could potentially impact the downstream fishery. A study by an independent consultant (Ariathurai 1980) predicted future sedimentation and erosion that may be quite severe in some areas of Lake of the Ozarks. Some bank and bottom erosion has already occurred and has been documented by changes in the bottom and bank contours of the upper 9 mi of Lake of the Ozarks. In addition, observations by Missouri Department of Conservation personnel indicate that the predicted erosion of sediment bars has begun. These bars previously supported extensive growth of smartweed and wild millet. An almost complete elimination of plant growth in these areas has occurred since 1981. The elimination of these nursery areas for fish and fish food production may impact the recruitment and growth of fish.

The Missouri Department of Conservation and the State of Missouri have recommended that the proposed operation of the Truman project be altered to reduce fluctuations in water level and velocity in Lake of the Ozarks, and to minimize other detrimental impacts. USACE has agreed to evaluate erosion, water quality, and flow velocities during three- and four-unit operation; the evaluation will be discussed with other agencies before each additional generating unit is put into operation and continued until full operation has been evaluated. USACE has also initiated downstream measures to minimize some of the adverse effects to the physical environment. In 1981 a $6.5 million plan to minimize flooding, erosion, and some of the safety hazards in the upper 4 mi of Lake of the Ozarks was approved (USACE 1981).

Other expected effects of operation that were outlined in the 1980 report include changes in the aquatic community of the upper Osage River Arm of Lake of the Ozarks and entrainment of tailwater fish during pumpback (USACE 1980). These and other problems associated with the Truman project are discussed in this report as they relate to the fishery of Lake of the Ozarks.

Observed Fishery Problems

Migration, Spawning, and Fish Concentrations

Truman Dam forms a barrier that concentrates many fish species that would otherwise migrate throughout the upper Osage River drainage. Beginning in the spring of 1978, annual spawning migrations of paddlefish (*Polydon spathula*), walleye (*Stizostedian vitreum*), white bass (*Morone chrysops*), and other species were blocked by the dam. Concentration of fish below the dam has been substantiated by creel surveys (Richards 1982, 1984).

The Osage River-Lake of the Ozarks paddlefish population supports the largest paddlefish sport fishery in Missouri and one of the largest in the United States (Russell et al. 1980). The spawning requirements of the paddlefish and the historical significance of this fishery were described by Purkett (1963). Russell et al. (1980) noted that there are no known spawning areas for paddlefish below Truman Dam, and that the potential for natural reproduction in the Osage River Basin above the dam is now limited because the reservoir has inundated all known spawning areas. The Lake of the Ozarks snag fishery for paddlefish is expected to decline and eventually disappear without maintenance stocking. At present, negotiations are underway to maintain this fishery through a cooperative stocking agreement.

Lake of the Ozarks has been noted as one of the best walleye fishing areas in Missouri (Pflieger 1975), a reputation largely based on fishing during the spring spawning season. Prior to the construction of Truman Dam, an extensive spawning run of walleye occurred each year out of Lake of the Ozarks into the Osage River. Richards (1982) reported that during the spring spawning season walleye now concentrate in the 40-acre discharge channel immediately below Truman Dam (Fig. 1), which provides an important fishery for Missouri anglers. Approximately 70% of the total walleye harvest reported for the upper 30 mi of Lake of the Ozarks during 1981 and 1982 occurred during the spring quarter in this channel (Richards 1982, 1984). Observations since the closure of the dam indicate walleye may successfully spawn in the tailwater channel when conditions are favorable; no other substantial walleye spawning runs have been documented in Lake of the Ozarks or its tributaries.

Richards (1982) noted two instances during which walleye spawning activity was adversely affected by Truman Dam operation. The first of these was related to gas supersaturation in 1978, and the second was due to erratic water releases in 1982. Other authors have discussed potential adverse effects of water level fluctuations and flow variation on spawning of walleye and other species (Fraser 1972; Hildebrand 1980; Walburg et al. 1981). During 1982 the average size of walleye harvested in the upper 30 mi of Lake of the Ozarks increased, while the numbers decreased, relative to the previous year (Richards 1984). A continuation of this trend would indicate problems with reproduction and/or recruitment.

Gas Supersaturation

Gas supersaturation occurs when excessive amounts of air are entrained in water plunging over a spillway into a deep stilling basin. Gas supersaturation caused massive fish kills in Lake of the Ozarks during the spring of 1978 and 1979, with a total loss exceeding 0.5 million fish. This was the first recorded incidence of serious gas supersaturation below a warmwater hydroelectric dam (Crunkilton et al. 1980). The spillway design for Truman Dam was similar to designs at hydroelectric projects in the Pacific northwest which caused major fish kills in the 1960's and 1970's; these were modified later to minimize supersaturation. Spillway deflectors similar to those described by Smith (1972, 1974) for Pacific northwest facilities were installed at Truman Dam following the 1979 fish kill. They function by deflecting water horizontally across the stilling basin and thereby reducing the deep vertical plunging.

The deflectors lose effectiveness during periods of high volume spillway releases (25,000–30,000 cfs), and gas saturation levels then reach unacceptable levels. This results in direct fish mortalities, or mortalities due to secondary infections resulting from long term exposure to sublethal supersaturation levels. Several fish kills occurred during July 1981 under these conditions. Saturation levels are now monitored during periods of high volume spillway releases with the intent of discontinuing or reducing spillway releases when saturation reaches excessive levels. However, mechanical problems with the pump-turbines, coupled with excessive inflow into Truman Reservoir, could necessitate future high-volume spillway releases.

Dissolved Oxygen

Under most circumstances dissolved oxygen levels below Truman Dam have been adequate. Exceptions were noted during July and August 1981, when surface water temperatures exceeded 80 F and water releases were discontinued for periods of up to 28 hours. Dissolved oxygen levels in the tailwater were recorded at less than 2.0 ppm, and significant fish kills resulted on two occasions. Low-volume spillway releases (250–500 cfs) were successfully used later in 1981 to increase downstream oxygen levels and will help to alleviate future fish mortalities of this nature.

Dewatering of Turbine Draft Tubes

The unique design of Truman Dam has required some innovations in construction of the facility and installation of the generators and turbines (Chapuran and Juett 1981). Frequent dewatering of the turbine draft tubes has been required for alignment and maintenance purposes. The downstream end of each draft tube extends into the tailwater area of upper Lake of the Ozarks, which supports dense concentrations of fish and an excellent tailwater fishery. When a generating unit is not in operation, fish in the tailwater can freely enter the downstream end of the draft tube.

Dewatering is accomplished by sealing the draft-tube intake and discharge openings, inserting bulkheads, and pumping out all water. Fish trapped within the draft tube must be removed manually through the 2 ft by 3 ft access hatches after nearly all of the water has been drained (Fig. 2). Each tube has a volume of nearly 2 million gallons, and a minimum of 6 hours is required to drain the water if the bulkheads form a tight seal; a large-volume leak can double the draining time.

Twelve dewatering operations occurred between August 1980 and February 1983. As few as 50 fish were found in a tube during one winter dewatering operating; however, an estimated 5,000 lb of fish were trapped during one late summer dewatering operation. Highest entrapment of fish has consistently occurred during the summer, when fish densities in the tailwater area, as indicated by creel survey estimates, are highest (Richards 1982, 1984). An estimated 2,400 lb of fish were killed during one dewatering in June 1982. A high proportion of the fish trapped and killed during dewatering operations have been channel catfish (*Ictalurus punctatus*) and blue catfish (*Ictalurus furcatus*). Flathead catfish (*Pylodictis olivaris*) up to 83 lb have also been killed. Catfishes, in particular, have exhibited an affinity for this dark, cave-like habitat.

Fish trapped in the draft tubes were not rescued effectively prior to September 1982, when a rescue procedure was developed by USACE and Missouri Department of Conservation personnel. The procedure includes flushing of the draft tubes prior to dewatering (when mechanically possible), screening the tube openings while sealing the bulkheads, oxygenation of the water in the tube during the draining period, use of specialized hauling tanks and cranes for fish removal, and transportation of fish from the powerhouse back to Lake of the Ozarks in an aerated hatchery truck. Screens are being constructed for each draft tube which can be set in place whenever a unit is not in operation. All of these measures should work together to minimize future fish kills due to dewatering operations.

Fish Passage

Downstream passage of fish through the turbines has been suggested by the entrapment of fish in the upstream portion of draft tubes during dewatering operations. The predominant species removed from this portion of the tubes have been bluegill (*Lepomis macrochirus*), gizzard shad (*Dorosoma cepedianum*), crappies (*Pomoxis* spp.), catfishes, and paddlefish. Field observations and reports from anglers have confirmed mortalities of some fish passing through the turbines during generation. In an effort to quantify fish passage through Truman Dam test netting was initiated in 1982. The methodology is still being refined.

Pumpback Operation

The Truman project is one of only 32 pumped storage hydroelectric facilities in operation in North America (IEEE 1976). It has a relatively small generating capacity, but the inclined-axis design, coupled with a low head differential, results in a high individual pump capacity of 5,500 cfs. Only two other pumped storage facilities have greater pump capacities.

Planned operation of the facility would result in over 400 unit-hours of pumpback each week from June through September, with each of five units pumping for up to 82.5 hours (USACE 1980). Early concerns that fish concentrated in the tailwater areas would be entrained and injured during pumpback were confirmed during pumpback testing in April 1982. One unit was tested for 2 hours on April 13 and a second unit was tested for 1 hour on April 14. These 3 unit-hours of pumpback testing resulted in the loss of 579 identifiable fish weighing approximately 2,000 lb; these were collected from the surface of Truman Reservoir. The number of totally-mutilated fish could not be determined, but the additional loss was considered to be at least equal to observed losses (Richards 1982). Liston (1979) found that surface counts of dead fish accounted for only a small portion of the pumpback-related mortality at the Ludington pumped storage facility in Michigan.

Catfishes made up the highest proportion of fish killed during the pumpback testing, at a time when catfish densities were considered to be relatively low. Creel survey results indicate that catfishes are most numerous in the tailwater channel during the summer period (Richards 1982); increased densities then, coupled with high susceptibility to pumpback-related mortality, could impact the catfish fishery of Lake of the Ozarks. Crappies and other centrarchids are also expected to be more abundant in the tailwater during summer months.

Richards (1982) reported that mechanical damage during pumpback testing appeared to be the primary cause of mortalities of large fish including paddlefish, walleye, catfishes, buffalos (*Ictiobus* spp.), and drum (*Aplodinotus grunniens*). Mortality resulting from internal damage associated with pressure changes was evident in white bass, walleye, and crappies. Six white bass, two walleye, and two white crappie (*Pomoxis annularis*) showing no external damage were dissected. All had bulging or ruptured air bladders in addition to obvious signs of hemorrhaging in the heart, liver, kidneys, and intestines. In many cases inflated air bladders became detached from fish and floated to the reservoir surface. During the second day of pumpback testing an attempt was made to recover floating air bladders in addition to picking up identifiable fish. Following the 1-hour test 61 air bladders were recovered. This probably represents a very low percentage of these organs as most would have ruptured and sunk. Foye and Scott (1965) reported ruptured air bladders and hemorrhagic kidneys in yellow perch (*Perca flavescens*) during pressure testing at the Pleasant Ridge pumped storage facility.

Research on fish mortality resulting from turbine passage during pumped storage operation of pumped storage facilities is limited. Studies reviewed by Turbak et al. (1980) indicate that mortalities at pumped storage facilities are species- and site-specific and that some facilities operate with negligible fish mortality. The Ludington pumped storage project in Michigan has been one of the most extensively documented cases. Liston (1979) estimated a 56% mortality rate for salmonids passing through the Ludington plant during pumpback. Heisey and Mathur (1980) reported an average mortality rate of 75% for adult fishes during testing at the Muddy Run pumped storage facility in Pennsylvania. However, it was concluded that the mortality estimates may have been biased because of the method of introduction of fishes into the Muddy Run intake. Miracle and Gardner (1980) cited cases in which fish mortalities were observed during initial testing and start-up of pumped storage operations but declined as operations continued. The location and design of the Truman Dam project make it likely that mortality during proposed summer pumpback would be more extensive than that observed during testing; however, this aspect remains to be proved.

The USACE has agreed to cease further pumpback operations at the Truman project until a solution to fish mortalities can be found. A consulting corporation has been contracted to evaluate fish protection technologies, and two types of protection systems have been identified as having the best potential (Stone and Webster Engineering Corporation 1983). The estimated costs of these systems are \$48–\$54 million. While it may be possible to develop a protection system that reduces losses of adult and large juvenile fishes it is unlikely that a system could be developed to protect small juveniles or larval fish. Liston et al. (1981) estimated high rates of juvenile and larval fish entrainment at Ludington and found mortality rates to be species-specific. Impingement of adult and juvenile fishes on a screening structure could also result in significant fish mortalities. Average water velocities measured during the April 1982 pumpback test were 2.5 ft/second immediately below Truman Dam. Bibko et al. (1974) reported fish impingement on screening structures when water velocities exceeded an average of 2.2 ft/second.

The potentially devastating effects of pumpback operation at Truman Dam have been discussed in detail by the Missouri Department of Conservation and USACE. Further pumpback testing is scheduled for 1985. Even if an effective fish protection system is developed, the pumpback capability will be held in reserve during the peak spawning months of March, April, and May. This should help reduce disruption of fish spawning in upper Lake of the Ozarks.

Summary

The Truman Dam project serves as an example of the wide range of fisheries and environmental problems that can develop at a conventional or pumped storage hydroelectric generating facility. The 1966 decision to increase the generation capacity of the project and to incorporate pumped storage capability increased the scope and magnitude of problems. The long-term effects of severe water level and velocity fluctuations are expected to be detrimental. The potential for substantial fish mortalities during pumpback operations has been recognized and solutions are being investigated. The pump-turbine/generator assembly, which was designed to accommodate pumped storage, has necessitated frequent dewaterings of turbine draft tubes for maintenance purposes, and thus has resulted in substantial fish mortalities. Downstream problems relating to supersaturation and low dissolved oxygen concentrations have been partially solved but could reoccur under certain operating conditions. Populations of paddlefish in Lake of the Ozarks will have to be maintained by stocking.

Solutions can be found for most of the problems associated with the Truman Dam project, yet the solutions are both time-consuming and costly. Hopefully, this review of the problems associated with this project will alert energy developers and government agencies to some of the important environmental concerns that should be thoroughly considered prior to authorization or construction of future hydroelectric projects. Hydropower production and quality fisheries can co-exist if such concerns are fully considered in the planning, design, construction, and operational phases of the hydroelectric project.

Acknowledgments

Thanks go to James P. Fry, Stanley M. Michaelson, Lee C. Redmond, Kenneth R. Perry, James R. Whitley, Thomas R. Russell, William M. Turner, John E. McPherson, Robert P. Vader, Melanie A. Benitz and other staff members of the Missouri Department of Conservation for their assistance with field data collections, manuscript preparation and manuscript review. The cooperation of U.S. Army Corps of Engineers personnel from the Kansas City District Office and the Harry S. Truman Project Office has been greatly appreciated.

References

ARIATHURAI, R. 1980. Erosion and sedimentation downstream from Harry S. Truman Dam as a result of hydropower operations. Contract report DACW41-79-C-0083. U.S. Army Corps of Engineers, Kansas City, Missouri, USA.

BIBKO, P. N., L. WIRTENAN, AND P. E. KUESER. 1974. Preliminary studies on the effects of air bubbles and intense illumination on the swimming behavior of the striped bass and gizzard shad. Pages 293–304 *in* L. D. Jensen, editor. Proceedings of the second entrainment and intake screening workshop. Electric Power Research Institute, Palo Alto, California, USA.

CHAPURAN, F. J., AND D. J. JUETT. 1981. Pumped storage using inclined-axis units at Truman Dam. Institute of Electrical and Electronic Engineers. Transactions on Power Apparatus and Systems 100:2452–2458.

CRUNKILTON, R. L., J. M. CZARNEZKI, AND L. TRIAL. 1980. Severe gas bubble disease in a warmwater fishery in the midwestern United States. Transactions of the American Fisheries Society 109:725–733.

FOYE, R. E., AND M. SCOTT. 1965. Effects of pressure on survival of six species of fish. Transactions of the American Fisheries Society 94:88–91.

FRASER, J. C. 1972. Regulated discharge and the stream environment. Pages 263–285 *in* R. T. Oglesby, C. A. Carlson, and J. A. McCann, editors. River ecology and man. Academic Press, New York, USA.

HANSON, W. D. 1975. Harvest of fish in Lake of the Ozarks, Missouri. Final report, Federal Aid Project F-1-R-23. Missouri Department of Conservation, Jefferson City, Missouri, USA.

HEISEY, P. G., AND D. MATHUR. 1980. Summary of ecological studies of fishes in Muddy Run Pumped Storage Pond, Pennsylvania. Pages 80–94 *in* J. P. Clugston, editor. Proceedings of the Clemson workshop on environmental impacts of pumped storage hydroelectric operations. FWS/OBS-80/28, U.S. Fish and Wildlife Service, Washington, D.C., USA.

HILDEBRAND, S. G., editor. 1980. Analysis of environmental issues related to small-scale hydroelectric development. III: Water level fluctuation. ORNL/TM-7453. Oak Ridge National Laboratory, Oak Ridge, Tennessee, USA.

IEEE (INSTITUTE OF ELECTRICAL AND ELECTRONIC ENGINEERS). 1976. Survey of pumped storage projects in the United States and Canada to 1975. Institute of Electrical and Electronic Engineers. Transactions on Power Apparatus and Systems 95:851–858.

LISTON, C. R. 1979. Estimates of salmonid fish mortalities occurring at the Ludington Pumped Storage Power Facility during 1975–1978, and related studies. 1978 Annual report, Volume I, Number 2. Consumers Power Co., Jackson, Michigan, USA.

LISTON, C. R., D. BRAZO, R. O'NEAL, J. BOHR, G. PETERSON, AND R. LIGMAN. 1981. Assessment of larval, juvenile, and adult fish entrainment losses at the Ludington Pumped Storage Power Plant on Lake Michigan. 1980 Annual report, Volume 1. Consumers Power Co., Jackson, Michigan, USA.

MIRACLE, R. D., AND J. A. GARDNER, JR. 1980. Review of the literature on the effects of pumped storage operations on ichthyofauna. Pages 40–53 *in* J. P. Clugston, editor. Proceedings of the Clemson workshop on environmental impacts of pumped storage hydroelectric operations. FWS/OBS-80/28. U.S. Fish and Wildlife Service, Washington, D.C., USA.

PFLIEGER, W. L. 1975. The fishes of Missouri. Missouri Department of Conservation, Jefferson City, Missouri, USA.

PURKETT, C. A., JR. 1963. The paddlefish fishery of the Osage River and the Lake of the Ozarks, Missouri. Transactions of the American Fisheries Society 92:239–244.

RICHARDS, K. R. 1982. Study of the fishery of the upper Osage Arm of Lake of the Ozarks, Missouri. Contract report DACW41-81-C-0103. U.S. Army Corps of Engineers, Kansas City, Missouri, USA.

RICHARDS, K. R. 1984. Study of the fishery of the upper Osage Arm of Lake of the Ozarks, Missouri. Contract report DACW41-82-C-0073. U.S. Army Corps of Engineers, Kansas City, Missouri, USA.

RUSSELL, T. R., L. K. GRAHAM, D. M. CARLSON, AND E. J. HAMILTON. 1980. Maintenance of the Osage River-Lake of the Ozarks paddlefish fishery. Final report. U.S. Army Corps of Engineers, Kansas City, Missouri, USA.

SMITH, H. A., JR. 1972. N_2 threat to Pacific northwest fisheries? Water Spectrum 4(2):41–48.

SMITH, H. A., JR. 1974. Spillway redesign abates gas supersaturation in Columbia River. Civil Engineering (New York) 144(9):70–73.

SPARROWE, R. D. 1977. Harry S. Truman Dam and Reservoir — A case history of problems in fish and wildlife coordination. Transactions of the North American Wildlife and Natural Resources Conference 42:42–55.

STONE AND WEBSTER ENGINEERING CORPORATION. 1983. Alternative fish protection systems evaluation for the Harry S. Truman Dam and Reservoir. Contract report DACW41-83-0017. U.S. Army Corps of Engineers, Kansas City, Missouri, USA.

TURBAK, S. C., D. R. REICHLE, AND C. R. SHRINER. 1980. Analysis of environmental issues related to small-scale hydroelectric development. IV: Fish mortality resulting from turbine passage. ORNL/TM-7521, Oak Ridge National Laboratory, Oak Ridge, Tennessee, USA.

USACE (U.S. ARMY CORPS OF ENGINEERS). 1980. Final report and final environmental impact statement II for downstream measures, Harry S. Truman Dam and Reservoir. U.S. Army Corps of Engineers, Kansas City, Missouri, USA.

USACE (U.S. ARMY CORPS OF ENGINEERS). 1981. Design Memorandum No. 52, Downstream measures, Harry S. Truman Dam and Reservoir. U.S. Army Corps of Engineers, Kansas City, Missouri, USA.

WALBURG, C. H., J. F. NOVOTNY, K. E. JACOBS, W. D. SWINK, T. M. CAMPBELL, J. NESTLER, AND G. E. SAUL. 1981. Effects of reservoir releases on tailwater ecology: A literature review. Technical report E-81-12. U.S. Army Corps of Engineers Waterways Experiment Station, Vicksburg, Mississippi, USA.

Pages 255–260 *in* G. E. Hall and M. J. Van Den Avyle, editors.
Reservoir Fisheries Management: Strategies for the 80's.
Reservoir Committee, Southern Division American Fisheries Society, Bethesda, Maryland. 1986.

Management of Largemouth Bass in a Perched Cooling Pond in Illinois

JAMES A. SMITHSON, KENNETH F. KURZAWSKI, AND THOMAS V. CLEVENGER

Illinois Power Company
Clinton Power Station
Box 678
Clinton, Illinois 61727

ABSTRACT

A population of largemouth bass (*Micropterus salmoides*) in a perched cooling pond in Illinois (Baldwin Cooling Pond) was monitored by electrofishing from 1979 through 1982. The population was characterized by seasonal movements in response to thermal discharges, rapid growth, and a high proportion of young-of-the-year (YOY) and Age-1 fish. During winter months, catch rates of largemouth bass were 90% higher in the discharge canal than at mid-lake or at the power plant intake. Catch rates declined in the discharge canal during the summer and increased again in the fall. Mean lengths attained by YOY (232 mm) and yearling largemouth bass (355 mm) by December were similar to the highest growth rates reported in the literature. Rapid growth is attributed to the general interaction of temperatures optimum for growth and establishment of an overwintering threadfin shad (*Dorosoma petenense*) population. The YOY and Age-1 fish accounted for more than 86% of the largemouth bass collected during August-December of each year. The combination of insufficient recruitment and overharvest of Age-2 and older fish contributed to the imbalanced age structure. Management strategies for largemouth bass are discussed in relation to unique characteristics of perched cooling ponds.

Sport fishing has become a viable secondary use of reservoirs that are initially constructed to dissipate heat created by electrical power generation. However, ecosystems in cooling reservoirs differ from those in other reservoirs, and successful fisheries management will recognize these differences and utilize beneficial aspects of altered thermal conditions. Continuous recirculation of water, elevated temperatures, and morphometry are the components of cooling reservoirs that have the most influence on the development of fisheries management strategies. Recirculation disrupts temperature stratification; hence, nutrients are recycled rather than trapped in the hypolimnion or bottom sediments. This recycling can partially alleviate declines in productivity and sport fish harvest associated with reservoir aging (Drew and Tilton 1970; Ploskey 1981).

Elevated temperatures influence many aspects of fish communities in cooling reservoirs. Growing seasons can be extended causing improved growth and earlier spawning of fishes (Busacker 1971; Joy and Tranquilli 1979; Tranquilli et al. 1981a). Ranges of desirable species such as threadfin shad (*Dorosoma petenense*) can be expanded into areas where cold temperatures would eliminate these species (Heidinger 1977). Some species are attracted to thermal discharges during winter and avoid these areas during the summer (Quinn et al. 1978; Tranquilli et al. 1981b), and these movements can affect harvest and growth of fishes in cooling reservoirs.

Much of the thermal influence depends on the thermal load and morphometry of the cooling reservoir. Morphometric features that lessen thermal influences include those that promote a variety of habitats, provide thermal refuges, and allow high exchange of water. Perched cooling ponds (those constructed by enclosure of an area by dikes) probably have less favorable morphometric features than cooling reservoirs constructed by damming a river. Perched cooling ponds usually are less than 1,200 hectares, relatively shallow, and have dikes that create a monotonous shoreline habitat.

A study was initiated in 1979 on Baldwin Cooling Pond, a perched cooling pond in Illinois, to assess influences of thermal discharges and management programs on the fish community. The entire community was assessed, but population dynamics of largemouth bass (*Micropterus*

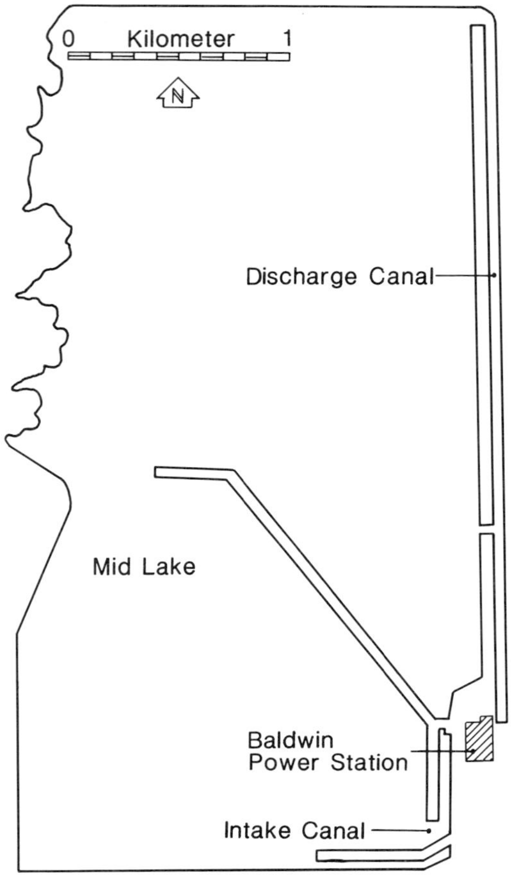

Figure 1. Map of Baldwin Cooling Pond indicating discharge canal, mid-lake, and intake canal sampling sites.

salmoides) were emphasized. The objective of this paper is to discuss management opportunities for largemouth bass in perched cooling ponds.

Study Site

Baldwin Cooling Pond is a 796-hectare perched pond located in southwestern Illinois. Mean depth is 3 m, and maximum depth is 15 m. The pond is rectangular with approximately 90% of the 24-km shoreline composed of riprapped levees. The reservoir was constructed by Illinois Power Company in 1968 to provide cooling water for three 600-megawatt coal-fired generating units which were phased into operation between 1969 and 1975. The entire volume of the reservoir can be recycled every 5 days with all units at full operation. This results in disruption of vertical stratification of temperature and dissolved oxygen. The reservoir has been leased to the Illinois Department of Conservation as a recreational fishing area since 1971.

Methods

Fishes were collected monthly (except November) from February 1979 through December 1982 using a 230-volt, alternating current electrofishing unit. Four 15-minute samples were taken at the intake canal, mid-lake region, and discharge canal (Fig. 1). Total lengths (mm) and weights (g) were measured for all fish collected. Information on mean lengths of largemouth bass during November 1979–1982 was obtained from administrative reports of the Illinois Department of Conservation (Division of Fisheries, Springfield, Illinois, USA). Adverse weather and equipment failure prevented collection of some scheduled samples. Surface water temperatures were taken with each 15-minute sample.

Ages of young-of-the-year (YOY), Age-1, and older bass were assigned using length-frequency distributions. Recaptures of fin-clipped YOY fish were used to validate these determinations. Accurate age and growth estimates for Age-2 and older fish could not be determined from length-frequencies; scales were not used because this method typically is inaccurate in cooling reservoirs (Tranquilli et al. 1981c).

Results and Discussion

Thermal discharges have influenced the movements, growth, and age structure of largemouth bass in Baldwin Cooling Pond. These factors are discussed in relation to appropriate management strategies.

Movements

Movements of largemouth bass in relation to the temperature gradient created by the thermal discharge were evident from electrofishing catch rates (Table 1). The wide seasonal variability in catch rates in the discharge canal was most likely due to water temperature. Largemouth bass were concentrated in the discharge canal during the winter and remained abundant there until June when water temperatures began to exceed the preferred range (30–32 C) reported for largemouth bass (Ferguson 1958; Smith and McNurney 1981; Tranquilli et al. 1981b). Catch of largemouth bass in the discharge canal was low throughout the summer months, and few bass were collected when temperatures exceeded 35 C. Catches began to increase in the discharge canal in September and October as the water temperature declined. Mean annual water temperatures in the discharge canal for August, September, and October were 37, 32, and 25 C, respectively.

Corresponding changes in the catch rates at mid-lake and intake sites were insufficient to determine the extent or direction of movements after largemouth bass left the discharge canal. This may have been partially due to sampling by electrofishing; for example, largemouth bass could have moved during summer to deeper areas of the pond where they were less susceptible to sampling. Catch data from Lake Sangchris, Illinois (Tranquilli et al. 1981b), indicated that largemouth bass made seasonal movements (up to 16.1 km) to preferred water temperatures. Their data were substantiated by recaptures of tagged fish and radiotelemetry observations.

Growth

Growth of largemouth bass through Age 2 was rapid (Tables 2 and 3). Average lengths attained by YOY (232 mm) and yearling largemouth bass (355 mm) by December exceeded the sizes attained by all similar-aged bass from all North American populations summarized by Carlander (1977) except for two from southerly locations. Maximum growth of largemouth bass occurs at 27–30 C for fry (Strawn 1961) and 27 C for subadults (Coutant 1975). The growth potential is affected by the amount of time spent at temperatures that are optimal for growth. Optimal temperatures for growth were available in Baldwin Cooling Pond for at least 4 months each year (Table 1). In comparison, the maximum ambient temperatures in a central Illinois reservoir (Lake Shelbyville) were less than 27 C (Joy and Tranquilli 1979). Elevated temperatures in Baldwin Cooling Pond increased the time when optimal growth temperatures were available.

Growth of largemouth bass is also affected by the length of the growing season. The growing season is determined by the number of days when temperatures are between 15.6 C (Bennett 1971) and 36 C (Strawn 1961; Coutant 1975). Joy and Tranquilli (1979) found that growth rates of largemouth bass from the heated area of Lake Sangchris were greater than those from nearby Lake Shelbyville and other unheated midwestern reservoirs. The increased growth was attributed

Table 1. Number of largemouth bass collected per hour by electrofishing from three locations in Baldwin Cooling Pond, 1979–1982. Values listed are from four 15-minute samples unless otherwise indicated. Surface water temperature (C) at time of collection is shown in parentheses.

	Discharge canal				Mid-lake				Intake canal			
Month	1979	1980	1981	1982	1979	1980	1981	1982	1979	1980	1981	1982
January	—	—	302 (18)	312 (10)	—	—	16 (9)	5 (4)	—	—	0 (8)	0 (3)[a]
February	368	525 (11)	375 (26)	528 (17)	12[a]	9 (7)	132 (16)	18 (9)	—	3 (7)	13 (13)	2 (8)
March	698[a]	311 (18)	147 (19)	275 (23)	76[a]	44 (13)	24 (13)[a]	32 (18)	17	6 (10)	—	25 (16)
April	494 (27)	286 (18)	210 (29)	449 (26)	56 (20)[b]	88 (14)	75 (22)	46 (20)	68 (18)	26 (13)	44 (22)	25 (17)
May	90 (34)	128 (32)	125 (23)	78 (34)	53 (26)	121 (26)	37 (20)	33 (26)	50 (26)	104 (27)	37 (22)	52 (25)
June	3 (35)	1 (37)	—	1 (37)	33 (29)[c]	36 (28)	18 (31)[a]	33 (31)	43 (28)	59 (29)	60 (30)[b]	40 (28)
July	1 (40)	0 (43)	0 (39)	0 (38)	69 (33)	81 (37)	19 (35)	19 (33)	72 (33)	178 (34)	16 (36)	47 (34)
August	2 (38)	1 (36)	—	—	56 (32)	76 (34)	—	12 (32)[a]	87 (29)	48 (30)	—	7 (30)
September	9 (34)	17 (31)	10 (29)	5 (34)	115 (27)	45 (28)	21 (25)	51 (29)	115 (26)	46 (30)	32 (22)	29 (27)
October	—	143 (25)	—	166 (25)	—	146 (20)	—	100 (20)	—	53 (20)	—	28 (17)
December	316 (15)	—	—	234 (17)	14 (6)	—	—	16 (8)	0 (4)	—	—	2 (7)

[a] Extrapolated from two 15-minute samples.
[b] Extrapolated from one 15-minute sample.
[c] Extrapolated from three 15-minute samples.

Table 2. Number (N) and mean total length (mm) of young-of-the-year largemouth bass collected from Baldwin Cooling Pond, 1979-1982.

	1979		1980		1981		1982		1979–1982
Month	N	Mean length	N	Mean length	N	Mean length	N	Mean length	combined mean length
August	82	133	37	126			6	147	132
September	136	169	57	166	29	169	48	169	168
October			175	210			150	226	217
November	406	205	226	223	317	178	180	215	203
December	249	222					124	248	232

to the extension of the growing season from 150 days in Lake Shelbyville to 245 days in Lake Sangchris. Growth of largemouth bass in Baldwin Cooling Pond exceeded that in Lake Sangchris, perhaps due to the extension of the growing season to 320 days. The elevated temperatures in Baldwin Cooling Pond exceeded the 36 C maximum for largemouth bass growth (Strawn 1961; Coutant 1975) at some locations; however, few bass were collected at these locations, and areas having cooler temperatures were available (Table 1).

Rapid growth at any temperature can only be achieved if there is adequate forage. Increased growth of largemouth bass followed the establishment of overwintering threadfin shad populations in some western and southern states (von Geldern and Mitchell 1975). Increased growth was not observed in West Virginia where annual introductions were required due to winterkills (von Geldern and Mitchell 1975). Thermal discharges into Baldwin Cooling Pond, after the final generating unit became operational in 1975, allowed establishment of an overwintering population of threadfin shad, and increased growth of YOY largemouth bass was observed thereafter. The mean total length of YOY largemouth bass in November 1972–1974 was 177 mm; in November 1975–1977, the average was 238 mm (Illinois Department of Conservation, Division of Fisheries, Springfield, Illinois, USA, unpublished data).

Age Structure

A favorable age structure in a population is the result of a balance between the rates of reproduction, growth, and mortality (Anderson 1975). The population of largemouth bass in Baldwin Cooling Pond was predominantly YOY and Age-1 fish, which accounted for 86–100% of the total electrofishing catch during August–December of each year. YOY largemouth bass first appeared in the samples in June or July and accounted for the largest percent of the samples by the end of the year (Table 4). These bass continued to dominate the catch until they were succeeded by YOY in September or October of the following year.

Table 3. Number (N) and mean total length (mm) of Age-1 largemouth bass collected from Baldwin Cooling Pond, 1979-1982.

	1979		1980		1981		1982		1979–1982
Month	N	Mean length	N	Mean length	N	Mean length	N	Mean length	combined mean length
January					136	244	286	193	209
February	337	232	419	229	311	252	413	225	233
March	387	232	248	227	99	268	303	217	230
April	551	255	303	237	259	274	478	233	248
May	173	267	289	257	150	288	133	253	265
June	53	260	74	276	23	280	71	272	271
July	131	284	207	281	33	296	50	277	283
August	46	301	73	302					302
September	93	311	43	334	32	364	33	332	328
October			159	353			136	330	343
November	100	366	238	355	97	369	122	342	357
December	63	371					104	345	355

Table 4. Percent of young-of-the-year (YOY), Age-1, and Age-2 or older largemouth bass in electrofishing samples from Baldwin Cooling Pond, 1979-1982.

	1979			1980			1981			1982		
Month	YOY	1	≥2	YOY	1	≥2	YOY	1	≥2	YOY	1	≥2
January								42	58		90	10
February		90	10		78	39 (22)		60	40		75	25
March		96	4		69	38 (31)		62	38		91	9
April		96	4		76	32 (24)		79	21		92	8
May		91	9		82	21 (18)		76	24		82	18
June	1	75	24	1	78	6 (21)		97	3		96	4
July	1	92	7	3	80	15 (17)	3	94	3	21	77	2
August	56	32	12	29	57	6 (14)				44	44	2
September	56	39	5	53	39	3 (8)	48	52		57	39	4
October				51	46	3 (3)				51	46	3
December	75	19	6							49	42	9

Management Implications

The high proportion of young age classes in Baldwin Cooling Pond suggests an unbalanced age structure. Management efforts to balance age structure would have to deal with the interrelated functions of reproduction, growth, and mortality.

Mortality probably accounted for the low proportions of Age-2 and older fish. Our data allowed no distinction between natural and angling mortality; however, there were no apparent reasons or evidence for excessive natural mortality. Because of rapid growth rates in Baldwin Cooling Pond, essentially all largemouth bass Age 2 or older were of legal harvestable size (356 mm minimum), and many Age-1 bass became subject to harvest during the later part of their second growing season.

Management strategies were implemented to decrease harvest. The length limit was raised from 356 to 457 mm in January 1982 to increase the abundance of larger fish available to anglers.

Regulation of angler access to warm-water discharges during winter might be useful for controlling harvest in some cooling lakes. At Baldwin Cooling Pond, anglers do not have access to the discharge canal itself, but most angling activity observed during winter and the highest catch rates occurred near the outlet of the discharge canal (IDOC 1980). Harvest probably would have been greater if the entire discharge canal was open to fishing.

Management techniques to increase spawning success and recruitment of largemouth bass in Baldwin Cooling Pond and similar reservoirs are limited. Morphometric features incorporated into the design of Baldwin Cooling Pond to increase heat dissipation may have contributed to poor habitat for spawning and recruitment. Features such as unrestricted wind flow and shallow depth contribute to heavy wave action, which is detrimental to spawning and early survival of largemouth bass (Miller and Kramer 1971; von Geldern 1971; Summerfelt 1975). Design of perched cooling ponds could incorporate features such as strategically located berms, wave dissipators, or protected enclosed bays to help reduce wave action.

Fingerling largemouth bass were supplementally stocked in an attempt to improve recruitment. In 1981, 6,783 fin-clipped fingerling largemouth bass were stocked during October. In 1982, these marked fish accounted for 40–45% of the Age-1 catch, indicating they had made a substantial contribution to the population. Supplemental stockings might also be used in other situations where spawning habitat is limited.

Acknowledgments

We thank Peter J. Paladino and Buz M. Imboden of the Illinois Department of Conservation for providing data from their November samples of Baldwin Cooling Pond. We also acknowledge Ronald L. Willmore, David W. Plunkett, and Linda M. Ruzevick for their assistance with sample collection and data handling.

References

Anderson, R. O. 1975. Factors influencing the quality of largemouth bass fishing. Pages 183–194 *in* R. H. Stroud and H. Clepper, editors. Black bass biology and management. Sport Fishing Institute, Washington, D.C., USA.

Bennett, G. W. 1971. Management of lakes and ponds, 2nd edition. Van Nostrand Reinhold Company, New York, USA.

Busacker, G. P. 1971. The age and growth of largemouth bass, *Micropterus salmoides* (Lacepede), in a thermally loaded reservoir. Master's thesis, University of Missouri, Columbia, Missouri, USA.

Carlander, K. D. 1977. Handbook of freshwater fishery biology, volume 2. Iowa State University Press, Ames, Iowa, USA.

Coutant, C. C. 1975. Responses of bass to natural and artificial temperatures regimes. Pages 272–285 *in* R. H. Stroud and H. Clepper, editors. Black bass biology and management. Sport Fishing Institute, Washington, D.C., USA.

Drew, H. R., and J. E. Tilton. 1970. Thermal requirements to protect aquatic life in Texas reservoirs. Journal of the Water Pollution Control Federation 42:562–572.

Ferguson, R. G. 1958. The preferred temperature of fish and their midsummer distribution in temperate lakes and streams. Journal of the Fisheries Research Board of Canada 15:607–624.

Heidinger, R. C. 1977. Potential of the threadfin shad as a forage fish in midwestern power cooling reservoirs. Transactions Illinois State Academy of Science 70:15–25.

IDOC (Illinois Department of Conservation). 1980. Baldwin Lake creel census summary, April 1979 through March 1980. Illinois Department of Conservation, Division of Fisheries, Springfield, Illinois, USA.

Joy, T., and J. A. Tranquilli. 1979. Annual growth of largemouth bass in Lake Sangchris, Lake Shelbyville, and other midwestern waters. Pages 10.1–10.14 *in* Evaluation of a cooling lake fishery. Vol. III. Electric Power Research Institute, Palo Alto, California, USA.

Miller, K. D., and R. H. Kramer. 1971. Spawning and early life history of largemouth bass (*Micropterus salmoides*) in Lake Powell. Pages 73–84 *in* G. H. Hall, editor. Reservoir fisheries and limnology. Special publication 8, American Fisheries Society, Bethesda, Maryland, USA.

Ploskey, G. R. 1981. Factors affecting fish production and fishing quality in new reservoirs, with guidance on timber clearing, basin preparation, and filling. Technical report E-81-11. U.S. Army Engineer Waterways Experiment Station, Vicksburg, Mississippi, USA.

Quinn, T., G. W. Esch, T. C. Hazen, and J. W. Gibbons. 1978. Long range movement and homing by largemouth bass (*Micropterus salmoides*) in a thermally altered reservoir. Copeia 1978:542–545.

Smith, V. J., and J. M. McNurney. 1981. Behavioral thermoregulation of largemouth bass and carp in an Illinois cooling lake. Pages 585–593 *in* R. W. Larimore and J. A. Tranquilli, editors. The Lake Sangchris study: case history of an Illinois cooling lake. Illinois Natural History Survey Bulletin, Volume 32, Article 4, Champaign, Illinois, USA.

Strawn, K. 1961. Growth of largemouth bass fry at various temperatures. Transactions of the American Fisheries Society 90:334–335.

Summerfelt, R. C. 1975. Relationship between weather and year-class strength of largemouth bass. Pages 166–174 *in* R. H. Stroud and H. Clepper, editors. Black bass biology and management. Sport Fishing Institute, Washington, D.C., USA.

Tranquilli, J. A., R. Kocher, and J. M. McNurney. 1981a. Population dynamics of Lake Sangchris fishery. Pages 413–499 *in* R. W. Larimore and J. A. Tranquilli, editors. The Lake Sangchris study: case history of an Illinois cooling lake. Illinois Natural History Survey Bulletin, Volume 32, Article 4, Champaign, Illinois, USA.

Tranquilli, J. A., D. W. Dufford, R. W. Larimore, R. Kocher, and J. M. McNurney. 1981b. Radiotelemetry observations on the behavior of largemouth bass in a heated reservoir. Pages 559–584 *in* R. W. Larimore and J. A. Tranquilli, editors. The Lake Sangchris study: case history of an Illinois cooling lake. Illinois Natural History Survey Bulletin, Volume 32, Article 4, Champaign, Illinois, USA.

Tranquilli, J. A., J. M. McNurney, and R. Kocher. 1981c. Results of a multiple-objective fish-tagging program in an artificially heated reservoir. Pages 536–558 *in* R. W. Larimore and J. A. Tranquilli, editors. The Lake Sangchris study: case history of an Illinois cooling lake. Illinois Natural History Survey Bulletin, Volume 32, Article 4, Champaign, Illinois, USA.

von Geldern, C. E. Jr. 1971. Abundance and distribution of fingerling largemouth bass, *Micropterus salmoides,* as determined by electrofishing at Lake Nacimiento, California. California Fish and Game 57:228–245.

von Geldern, C. E. Jr., and D. F. Mitchell. 1975. Largemouth bass and threadfin shad in California. Pages 436–449 *in* R. H. Stroud and H. Clepper, editors. Black bass biology and management. Sport Fishing Institute, Washington, D.C., USA.

Pages 261–269 *in* G. E. Hall and M. J. Van Den Avyle, editors.
Reservoir Fisheries Management: Strategies for the 80's.
Reservoir Committee, Southern Division American Fisheries Society, Bethesda, Maryland. 1986.

Ash Basin Effluents as a Concern of Fisheries Managers: A Case History and Perspective

LARRY L. OLMSTED, DONALD J. DEGAN, JOHN S. CARTER
Duke Power Company
Route 4, Box 531
Huntersville, North Carolina 28078

PETER M. CUMBIE
Duke Power Company
P.O. Box 33189
Charlotte, North Carolina 28242

ABSTRACT

Belews Lake is a 1,564-hectare impoundment for the fossil-fueled Belews Creek Steam Station in North Carolina. Ash from the plant is sluiced to an ash basin, the overflow of which began discharging into the lake in 1975. Water chemistry of the lake is substantially influenced by its long retention time (about 1,500 days), its circulation pattern, and the ash basin discharge. The fish community in Belews Lake has reflected changes in water quality from 1976 to the present. Annual rotenone samples prior to power plant operation (1972-1975) revealed a diverse community, dominated by centrarchids, with an average total standing stock of 98 kg/hectare. Samples after initiation of ash basin discharges (1976-1981) showed a progressively less-diverse fish community dominated by non-game fish, with an average standing stock of only 17 kg/hectare. These trends were accompanied by reduced reproduction of most species and a high incidence of blindness in some. Selenium in the ash basin effluent has been implicated as the primary cause of alterations of the fish community.

The current interest in using impoundments for cooling steam stations while minimizing environmental impact coincides with a period of decline in the construction of multiple-use reservoirs. This presents a valuable opportunity for fisheries managers to meet increasing public demand for fishing on reservoirs. However, reservoirs constructed specifically for heat dissipation can pose unique management problems. To date, most management concern has focused on problems related to the potential impacts of rejected heat. Recently, impacts of other operational features such as ash basin effluent discharges on the ecology of cooling impoundments have gained increasing attention.

Belews Lake is located in north-central North Carolina approximately 24 km northeast of Winston-Salem. The lake was built in the early 1970's by Duke Power Company as a cooling impoundment for Belews Creek Steam Station, a two-unit (1,120 megawatt each), coal-fired electric generating station. At full pool, the reservoir covers 1,564 hectares and has a mean depth of 14.6 m. Average retention time is approximately 1,500 days. Beginning in 1975, fly ash from the plant was wet sluiced to a settling basin, and the excess water was returned to Belews Lake. Return of this sluice water to the lake is important to maintain adequate water levels in Belews Lake for plant operation. A counter-clockwise circulation pattern of the surface waters in the main reservoir basin is induced by the plant intake and discharge.

Monitoring of physical, chemical, and biological characteristics of the streams to be impounded was initiated in the summer of 1970, approximately three years prior to Belews Lake reaching full pool level (Weiss and Anderson 1978). The North Carolina Wildlife Resources Commission, Division of Inland Fisheries, evaluated the fish populations of Belews Lake from 1971 through 1976 to determine the sport-fishery potential of an industrial cooling lake. Duke Power Company began fish population studies of Belews Lake in 1976. From the initial filling through 1975, the reservoir was considered a typical piedmont reservoir with respect to its water chemistry and biotic communities (Weiss et al. 1978). In 1976, however, North Carolina Wildlife Resources Commission and Duke Power Company biolo-

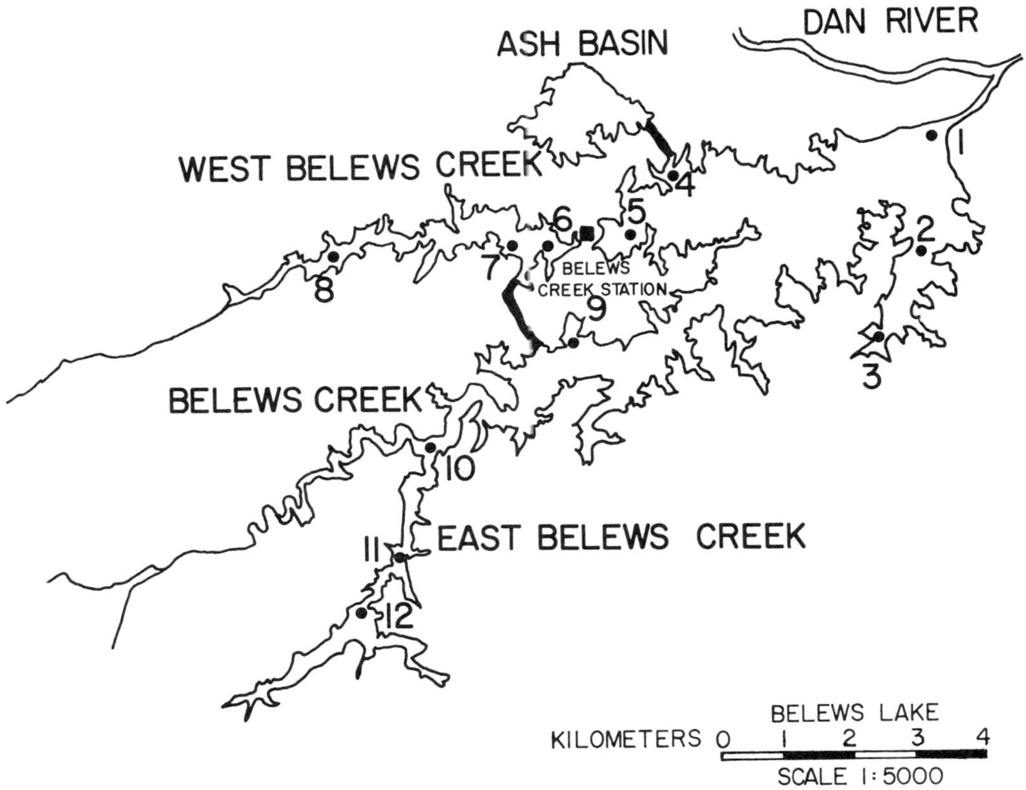

Figure 1. Map of Belews Lake, North Carolina, showing sampling stations.

gists noted a substantial decline in the abundance of adult fish in the main reservoir basin accompanied by the absence of young-of-the-year fish (Cumbie 1978; Harrell et al. 1978; Van Horn 1978). Concern about the declining fishery prompted further studies of chemical and biological factors judged to offer the greatest potential for delineating the causes of the degraded fishery.

Specific objectives of studies conducted on Belews Lake between 1976 and 1982 were (1) to document temporal and spatial trends in water quality, fishery parameters, and plant operating features; (2) to determine the cause(s) of any changes noted in the fish community; and (3) to recommend appropriate corrective actions.

Methods

Sampling stations and methods used to monitor water quality of Belews Lake from 1970 through 1976 were summarized by Weiss et al. (1978). Water quality monitoring after 1976 was performed by the Environmental Laboratories of Duke Power Company at 12 locations throughout the reservoir (Fig. 1). Hydraulic patterns caused most sampling stations to have similar characteristics. However, one station (12) was separated from the remainder of the lake by a causeway and was considered to be a reference station because it was isolated from the ash-basin effluent. Hereafter, Stations 1–11 are referred to collectively as main lake stations.

Data on fish standing crop for 1972–1976 were reported by Van Horn (1978), and estimates for 1977 were summarized by Harrell et al. (1978). Additional rotenone sampling was performed in July 1978–1981 at two or three main lake stations and at Station 12. A block net with 6.4-mm mesh was placed across the sample cove and rotenone operations performed following guidelines published by Grinstead et al. (1978). Coves averaged 0.4 hectare in surface area. Fish were sorted by species into 2-cm length groups. Total weight of each length group was determined for fish picked up the first day, and these relation-

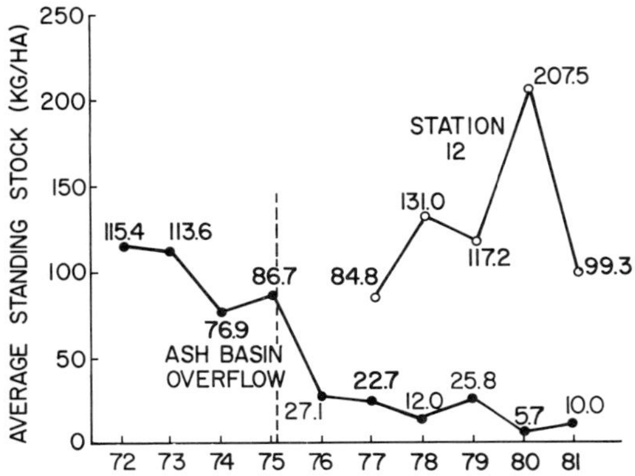

Figure 2. Average standing stock in main lake coves (lower line) and at Station 12 (upper line) in Belews Lake, 1972-1981. The vertical dashed line indicates when ash-basin effluents first entered the reservoir in early 1975.

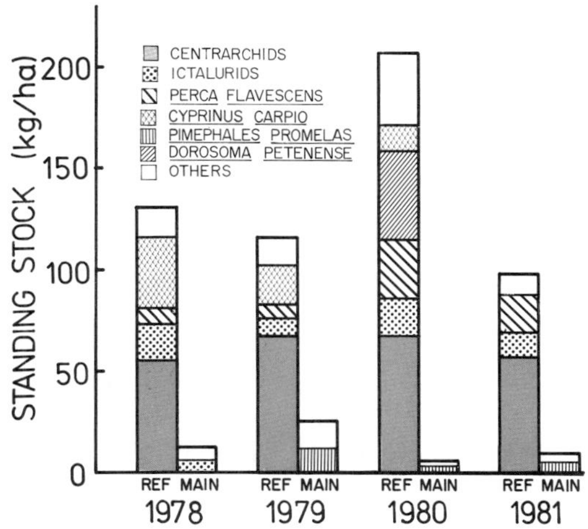

Figure 3. Standing stock of fishes collected in reference (Station 12) and main lake (Stations 3, 5, 7, 9, and 11) rotenone samples from Belews Lake, 1978-1981.

ships were used to calculate weights for the second-day fish.

Larval fish were sampled in shoreline areas at night using a 0.5-m diameter, 0.710-mm mesh push net. Samples were collected monthly at three or four main lake sites and at the reference site during May–August 1977, May–June 1979, May and July 1980, and April–June 1981. Adult and juvenile fish were sampled monthly from July 1976 through July 1977 by electrofishing four 100-m shoreline transects at each of five main lake stations and in the reference area.

Results and Discussion

Fish Populations

Fish abundance and diversity in the main lake declined substantially after ash-basin effluents were first discharged in 1975 (Fig. 2). Rotenone samples from 1972 through 1975 yielded 24 species and an average standing stock of 98 kg/hectare (Van Horn 1978). The fish community was dominated by centrarchids, primarily bluegills (*Lepomis macrochirus*), pumpkinseed sunfish (*L. gibbosus*), and largemouth bass (*Micropterus salmoides*), and cyprinids, primarily carp (*Cyprinus carpio*), and golden shiner (*Notemigonus crysoleucas*). After 1975, diversity and standing stock generally declined. By 1981, only seven species (10 kg/hectare) were collected, and fathead minnows (*Pimephales promelas*) and mosquitofish (*Gambusia affinis*) accounted for 82% of the standing stock (Fig. 3). Young-of-the-year fish were virtually absent from rotenone samples at main lake locations after 1975. Blindness, characterized by opaque lenses, occurred frequently in centrarchids (especially warmouth, *Lepomis gulosus*) during 1976 and 1977.

Comparisons of fish communities in the main lake and the reference station provide further evidence of impacts of the ash-basin operation. During 1977–1981, annual rotenone catches at Station 12 were about 4 to 35 times higher than in main lake areas (Fig. 2). Standing stock estimates at the reference station ranged from 85 to 208 kg/hectare, which is typical of many piedmont reservoirs. Centrarchids (primarily *Lepomis* spp.) were the dominant taxa each year and contributed 47% of the standing stock over the four-year period. Length-frequency data indicated that successful centrarchid reproduction occurred at the reference station each year. Electrofishing catch rates during 1976 and 1977 in the main body of the lake averaged 2.9 fish/100 m of shoreline, whereas catch rates at the reference area averaged 36.9 fish/100 m. Centrarchids and black bullheads (*Ictalurus melas*) dominated the catch (Harrell et al. 1978).

Shifts in fish abundance and species composition over time and differences between reference and main lake samples reflected the poor spawning success of several species after 1975. This was substantiated by results of the push net sampling. In 1977, larval fish were much more abundant in the reference area than at the main lake locations, averaging 2,001 and 348 larvae/ 1,000 m^3, respectively. Similar patterns occurred in subsequent years. Species composition in larval fish samples also varied substantially between reference and main lake areas. Reference area samples in 1977 included clupeids (89.0% by number), *Lepomis* spp. (8.9%), cyprinids other than carp (0.2%), and unidentifiable larvae (1.0%); similar results were found in 1979–1981. The main lake locations included carp (96.0%), mosquitofish (0.9%), and unidentifiable larvae (2.7%) in 1977 (Harrell et al. 1978). In following years, fathead minnows replaced carp so that the main lake was dominated by fathead minnows and mosquitofish. Shad (*Dorosoma* spp.) and *Lepomis* spp. reappeared in larval samples in low numbers in 1980. Samples collected in 1982 with a bow-mounted Tucker trawl described by Siler (1983) suggested that shad density in the main lake area may have been higher than predicted from push net samples, but these samples also revealed that about 60% of the threadfin shad (*Dorosoma petenense*) from the main lake showed signs of emaciation and appeared white rather than silvery in color. Similar abnormalities occurred in about 2% of the larval threadfin shad from the reference area.

Special Investigations

Many feasible environmental concerns were examined and quickly dismissed as possible causes for the rapid decline of the Belews Lake fish community. Thermal loading did not appear to be a possible cause, since numerous impoundments exhibit much higher temperatures yet support healthy fish communities (McNeely and Pearson 1970; Siler and Clugston 1975). Fluctuating water levels were also dismissed because water levels were relatively stable, fluctuating less than 1.5 m between January 1975 and December 1977. Impingement sampling performed during 1976 yielded estimates similar to those of other cooling impoundments supporting healthy fisheries, and entrainment was considered an unlikely cause of the magnitude of impacts noted. Moribund fishes were examined by Duke Power Company fisheries biologists to determine whether disease or parasitism could be considered the cause of the reduced standing stock, altered community structure, or blindness. Additional moribund fish were sent to fish pathologists at Auburn University and North Carolina State University for examination. No incidence of parasitism or disease sufficient to explain the trends was found.

The influence of chemical factors was then hypothesized to be the most likely explanation of the degraded fishery (Cumbie 1978; Harrell et al. 1978). Pesticides and related organic chemicals were considered possible causes of the depressed fishery since it has been shown that these compounds can interfere with reproduction through toxic effects on spawning adults, gametes, developing embryos, or fish larvae (National Academy of Sciences 1972). Cumbie (1978) found that pesticide levels in water samples were below detection limits. DDE and PCB residues were 1 to 2 ng/g in fish, but these concentrations are too low to cause reproductive impairment (National Academy of Sciences 1972). Furthermore, the fish with detectable levels were from the reference area rather than the main lake, where residues were below detection limits. Hence, accumulation of PCB and DDE residues did not appear to cause reproductive impairment in Belews Lake fish (Cumbie 1978).

Intermittent exposure of larval fish to inputs of toxic, non-persistent pesticides (such as organophosphates) from agricultural sources could cause mortality. Water samples were collected following heavy rainfall in 1977 from two stations in the main lake, the reference area, and the three main tributary streams. This rainfall followed a protracted period of near-drought conditions during the spring planting season, and runoff at this time was expected to carry any pesticides and herbicides that might have been present into the lake. No pesticides or herbicides were detected, and investigation of them as a possible cause of reproductive failure was discontinued.

Analysis of water chemistry data showed low nutrient levels, hypolimnetic oxygen depletion, and other trends characteristic of recently impounded piedmont reservoirs. However, several parameters demonstrated trends different from those commonly observed (Table 1). Conductivity in the lake increased from 71 μmho/cm in 1974, to 92 in July 1976, and to 145 at Station 1 in June 1977 (Weiss and Anderson 1978). Accompanying the increased conductivity were elevated values for selenium, arsenic, and zinc at the main lake stations (Table 1). Fish tissue samples collected from Belews Lake in October and November 1976 were analyzed for trace ele-

Table 1. Representative water chemistry data for selected years in 1973-1982 at the reference site (Station 12) and main lake (Station 1) of Belews Lake, North Carolina. Dashes indicate that no samples were collected.

	1973–1974		1976–1977		1979		1982	
	Reference	Lake	Reference	Lake	Reference	Lake	Reference	Lake
Sulfate (mg/l)	4.7	4.9	7.0	17.5	8.9	24.0	11.6	38.0
Calcium (mg/l)	2.2	2.9	6.0	6.6	5.1	12.0	8.4	14.5
Specific conductance (μmho/cm)	64.8	71.2	—	127.6	58.0	115.0	81.0	156.0
Arsenic (μg/l)	—	—	1.6	6.6	1.1	4.3[a]	1.3	3.1[a]
Selenium (μg/l)	—	—	4.1	12.6	1.7	9.5[a]	1.1	8.8[a]
Zinc (μg/l)	13.5	—	17.5	—	4.5	1.8	4.6	4.3
Turbidity (NTU)	18.3	4.2	16.9	1.3	31.1	4.4	13.6	1.8
Ortho-phosphate (mg/l)	0.005	0.005	0.002	0.002	—	—	—	—
Total phosphorus (mg/l)	0.027	0.013	0.010	0.010	—	—	—	—
(NO_2 + NO_3) Nitrogen (mg/l)	0.084	0.061	0.151	0.023	—	—	—	—
Ammonia Nitrogen (mg/1)	0.059	0.042	0.046	0.032	—	—	—	—

[a] Data from Station 5.

ments and heavy metal concentrations. Cumbie (1978) reported that 16 elements were detected in composited skeletal muscle tissue, and two of them, selenium and zinc, were present at concentrations which differed greatly between the reference and main lake stations. Further examination, however, showed that the zinc analyses were apparently in error due to contamination or analytical interferences. Hence, selenium was identified as the factor most likely impacting the fish populations in Belews Lake.

Further research was done to determine the distribution of selenium in the food chain. Selenium concentrations in muscle tissue from fish collected in 1977 were much higher in the main lake than in the reference area (Table 2). Mean

Table 2. Selenium concentrations (μg/g wet weight) in Belews Lake fish collected from the main lake (Stations 1, 5, 6, and 10) and reference location (Station 12), March-September 1977.

	Main lake			Reference area		
Species	Number of fish	Average concentration	Range	Number of fish	Average concentration	Range
March 1977						
Lepomis auritus	—	—	—	2	4.31	2.63–5.99
L. cyanellus	3	9.62	3.40–15.1	6	2.11	1.42–2.69
L. gulosus	—	—	—	8	3.90	3.06–5.03
L. macrochirus	—	—	—	2	3.40	2.34–4.46
Ictalurus spp.	10	6.58	1.76–12.6	10	0.57	<0.10–1.16
Cyprinus carpio	5	12.00	6.77–14.2	4	2.95	1.46–5.49
May 1977						
Ictalurus spp.	11	7.11	0.31–15.5	—	—	—
June 1977						
Lepomis gulosus	3	9.64	4.28–12.6	—	—	—
L. macrochirus	3	8.79	1.93–17.9	—	—	—
L. microlophus	2	31.50	26.5–36.5	—	—	—
Ictalurus punctatus	3	26.10	22.3–30.6	—	—	—
Ictalurus spp.	7	11.25	3.46–15.3	—	—	—
September 1977						
Lepomis cyanellus	—	—	—	8	2.94	2.14–3.63
L. gulosus	4	19.60	13.9–25.5	12	6.71	3.6–10.2
Lepomis spp.	3	54.60	41.0–77.1	—	—	—
L. macrochirus	4	17.20	8.9–25.9	12	6.98	2.90–13.4
L. microlophus	1	21.10	—	5	2.21	1.30–2.70
Ictalurus spp.	6	23.60	10.1–36.2	—	—	—
Cyprinus carpio	4	29.30	14.1–36.7	—	—	—

selenium concentrations in the reference area ranged from 0.57 to 6.98 μg/g, whereas concentrations in fish from the main lake ranged from 6.58 to 54.6 μg/g.

Sunfish were collected from two other water bodies to check for background selenium levels in fish in the region. Green sunfish (*Lepomis cyanellus*) were collected from a pond near Belews Lake, and bluegills were collected from Lake Norman, an impoundment on the Catawba River that supports a diverse fish community (Siler et al. 1982). Analysis of skeletal muscle from these fish showed means of 1.9 μg/g selenium for green sunfish and 0.7 μg/g for bluegills (Cumbie 1978). These values were comparable to the reference area of Belews Lake.

To provide a measure of the possible movement of selenium into the food chain in Belews Lake, plankton were collected from reference and main lake areas during March, June, July, August, and November 1977 (Cumbie 1978). Plankton were collected using an 80-μm mesh net hauled vertically from 10 m to the surface. No attempt was made to separate zooplankton from phytoplankton; however, mesh size selectivity excluded all but the largest phytoplankton. Plankton in the main lake had higher selenium concentrations than those in the reference area. Reference area plankton contained 4–20 μg/g selenium (dry weight), while plankton from the main lake area contained 40–100 μg/g selenium.

Plankton have been found to be efficient accumulators of selenium. Concentration factors of 125×, 190×, and 7,230× relative to water concentrations have been reported by Sandholm et al. (1973), Cumbie (1978), and Copeland and Ayers (1972), respectively. Long-term ingestion of as little as 5 μg/g selenium (dry weight) in the diet of domestic livestock or fish will result in selenium poisoning and associated disorders (Oldfield et al. 1974). Fish which feed on zooplankton in the main body of Belews Lake are apparently exposed to dietary levels of selenium from 40 to 100 μg/g (dry weight). Bioaccumulation of selenium by Belews Lake fish to toxic levels appears to be a valid concern. The role of food organisms in the accumulation of selenium in fish is presently being investigated.

In December 1977, a SO_3 flue gas conditioning system was installed on Unit 1 of Belews Creek Steam Station to increase efficiency of the electrostatic precipitators. This system was modified to also service Unit 2 in August 1978, and a separate system was installed on Unit 2 in 1979. During 1978, use of the SO_3 conditioning system acidified the fly ash slurry, which led to a reduction of the pH in the ash basin effluent. The acidic conditions of the fly ash slurry also reduced selenium and arsenic levels in the effluent (Cumbie 1980). Following the reduction in selenium discharges from the ash basin, selenium concentrations in the lake water declined to lower levels. This prompted North Carolina Wildlife Resources Commission officials and Duke Power Company personnel to evaluate the feasibility of stocking fish into Belews Lake in early 1979. It was agreed that a closely monitored trial stocking might prove insight into the lake's ability to sustain fish populations as selenium concentrations declined. On 25 April 1979, about 200,000 largemouth bass fingerlings were stocked into Belews Lake. Two main lake coves were isolated with block nets to facilitate observation of the stocked bass. None of the stocked bass were recovered from the coves one week after stocking, indicating no survival.

A second cove stocking study was performed by Duke Power Company to determine survival and short-term selenium accumulation rates for bluegill. Bluegill were selected because they are native to the drainage and are a commonly used bioassay organism. Coves of approximately 0.4 hectare surface area were selected at two main lake stations, at mid lake, and in the reference area. The two main lake coves were isolated with a block net, and resident sunfish were removed by electrofishing. Two 0.61 × 0.61 × 0.61-m cages with 3.2-mm nylon mesh were suspended 1 m below the surface in each of the four coves. On 23 July 1978, the eight cages and two main lake coves were stocked with hatchery-reared bluegill fingerlings (50–75 mm total length). Each cage was stocked with at least 100 fingerlings, and the two coves were stocked at a rate of 2,500/hectare. Survival of bluegills in the cages was monitored daily during the first 10 days of the experiment and then periodically until 20 September 1978 (59 days). Dead and moribund fish in cages and coves were counted and preserved. Accumulation of trace metals by bluegills was monitored by analyzing muscle and viscera concentrations prior to and 8 days after (31 July 1978) exposure. About 20 fingerlings were randomly chosen for baseline evaluation and about 10 were collected from the cages and coves at the two main lake stations during each obser-

Table 3. Arsenic, selenium, and zinc concentrations (μg/g wet weight) in composite samples of muscle and viscera from bluegills prior to and after placement in cages and isolated coves in the main lake portion of Belews Lake.

	Arsenic		Selenium		Zinc	
	Muscle	Viscera	Muscle	Viscera	Muscle	Viscera
Pre-exposure	0.03	<0.01	<0.5	0.5	27	20
Station 4						
Cove	0.03	0.04	3.4	5.5	42	33
Cages	<0.01	0.03	0.6	3.6	29	56
Station 1						
Cove	0.01	<0.02	2.0	7.5	57	33
Cages	<0.01	0.20	2.0	5.5	24	36

vation period. Appropriate tissues from each collection were composited prior to chemical analyses. Benthic organisms were collected from the ash basin discharge cove (Station 4) and the reference area with a sweep net and Petersen grab sampler, and organisms were live-picked in the field and classified as *Hexagenia* and non-*Hexagenia*. These samples were freeze-dried and analyzed for trace elements.

Selenium concentrations in surface water during the stocking experiment in the lake and reference areas averaged 10.0 and <1.0 μg/l, respectively. Mortality of caged fish reflected the differences in selenium exposure. Generally, fish stocked in both the reference and main lake cages remained healthy throughout the study, with only a small amount of mortality that was attributable to confinement stress. However, fish held free in coves in the main lake exhibited increasing stress and mortality after the first 3 days of exposure. On day four, these fish began swimming erratically at the surface and exhibited exophthalmia and extreme abdominal distention. When dissected, the abdominal cavity was found to be filled with fluid. These symptoms are identical to those reported in channel catfish (*Ictalurus punctatus*) intraperitoneally injected with acutely lethal doses of sodium selenite (Ellis et al. 1937).

After 8 days exposure, main lake fish showed a higher accumulation of selenium than arsenic or zinc (Table 3). Selenium accumulation ratios ranged from about 1.0 to 15.0, whereas arsenic and zinc accumulation ratios ranged from <1.0 to 2.8. Selenium accumulation was greater in viscera than in muscle, indicating that visceral organs were a primary accumulation site. Adams (1976) noted that selenium appeared to accumulate in the viscera until an equilibrium was reached between the fish and its surroundings. Selenium concentrations in viscera of fish stocked in coves were greater than levels in the corresponding caged fish, suggesting that differences in accumulation may have been due to differences in prey consumption.

Stomach content analyses of fish from the stocking experiment revealed the primary food source to be ephemeropterans, especially *Hexagenia*. Elemental analyses of benthic organisms revealed that *Hexagenia* and non-*Hexagenia* organisms from the ash basin discharge cove contained much higher concentrations of selenium and arsenic than observed in the reference area (Table 4).

Resident fishes were examined histologically to determine the mode of toxicity of selenium as well as the possibility of recovery of fish populations. Sorensen et al. (1984) performed a histopathological and hematological study of green sunfish from main lake (Stations 10 and 11) and reference (Station 12) areas in 1981. Levels of selenium in green sunfish differed significantly between the two sites. The mean concentration at the reference area was 2.3 μg/g, while the fish from the downlake station had a mean concentration of 12.9 μg/g in muscle tissue. The downlake fish exhibited a number of conditions associated with selenium poisoning, including general edema, swollen gill lamellae, reduced hematocrit, necrosis of ovarian follicles, glomerulonephritis, pericarditis, and liver cell degeneration. Similar conditions were induced experimentally in fingerling bluegill maintained in the Environmental Laboratory at Duke Power Company on a diet of selenium-contaminated *Hexagenia* collected from Belews Lake. These observations emphasize the importance of the

Table 4. Arsenic, selenium, and zinc concentrations (μg/g wet weight) in benthic invertebrates from the ash basin discharge area (Station 4) and reference area (Station 12) of Belews Lake.

	Arsenic		Selenium		Zinc	
	Hexagenia	Non-*Hexagenia*	*Hexagenia*	Non-*Hexagenia*	*Hexagenia*	Non-*Hexagenia*
Station 4 cove (ash basin discharge area)	3.05	2.02	8.36	6.17	11.25	6.20
Station 12 (reference area)	0.11	0.12	0.99	0.39	21.55	7.60

food chain as the pathway by which selenium accumulated in Belews Lake fish.

The possibility that selenium and arsenic are removed from the water column by natural physical and chemical processes and deposited in bottom sediments (Cumbie 1978; Holland 1979) prompted more detailed investigation of selenium and arsenic dynamics in sediments of Belews Lake (Cumbie 1984). Based on sediment dredge samples, mean selenium concentrations near the ash basin discharge were higher (11.5 ± 8.6 μg/g) than those in the reference area (1.5 ± 0.3 μg/g). These results compared well with the earlier observations of Cumbie (1978) and Holland (1979). Selenium levels in the reference area are comparable to those reported in the literature from uncontaminated locations (Adams 1976; Wiersma and Lee 1971). In core samples taken to investigate vertical distribution of contaminants in sediments, it was found that selenium and arsenic were limited to the surface materials (to depths of 5 cm or less). Selenium and arsenic concentrations tended to increase in the core sections with increasing overlying water column depth. Cumbie (1984) attributed this to "sediment focusing," where the selenium-contaminated fine sediments are transported from littoral areas to deeper locations due to wave action and wind-induced currents. Sedimentation in Belews Lake is very slow and was estimated by Cumbie (1984) to average 0.1–1.0 cm/yr.

Remedial and Future Actions

Evaluation of the various studies conducted on Belews Lake and the Belews Creek ash-handling system has led to the conclusion that the ash basin discharge into Belews Lake must be eliminated to avoid continuing adverse impacts and to allow recovery of the ecosystem. Analyses of alternatives have led to a decision to convert the present wet bottom ash-fly ash sluicing system to a combined wet bottom ash-dry fly ash system. This will eliminate the introduction of fly ash sluice water to the ash settling pond. Concurrently, the discharge point for the ash settling pond will be relocated so that the bottom ash sluice water will enter the Dan River rather than Belews Lake. Then, all ash sluice water discharges to Belews Lake will be eliminated under normal operating conditions. During drought, the ash basin discharge could be temporarily returned to the reservoir to maintain operating water levels.

Studies to document recovery of Belews Lake from selenium contamination will be performed in conjunction with the ash-handling system modifications. These studies will include evaluation of the rate at which sediment focusing and deposition processes result in sediment conditions that will allow the Belews Lake food chain to purge itself of selenium. Assessment of the time required for recovery of fish populations and the need for further mitigative actions will be a prime objective of these studies.

Many cooling impoundments that receive ash basin effluent discharges support excellent fisheries. Creel surveys conducted on Duke Power Company cooling impoundments found the highest fishing pressure, success, and harvest occur in proximity to the steam stations (Siler et al. 1986). The environmental problems encountered at Belews Lake resulted from a combination of factors. The most important factors were those relating to the loading and retention of selenium. Loading of selenium to the lake was influenced by chemical characteristics of coal, design and operation of electrostatic precipitators, and treatment effectiveness of the ash basin. However, factors controlling the retention of selenium in Belews Lake were of overriding importance. The thermal regime, small surface area, and long retention time of Belews Lake combined to exacerbate bioaccumulation of elements that might otherwise have been of little concern.

The Belews Lake experience offers valuable lessons on the fisheries management implications of ash basin effluent discharges into cooling ponds.

References

ADAMS, W. J. 1976. The toxicity and residue dynamics of selenium in fish and aquatic invertebrates. Doctoral dissertation, Michigan State University, East Lansing, Michigan, USA.

COPELAND, R. A., AND J. C. AYERS. 1972. Trace element distributions in water, sediment, phytoplankton, zooplankton, and benthos of Lake Michigan: A baseline study with calculations of concentration factors and buildup of radioisotopes in the food web. Special report 1. Environmental Research Group, Inc., Ann Arbor, Michigan, USA.

CUMBIE, P. M. 1978. Belews Lake environmental study report: Selenium and arsenic accumulation. Technical report 78-04. Duke Power Company, Charlotte, North Carolina, USA.

CUMBIE, P. M. 1980. Effects of pH adjustment and SO_3 fly ash conditioning on the Belews Creek Steam Station Ash Basin. Technical report 80-01. Duke Power Company, Charlotte, North Carolina, USA.

CUMBIE, P. M. 1984. Deposition of selenium and arsenic in sediments of Belews Lake, North Carolina. Pages 1–27 *in* B. J. Ward, editor. Workshop proceedings: The effects of trace elements on aquatic ecosystems. Report EA-3329. Electric Power Research Institute, Palo Alto, California, USA.

ELLIS, M. M., H. L. MOTLEY, M. D. ELLIS, AND R. O. JONES. 1937. Selenium poisoning in fishes. Proceedings of the Society for Experimental Biology and Medicine 36:519–522.

GRINSTEAD, B. G., R. M. GENNINGS, G. R. HOOPER, C. A. SCHULTZ, AND D. A. WHORTON. 1978. Estimation of standing crop of fishes in the predator-stocking-evaluation reservoirs. Proceedings of the Annual Conference of the Southeastern Association of Fish and Wildlife Agencies 30:120–130.

HARRELL, R. D., R. L. FULLER, AND T. J. EDWARDS. 1978. An investigation of the fish community of Belews Lake, North Carolina. Technical report 78-07. Duke Power Company, Charlotte, North Carolina, USA.

HOLLAND, E. A. 1979. Arsenic and selenium in the water, sediments, and biota near a coal-fired power-plant—Belews Lake, North Carolina. Master's thesis, University of North Carolina, Chapel Hill, North Carolina, USA.

MCNEELY, D. L., AND W. D. PEARSON. 1974. Distribution and condition of fishes in a small reservoir receiving heated waters. Transactions of the American Fisheries Society 3:518–530.

NATIONAL ACADEMY OF SCIENCES, U.S. 1972. Water quality criteria, 1972. National Academy of Sciences, Washington, D.C., USA.

OLDFIELD, J. E., W. H. ALLAWAY, H. A. LAITENEN, H. W. LAKIN, AND O. H. MUTH. 1974. Selenium. Pages 57–63 *in* U.S. National Committee for Geochemistry, National Reserach Council. Geochemistry and the environment. Vol. 1. The relation of selected trace elements to health and disease. National Academy of Sciences, Washington, D.C., USA.

SANDHOLM, M., H. E. OKSANEN, AND L. PESONEN. 1973. Uptake of selenium by aquatic organisms. Limnology and Oceanography 18:496–499.

SILER, J. R. 1983. A description of a trawl handling structure for a bow-fished Tucker Trawl. Progressive Fish Culturist 45:217–219.

SILER, J. R., AND J. P. CLUGSTON. 1975. Largemouth bass under conditions of extreme thermal stress. Pages 333–341 *in* R. H. Stroud and H. Clepper, editors. Black bass biology and management. Sport Fishing Institute, Washington, D.C., USA.

SILER, J. R., W. J. FORIS, AND M. C. MCINERNY. 1986. Heterogeneity in fish populations within a reservoir. Pages 122–136 *in* G. E. Hall and M. J. Van Den Avyle, editors. Reservoir fisheries management: Strategies for the 80's. Reservoir Committee, Southern Division American Fisheries Society, Bethesda, Maryland, USA.

SILER, J. R., R. E. LEWIS, AND B. K. BAKER, R. A. HANSEN, AND G. E. VAUGHAN. 1982. Fish. Chapter 10 *in* J. E. Hogan and W. D. Adair, editors. Lake Norman summary, Volume 11. Technical report 82-02. Duke Power Company, Charlotte, North Carolina, USA.

SORENSEN, E. M. B., P. M. CUMBIE, T. L. BAUER, J. S. BELL, AND C. W. HARLAN. 1984. Histopathological, hematological, condition factor, and organ weight changes associated with selenium accumulation in fish from Belews Lake, North Carolina. Archives of Environmental Contamination and Toxicology 13:153–162.

VAN HORN, S. L. 1978. Piedmont Fisheries Investigation. Study I. Develoment of the sport fishing potential of an industrial cooling lake. Final report. Federal Aid Project F-23. North Carolina Wildlife Resources Commission, Division of Inland Fisheries, Raleigh, North Carolina, USA.

WEISS, C. M., AND T. P. ANDERSON. 1978. Belews Lake: A summary of a seven-year study (August 1970–June 1977) to assess environmental effects of a coal-fired power plant on a cooling pond. Department of Environmental Sciences and Engineering, University of North Carolina at Chapel Hill, Chapel Hill, North Carolina, USA.

WEISS, C. M., P. H. CAMPBELL, AND D. R. LENAT. 1978. Environmental effects of power plant operation, Belews Lake, Years V and VI. July 1974–June 1976. Department of Environmental Sciences and Engineering, University of North Carolina at Chapel Hill, Chapel Hill, North Carolina, USA.

WIERSMA, J. H., AND G. F. LEE. 1971. Selenium in lake sediments: Analytical procedure and preliminary results. Environmental Science and Technology 5:1203–1206.

Pages 270–277 *in* G. E. Hall and M. J. Van Den Avyle, editors.
Reservoir Fisheries Management: Strategies for the 80's.
Reservoir Committee, Southern Division American Fisheries Society, Bethesda, Maryland. 1986.

MANAGEMENT OF RESERVOIR RELEASES

Improving the Downstream Environment by Reservoir Release Modifications

RICHARD J. RUANE, CHARLES E. BOHAC, WILLIAM M. SEAWELL, RICHARD M. SHANE

Tennessee Valley Authority
Division of Air and Water Resources
Knoxville, Tennessee 37902

ABSTRACT

Tailwater fisheries can be adversely impacted by reservoir releases, but engineering modifications can significantly improve them. Physical and chemical factors that can affect tailwater fisheries include low dissolved oxygen, changes in temperature, supersaturation of total dissolved gases, dissolved chemical substances such as iron, manganese, hydrogen sulfide, and ammonia, and low streamflow. Tailwaters usually are affected by only a few of these factors. Each tailwater is uniquely affected depending on upstream reservoir operations, the water quality of the reservoir releases, and tailwater characteristics. Methods that have been demonstrated to improve tailwater fisheries include aeration of the hypolimnion in the upstream impoundment, destratification of the upstream impoundment, hydroturbine aeration, tailrace aeration, multi-level intakes for the upstream reservoir, sluicing, modified reservoir operations, and reregulation of the present hydropower releases. Few of these methods have been applied on a widespread basis; therefore, experience with these techniques is limited, and associated fishery benefits have not yet been adequately documented.

Numerous scientific investigations have documented changes in water quality that take place when a river is impounded. While such studies generally indicate suitable water quality, some physical and chemical constituents deteriorate, particularly in hypolimnetic releases from stratified reservoirs with long water storage or retention times. Dissolved oxygen concentrations can be lower than desirable, and temperature, dissolved iron and managenese, hydrogen sulfide, ammonia, and turbidity often reach levels that adversely impact downstream uses. Downstream uses are also impacted by the intermittent streamflows related to reservoir releases. These uses include recreational fishing, municipal and industrial water supplies, and the ability of the river to assimilate treated wastewaters.

This paper reviews the characteristics of environmental impacts associated with reservoir releases and summarizes various structural and operational methods that may be employed to alleviate adverse impacts.

CHARACTERISTICS OF ENVIRONMENTAL IMPACTS ASSOCIATED WITH RESERVOIR RELEASES

Water Quality

Most water quality problems associated with reservoirs stem from stratification, which isolates the hypolimnion from reoxygenation processes. Oxygen-demanding processes continue to work within the hypolimnion depressing the dissolved oxygen (DO) content. Sometimes oxygen is depleted entirely resulting in anoxic conditions, which can lead to resolubilization of chemical substances that impact downstream uses.

Although stratification provides coldwater storage in reservoirs, the water may not be at an optimal temperature or be adequate to provide for a cold tailwater fishery. Conversely, for river reaches supporting warmwater and coolwater fisheries, reservoir releases may be too cold for an extended period each year to permit adequate reproduction or growth. Additionally, reservoir releases affected by hydropower are subject to sudden flow fluctuations, causing rapid temperature changes.

Several studies have evaluated environmental impacts associated with reservoir releases (TVA 1978, 1981; Walburg et al. 1983). Following is a summary of typical consequences of impounded waters for various physical and chemical constituents.

Dissolved oxygen.—Low concentrations of DO can adversely affect fish and benthic invertebrates in tailwaters (Pearson et al. 1968; Hill 1980). These effects, particularly on the salmonid fishes, have been extensively studied in laboratory situations (Doudoroff and Shumway 1970; Davis 1975; USEPA 1983). Although the concentrations that adversely affect fish have been found to vary widely, general effects include depressed feeding and growth rates, impaired reproduction, anoxia, and death. Concentrations producing effects on fish vary widely under apparently similar conditions, and interpretation of these contradictory findings is quite subjective. The body of knowledge on warmwater fish species is much more limited than that on salmonids, and applicable criteria, for the most part, are based on incomplete data. Application of laboratory results to stream conditions, including tailwaters, is an even greater concern with respect to defining the DO conditions required to provide a desired quality of fishery. The comparison of laboratory results with observations in field situations has not been defined adequately (USEPA 1983).

Davis (1975) concluded that invertebrates would probably be adequately protected by DO standards established for fish. Low DO may drastically affect species composition of benthic invertebrates in tailwaters, resulting in less diverse communities of low oxygen-tolerant species (chironomids, isopods, and amphipods) dominating these waters. The effects of low DO concentrations on tailwater biota are often difficult to separate from the accompanying conditions of altered temperatures and flows that affect organisms in the same manner.

Dissolved oxygen concentrations in tailwaters can change rapidly. Photosynthetic activity during shutoff periods, combined with release of deoxygenated water during hydropower peaking, can change tailwater dissolved oxygen concentrations from as high as supersaturation to almost complete depletion within minutes. The effects of such rapid changes on fish are not well known.

One case history was reported by Weithman and Haas (1982), who reported a drastic decline in rainbow trout catch rates below Table Rock Dam, Missouri (Lake Taneycomo), during periods of low DO.

Temperature.—The temperature of reservoir releases is a critical factor affecting fish and benthic invertebrates. The low temperature of some hypolimnetic releases can produce beneficial effects; trout fisheries are maintained in many large tailwaters where they would not exist under conditions of natural streamflow (Axon 1976). The temperature of releases also can adversely affect native stream fishes, particularly warmwater species. Major changes from the temperature pattern of an unregulated stream include warmer temperatures during the autumn, slower rates of warming and cooling in the spring and fall, respectively, and depressed temperature during the summer months. Such changes can inhibit reproduction and growth of warmwater fish species (Dendy and Stroud 1949) and cause major changes in the fish community (Edwards 1978).

Sudden decreases in water temperature that occur at the onset of power generation can induce cold shock reaction in fish in the tailwater. Effects of cold shock include impairment of osmoregulation, changes in behavior, disorientation, loss of equilibrium, immobilization, and death (Block 1974). When the magnitude of temperature fluctuation is not great enough to produce drastic effects, more subtle effects may occur such as increased vulnerability to predation (Coutant 1975). Information on the potential impact of thermal shock on the total tailwater fishery is not available.

Effects of an altered temperature regime on benthic organisms in tailwaters receiving cold hypolimnetic releases include delayed emergence, altered species composition, and lower diversity (Pearson et al. 1968; Ward 1976a). Interaction with water temperature is also significant in determining the impact of low dissolved oxygen concentrations on fish and benthic invertebrates. Depressed temperatures may permit some aquatic life to tolerate lower DO levels than usual.

Total dissolved gases.—Supersaturation of total dissolved gases (TDG), consisting primarily of nitrogen and oxygen, can be a serious problem in reservoir releases under certain conditions where air is induced under pressure. Supersaturation below dams is usually associated with release of water over spillways. MacDonald and Hyatt (1973) and Ebel (1969) reported supersaturation of nitrogen in water caused by turbine venting. However, the Tennessee Valley Authority (TVA) has not experienced any values of TDG that exceed criteria when using turbine aeration.

Excess gases can produce eye damage and other sublethal effects while higher levels can produce death. Although high partial pressures of either nitrogen or oxygen can produce gas bubble disease symptoms, TDG pressure has been reported to be the primary factor. Levels as low as 105% continuous concentrations of TDG can damage hatchery trout (Bouck 1980), while higher continuous concentrations (125%) can cause mortality (Bouck et al. 1976). For wild fish, Ebel et al. (1979) recommended a "safe" maximum concentration of 115% for rivers. Young fish are much more sensitive than adults, and intermittent exposures are tolerated better than continuous exposures. Relatively little is known about the TDG requirements for nonsalmonids, but studies have indicated that most warmwater fishes are more tolerant than salmonids (Weithkamp and Katz 1980). A few studies have indicated an avoidance reaction to supersaturated TDG concentrations by some fishes; however, there have been conflicting results (Weithkamp and Katz 1980), and more data are needed to resolve this issue.

Gas bubble disease can also affect other aquatic life. Nebeker et al. (1976) reported that supersaturated gases affect *Daphnia,* crayfish, and stoneflies, but specific effects are not well documented. A maximum saturation value of 110% TDG is generally considered adequate to protect salmonids, but more work is needed on warmwater species and invertebrates.

Dissolved chemical substances. —Thermal stratification of reservoirs followed by depletion of the hypolimnetic DO provides potential for the chemical reduction of metals, such as iron and manganese, trapped in the sediments in their oxidized and insoluble form. Fe^{+++}, while almost completely insoluble except at low pH, is easily reduced to Fe^{++} and dissolved as ferrous bicarbonate under anoxic conditions; it can reach high values expecially in meromictic lakes with values of up to 40 mg/liter having been reported (Ruttner 1963). When oxygen is reintroduced into the environment, even at very low levels such as during lake turnover, the iron is quickly oxidized and precipitated as ferric hydroxide. Most of the iron then settles back into the sediments before it is released through the reservoir outlets. Manganese reacts similarly to iron except that it is somewhat more easily reduced and does not readily oxidize like iron when pH is at common levels (5.5–7.5). Therefore, dissolved manganese can occur with higher oxygen levels than dissolved iron (Ruttner 1963).

Reduced substances in hypolimnetic releases, including iron, manganese, sulfur, and ammonia which can be produced by anaerobic processes in the reservoir, can significantly impact tailwater biota. Hydrogen sulfide is toxic at very low concentrations (0.002 mg/liter), and ammonia toxicity increases with pH (EPA 1976). The effects of manganese and iron on aquatic life are significant only at concentrations greater than about 1 mg/liter (Gordon 1983). Synergistic effects of reduced substances may be involved in some cases (Grizzle 1981). The portion of the tailwater impacted by reduced substances can be decreased by reaeration measures, but the effects of precipitated iron and manganese on tailwater biota can still be a concern.

Streamflow in Tailwaters

Effective operation for hydropower at tributary storage reservoirs can result in periods when turbines are idle; at those times, downstream flows are reduced to minimum levels produced by leakage from the dam and local inflows. Low-flow periods can last from several hours each day to periods of one or two months during drought conditions. Low-flow periods are often a direct result of attempts to maximize hydropower production and capacity to replace expensive thermal generation that would otherwise be needed during peak power demand periods.

Daily low-flow periods result from hourly load variations. For example, in the TVA system summer peaking releases normally begin about midmorning and continue for the remainder of the daylight hours. Winter load peaks tend to occur in the morning and again in the evening. Unless there is an above-normal amount of water in storage, the turbines do not usually operate during offpeak hours.

Annual low-flow periods, where turbines are idle for days, weeks, or sometimes even months, result from seasonal load variations. During low-load seasons, normal TVA reservoir operations require that hydrogeneration be reduced in spring so that reservoir pool levels can be raised and sometimes in fall so that water can be saved for future use. Extended periods with no generation are most likely to occur in dry years, when limited reservoir inflows make economy of hydropower generation especially critical.

Studies actually documenting changes in fish and invertebrate populations after increased minimum flows are not available to more accurately predict effects and validate models. Most of the direct information available is from comparisons of tailwaters with different minimum flow conditions (Blanz et al. 1970; Hill 1980). In general, these studies show greater fish and benthic invertebrate diversity in tailwaters with higher minimum flows.

Although low flows can adversely affect fish, they can also contribute to higher angler catch rates. Deppert (1978), reported 86% of trout harvested in tailwaters were caught during low water release periods. Aggus et al. (1979) showed that effort and harvest in the Bull Shoals Reservoir tailwater increased with lower daily reservoir releases. Thus, it could be inferred that increasing minimum flows in tailwaters can also decrease angler effectiveness.

Procedures have been developed for assessing requirements for instream flows (Orsborn and Allman 1976; Stalnaker and Arnette 1976). This methodology uses water depth and velocity requirements for various life stages of a particular fish species to arrive at minimum acceptable streamflows. One problems with this methodology lies in the validity of assumptions about requirements for individual species. These assumptions often involve compromises and are quite arbitrary for some species. A simplified method for estimating required flows, such as that proposed by Newcombe (1981), may be more useful in quantifying needs for tailwater conditions. However, even this simplified method requires decisions about depth and velocity requirements for fish species, many of which are poorly supported in the literature. Hence, "reasonable" minimum flows in tailwaters need to be defined on the basis of costs, practical feasibility, and subjective analysis.

The gross effects of fluctuating flows and periodic shutoffs on tailwater invertebrates have received some attention (Ward 1976b). However, most methods for assessing instream flow requirements are keyed to fish, and needs of benthic invertebrates have been largely ignored. The assumption that providing directly for the fish's habitat needs and food requirements will provide suitable conditions for invertebrates is probably generally valid, but this area needs additional investigation.

Alternative Methods of Water Quality Improvement

Essentially seven alternatives can be considered for improving the quality of reservoir releases. These methods include various schemes of aerating the reservoir, destratifying the reservoir, aerating the releases, or aerating water in the tailrace.

Hypolimnetic Aeration

High-purity oxygen, or in some cases air, injection into the hypolimnion has been investigated at a number of reservoirs. Studies of particular importance include those at Fort Patrick Henry, Tennessee (Fain 1978), and Clark Hill, Georgia and South Carolina (Speece et al. 1982). Pure oxygen has the advantage of minimizing potential difficulties arising from excess total gas supersaturation and, when used in combination with small-pore diffusers located near the reservoir outlets, is highly efficient. Oxygen costs tend to be the major expense whether the oxygen is trucked to the site or generated onsite. Hypolimnetic aeration can be achieved using pneumatic systems as well as high-purity oxygen systems. Several of these systems were compared by Fast and Lorenzen (1976a) for the San Vicente Reservoir, California.

Destratification

A bubble column produced with compressed air will create an upwelling in a reservoir that, in combination with wind energy, can be used to prevent stratification from occurring or to destratify if that has already occurred (Pastorok et al. 1982). Compressed air has been investigated in a number of studies, but mechanical mixing has been investigated on a more limited basis. Mechanical mixing devices include high-capacity, low-head pumps to exchange water between the hypolimnion and the epilimnion (Dortch 1979) and airplane propellers attached to floats (Garton pumps) and powered through gear reducers to force the epilimnetic water into the hypolimnion (Garton and Punnett 1981).

Fast and Lorenzen (1976b) compared the costs of methods of hypolimnetic aeration to methods of destratification. They concluded that each system has unique properties that must be considered for each application. Only then can cost advantages be determined.

Destratification methods can result in the increase of quantities of seston into the releases and can be expected to increase production of filter feeding invertebrates (Webster et al. 1979).

Hydroturbine Aeration

Aeration within the turbines at hydroelectric plants has received considerable interest recently (Bohac et al. 1983). Turbines are attractive for aeration systems because the pressure in the draft tube immediately below the turbine often is be-

Table 1. Low flow augmentation alternatives for Norris Dam, Tennessee.

Alternative	Minimum flow (ft^3/second)		Estimated costs (millions of 1980 dollars)	
	Preferred[a]	Guaranteed	Initial	Annual
Rock weir	—	200	0.5	0.04
Sluices	300	200	—	1.13
Collapsible barrier dam	300	200	5.8	0.67
Sluices	600	200	—	1.40
Collapsible barrier dam	600	200	5.8	0.79
Small hydro unit[b]	600	200	14.1	0.47
Sluices	1,000	200	—	1.70
Collapsible barrier dam	1,000	200	5.8	0.94

[a] Daylight hours (6 a.m.–8 p.m.) during warm months (April–September); could be less during moderate to severe droughts.
[b] Normal flow through each of the existing turbines ranges from about 3,000 to 4,500 ft^3/second. Six hundred ft^3/second would be such a small flow that generation would be inefficient. Therefore, a small turbine designed for 600 ft^3/second could be added to the power house.

low atmospheric pressure, and draft tubes downstream from turbines usually provide long contact times, high turbulence, and high water pressures that greatly increase the transfer of oxygen from air bubbles into the water. However, it is unlikely that a given turbine will aspirate at all operating conditions. Aspiration can be enhanced by using baffles or deflector plates within the turbine system to create a localized vacuum immediately downstream from the baffles. In lieu of baffles, compressed air also has been used (Bohac et al. 1983).

Turbine aeration is quite efficient because comparatively less energy is required to aspirate or inject air into the water immediately downstream from the turbine wheels. Travel time in the draft tube is significant, and the pressure after injection usually exceeds atmospheric pressure throughout the draft tube system. Oxygen transfer efficiencies range from 40 to 70%. The presence of air and baffles does, however, cause the turbine to operate less efficiently, and the cost of lost power generation can be significant (Bohac et al. 1982).

Tailrace Aeration

Methods of tailrace aeration include diffused air, mechanical aeration, and high-purity oxygen systems. Ruane (1972) evaluated these for possible use at Fort Patrick Henry Dam (Tennessee). The efficiency of such systems is low (8 to 12%) because of shallow water depth, and other aeration methods are deemed more cost-effective.

Multi-level Intakes

Inlet structures which enable operators to select the reservoir strata from which withdrawals are to be made can be an effective method of meeting downstream water quality goals. Generally, multi-level intakes have been used on nonpower dams. However, in cases where temperature manipulation was required at hydropower dams, deflectors have been placed in front of the penstock and trash racks to draw water from a higher and warmer stratum. Labadie et al. (1981) combined simulation and optimization techniques for determining operational guidelines for meeting downstream temperature objectives.

Sluicing

Releasing water through dam sluices can be used to aerate releases from reservoirs. The sluices themselves can provide adequate aeration in some situations (Wilhelms and Smith 1981). In other cases, aerating cone-sleeved valves have been used.

Release Manipulation

Releases from reservoirs that stratify can be sequenced to store as much or as little cold water as practicable. This method has been investigated by TVA to maximize the duration of releases below 20 C throughout the summer (TVA 1981).

Alternative Methods of Flow Enhancement

Preliminary feasibility studies of alternatives for low flow augmentation have been conducted by TVA (1981) for three large tributary reservoirs: Norris, Cherokee, and Douglas (all in Tennessee). Alternatives investigated for Norris Reservoir are summarized in Table 1. Only sluicing alternatives have been considered at Cherokee and Douglas Reservoirs, and they are similar to the three sluicing alternatives shown for Norris Reservoir. Table 1 presents a comparison of the minimum flows provided with each alternative along with estimated initial and annual costs. Initial costs reflect construction-related costs. Annual costs are primarily power losses associated with head loss at the dam and pulsing requirements during periods when generation is of minimum value.

Among the alternatives considered for flow regulation, the only one to reach the experimental phase is the rock weir to be located 2 mi below Norris Dam. This is a low dam (5 ft high) constructed of layers of wire baskets filled with rocks (gabions). Discharge is controlled by 54, 1-ft diameter steel discharge pipes located near the bottom of the weir. Twenty of the pipes have float-activated controls that open as the water level drops, permitting a constant minimum discharge of about 200 ft^3/second at all times. With the pipe controls, channel storage behind the weir is adequate to maintain 200 ft^3/second on a normal day of peaking, with no off-peak pulses required. If the turbines are to be idle for one day or more, then two, 30-minute, one-turbine pulses would be required each day.

Extensive testing of alternate weir shapes and widths was conducted with a 1:7 scale physical model. Of major concern was the nature of overflow conditions and the degree of risk to boaters who might pass over the weir. Model results showed that the 5-ft high weir should be at least 21 ft in length and be stepped down gradually a total of 2 ft on the downstream side to avoid the hazard of standing waves and rollers below the weir.

Post construction monitoring of the Norris Dam tailwater is planned for 1984 and 1985. This will include daily creel surveys, monitoring of stress-related variables such as water temperature and dissolved oxygen, and fishery measurements (population, growth, and food supplies).

Conclusions and Recommended Research

Evaluations of specific reservoir projects generally indicate that benefits are less than cost when engineering methods are designed to achieve water quality standards and streamflow criteria. However, it is possible that presently accepted water quality and streamflow criteria are not required for maintenance of aquatic life in certain tailwaters, allowing the possibility that fishery benefits can be attained with lower-cost engineering measures that will provide less than the optimum desired water quality and streamflow levels.

Advances in our ability to control downstream effects of reservoir releases have been slowed by the lack of evaluations of specific techniques and engineering designs. Various demonstration efforts should be implemented at representative sites around the country to evaluate full-scale methods to alleviate low DO conditions, enhance intermittent streamflows, and modify inappropriate water temperatures. During the demonstration period, releases from selected reservoirs should be regulated to increase DO to various concentrations up to State and Federal criteria. Various ranges of minimum flows should also be maintained for evaluating fishery benefits.

Demonstrations are also needed to increase DO and reduce concentrations of iron, manganese, hydrogen sulfide, and ammonia in low-level reservoir releases from nonpower as well as hydroelectric power projects, and in both coldwater and warmwater releases. Such demonstrations will have to consider oxygenation of hypolimnetic waters, or mixing of hypolimnetic waters with epilimnetic waters in the reservoir, using means that will not increase the total dissolved gas concentrations to levels that impact aquatic life in the tailwater. This approach should allow precipitation of the undesirable water quality constituents in the reservoir instead of releasing them downstream.

One reaeration method having great promise is with turbines designed specifically for that purpose. Because many of the existing turbine wheels have been in operation for 30 to 50 years, a large number of these will be replaced during the next 10 to 20 years. A demonstration project is needed to develop and test a prototype "ventable turbine" i.e., a replacement turbine wheel designed to achieve maximum air aspiration at minimum power losses. Installation and field testing of a suitable model should be accomplished at an existing hydropower facility and include measuring water quality, fishery and benthos responses, power losses, and turbine cavitation.

DO criteria based on laboratory tests probably are not representative of the actual field environment, and tests in large-scale microcosms (mesocosms) are needed to provide the linkages between laboratory and field conditions. Research on aquatic life under actual field conditions often is impractical and difficult to perform with scientific credibility; therefore, large-scale evaluations in mesocosms are proposed for various species of fish important to sport and commercial interests. Such tests are especially needed for DO conditions that often occur downstream from reservoir projects, i.e., the effects of low

DO for various periods each day, over three- to six-month periods. Similarly, research is needed to better determine the fishery effects of rapid temperature decreases associated with reservoir releases.

The instream-flow analysis procedures proposed to date result in a wide range of desirable streamflow levels, some which cannot be achieved without impacting other reservoir uses. Further research is needed to determine the fishery benefits of streamflow levels that can be provided without significantly affecting other reservoir project uses and can be provided in a cost-effective manner.

References

Aggus, L. R., D. I. Morais, and R. F. Baker. 1979. Evaluation of the trout fishery in the tailwater of Bull Shoals Reservoir, Arkansas, 1971–73. Proceedings of the Annual Conference of the Southeastern Association of Fish and Wildlife Agencies 31:565–573.

Axon, J. R. 1976. Review of coldwater fish management in tailwaters. Proceedings of the Annual Conference of the Southeastern Association of Game and Fish Commissioners 29:351–355.

Blanz, R. E., C. E. Hoffman, R. U. Kilambi, and C. R. Liston. 1970. Benthic macroinvertebrates in cold tailwaters and natural streams in state of Arkansas. Proceedings of the Annual Conference of the Southeastern Association of Game and Fish Commissioners 23:281–292.

Block, R. M. 1974. Effects of acute cold shock on the channel catfish. Pages 109–118 *in* J. W. Gibbons and R. R. Sharitz, editors. Thermal ecology. Symposium series CONF-730505, Office of Information Services, U.S. Atomic Energy Commission, Oak Ridge, Tennessee, USA.

Bohac, C. E., E. D. Harshbarger, J. L. Davis, R. J. Ruane, and S. Vigander. 1982. Methods of reservoir release improvement. Proceedings of the 37th Industrial Waste Conference, Purdue University, West Lafayette, Indiana, USA.

Bohac, C. E., E. D. Harshbarger, J. W. Boyd, and A. W. Lewis. 1983. Techniques for reaeration of hydropower releases. Technical report E-83-5. U.S. Army Corps of Engineers Waterways Experiment Station, Vicksburg, Mississippi, USA.

Bouck, G. R. 1980. Etiology of gas bubble disease. Transactions of the American Fisheries Society 109:703–707.

Bouck, G. R., G. A. Chapman, P. W. Schneider, Jr., and D. G. Stevens. 1976. Observations on gas bubble disease among wild adult Columbia River fishes. Transactions of the American Fisheries Society 105:114–115.

Coutant, C. C. 1975. Temperature selection by fish—a factor in power plant impact assessments. *In* Environmental effects of cooling systems at nuclear power plants. Publication 701, Environmental Sciences Division, Oak Ridge National Laboratory, Oak Ridge, Tennessee, USA.

Davis, J. C. 1975. Minimal dissolved oxygen requirements of aquatic life with emphasis on Canadian species: a review. Journal of the Fisheries Research Board of Canada 32:2295–2332.

Dendy, J. S., and R. H. Stroud. 1949. The dominating influence of Fontana Reservoir on temperature and dissolved oxygen in the Little Tennessee River and its impoundments. Journal of the Tennessee Academy of Science 24(1):41–51.

Deppert, D. L. 1978. The effect of striped bass predation and water quality, the rainbow trout fishery of the lower Illinois River. Master's thesis, University of Oklahoma, Norman, Oklahoma, USA.

Dortch, M. S. 1979. Artificial destratification of reservoirs. Technical report E-79-11. Office of the Chief of Engineers, U.S. Army Corps of Engineers, Washington, D.C., USA.

Doudoroff, P., and D. L. Shumway. 1970. Dissolved oxygen requirements of freshwater fishes. Fisheries technical paper 86, United Nations Food and Agricultural Organization, Rome, Italy.

Ebel, W. J. 1969. Supersaturation of nitrogen in the Columbia River and its effect on salmon and steelhead trout. U.S. National Marine Fisheries Service Fishery Bulletin 68:1–11.

Ebel, W. J., K. T. Beiningen, G. R. Bouck, W. R. Penrose, and D. E. Weitkamp. 1979. Gases, total dissolved. Pages 113–118 *in* R. V. Thurston, R. C. Russo, C. M. Fetterolf, T. A. Edsall, and Y. M. Barber, Jr., editors. A review of the EPA redbook: quality criteria for water. Water Quality Section, American Fisheries Society, Bethesda, Maryland, USA.

Edwards, R. J. 1978. The effect of hypolimnion reservoir releases on fish distribution and species diversity. Transactions of the American Fisheries Society 107:71–77.

Fain, T. G. 1978. Evaluation of small-pore diffuser technique for reoxygenation of turbine releases at Fort Patrick Henry Dam. WK28-1-32-100, Division of Water Resources, Tennessee Valley Authority, Norris, Tennessee, USA.

Fast, A. W., and M. W. Lorenzen. 1976a. Synoptic survey of hypolimnetic aeration. Journal of the Environmental Division Proceedings of the American Society of Civil Engineers 102:1161.

Fast, A. W., and M. W. Lorenzen. 1976b. Comparative study with costs of hypolimnetic aeration. Journal of the Environmental Division Proceedings of the American Society of Civil Engineers 102:1175.

Garton, J. E., and R. F. Punnett. 1981. Quality improvement of releases from reservoirs. Pages 782–791 *in* H. G. Stefan, editor. Symposium on surface water impoundments. American Society of Civil Engineers, Minneapolis, Minnesota, USA.

Gordon, J. A. 1983. Iron, manganese, and sulfide mechanics in streams and lakes. Report TTU-CE-83-2. Department of Civil Engineering, Tennessee

Technological University, Cookeville, Tennessee, USA.

GRIZZLE, J. M. 1981. Effects of hypolimnetic discharge on fish health below a reservoir. Transactions of the American Fisheries Society 110:29–43.

HILL, D. M. 1980. Characteristics and determinants of the fisheries resources of three cold tailwaters in Tennessee. Journal of the Tennessee Academy of Sciences 55(4):130–136.

LABADIE, J. M., D. G. FONTANE, AND B. LOFTIS. 1981. Optimal control of reservoir discharge quality through selective withdrawal. Pages 772–781 *in* H. G. Stefan, editor. Symposium on surface water impoundments. American Society of Civil Engineers, Minneapolis, Minnesota, USA.

MACDONALD, J. R., AND R. A. HYATT. 1973. Supersaturation of nitrogen in water after passage through hydroelectric turbines at Mactaquac Dam, Saint Johnson River, New Brunswick. Journal of the Fisheries Research Board of Canada 30:1392–1394.

NEBEKER, A. V., D. G. STEVENS, AND J. R. BRETT. 1976. Effects of gas bubble supersaturated water on freshwater aquatic invertebrates. Pages 51–65 *in* D. H. Fickerson and M. J. Schneider, editors. Gas bubble disease. Technical Information Center, Energy Research and Development Administration, Oak Ridge, Tennessee, USA.

NEWCOMBE, C. 1981. A procedure to estimate changes in fish populations caused by changes in stream discharge. Transactions of the American Fisheries Society 110:382–390.

ORSBORN, J. F., AND C. H. ALLMAN, editors. 1976. Instream flow needs, Vol. 1. Western Division American Fisheries Society, Bethesda, Maryland, USA.

PASTOROK, R. A., M. W. LORENZEN, AND T. C. GINN. 1982. Environmental aspects of artificial aeration and oxygenation of reservoirs: a review of theory, techniques, and experiences. Technical report E-82-3. U.S. Army Corps of Engineers Waterways Experiment Station, Vicksburg, Mississippi, USA.

PEARSON, W. D., R. H. KRAMER, AND D. R. FRANKLIN. 1968. Macroinvertebrates in the Green River below Flaming Gorge Dam, 1964–65 and 1967. Proceedings of the Utah Academy of Science, Arts, and Letters 45:148–167.

RUANE, R. J. 1972. Investigations of methods for increasing dissolved oxygen concentrations downstream from reservoirs. 11th Water Resources Engineering Conference, Nashville, Tennessee, USA.

RUTTNER, F. 1963. Fundamentals of limnology. Third edition. University of Toronto Press, Toronto, Canada.

SPEECE, R. E., C. GIVLER, R. AUBERT, J. CRATE, R. CAIRE, AND R. H. SIDDIQI. 1982. Hypolimnion oxygenation studies in Clark Hill Lake. Journal of the Division of Hydraulics Proceedings of the American Society of Civil Engineers 108:225.

STALNAKER, C. B., AND J. L. ARNETTE. 1976. Methods for the determination of stream resource flow requirements: an assessment. FWS/OBS-03. U.S. Fish and Wildlife Service, Fort Collins, Colorado, USA.

TVA (TENNESSEE VALLEY AUTHORITY). 1978. Impact of reservoir releases on downstream water quality and uses. Special executive report. Division of Environmental Planning, Tennessee Valley Authority, Chattanooga, Tennessee, USA.

TVA (TENNESSEE VALLEY AUTHORITY). 1981. Improving reservoir releases. Publication ONR/WR-83/6. Office of Natural Resources, Division of Water Resources, Tennessee Valley Authority, Knoxville, Tennessee, USA.

USEPA (U.S. ENVIRONMENTAL PROTECTION AGENCY). 1976. Quality criteria for water. U.S. Environmental Protection Agency, Washington, D.C., USA.

USEPA (U.S. ENVIRONMENTAL PROTECTION AENCY). 1983. Ambient water quality criteria for dissolved oxygen. U.S. Environmental Protection Agency, Office of Water Regulations and Standards, Washington, D.C., and Office of Research and Development, Corvallis, Oregon, USA.

WALBURG, C. H., J. F. NOVOTRY, K. E. JACOBS, AND W. D. SWINK. 1983. Effects of reservoir releases on water quality, macroinvertebrates, and fish in tailwaters: field study results. Technical report E-83-6. Office of Chief of Engineers, U.S. Army Corps of Engineers, Washington, D.C., USA.

WARD, J. V. 1976a. Effects of thermal constancy and seasonal temperature displacement on community structure of stream macroinvertebrates. Pages 302–307 *in* G. W. Esch and R. W. McFarlane, editors. Thermal ecology II. Energy Research and Development Administration, Washington, D.C., USA.

WARD, J. V. 1976b. Effects of flow patterns below large dams on stream benthos: a review. Pages 235–253 *in* J. F. Orsborn and C. H. Allman, editors. Instream flow needs, Vol. 2. Western Division American Fisheries Society, Bethesda, Maryland, USA.

WEBSTER, J. R., E. F. BENFIELD, AND J. CAIRNS, JR. 1979. Model predictions of effects of impoundment on particulate organic matter transport in a river system. Pages 339–364 *in* J. V. Ward and J. A. Standord, editors. The ecology of regulated streams. Plenum Press, New York, New York, USA.

WEITHMAN, S. A., AND M. A. HAAS. 1980. Impacts of varying levels of dissolved oxygen on the Lake Taneycomo trout fishery. Missouri Department of Conservation, Jefferson City, Missouri, USA.

WEITKAMP, D. E., AND M. KATZ. 1980. A review of dissolved gas supersaturation. Transactions of the American Fisheries Society 109:659–702.

WILHELMS, S. C., AND D. R. SMITH. 1981. Reaeration through gated-conduit outlet works. Technical report E-81-5. U.S. Army Corps of Engineers Waterways Experiment Station, Vicksburg, Mississippi, USA.

Pages 278–285 *in* G. E. Hall and M. J. Van Den Avyle, editors.
Reservoir Fisheries Management: Strategies for the 80's.
Reservoir Committee, Southern Division American Fisheries Society, Bethesda, Maryland. 1986.

Enhancing Tailwater Fisheries

JOHN C. PETERS

Environmental Affairs Office
Bureau of Reclamation
Engineering and Research Center
Denver, Colorado 80225

ABSTRACT

The state-of-the-art for managing tailwater fisheries is assessed and deficiencies are identified in this paper. The physical, legal, biological, and political constraints confronting biologists, as well as opportunities for achieving effective management for tailwater fishery resources, are discussed. Recommendations for future action include setting standards, developing guidance documents, and carrying out environmental quality assurance programs. Needs identified include improved life history information for warm and cool water fishes in rivers, more effective fish sampling techniques in tailwaters, increased knowledge on interspecific competition, published case histories documenting successful management of tailwater fishes, and participation by biologists in developing and implementing project operational criteria affecting tailwater fisheries.

An increased recognition of the value of fisheries resources in rivers and streams occurred in the 1970's. Books were written (Hynes 1970) and symposia dealing with the ecology and management of fisheries resources in streams and rivers were sponsored by the American Fisheries Society and North American Benthological Society (Oglesby et al. 1972; Orsborn and Allman 1976; Ward and Stanford 1979; Krumholz et al. 1981). Recently, a comprehensive technical literature review of the effects of reservoir releases on tailwater ecology was prepared by Walburg et al. (1981). Often these publications were the product of multidisciplinary interaction with contributions made by fishery biologists, zoologists, and botanists; water quality, design, and construction engineers; social scientists and economists; political scientists, lawyers, and elected officials; and public environmental interest groups. Federal water resource agencies became more aware that quality fisheries in rivers below dams often were the result of planned project design and/or operation.

Advances in state-of-the-art for the management of tailwater fisheries in the past decade can be attributed to increased knowledge resulting from studies throughout all regions of North America. Biologists were given the opportunity to learn by doing and have responded by improving sport and commercial tailwater fishery resources on certain projects. The newfound public support during the "environmental decade" of the 1970's permitted biologists to contribute in a more meaningful manner in water resource development decisions. Biologists came to be equal partners with water resource economists, engineers, and planners in water resource programs. New water projects were built or operations of existing projects modified to enhance tailwater fisheries. Finally, a number of environmental laws (National Environmental Policy Act, Clean Water Act, and Endangered Species Act) enacted in the 1970's directly and indirectly strengthened the biologists' participation in developing and managing water resource projects as they affected tailwater fishery resources.

Water resource projects are now planned and operated with confidence in predicting the fate of future fishery resources that is based on science rather than experience or intuition. Each water project affects tailwater fisheries, and those in various regions of North America will result in both similar and different tailwater fishery effects, with the biological constraints limiting practical management choices. Fishery resources have benefited from the interaction between biologists, water resource engineers, economists, political scientists, etc.

Dams store and release water for a variety of uses, including power, flood control, irrigation, navigation, municipal and industrial water supply, water quality control, recreation, and combinations thereof. Water releases for fishery purposes are rarely a legal requirement of existing water projects, especially for instream use to maintain or enhance fish habitat. Most states and provinces now recognize fishery protection as a beneficial use of water, but the quantification, operational, and legal requirements may not be sanctioned enough to be useful for adequately protecting tailwater fishery resources (Peters 1982). This paper assesses the state-of-the-art, identifies deficiencies, and recommends future needs for tailwater fishery resources in North America. To achieve this purpose, the physical, legal, biological, and political constraints confronting biologists are discussed as well as opportunities in achieving effective management of tailwater fishery resources.

Project Constraints and Opportunities

The importance of tailwater fisheries is well known in North America. In some tailwaters, fish populations and harvest are greater than those in the reservoir upstream especially when considered on the basis of surface area. Knowledge of reservoir limnology is essential to understanding and managing tailwater fisheries. Whether or not a reservoir stratifies and the location of the dam intake dictates the quality and quantity of tailwater fisheries.

Efforts to manage a tailwater fishery to reflect conditions in an unregulated river are often impossible because of constraints imposed by a water development project. The quantity, quality, and timing of project releases creates an environment that may differ measurably from an unregulated stream. The impacts of project releases on the tailwater environment are controlled by overall project design and operation. Water development has occurred basin-wide in many areas; therefore, river basin decisionmaking must be evaluated for multiple dams on river basins as the Columbia, Colorado, Arkansas, Tennessee, Ohio, Missouri, and Rio Grande. In these basins, management must focus on system evaluation as well as on single projects.

The biologist's role in achieving effective management for tailwater fisheries can be carried out by first identifying rational objectives or goals for fisheries management. To be effective, the biologist must clearly recognize the physical, biological, legal, political, and economic constraints imposed during the development of water projects from a basin-wide perspective. The biologist must then identify opportunities for developing tailwater fisheries that are project specific, and encourage operating agencies to internalize into their decisionmaking all the effects of their project on other resources. Finally, the biologist must be aware of the tools available for identifying impacts and how to use these tools effectively.

Physical Constraints

Typically, the more important physical constraints on tailwater fisheries are hydrology, land use, geomorphology, and water quality. The amount of water released constitutes the most fundamental restriction on the range of management and allocation options available to decisionmakers. Timing of water releases must also be evaluated. The biologist must recognize that competition for releases for other purposes may limit prospects of a complete program of instream flows for fish.

Land use should be evaluated in terms of nutrient yield from the watershed, the significance of allochthonous sources of organic material to aquatic environments, and their relationship to the productivity of fishery resources. In streams, metabolic essentials are not recycled as they are in lakes and reservoirs; they must be supplied constantly from upstream sources or stream-side vegetation. Options for influencing water quality are available to managers, but land management programs are difficult to implement. In addition, tailwaters represent a flowing stream that has been altered by the passage of water through an impoundment.

From a geomorphic perspective, aquatic organisms are not randomly distributed in a tailwater or unregulated channel but rather are part of a complex mosaic pattern based on habitat needs. This pattern of distribution is related to channel morphology and the interspersion of pools, riffles, and runs, the nature of the streambed and streambank material, cross-sectional channel shape, and the meandering nature of the channel itself. Fish seek different habitats within a channel, and at different times of the year, they may occupy different places within a channel. This process has evolved over geologic time through natural selection, and needs are often specific and limiting.

Fish are able to carry out their life processes in these dynamic, pulse-stabilized systems and have established self-sustaining populations in stream channels. The geomorphic physical constraint is not fixed, and options for management are available but not widely used. For example, manmade spawning channels have been built and successfully operated to produce salmonid fry (MacKinnon 1960).

Water quality evaluation in tailwaters should be undertaken from both a physical and chemical perspective. Changes in temperature, dissolved gases, reduced compounds (e.g., iron, manganese, sulphur, hydrogen sulfide, and ammonia), availability of nutrients, and other factors can be modified if appropriate management options are determined during planning. Costs associated with water quality control are generally higher if added after construction than if incorporated when a project is being planned and designed.

Legal Constraints

In multiple-purpose water projects, there are several legally sanctioned water-use obligations a project must satisfy before managers can begin to exercise options to increase the value of the tailwater fishery and associated recreation. These include withdrawals of water from the tailwater for municipal and industrial use and irrigation, which are protected by the "first in time, first in use" concept governing water laws in the western United States. In the Pacific Northwest, they also may include treaty obligations covering the level of different fish runs and an Indian tribe's access to the runs at specific locations in a river basin. In other geographic areas of North America, water releases are managed primarily for navigation, flood control, or hydropower.

Most water projects have been authorized and managed for multiple purposes. Federal or State legislation authorizing water projects lists these purposes, but water agencies allocate and manage project-developed water. Private hydropower projects must be licensed by the Federal Energy Regulatory Commission. These licenses are reviewed periodically and may be modified. In any case, whether a water project is private or public, there are legal requirements that must be met either by law or by license. It is within this legal framework that water is allocated to various users.

A number of states legally recognize instream flows for fish as a beneficial use. However, the quantification, operation, and legal requirements continue to cause administrative problems associated with adequate protection of tailwater fisheries in North America. Since 1977, the U.S. Fish and Wildlife Service has published a series of reports that detail opportunities and constraints which exist under existing State laws. The latest report intended to help fish and wildlife managers covers the 26th state in the series to protect instream uses of water (White 1983).

Biological Constraints

The natural processes governing the life cycle of fish include waterflow and water quality requirements for specific times and places in all streams. Some requirements are absolute; failure to meet them will lead to local extirpation of a species. However, the biological requirements for developing quality tailwater fisheries may be less restrictive on management options than for other constraints.

In regulated and unregulated rivers, biota reflect the interactions between native and introduced species. Therefore, the consequences and benefits of an introduction of a new fish species should be carefully evaluated before such an action occurs.

Many biological changes such as the blockage of upstream migration or the inundation of critical spawning grounds are a direct result of the construction of a water project. In tailwaters it may not be feasible to manage for some fish species because it would not be practical to maintain essential life history requirements. However, there are a wide range of opportunities available to the fishery biologist provided a process exists for consideration of biological resources. The ability to predict biological responses is frequently limited by the biologist's understanding of the tailwater community.

Political and Economic Constraints

These constraints are the most flexible and negotiable. If a water project or river basin institutes a new management goal for tailwater fishery resources aimed at maximizing social benefits, political negotiation over economic issues intensifies. However, it is through this process that important changes are often made, either during water project planning or after a water project is fully operational. In many cases, water projects do not fully allocate all water resources set aside by development. Therefore, agencies operating dams may end up with de facto ownership of the river and resulting tailwater fishery.

Because of their authority for controlling flows to achieve project purposes, and the principal mandate to carry out these purposes, agencies have generally used the dams for only these statutory purposes. It is often quite by accident that quality tailwater fisheries develop for many existing projects.

Project Opportunities

Though each water project has constraints, opportunities also exist for achieving effective management of tailwater fishery resources. Significant opportunities occur during project location, design, and operation. However, unless all resources are considered and objectives and goals defined, many options may be overlooked by decisionmakers. Selection of the proper location for a project can produce significant tailwater fishery benefits often with minimum social and/or economic costs.

Design features such as multilevel water intakes, screens, and fish passage facilities should be carefully evaluated because of their costs. Proper design of spillways and discharge pools can prevent gas saturation and minimize other water quality problems to tailwaters. Recently, the U.S. Bureau of Reclamation took action to correct gas supersaturation below Navajo and Yellowtail dams and to improve water temperature below Flaming Gorge Dam. However, the structural changes were made to correct current tailwater fishery problems after the dams were built.

Operations should include not only water quantity and consideration of seasonal and daily variation for tailwater fishery resources but also water quality needs as well as possible problems with temperature, nutrients, and toxic levels of reduced substances. In any case, the tailwater flow regime necessary to maintain the desired habitat should be evaluated. The U.S. Bureau of Reclamation's Fryingpan-Arkansas Project was authorized by Congress with instream flow needs recognized in project operation principles set forth in the enabling legislation (USBR 1975).

Fortunately, predictive tools are available that allow for more rational decisionmaking regarding tailwater fisheries. Models for predicting water temperature for tailwater fisheries and the downstream extent of the fishery and water quality are available for individual projects (USACE 1982) and for river basins (Smith 1978).

Recent predictive tools for stream fisheries habitat needs include the instream flow methodology, which integrates hydraulic simulation and habitat suitability in a computer library of models called the PHABSIM (Physical Habitat Simulation System) (Milhous et al. 1981). The PHABSIM considers the physical habitat as a function of flow and the variation of physical habitat over time. An earlier version of the model (Stalnaker 1978) allows for the quantification of the amount of physical habitat for fish species and life history phases in a given reach of stream, at different stream-flow regimes, and with different channel configurations and slopes.

Advances in computer technology in recent years have resulted in applying operations research techniques to the difficult problems of effectively managing tailwater resources. Practical applications of implementing this rapidly developing body of knowledge for resolving conflicts between competing uses should be undertaken. In a recent paper discussing hydroelectric operation at the river basin level, Shane (1981) summarized the state-of-the-art in developing models for multipurpose multiple reservoir systems and concluded that modeling could help the manager by evaluating new operating requirements, providing long-range guidance and forecasts, making more efficient use of historical data, and documenting of operating policy. Models that address the major ecological issues related to the operation of dam/impoundment systems are a major new tool for achieving effective tailwater fisheries management.

Recommendations

All water projects should be planned and operated with development of the tailwater fisheries receiving full consideration. Further, opportunities for using unallocated stored water in existing projects for enhancing tailwater fisheries should be vigorously pursued. To implement these recommendations, I offer three integrated courses of action: (1) develop and set standards; (2) prepare guidance manuals; and (3) establish auditing procedures. Fisheries professionals must become more effective in communicating with water resource decisionmakers; these techniques have been found to be useful for that purpose.

Standards for Tailwater Fisheries

Much effort has been expended in the past 10 years to identify instream-flow needs for several species of fish, and most biologists now recognize that different species of fish have different flow needs. Decisionmakers responsible for allocating flows seem to accept the basic idea of maintain-

Table 1. Environmental concerns related to setting standards for fishery uses in streams below dams.

Instream flow	Water quality	Migratory fish
Flows required to maintain acceptable hydraulic and temperature conditions during critical life history stages in order to sustain suitable habitat for aquatic biota	Changes in nutrient transport and cycling and altered temperature regimes	Temperatures or water quality barriers to migration through large reservoirs
Flows needed to initiate and sustain fish migrations	Gas supersaturation	Fish mortality resulting from upstream and downstream passage
Flow needed to minimize effects of peaking operations on the rapid fluctuation and lateral movement of usable microhabitats of aquatic biota	Altered material cycles including trace elements, heavy metals, and organics	Passage of fish out of reservoir, which enhances the tailwater fishery
Effects of changing flows on community structure and productivity of aquatic biota		Cumulative and synergistic effects due to multiple dam systems
Effects of changing flows on food chains that support fish	Altered dissolved oxygen regimes	Seasonal alteration of downstream fishes to tailwater area

ing releases for fish. Presently, instream-flow needs relate primarily to minimum and optimum flow patterns required to protect tailwater fishery resources. Tailwater impacts that can result from project operation range from those that are easily described (stranding of fish and loss of habitat at low discharges) to those that are very complex (decreased habitat quality and reduction in productivity at all trophic levels).

The work done through the Cooperative Instream Flow Group of the U.S. Fish and Wildlife Service (Stalnaker 1978; Milhous et al. 1981) for establishing instream-flow needs on a fish species basis should be used and refined when necessary to reflect local or regional conditions. However, this work should also be expanded to include the effects of water quality, behavior, productivity of various trophic levels and their relationships with tailwater fisheries. Table 1 lists major ecological considerations that should be an integral part of setting standards; they present a skeletal framework that needs further development and refinement. These standards must then be available for all users and for presentation to decisionmakers. The difficult task is to gain acceptability of such standards by water development agencies and put them into practice.

Guidance Manuals for Tailwater Fisheries

Dams are designed and water projects operated to achieve recognized project needs. For example, the location and design of an intake structure are engineering decisions based primarily on costs related to identified project needs. Subsequently, the operation of a water project is a function of project purposes. Ordinarily, water is released to produce power, to provide water for irrigation, for municipal and industrial users, or navigation. Conversely, water may be stored in a reservoir to prevent flooding. In certain cases, stored water could be released for instream flow for tailwater fisheries. Identifying fishery needs and effective implementation of actions for these needs can be facilitated by developing and using guidance manuals for tailwater fisheries.

Water quality and water temperature for tailwater fisheries are directly related to the location and design of an intake structure. The U.S. Fish and Wildlife Service (Neitzel et al. 1982) prepared a guidance manual concerned with the input of biological information useful for intake structure design. This manual was developed for pumping operations, intake and discharge structures, and other water facilities required for steam-generated electrical powerplants. Construction, location, and operation of intake and discharge facilities for these plants are regulated under the Clean Water Act of 1977. Section 316(b) of the Act provides protection for aquatic organisms that are affected by water intake structures. The concepts developed in this guidance manual are applicable to all water intake structures, including those used for dams and canals.

The purpose of a guidance manual would be to facilitate input from biologists to engineers on the design and operation of a project as related to fisheries. For tailwater fisheries, the manual would consider the ability of the project to control instream water temperature and water quality, as well as daily or seasonal water level fluctuations. Biological information would be available to the design engineer to meet project purposes and pertinent features could be designed accordingly. Guidance manuals would provide opportunities for maximizing uses of aquatic resources through appropriate design analyses, a program activity not often used in the past.

Environmental Auditing Program

The main object of environmental impact assessment under the National Environmental Policy Act of 1969 has been to improve decision-maker sensitivity to environmental factors. However, little attention has been paid to ensure that mitigation and enhancement measures identified during impact assessment were implemented. Recently, regulations of the Council on Environmental Quality introduced the Record of Decision as a legally enforceable public statement of an agency's intention (CEQ 1978, section 1505.2). This provides a mechanism for environmental quality assurance to determine if mitigation measures have been implemented. It also provides an opportunity for feedback to determine if mitigation or enhancement measures have been effective.

On a new water project, the Record of Decision may mandate a commitment from management to implement a tailwater fishery program as part of the project. Putting together the administrative fabric to modify existing projects to consider enhancing tailwater fisheries represents a real challenge for biologists. Nevertheless, the concept of auditing environmental commitments for quality assurance is a valuable management tool. Equally important is the notion that a measure of the effectiveness of the commitment should be part of this process. This allows for refinements to be made on existing projects and improves the capability to make better decisions on new projects.

Needs and Trends

Stream habitat requirements are poorly understood for warm water and cool water fishes compared to cold water fishes. Lack of habitat data for warm and cool water fishes has limited the fisheries biologist's effectiveness to negotiate tailwater flows necessary for them. Biologists should shift their priorities so that stream surveys and other ongoing activities at the State level are devoted to gathering basic fish life history information in streams. More effective sampling techniques are needed to determine fish populations in large rivers. While some progress has been made toward better sampling of trout and salmon populations in large rivers with electrofishing gear (Mahon 1980), there is a continuing need to improve population sampling with electrofishing gear for warm and cool water fishes in large, turbid tailwaters. Recent refinements in stream rotenone sampling (Ober 1981) offer new opportunities to quantitatively measure stream fish populations.

The effects of interspecific competition among fishes in tailwaters is not well known. Tailwater fishery management is often limited to introducing new species of fishes and hoping for the best. Little thought is given to the effects of the introduced species to waters downstream of the tailwaters, either from the intra- or interspecific view. Fishery biologists need to more carefully assess the possible impacts of such introductions before they occur. The collection and use of additional fish behavioral data could prevent unwise introductions in tailwaters.

Stocking is a useful fisheries management tool in certain tailwaters. Stocking of rainbow trout (*Salmo gairdneri*) and brown trout (*Salmo trutta*) in tailwaters of reservoirs with deep outlets in the southeastern United States has provided some outstanding fisheries (Axon 1975). Recently, Vincent (1980) concluded that annual stocking of catchable-sized rainbow trout depressed wild rainbow and brown trout populations in the Madison River, Montana. There is a need to further evaluate annual stocking in tailwaters, as it might affect established populations of both the stocked and native species.

Case histories that document the enhancement of tailwater fisheries by structural or operational actions are lacking. These should include documentation of specified design or operational modifications, with adequate monitoring to assess impacts on tailwater communities. Table 2 summarizes recent work completed by the Bureau of Reclamation to correct problems affecting tailwater trout fisheries. Monitoring to measure the long-term effectiveness of these actions is underway, so detailed case histories are not yet available for these projects.

Table 2. Modifications made at U.S. Bureau of Reclamation dams to enhance tailwater fisheries.

Dam	River	Problem	Action and date
Flaming Gorge	Colorado	Water temperature	Installed multi-level intakes, 1978
Navajo	San Juan	Gas supersaturation	Modified discharge valve auxiliary outlet, 1977
Yellowtail afterbay	Big Horn	Gas supersaturation	Installed fliplip 1982

Fisheries biologists need to work further with water development agencies to document similar actions to enhance tailwater fisheries and report on their effectiveness. Documentation is needed on actions taken to prevent problems with new projects as well as to correct existing problems, including changes in operations and structural modifications.

Fishery biologists have a responsibility to see that appropriate language, rule curves, and criteria for operations affecting tailwater fisheries are incorporated into each water project operating manual. However, biologists must also maintain continual involvement with the project. No set of rules will work well all the time, but must be modified periodically. Other water users understand this and are ready to provide inputs and negotiate particular operations when infrequent events, such as droughts or floods, occur. Biologists have to do more than just lay down some rules; they need to be aware of the constraints and opportunities available for each project and to use them as a roadmap toward achieving management goals for tailwater fisheries on a continuing basis.

Engineers must be convinced that biologists are knowledgeable in terms of a fish species' responses to altering streamflow and habitat conditions. Although biologists have made some progress in communicating their needs to engineers, further improvements in communication between the two are still needed. Furthermore, fisheries agencies have not been very successful in convincing water development agencies that the project should pay for the data gathering and analysis necessary to describe fish habitat preferences; this fundamental impasse needs to be broken before much further progress can occur.

Conclusions

The previous discussion focused on problems associated with tailwater fisheries and recommendations for more effective techniques to better manage these resources. Fundamental to this process must be a balancing between the need for healthy environments for reservoir and tailwater fisheries. Although there is little information to support this conclusion, such an equitable balance should be possible for most water development projects. Where that is not so, managers will certainly have to consider trade offs; however, their choices should be based on the biological facts.

Fishery biologists must (1) document where there have been successes on new projects or where existing projects have been successfully modified and (2) identify legislative needs to correct longstanding problems related to tailwater fisheries. The profession would be well served by the development of a list of case histories where successful tailwater fisheries have occurred and by documentation of what happened specifically to allow for such success. This should be done for all regions in North America and special efforts made in those instances where a river basin system has provided milestones in tailwater fisheries management.

New legislation, such as the recently enacted Northwest Power Act of 1980, or modification of existing legislation, is needed. Perhaps the Fish and Wildlife Coordination Act could be modified and strengthened to serve this purpose. Opportunities exist at both the Federal and State level to correct legislative deficiencies. How the Columbia River Basin fisheries respond to the mandates of the Northwest Power Act should be watched closely. Its fish and wildlife program may serve as a model for similar programs in other developed river basins.

Finally, water projects have benefited from the input of fisheries biologists, especially where the water resource agencies have been willing to try new, innovative ideas based on biological needs. There have been failures as well as successes; downstream passage of anadromous fish continues to be a problem on certain projects; recognition of instream flows for tailwater fisheries is an institutional and legal constraint in the western United States. Biologists have taken advantage of opportunities in the past 10 years and can point to a number of successes. However, additional research and development opportunities are still greatly needed for tailwater fisheries, as are enlightened managers and decisionmakers to allow for changes in existing water development projects and innovation on new projects.

References

AXON, J. R. 1975. Review of coldwater fish management in tailwaters. Proceedings of the Annual Conference of the Southeastern Association of Game and Fish Commissioners 28:351–355.

CEQ (COUNCIL ON ENVIRONMENTAL QUALITY). 1978. Regulations for implementing the procedural provisions of the National Environmental Quality Act. 43 FR 55978-56007, U.S. Congress, Washington, D.C., USA.

HYNES, H. B. N. 1970. The ecology of running waters. University of Toronto Press, Toronto, Canada.

KRUMHOLZ, L. A., C. F. BRYAN, G. E. HALL, AND G. B. PARDUE, editors. 1981. The warmwater streams symposium. Southern Division American Fisheries Society, Bethesda, Maryland, USA.

MACKINNON, D. 1960. A successful transplant of salmon eggs in the Robertson Creek Spawning Channel. Canadian Fish Culturist 27:25–31.

MAHON, R. 1980. Accuracy of catch–effort methods for estimating fish density and biomass in streams. Environmental Biology of Fishes 5:343–360.

MILHOUS, R. T., D. L. WAGNER, AND T. WADDLE. 1981. User's guide to the physical habitat simulation system (PHABSIM). FWS/OBS-78/33, U.S. Fish and Wildlife Service, Washington, D.C., USA.

NEITZEL, D. A., M. A. SIMMONS, AND D. H. MCKENZIE. 1982. A guidance manual for the input of biological information to water in the structure design. FWS/OBS-82/13.1, U.S. Fish and Wildlife Service, Washington, D.C., USA.

OBER, R. D. 1981. Operational improvements for sampling large streams with rotenone. Pages 364–369 *in* L. A. Krumholz, C. F. Bryan, G. E. Hall, and G. B. Pardue, editors. The warmwater streams symposium. Southern Division American Fisheries Society, Bethesda, Maryland, USA.

OGLESBY, R. T., C. A. CARLSON, AND J. A. MCCANN, editors. 1972. River ecology and man. Academic Press, New York, New York, USA.

ORSBORN, J. F., AND C. H. ALLMAN, editors. 1976. Proceedings of the symposium and Specialty Conference on instream flow needs, volumes I and II, Boise, Idaho, May 3–6, 1976. American Fisheries Society, Bethesda, Maryland, USA.

PETERS, J. C. 1982. Effects of river and streamflow alteration on fishery resources. Fisheries (Bethesda) 7(2):20–22.

SHANE, R. M. 1981. Scheduling model development for multipurpose multiple reservoir releases: The state of the art. Pages 2.1–2.14 *in* Proceedings of hydroelectric operation at the river basin level: research needs to include ecological issues in basin-level hydropower planning. Electric Power Research Institute, Palo Alto, California, USA.

SMITH, D. J. 1978. Water quality for river-reservoir systems. Hydrologic Engineering Center, University of California–Davis, Davis, California, USA.

STALNAKER, C. B. 1978. Methodologies for preserving flows. Pages 1–9 *in* The incremental method in instream flow management—state of the art. Upper Mississippi River Basin Commission, Minneapolis, Minnesota, USA.

USACE (U.S. ARMY CORPS OF ENGINEERS) 1982. CE-QUAL-R1: A numerical one-dimensional model of reservoir water quality; user's manual. Instruction report E-82-1. U.S. Army Corps of Engineers Waterways Experiment Station, Vicksburg, Mississippi, USA.

USBR (U.S. BUREAU OF RECLAMATION). 1975. Environmental Impact Statement, Fryingpan-Arkansas Project, Colorado. FES 75-43, Appendix B, Volume 1. U.S. Bureau of Reclamation, Denver, Colorado, USA.

VINCENT, E. R. 1980. Effects of stocking catchable trout on wild trout populations. Pages 88–91 *in* Proceedings of wild trout symposium. Trout Unlimited, Denver, Colorado, USA.

WALBURG, C. H., J. F. NOVOTNY, K. E. JACOBS, W. D. SWINK, T. M. CAMPBELL, J. NESTLER, AND G. E. SAUL. 1981. Effects of reservoir releases on tailwater ecology: a literature review. Technical report E-81-12. U.S. Army Corps of Engineers Waterways Experiment Station, Vicksburg, Mississippi, USA.

WARD, J. V., AND J. A. STANFORD, editors. 1979. The ecology of regulated streams. Plenum Press, New York, New York, USA.

WHITE, M. 1983. Opportunities to protect instream flows in Texas, Oklahoma, and Arkansas. FWS/OBS-83.22, U.S. Fish and Wildlife Service, Washington, D.C., USA.

Pages 286–293 *in* G. E. Hall and M. J. Van Den Avyle, editors.
Reservoir Fisheries Management: Strategies for the 80's.
Reservoir Committee, Southern Division American Fisheries Society, Bethesda, Maryland. 1986.

Effect of a Hypolimnetic Discharge on Reproductive Success and Growth of Warmwater Fish in a Downstream Impoundment

GARY D. HICKMAN AND KERRY W. HEVEL

Tennessee Valley Authority
Office of Natural Resources
Division of Natural Resource Operations
Norris, Tennessee 37828

ABSTRACT

Cove rotenone samples (1972-1982), electrofishing samples (1975-1977), and ichthyoplankton samples (1975-1976) in Melton Hill Reservoir were compared with water release rates from upriver Norris Dam to determine impacts of a hypolimnetic discharge on fish reproductive success and growth of largemouth bass (*Micropterus salmoides*) in a downstream impoundment. A significant inverse relationship was found between water volume discharge during spring and early summer from Norris Dam and same-year estimates of young-of-year (YOY) numbers from cove rotenone samples. Larval fish samples revealed low diversity and densities in 1975, when spring discharges were high, and much higher levels in 1976, when spring discharges were low. Cove rotenone data showed similar increases of 2,938% in clupeid numbers and 309% in centrarchids from 1975 to 1976. Largemouth bass (Age 1) grew significantly faster and had higher YOY production during years of low spring releases from Norris Dam and slower growth and low YOY production during years of high spring discharges. Variable reservoir temperature regimes that resulted from yearly differences in spring and early summer discharge rates produced significant variations in spawning success of warmwater species and growth of piscivorous sport fish.

The fish populations of Melton Hill Reservoir were intensively sampled from 1972 through 1980 to satisfy requirements of the Clean Water Act for operation of the Bull Run Steam-Electric Plant (TVA 1976a, 1976b; 1981). Results of these investigations indicated Norris Dam discharge rates had a greater impact on the fishery of Melton Hill Reservoir than did operation of the Bull Run plant. It was concluded that the heated discharge from Bull Run had a beneficial impact by increasing water temperatures in the vicinity of the plant to more closely approximate a normal temperature regime of a southeastern impoundment.

While the effects of hypolimnetic discharges from dams on tailwater fisheries have been well documented, few investigations have addressed the impacts of such discharges on the fisheries of downstream reservoirs. Fry and Hanson (1968) reported a shift from a self-sustaining warmwater fishery to a coldwater put-and-take rainbow trout (*Salmo gairdneri*) fishery in 700-hectare Lake Taneycomo, Missouri, after construction of Table Rock Dam. Coldwater releases through Table Rock Dam virtually eliminated reproduction and severely reduced growth rates of warmwater fish species in Lake Taneycomo. In this case, reservoir releases of the upstream impoundment determined the thermal conditions of the entire downstream reservoir. Dendy and Stroud (1949) found similar effects of hypolimnetic discharges from Fontana Dam, North Carolina, on fish in Cheoah (250 hectares) and Calderwood (215 hectares) reservoirs. Where smallmouth bass (*Micropterus dolomieui*) and walleye (*Stizostedion vitreum*) dominated the fisheries before construction of Fontana Dam, only put-and-take trout fisheries existed after its closure.

The objective of this study is to evaluate the impacts of coldwater releases from a large reservoir (Norris) on warmwater fish reproductive success and on growth of largemouth bass (*Micropterus salmoides*) in a downstream reservoir (Melton Hill).

STUDY AREA

Melton Hill Dam, located at Clinch River mile (CRM) 23.3, was completed in 1960, and inundation of the basin began in 1963. The reservoir has a river channel length of 70.8 km and

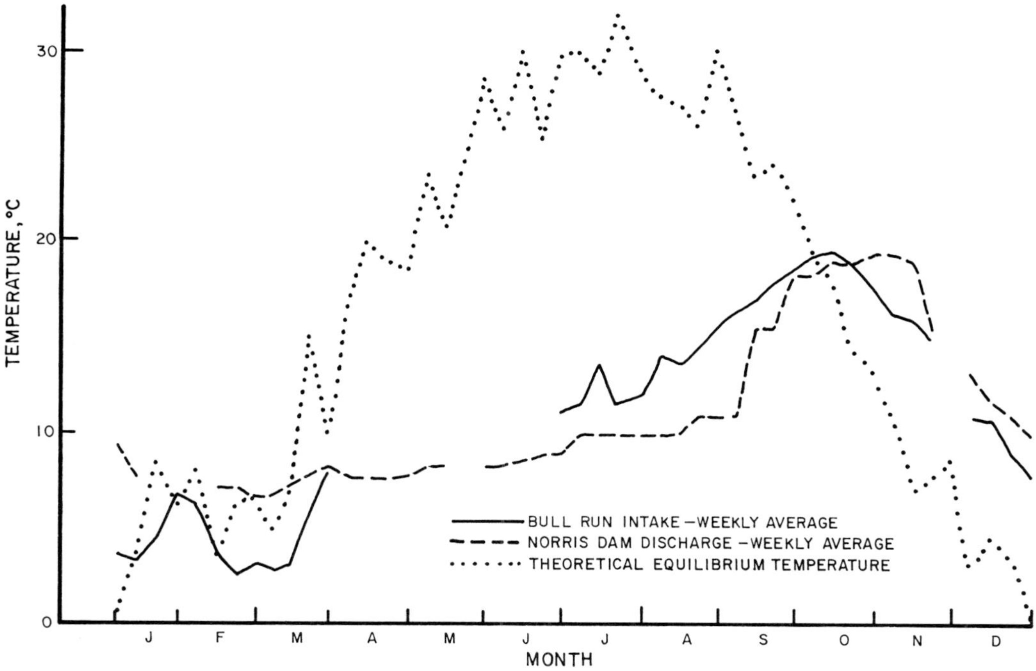

Figure 1. Comparison of water temperatures recorded from Norris Dam discharge and the Bull Run intake channel with theoretical equilibrium water temperatures for Melton Hill Reservoir in 1969.

a shoreline length of 231.7 km at full pool. Reservoir volume at the top of the spillway gates is 15,547 hectare-meters, and surface area at full pool is 2,315 hectares. Melton Hill Reservoir has a short water retention time (average 8.5 days), with water turnover rates related to releases from Norris and Melton Hill dams.

Bull Run Steam-Electric Plant is a single unit, fossil-fueled facility located within the transition zone between riverine and lacustrine conditions on Melton Hill Reservoir (CRM 48.0). The plant has a skimmer wall, and cooling water is withdrawn from the lower 1.5 m of the water column. At peak operation, the facility has a generating capacity of 950,000 kilowatts and uses 25.2 m^3 of cooling water per second. The water is heated a maximum of 10.6 C above ambient and returned via a canal to Melton Hill Reservoir.

Norris Dam is located at CRM 79.7, approximately 20 km above the headwaters of Melton Hill Reservoir. Surface area of Norris Reservoir at full pool is 13,840 hectares. Norris Reservoir water is discharged through two hydroelectric generators that withdraw water from a depth of 46.6 m at full pool. Since Norris Reservoir is thermally stratified during summer, the discharge is typically cold. Water is released irregularly through Norris Dam depending on electric power demands and flood control requirements. A 24-hour release cycle usually varies between no turbine discharge, 1.8 m^3/second, and two-unit discharges of 170 m^3/second. Under flood control regimes, however, releases of 565 m^3/second utilizing sluice gates have been recorded.

The thermal regime of Melton Hill Reservoir is characterized by lower spring and summer water temperatures than theoretical equilibrium water temperatures, i.e., those that would exist in the absence of Norris Dam or Bull Run Steam-Electric Plant (Fig. 1). During high-flow years, large volumes of cold hypolimnetic water released through Norris Dam result in colder temperatures in Melton Hill Reservoir; during low-flow years solar heating warms Melton Hill Reservoir to a greater extent (Fig. 2). Peak operation of Bull Run Steam-Electric Plant elevates the water temperature in the immediate vicinity of the discharge canal up to 10.6 C. However, this heat dissipates rapidly in the main body of the reservoir.

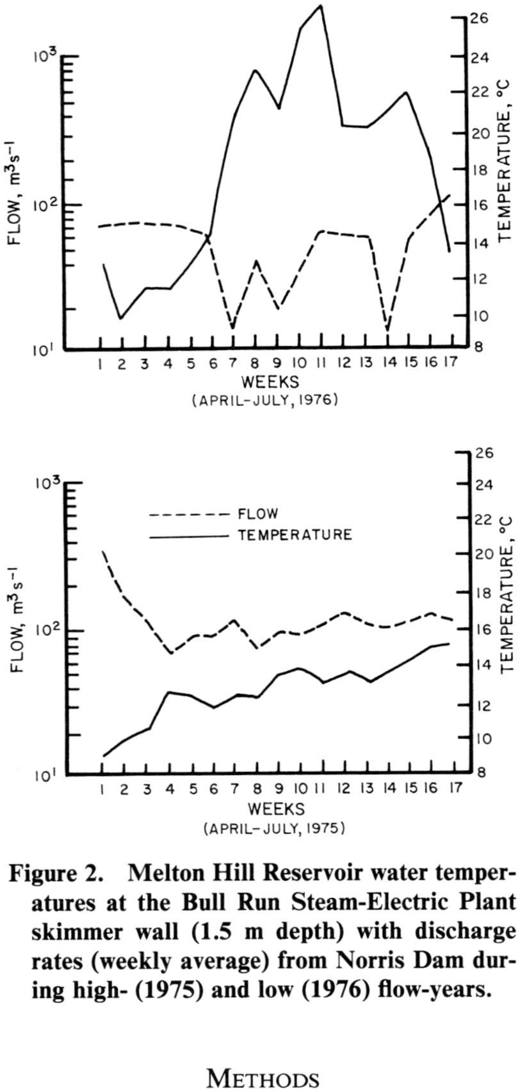

Figure 2. Melton Hill Reservoir water temperatures at the Bull Run Steam-Electric Plant skimmer wall (1.5 m depth) with discharge rates (weekly average) from Norris Dam during high- (1975) and low (1976) flow-years.

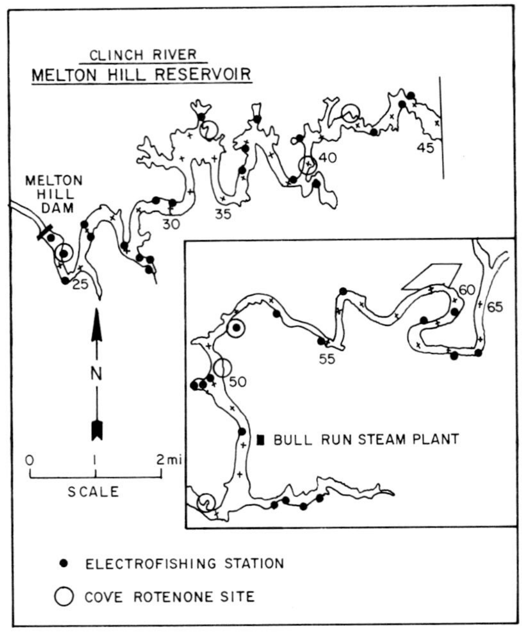

Figure 3. Location of cove rotenone and electrofishing sample stations on Melton Hill Reservoir.

Methods

Data presented in this report were obtained from cove rotenone surveys, electrofishing samples, and ichthyoplankton tows. Since 1972, 37 cove rotenone samples have been collected along the length of Melton Hill Reservoir (Fig. 3). To allow comparisons with fish communities in adjacent reservoirs, data from cove rotenone samples collected since 1970 from Norris (39 coves) and Watts Bar (38 coves) reservoirs were also included in the analysis. Standard cove rotenone sampling procedures were utilized in all rotenone studies (TVA 1982). Cove areas, usually 0.4 hectare in surface area, were blocked off before rotenone was applied. Students t-tests were used to compare densities of sport, commercial, and prey species from Melton Hill, Watts Bar, and Norris reservoirs.

Norris Dam discharge rates and temperatures of released water were obtained from the Tennessee Valley Authority (Reservoir Operations Branch, Knoxville, Tennessee, USA). Regression analysis was done on water volume discharged during spring and early summer (April–July) from Norris Dam and estimates of young-of-year fish (YOY) numbers from cove rotenone samples in the same year. Production of YOY in Melton Hill Reservoir was also compared with same-year abundance of YOY in Norris and Watts Bar reservoirs.

Scale samples were taken from Melton Hill largemouth bass collected from 1970 through 1977 for analysis of age and growth. Comparison of growth increments for each age class during high and low spring-early summer discharge years was made using a t-test. In addition, a regression analysis was done on growth increments for each age class and spring-early summer discharge rates from Norris Dam.

Daytime electrofishing samples (50-m runs) were collected for one week each month throughout Melton Hill Reservoir (Fig. 3) from October

Table 1. Mean number of fish per hectare in rotenone samples from Melton Hill and adjacent reservoirs. Number of samples in parentheses.

Species	Abundance (mean number/hectare)[a]		
	Melton Hill 1972–1982 (37)	Norris 1970–1980 (39)	Watts Bar 1970–1980 (38)
Sport fishes			
Esox lucius	tr	0	0
Esox masquinongy	0	tr	0
Morone chrysops	7	27	49
Morone mississippiensis	6	0	30
Morone saxatilis	tr	2	tr
Morone hybrid	tr	0	tr
Lepomis spp.	53	17	38
Lepomis gulosus	95	5	67
Lepomis auritus	57	tr	97
Lepomis cyanellus	0	0	tr
Lepomis humilis	0	0	tr
Lepomis macrochirus	1,160	1,675	1,641
Lepomis megalotis	1	1	28
Lepomis microlophus	7	0	4
Micropterus dolomieui	8	197	52
Micropterus punctulatus	44	97	21
Micropterus salmoides	112	84	176
Pomoxis spp.	15	159	tr
Pomoxis annularis	79	85	53
Pomoxis nigromaculatus	8	21	2
Perca flavescens	tr	0	12
Stizostedion canadense	1	2	6
Stizostedion vitreum	0	21	tr
Subtotal	1,653	2,442	2,276
Commercial fishes			
Lepisosteus oculatus	0	tr	tr
Lepisosteus osseus	tr	1	2
Alosa chrysochloris	2	0	16
Hiodon tergisus	0	5	1
Carassius auratus	3	0	0
Cyprinus carpio	262	8	39
Carpiodes carpio	4	0	1
Carpiodes cyprinus	51	1	tr
Carpiodes velifer	tr	0	0
Catostomus commersoni	1	tr	tr
Hypentelium nigricans	1	tr	3
Ictiobus bubalus	65	0	12
Ictiobus cyprinellus	1	0	1
Ictiobus niger	5	0	tr
Minytrema melanops	0	0	9
Moxostoma anisurum	1	1	tr
Moxostoma macrolepidotum	tr	1	tr
Moxostoma carinatum	tr	0	tr
Moxostoma duquesnei	1	1	4
Moxostoma erythrurum	6	8	13
Ictalurus furcatus	tr	0	0
Ictalurus melas	1	0	tr
Ictalurus natalis	5	1	1
Ictalurus nebulosus	tr	0	tr
Ictalurus punctatus	1	56	46
Pylodictis olivaris	tr	35	6

Table 1. Continued.

Species	Abundance (mean number/hectare)[a]		
	Melton Hill 1972–1982 (37)	Norris 1970–1980 (39)	Watts Bar 1970–1980 (38)
Aplodinotus grunniens	68	50	315
Subtotal	478	163	469
Prey fishes			
Dorosoma cepedianum	2,789	1,553	2,088
Dorosoma petenense	2,150	2,778	2,352
Campostoma anomalum	tr	1	1
Hybopsis storeriana	0	0	1
Notemigonus crysoleucas	2	tr	tr
Notropis atherinodes	17	4	43
Notropis cornutus	0	tr	tr
Notropis galacturus	3	tr	tr
Notropis photogenis	0	0	3
Notropis rubellus	0	tr	0
Notropis spilopterus	85	58	51
Notropis stramineus	0	tr	0
Notropis volucellus	0	10	0
Notropis whipplei	33	65	27
Pimephales spp.	34	3	20
Pimephales notatus	292	1,382	246
Pimephales promelas	12	0	7
Pimephales vigilax	85	126	351
Fundulus catenatus	0	tr	0
Gambusia affinis	58	0	tr
Labidesthes sicculus	42	49	14
Etheostoma blennioides	1	0	tr
Etheostoma simoterum	2	0	0
Percina caprodes	17	107	52
Cottus carolinae	tr	0	tr
Unidentified minnows	1	22	943
Subtotal	5,623	6,158	6,199

[a] tr indicates an average of <0.5 fish/hectare.

1975 to September 1977 using a boat-mounted direct current electrofishing unit.

Ichthyoplankton samples were collected biweekly from March 19 to August 6, 1975, and March 31 to August 18, 1976, at a transect adjacent to the Bull Run Steam-Electric Plant. Samples were collected with a towed beam net (0.5 m square, 1.8 m long, 0.51-mm mesh). Towed net samples were of 10 minutes duration at a velocity of approximately 1 m/second. A tow consisted of lowering the net to the bottom, accelerating the boat to 1 m/second, and raising the net one-tenth of the overall depth at 1-minute intervals (nine lifts during the tow), as described by Graser (1977). Volumes filtered were calculated using water velocities obtained from a flowmeter suspended in the mouth of each net. Dur-

Table 2. Total numbers per hectare for selected young-of-year fish species collected in rotenone samples from Melton Hill Reservoir, 1972-1982.

	Abundance (number/hectare)								
Species	1972	1974	1975	1976	1977	1978	1979	1980	1982
Dorosoma cepedianum	0	146	5	3,446	3,651	1,805	0	14	479
Dorosoma petenense	15	88	256	2,478	603	7,707	104	5,180	5,677
Morone chrysops	0	0	1	9	4	10	0	0	2
Lepomis macrochirus	289	203	246	1,277	941	888	77	338	1,750
Micropterus dolomieui	0	0	1	2	10	27	0	8	7
Micropterus punctulatus	1	0	26	108	23	117	7	4	3
Micropterus salmoides	5	34	46	166	56	78	9	90	52
Pomoxis annularis	1	19	19	111	32	13	2	17	59
Number of coves sampled	2	6	6	10	6	4	2	3	2
Total YOY collected	401	981	1,829	10,344	6,249	11,318	382	6,151	9,389

ing each excursion, samples were collected four times per day (dawn, noon, dusk, and night).

Results

Fish Community Structure

Five species contributed more than 83% (by number) of all fish collected in cove rotenone samples in Melton Hill Reservoir (Table 1). These were threadfin shad (*Dorosoma petenense*), gizzard shad (*Dorosoma cepedianum*), bluntnose minnow (*Pimephales notatus*), bluegill (*Lepomis macrochirus*), and common carp (*Cyprinus carpio*). The most abundant prey species were threadfin and gizzard shad (64% combined), the most common sport fish was bluegill (15%), and the most abundant commercial species was common carp (3%).

A comparison of Melton Hill fish densities with those of Norris and Watts Bar reservoirs revealed that numbers of sport fish were significantly greater ($P \leq 0.05$) in both Watts Bar and Norris than in Melton Hill Reservoir (Table 1). Numbers of commercial species were significantly higher in Melton Hill than in Norris ($P \leq 0.01$) but very similar to those in Watts Bar. Prey numbers were not significantly different among reservoirs.

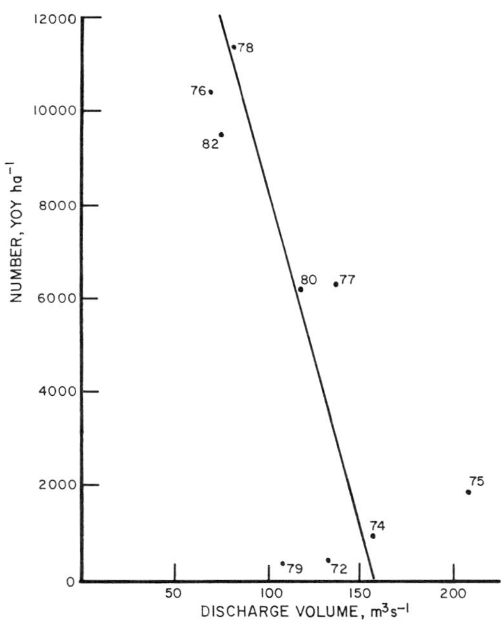

Figure 4. Relationship between number of young-of-year (YOY) fish per hectare in rotenone samples and water volume discharged during spring and early summer (April-July) from Norris Dam in 1972-1982.

Juvenile Fish Abundance

Considerable fluctuation in numbers of YOY fish was observed in cove rotenone samples collected in Melton Hill Reservoir (Table 2). A correlation analysis of same-year numbers of YOY in Melton Hill Reservoir with those in Norris and Watts Bar reservoirs revealed no significant relationships (Melton Hill vs Norris, $r = -0.40$; Melton Hill vs Watts Bar, $r = 0.11$). This suggests that factors influencing numbers of YOY in Melton Hill Reservoir did not have similar influence on Norris and Watts Bar numbers of YOY. Regression analysis of water volume discharged during spring and early summer from Norris Dam and estimates of YOY numbers from cove rotenone samples (Fig. 4) revealed a significant inverse relationship ($r = -0.71$, $P \leq 0.05$). That is, as volume of water released through Norris

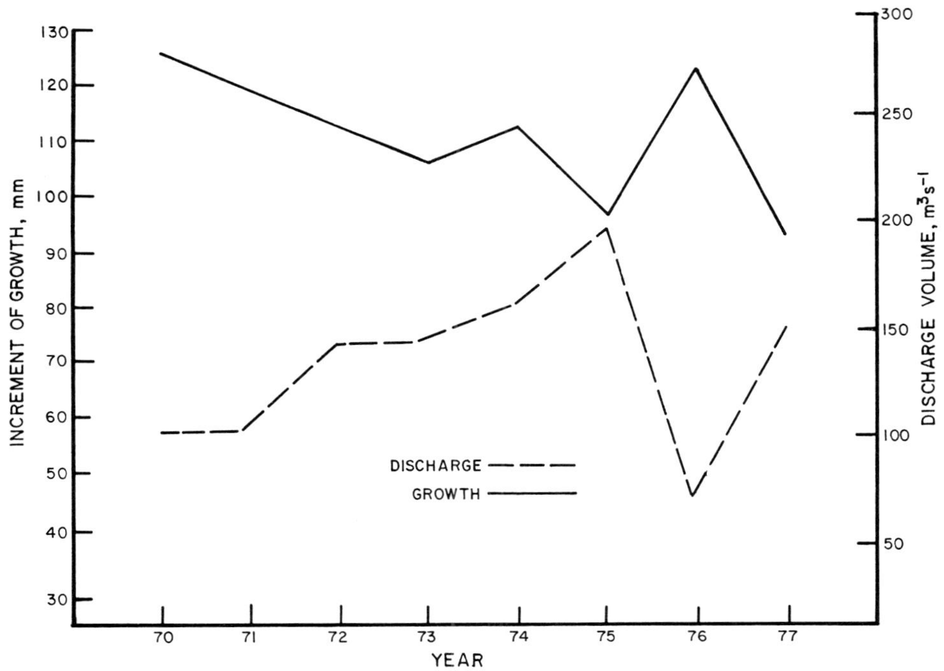

Figure 5. Growth of Age-1 largemouth bass in Melton Hill Reservoir and April-July average daily discharge water volumes from Norris Dam for 1970-1977.

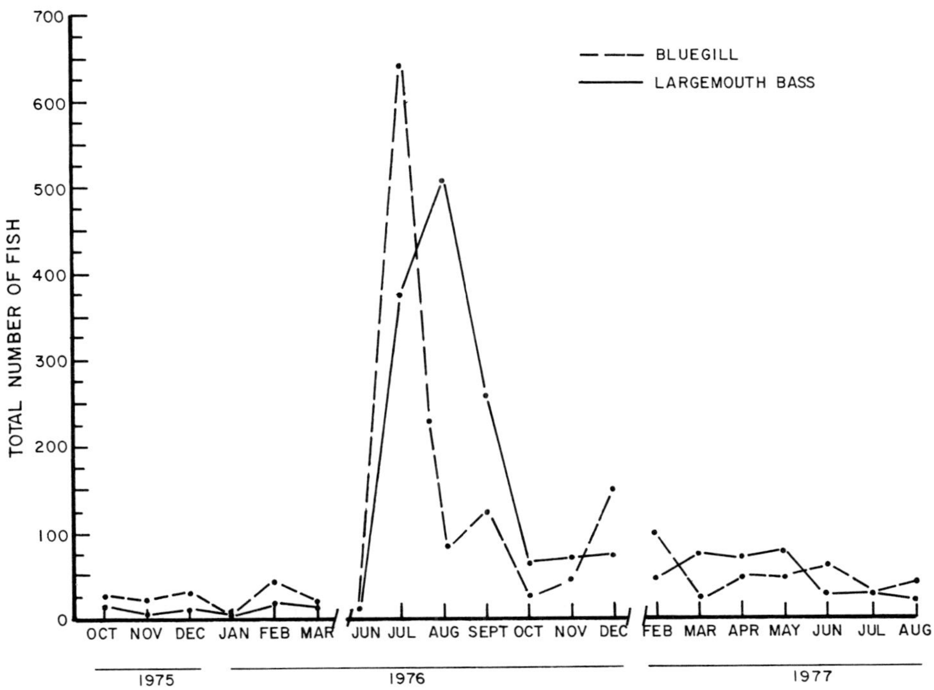

Figure 6. Numbers of largemouth bass and bluegill collected in electrofishing samples from Melton Hill Reservoir, October 1975-August 1977.

Table 3. Number of larval fish per 1,000 m³ collected from April-August near Bull Run Steam-Electric Plant, Melton Hill Reservoir, 1975-1976.

Taxon	Number collected 1975	Number collected 1976
Clupeidae		
Unspecified clupeids	3	55,078
Dorosoma spp.		20
D. cepedianum		71
D. petenense		2,128
Cyprinidae		
Unspecified cyprinids		13
Cyprinus carpio	6	49
Catostomidae	71	11
Atherinidae		
Labidesthes sicculus		6
Percichthyidae		
Morone spp.		225
Centrarchidae		
Lepomis spp.		1,237
Pomoxis spp.	1	51
Percidae		
Unspecified darter		2
Percina caprodes	4	1
Stizostedion canadense	1	
Sciaenidae		
Aplodonotus grunniens		4
Cottidae		
Cottus carolinae	1	2

Dam increased, overall numbers of YOY fish in Melton Hill Reservoir decreased.

Such fluctuations in available fish prey could result in vacillating growth of piscivorous species. Largemouth bass, the major piscivore in Melton Hill Reservoir, reflected this pattern. Age-1 fish grew significantly faster ($P \leq 0.05$) during years with low releases from Norris Dam than during years of high discharges. This pattern was also suggested for ages 2 through 5; however, relationships were not statistically significant. Regression analysis revealed a highly significant ($P \leq 0.01$) negative correlation ($r = -0.50$) between Age-1 bass growth and discharge rate (Fig. 5).

Many largemouth bass and bluegill (mostly YOY) were present in Melton Hill Reservoir electrofishing samples during July through September 1976 (Fig. 6). Numbers of both species decreased by October 1976 and remained near these levels until September 1977 when electrofishing sampling was terminated. While inconclusive, this information supports other observations that favorable spawning and survival conditions existed in spring and early summer of 1976 and not in 1977 in Melton Hill Reservoir.

Table 4. Abundance of young-of-year fishes in 1975 and 1976 cove rotenone samples from Melton Hill Reservoir.

Family	Abundance (number/hectare) 1975	Abundance (number/hectare) 1976
Clupeidae	250	7,592
Cyprinidae	1,001	788
Catostomidae	14	1
Ictaluridae	2	6
Poeciliidae	113	112
Atherinidae	22	145
Percichthyidae	1	4
Centrarchidae	405	1,654
Percidae	21	34
Total	1,830	10,344

Larval Fish Abundance

Larval fish collections in 1975 and 1976 also demonstrated the effect of Norris Dam discharge volumes on fish reproduction in Melton Hill Reservoir. As discussed previously, discharges from Norris Dam were high in 1975 and relatively low in 1976. Larval fish samples reflected considerable differences in diversity of taxa and densities between the two years (Table 3). Fish larvae of 13 taxa representing nine families were collected in 1976 compared to seven taxa from six families in 1975. Clupeids reflected the greatest difference in density, but most of the other taxa also showed significant increases in numbers taken in the low water year (1976).

Further evaluation of reproductive success during 1975 and 1976 was made by comparing numbers of YOY collected in cove rotenone samples (Table 4). As with larval fish, numbers of YOY were substantially higher in 1976, with clupeids increasing drastically (2,938%) followed by centrarchids (309%).

Conclusions

Eschmeyer and Smith (1943) reported that hypolimnetic discharges from Norris Dam probably eliminated successful reproduction of warmwater species in the area immediately below the dam. Present studies indicate that coldwater releases also impact the fish community in the main body of Melton Hill Reservoir (20.9 to 90.7 km downstream). High discharge rates from Norris Dam severely lower reservoir water temperatures during spring and summer, a period critical to successful spawning and growth of fish. Conversely, low rates of discharge from Norris Dam allow warmer reservoir temperatures and a corresponding increase in YOY fish production and growth rate of largemouth bass. These coldwater discharges resulted in critically-reduced thermal conditions for spawning of warmwater species and limited growth of piscivorous sport fish.

The most critical months for fish spawning in Melton Hill Reservoir are April to July. Discharge rates during this period would have to be minimized for maximum spawning of warmwater fish. However, considering the multipurpose functions of Norris Reservoir (power production, flood control, recreation, etc.), the effects and cost of reducing flows during spring and early summer would need to be carefully evaluated in conjunction with potential benefits to the Melton Hill Reservoir warmwater fish community. Agencies regulating dam discharges should be aware of this need for reduced discharges in the downstream reservoir during the fish spawning and growing period, and flows should be reduced whenever power demands and water levels make it possible. Through coordinated efforts, adequate fish reproduction should be obtainable in coldwater receiving reservoirs, at least on a periodic basis.

References

DENDY, J. S., AND R. H. STROUD. 1949. The dominating influence of Fontana Reservoir on temperature and dissolved oxygen in the Little Tennessee River and its impoundments. Journal of the Tennessee Academy of Sciences 24:41–51.

ESCHMEYER, R. W., AND C. G. SMITH. 1943. Fish spawning below Norris Dam. Journal of the Tennessee Academy of Sciences 23:4–5.

FRY, J. P., AND W. D. HANSON. 1968. Lake Taneycomo: a cold-water reservoir in Missouri. Transactions of the American Fisheries Society 97:138–145.

GRASER, L. F. 1977. Selectivity of larval fish gear and some new techniques for entrainment and open water larval fish sampling. Pages 56–71 *in* L. L. Olmsted, editor. Proceedings of the first symposium on freshwater larval fish. Duke Power Company, Charlotte, North Carolina, USA.

TVA (TENNESSEE VALLEY AUTHORITY). 1976a. 316(a) demonstration, Bull Run Steam Plant; Effects of thermal discharges from Bull Run Steam Plant on the fish populations of Melton Hill Reservoir, Vol. 4. Division of Natural Resource Operations, Tennessee Valley Authority, Norris, Tennessee, USA.

TVA (TENNESSEE VALLEY AUTHORITY). 1976b. Estimates of entrainment of fish eggs and larvae by Bull Run Steam Plant, 1975, and assessment of the impact on the fish populations of Melton Hill Reservoir. Division of Natural Resource Operations, Tennessee Valley Authority, Norris, Tennessee, USA.

TVA (TENNESSEE VALLEY AUTHORITY). 1981. Update report of the potential for Bull Run Steam-Electric Plant to impact the fish community of Melton Hill Reservoir, Tennessee. Division of Natural Resource Operations, Tennessee Valley Authority, Norris, Tennessee, USA.

TVA (TENNESSEE VALLEY AUTHORITY). 1982. Reservoir cove rotenone standard procedures. *in* Biological resources procedures manual. Division of Natural Resource Operations, Tennessee Valley Authority, Knoxville, Tennessee, USA.

Pages 294–300 *in* G. E. Hall and M. J. Van Den Avyle, editors.
Reservoir Fisheries Management: Strategies for the 80's.
Reservoir Committee, Southern Division American Fisheries Society, Bethesda, Maryland. 1986.

Walleye Migration through Tygart Dam and Angler Utilization of the Resulting Tailwater and Lake Fisheries

FRANK JERNEJCIC
West Virginia Department of Natural Resources
Division of Wildlife Resources
Fairmont, West Virginia 26554

ABSTRACT

Fish populations in 1,740-acre Tygart Lake and its tailwater were sampled to provide information needed to evaluate impacts associated with the addition of hydropower facilities to the Tygart Lake project. Walleyes (*Stizostedium vitreum vitreum*) dominate the sport fishery of the lake and migrate through the dam, providing a major tailwater fishery. Anglers caught 6,042 walleyes from the lake and 8,724 from the tailwater during a 1-year period. Walleye fishing success (C/f) was higher in the tailwater than in the lake (0.56 vs. 0.32 caught per hour). C/f was highest during the fall in the lake but during the spring in the tailwater. Tag returns indicated a 6% exploitation rate for lake walleyes during a 15-month period. Tailwater walleyes experienced a 25% exploitation rate during a 7-month period. Nine percent of walleyes tagged in the lake were caught by anglers in the tailwater from December through March. Age-0 and -1 walleyes migrated through the dam more readily than older walleyes. Walleye migration occurred during the winter, December through April, at times when the pool elevation was decreasing at a rate of at least 6-ft per 24 hours.

Addition of hydropower facilities to Tygart Dam by the U.S. Army Corps of Engineers will impact the existing lake and tailwater fisheries. Hydropower alternatives that are being considered involve lake level increases and fluctuations, variations in daily releases, and construction of a reregulating dam 9 mi downstream that would cause the loss of the existing tailwater and creation of a new one.

This study was conducted to help predict effects of possible changes in reservoir operation on lake and tailwater fishing and fisheries. Aspects evaluated were angler utilization of the lake and tailwater fisheries and migration of walleyes through the dam during simulated hydropower flows at proposed summer and winter pool elevations. Although information was obtained for all species, walleyes (*Stizostedion vitreum vitreum*) are emphasized in this paper because they provide the most valuable fishery in this system.

A reproducing walleye population was established in Tygart Lake following fry stockings from 1973 through 1976. Walleyes now comprise a major fishery at Tygart Lake and migrate downstream through Tygart Dam providing a substantial tailwater fishery. Reproduction does not occur in the Tygart watershed outside of the lake.

STUDY AREA

Tygart Lake is a 1,740-acre impoundment located on the Tygart River at mile 22.7. The dam was completed in 1938, and the reservoir is operated for flood control and low flow augmentation for navigation on the Monongahela River. Maximum depth is 134 ft and mean depth is 63 ft. Lake elevations vary with precipitation, flood control, and low flow augmentation needs in the basin, resulting in a possible winter pool level 84 ft below the initial summer pool elevation of 1,094 ft mean sea level (msl). Daily fluctuations in lake elevations can exceed 10 ft, and total fluctuations exceeding 75 ft within one week are not uncommon because of the relatively low storage volume (111,200 acre-ft at 1,094 msl).

The lake is infertile, and nutrients do not accumulate because of the short retention time (20 days at 1,094 msl) and hypolimnetic withdrawal (USACE 1976; 1981). Inlet invert elevation of the eight sluice gates is 995.0 msl. The average annual discharge is 2,358 ft^3/second (cfs). Since 1975, the minimum annual discharge has ranged from 124 to 340 cfs and the maximum from 11,600 to 16,000 cfs. The lake is weakly stratified during some years, but dissolved oxygen is always adequate to sustain fish throughout the impoundment and in the tailwater.

A stilling basin extends 223 ft downstream from the dam. The width, area, and depth are 488 ft, 2.5 acres, and 30 ft, respectively. Bottom composition is boulders and bedrock. Fish cannot enter the basin from the tailwater, only through the dam; fishing is not permitted in the basin.

For this study, the tailwater was considered to extend 1 mi downstream (50 acres) from the stilling basin. Average width and depth of the tailwater were 430 and 10 ft, respectively. A falls 9 mi downstream prevents upstream fish movement.

Methods

Angler effort, catch, and harvest were evaluated with a roving clerk, day-time creel survey that randomly sampled one weekday and an alternate weekend day per week in both the lake and tailwater (50 acres). The lake was surveyed from ice-out on 1 March through 28 December 1981, when the lake froze. The tailwater area was surveyed from 1 March 1981 through 27 February 1982.

Walleye migration through the dam and distribution in the lake and tailwater were evaluated in several ways. Walleyes were tagged with Floy anchor tags in the lake and tailwater; recoveries by anglers and electrofishing were used to document migration and estimate exploitation rates.

The stilling basin was sampled using rotenone before and after simulated hydropower flows at summer and winter pool elevations to evaluate downstream migration through the dam. Samples were obtained on 15–16 July 1981 and 16–18 February 1982. Fish were removed from the basin during the first day of each sample period. A simulated hydropower release was made for 8 hours, and a second sample, assumed to represent fish that moved through the dam, was collected. The July 1981 simulated hydropower flow was during the night while the February 1982 simulation was during the day.

Sampling of the stilling basin required complete control of discharges during the sample period. First, valves were opened to drain the basin from a depth of 30 ft to 7 ft. Rotenone was then pumped into the bottom-half of the water column. After a minimum of 3 hours was allowed for the rotenone to be effective, a gate in the dam was briefly opened, causing a 3 ft to 5 ft rise in the water level. This agitated the stilling basin and displaced dead fish from the bottom so they could be collected. A 2.5% emulsified rotenone formulation (Nu-Syn Noxfish, S. B. Penick and Co.) was used at rates of 0.86 ppm during July (71 F) and 5.17 ppm during February (38 F). An earlier bioassay using nine walleyes (9.2–21.4 in long) revealed that 5 ppm rotenone was required to be effective at 36–40 F temperatures expected in February. During the February samples, $KMnO_4$ sufficient to neutralize 70% of the rotenone was used to prevent a fish kill downstream.

The tailwater was electrofished in conjunction with the stilling basin rotenone samples and following certain high discharge periods to determine differences in walleye abundance before and after hydropower flows. A Smith-Root electrofishing boat with a variable voltage pulsator was used to produce 220–400 volts AC at 4–5 amps. Sampling was conducted after sunset for about 2 hours per trip. Actual electrofishing time was measured by a timer.

The creel census clerk routinely observed the tailwater and counted dead or distressed fish. Additional observations were made when combinations of pool elevation and flow were expected to cause fish movement through the dam and/or resulting mortality. Daily flushing rates were calculated to relate lake elevation and discharge.

Scales were taken to determine ages of walleyes collected and to categorize age groups by length range. Back-calculation of length used the Fraser model (Tesch 1968) with an *a* value of 1.5 in (Priegel 1964). Mean total lengths (in) of walleyes from Tygart Lake at Ages 1 through 6 were 8.6 (136 fish), 13.7 (123), 16.6 (71), 19.1 (36), 21.3 (12), and 24.6 (2). Age groups were then assigned the following length ranges: Age 0 $<$ 9.0 in; Age 1, 9.0–13.9 in; Age 2, 14.0–16.9 in; and Age 3+, $\geq$ 17.0 in. Growth in Tygart Lake is comparable to other West Virginia reservoirs and is considered rapid in relation to growth of the species throughout its range (Anderson and Weithman 1978). Growth was assumed to be similar in the tailwater and lake.

Results

Stilling Basin Samples

The first July stilling basin sample, used to eliminate all fish in the basin prior to the simulated hydropower flows, produced nine species

Table 1. Comparison of walleye captures by electrofishing in Tygart Dam tailwaters following intervals having different lake elevations and tailwater flows, February-April 1982.

	Sampling dates			
	February 15–16	February 17–18	March 1-3	April 12–13
Electrofishing time (sec)	5,900	5,250	6,150	5,475
Average flow (cfs)	1,050	1,300	1,450	1,100
Total number walleye	116	55	159	225
Sec/walleye	51	95	39	24
During interval between samples:				
Maximum flow (cfs)		7,640	9,500	14,550
Maximum lake elevation (msl)		1,043	1,072	1,116
Lake level fluctuations (ft)				
Increase		5	33	84
Decrease		0	37	80

Table 2. Lake elevations, daily water level fluctuations, and maximum tailwater flows associated with walleye migration through Tygart Dam.

Date	Lake elevation (msl)		Maximum daily drop (ft)	Number of dead walleyes observed	Maximum daily flows (cfs)
	Maximum	Minimum[a]			
26–30 Dec 81	1,078	1,042	9	3	9,660
26–27 Jan 82	1,083	1,073	7	2	9,040
23 Feb 82	1,063	1,057	6	1	9,031

[a] Following maximum elevation.

weighing 16 lb. Following the simulated hydropower release of 6,200 cfs (8 hours, 1,096 msl), the rotenone sample yielded four species of fish weighing 1 lb. Walleyes were the dominant species in the pre-flow sample by weight (75%) and number (41%) but were not found in the post-flow sample.

The first February stilling basin sample produced six species weighing 2 lb. Following the simulated hydropower release of 7,640 cfs (8 hours, 1,043 msl), only two species weighing less than 1 lb were collected. No walleyes were found; one fingerling channel catfish (*Ictalurus punctatus*) was the only game fish collected.

The lake elevation decreased 2 ft during the July hydropower flow but increased 5 ft during the February flow. Daily flushing rates were 9.3 and 2.6 days during the July and February flows, respectively.

Tailwater Samples

Walleyes were relatively less abundant following the July simulated hydropower flow than before. An average of 247 seconds of electrofishing was required to capture a walleye on 12 and 13 July before the hydropower flow (total of 17 collected); after the release, only one walleye was caught during 4,560 seconds of shocking.

During 1982 fewer walleyes (55) were collected after the February hydropower release than before (116), and the time required to capture each walleye nearly doubled (Table 1). It appeared that few, if any, walleyes had moved through the dam during either the summer or winter simulated hydropower flows.

Major lake elevation fluctuations and increased releases subsequent to the February sampling prompted additional sampling during March and April (Table 1). Total numbers of walleye collected increased in the March and April samples, and times required to collect each walleye decreased. There were obviously more walleyes in the tailwater during the March and April samples than during February, indicating an increase in walleye migration during that interval.

Dead Fish in Tailwater

Physical trauma from rapid pressure changes and impact undoubtedly resulted in injury or mortality to a portion of fish moving through the dam. Thus, the presence of dead and/or distressed fish in the tailwater (not resulting from angling mortality) is indicative that fish migration is occurring.

Dead walleyes were observed in the tailwater during three periods when lake elevations were decreasing at least 6 ft in 24 hours (Table 2). Flows during these times ranged from 5,280 to 9,660 cfs, resulting in flushing rates of 3.8 to 4.5 days.

Creel Census

Walleyes accounted for 41% of the 14,806 fish caught from Tygart Lake, followed by largemouth bass (*Micropterus salmoides*), smallmouth bass (*M. dolomieui*) (26%), and sunfish (*Lepomis* spp.) (23%). In the tailwater, walleyes comprised 63% of the 13,880 fish caught, followed by stocked rainbow trout (*Salmo gairdneri*) (33%). Walleye fishing success (C/f) from the lake was 0.32 caught per hour and 0.17 har-

Table 3. Estimated number of walleyes caught, angler hours for walleyes, and number caught per hour (C/f) in Tygart Lake and tailwater, 1981-1982.

Month	Lake			Tailwater		
	Number	Hours	C/f	Number	Hours	C/f
March	384	2,065	0.19	1,528	3,741	0.41
April	358	4,305	0.08	2,524	3,576	0.71
May	707	2,192	0.32	1,500	2,218	0.68
June	285	1,122	0.25	892	1,184	0.75
July	317	586	0.54	1,033	1,610	0.64
August	341	1,340	0.25	3	393	0.01
September	716	2,362	0.30	9	144	0.06
October	2,028	3,307	0.61	39	221	0.18
November–December	906	1,772	0.51	73	374	0.20
January–February		Ice cover		1,119	1,987	0.56
Total number caught	6,042		0.32	8,724		0.56
Total number harvested	3,305		0.17	3,061		0.20
Total hours		19,051			15,448	
Hours per acre		11			309	

vested; from the tailwater 0.56 were caught and 0.20 harvested per hour (Table 3).

The mean length of walleyes caught was 12.2 in from the lake and 10.4 in from the tailwater (Table 4). Anglers released fish averaging 9.7 in from the lake vs 8.7 in from the tailwater and kept 14.2 in walleyes from the lake vs 13.8 in from the tailwater.

Walleye fishing pressure was greatest in the lake during the spring and fall, but C/f was highest during July, and October through December (Table 3). More walleyes were caught in Tygart Lake during October (35% of total) than in any other month. In the tailwater, the majority of walleye fishing pressure (93%) occurred from February through July, when walleye C/f was greatest. More walleyes were caught in the tailwater during April (29% of total) than in any other month. Only 39 walleyes were caught during October when the greatest number (1,062) of trout were caught.

Tag Returns

Returns of tagged fish by anglers confirmed that walleyes migrated through Tygart Dam to the tailwater and further downstream. Two tagged in the tailwater were caught 10 mi downstream and one 30 mi downriver in the Monongahela River. Numbers of tags returned may have been minimal because returns were voluntary, no rewards were offered, and the program was not extensively publicized.

Six percent of the walleyes tagged in Tygart Lake during 1981 were recaptured by anglers in

Table 4. Estimated numbers of walleyes harvested and released by anglers from Tygart Lake and tailwater, March 1981-February 1982.

Age	Inch class	Lake		Tailwater	
		Released	Harvested	Released	Harvested
0	4	7	0	0	0
	5	0	0	154	0
	6	177	4	712	6
	7	150	0	768	4
	8	509	33	1,304	14
	Subtotal	843	37	2,938	24
1	9	342	9	681	10
	10	708	80	1,508	53
	11	328	68	285	180
	12	363	480	369	387
	13	151	649	62	568
	Subtotal	1,892	1,286	2,905	1,198
2	14	37	707	88	783
	15	13	621	7	677
	16	7	235	0	257
	Subtotal	57	1,563	95	1,717
3+	17	0	180	0	71
	18	2	91	0	32
	19	0	28	0	3
	20	0	39	0	3
	21	0	31	0	4
	22	0	9	0	6
	≥23	0	43	0	3
	Subtotal	2	421	0	122
Total		2,794	3,307	5,938	3,061
Mean length (in)		9.7	14.2	8.7	13.8
Mean length caught		12.2		10.4	

Table 5. Numbers of walleyes tagged in Tygart Lake during 1981 and recaptured by anglers in the lake and tailwaters, April 1981-May 1982.

	Age group (years)				
Length range (in)	0 (<9 in)	1 (9–13 in)	2 (14–16 in)	3 (≥17 in)	All ages
Number tagged	4	54	13	15	86
Number recaptured in lake	0	1	1	3	5
Number recaptured in tailwater	0	8	0	0	8
Overall percent recaptured	0	17	8	20	15

Table 6. Dates of tags returned by anglers from walleyes tagged in Tygart Lake during 1981.

Period tagged in 1981	Number tagged	Number of recaptures	Recapture location	Date recaptured
April	23	3	Lake	April, August, December 81
September–October	30	5	Tailwater	February–May 82
		1	Lake	May 82
November	33	3	Tailwater	February–April 82
		1	Lake	May 82

the lake and 9% were caught in the tailwater (Table 5). These tailwater recaptures represent a minimum estimate of the proportion of the lake walleye population migrating to the tailwater. At least 25% of those tagged in the tailwater during 1982 (February–April) were recaptured by anglers in the tailwater from February through August.

Recaptures in the tailwater of fish tagged in the lake during 1981 occurred from February through May 1982 (Table 6). The majority of lake-tagged fish (79%) were released 3–7 mi above the dam. No fish released within 1 mi of the dam (21%) were recaptured in the tailwater.

A comparison of the ages of walleyes caught by anglers in the lake and tailwater revealed that Age-0 fish were three times more abundant in the tailwater creel than in the lake (Table 4). In contrast, Age-3 and older walleyes were three times more numerous in the lake creel than in the tailwater creel. The relative proportions for Age-1 and Age-2 walleyes also were greater in the lake than in the tailwater, and the differences in abundance of age groups were statistically significant (Chi-square, $P \leq 0.01$).

Age-1 walleyes were the only tagged walleyes from the lake that were recaptured in the tailwater (Table 5). Eleven of the 28 Age-2 and older tagged walleyes were released 0.1 mi upstream of the dam on 2 April 1981, and the only recapture from that group was on 23 August 1981 near the head of the lake, 10 mi upstream. Three (18%) of the remaining 17 Age-2 and older walleyes were tagged and released 4–6 mi above the dam and recaptured in the lake. These recaptures suggest a reduced tendency for older walleyes to move downstream through the dam.

Discussion

Walleyes dominate the sport fishery of Tygart Lake and migrate through the dam to provide a major tailwater fishery. Other species common in the lake were not caught or sampled in the tailwater, which indicated a selective movement by walleyes. Fisheries at two Missouri tailwaters, Lake of the Ozarks and Pomme de Terre, were supported by fish passing from the lakes (Hanson 1979). White crappie (*Pomoxis annularis*), channel catfish, carp (*Cyprinus carpio*), and bluegills (*Lepomis macrochirus*), which were abundant in these Missouri tailwater fisheries, were also common in Tygart Lake but not the tailwater.

Only 6% of the walleyes tagged in Tygart Lake were caught in the lake during a 15-month period compared with 25% tagged and caught in the tailwater in 7 months. Since walleye exploitation rates generally average between 20% and 30% annually (Colby et al. 1979), the Tygart Lake walleye population would be considered lightly exploited by anglers and the tailwater population heavily exploited.

Walleye fishing pressure on Tygart Lake was greatest during the spring and fall although the highest success rates occurred during the fall drawdown. In Tygart tailwater, fishing pressure was concentrated from February through July, and walleye fishing success remained high throughout that period. Fishing success declined only because migration from the lake had ceased and the walleyes initially present had been harvested or had moved downstream. In contrast to Tygart Lake and tailwater, Colby et al. (1979) cited several authors who indicated walleye fishing was best during the first month of the spring season and decreased thereafter.

A minimum of 9% of walleyes tagged in the lake during 1981 migrated to the tailwater from

December 1981 through March 1982. This loss from the lake was considered a net benefit because of the heavy angler use of the resulting tailwater fishery. In contrast, loss of walleyes, particularly brood stock, from Kansas impoundments was considered detrimental because of significant declines in the lake population (Groen and Schroeder 1978). Sixty percent of walleyes tagged in Pomona Reservoir, Kansas, were recaptured in the tailwater.

The scarcity of larger walleyes creeled in the tailwater, in conjunction with tag returns from Age-1 walleyes, indicated that younger walleyes (Age 0 and Age 1) migrated through Tygart Dam more readily than those Age 2 and older. Walleyes caught in the lake were larger than those caught in the tailwater. Seven percent of the walleyes caught in the lake were Age 3 or older as opposed to 1% of the tailwater walleyes. Conversely, 14% of the lake walleyes were Age 0 compared with 33% of the tailwater walleyes. Serns and Kempinger (1981) found that angling vulnerability decreased with increasing walleye age and length in Escanaba Lake, Wisconsin. Likewise, Elsey and Thomson (1977) found angling selective for younger walleyes in Lac des Mille Lacs, Ontario.

The scarcity of larger walleyes creeled in the tailwater cannot be attributed to angling vulnerability. The smaller area, shallower depth, and increased velocities would concentrate fish and increase their vulnerability to anglers. Food is also less abundant in the Tygart tailwater than in the lake, which should increase angler success (Forney 1967; Kempinger et al. 1975).

Observations of dead fish, angler success rates, and numbers of walleye caught each month from the tailwater revealed that migration started during December or January with releases of flood storage. Numbers of walleyes caught by tailwater anglers were greatest from March through May then decreased substantially during June. It appears that essentially all tailwater walleyes, even those harvested after April, were fish that migrated through the dam during the period of winter pool drawdowns, December through March. Tailwater recaptures of walleyes tagged in the lake also substantiated their winter and early spring migration.

Groen and Schroeder (1978) reported walleye loss from Kansas reservoirs occurred during late March and April when walleyes concentrated along dam rip rap to spawn. However, walleye spawning habitat is prevalent throughout Tygart Lake, so walleyes do not concentrate near the concrete dam for spawning.

Migration did not occur with high flows from summer pool elevations during July or from a rising winter pool during February. Increases in numbers of tailwater walleyes were documented following high discharges accompanied by major decreases in reservoir elevations during February, March, and April. Pool elevations during this period when rapid reservoir withdrawals were made to ensure adequate storage capacity for subsequent winter and spring floods even exceeded summer pool elevations.

It appears that a combination of two reservoir operations are necessary for walleyes to migrate through Tygart Dam. First, the lake must be in the winter pool operating schedule, December through April. Secondly, withdrawal must be sufficiently rapid to decrease the reservoir elevation at least 6 ft per 24 hours, equivalent to flushing rates of 3.8 to 4.5 days.

Hydropower schemes requiring alteration of historical flows could prevent walleye migration or cause excessive mortality to migrating fish. Spillage of adequate flows, 5,000 to 10,000 cfs, resulting in suitable flushing rates for walleye migration will be necessary from December through April. Discharge records for 1981 indicated suitable conditions for migration occurred for three or four days each month during that period. Power production may have to be sacrificed if sufficient spillage is not available from December through April to sustain adequate walleye migration for maintenance of the existing tailwater fishery.

Acknowledgments

This project could not have been completed without the assistance of many West Virginia Department of Natural Resources and U.S. Army Corps of Engineers personnel. Denzil Courtney, Pat Dockerty, Diane Dumont, Ed Godfrey, Bert Pierce, and Craig Stihler contributed substantially to the project. The major part of this study was funded by the Pittsburgh District of the U.S. Army Corps of Engineers.

References

Anderson, R. O., and A. S. Weithman. 1978. The concept of balance for coolwater fish populations. Pages 371–381 *in* R. L. Kendall, editor. Selected coolwater fishes of North America. Special publication 11, American Fisheries Society, Bethesda, Maryland, USA.

Colby, P. J., R. E. McNicol, and R. A. Ryder. 1979. Synopsis of biological data on the walleye *Stizostedion vitreum vitreum* (Mitchell 1818). Fisheries Synopsis 119, Food and Agriculture Organization of the United Nations, Rome, Italy.

Elsey, C. A., and R. T. Thomson. 1977. Exploitation of walleye (*Stizostedion vitreum vitreum*) in Lac des Mille Lacs, northern Ontario, by commercial and sport fisheries, 1958–75. Journal of the Fisheries Research Board of Canada 34:1769–1773.

Forney, J. L. 1967. Estimates of biomass and mortality rates in a walleye population. New York Fish and Game Journal 14:176–192.

Groen, C. L., and T. A. Schroeder. 1978. Effects of water level management on walleye and other coolwater fishes in Kansas reservoirs. Pages 278–283 *in* R. L. Kendall, editor. Selected coolwater fishes of North America. Special publication 11, American Fisheries Society, Bethesda, Maryland, USA.

Hanson, W. D. 1979. Tailwater fisheries of Lake of the Ozarks and Pomme de Terre Lake, Missouri. Proceedings of the Southeastern Association of Fish and Wildlife Agencies 31:505–513.

Kempinger, J. J., W. S. Churchill, G. R. Priegel, and L. M. Christenson. 1975. Estimates of abundance, harvest, and exploitation of the fish population of Escanaba Lake, Wisconsin, 1946–69. Technical Bulletin 84, Wisconsin Department of Natural Resources, Madison, Wisconsin, USA.

Priegel, G. H. 1964. Early scale development in the walleye. Transactions of the American Fisheries Society 93:199–200.

Serns, S. L., and J. J. Kempinger. 1981. Relationship of angler exploitation to the size, age and sex of walleyes in Escanaba Lake, Wisconsin. Transactions of the American Fisheries Society 110:216–220.

Tesch, F. W. 1968. Age and growth. Pages 93–123 *in* W. E. Ricker, editor. Methods for assessment of fish production in fresh water. IBP Handbook Number 3, Blackwell Scientific Publications, Oxford, England.

USACE (U.S. Army Corps of Engineers). 1976. Tygart River Lake water quality report. Pittsburgh District, U.S. Army Corps of Engineers, Pittsburgh, Pennsylvania, USA.

USACE (U.S. Army Corps of Engineers). 1981. Tygart Lake West Virginia hydropower, stage 2 intermediate evaluation. Pittsburgh District, U.S. Army Corps of Engineers, Pittsburgh, Pennsylvania, USA.

Pages 301–305 *in* G. E. Hall and M. J. Van Den Avyle, editors.
Reservoir Fisheries Management: Strategies for the 80's.
Reservoir Committee, Southern Division American Fisheries Society, Bethesda, Maryland. 1986.

NEEDS OF RESERVOIR USER-GROUPS

S. L. SPENCER
Alabama Department of Conservation and Natural Resources
Montgomery, Alabama 36130

The organizers of this Symposium recognized that future policy and direction for reservoir fisheries management must incorporate the concerns and interests of user-groups and that their needs could best be communicated by outdoor writers, the fishing tackle industry, environmental interests, and representatives of State and Federal agencies who deal routinely with reservoir user-groups. Providing a link between the technical portion of the Symposium and the primary users of reservoir fishery resources was considered essential.

To obtain the perspectives of user interests and concerns, the organizers invited presentations by user-group representatives at the opening session of the Symposium. A special evening Panel Session was also organized to facilitate the exchange of information between the user-interests and technical attendees. Invited presentations at the opening session were provided by Mr. Grits Gresham, a distinguished outdoor writer and sportsman, Mr. Richard Kotis, a tackle manufacturer and recognized leader in fish conservation, and Mr. Robert Jantzen, Director of the U.S. Fish and Wildlife Service, who presented the Keynote address. A panel of respected representatives from major sportsmen's organizations, outdoor writers, and recreational fishery managers was also convened to provide an open discussion of user needs and interests from technical and user-group perspectives. This panel included the following individuals:

Mr. Sam L. Spencer, Panel Moderator, Alabama Department of Conservation and Natural Resources
Mr. Aron Pass, Senior Associate Editor, Game and Fish Publications
Mr. Stu Tinney, Publisher, Striper Magazine
Dr. David Philipp, Illinois Natural History Survey
Mr. Bennie Fontenot, Louisiana Department of Wildlife and Fisheries
Mr. Darryl Lewis, Natural Resources Branch, U.S. Army Corps of Engineers
Mr. Dennis Holcomb, Florida Game and Fresh Water Fish Commission
Mr. Richard Kotis, Fred Arbogast Company

This paper is a synopsis of comments presented by the user-group participants during the Opening Session of the Symposium and the User Group Panel Discussion.

Opening Session

In his opening presentation, Mr. Gresham summed up his perception of the angler's view of reservoir fisheries management as "confused." He postulated that the fisheries profession was presently at about the same point as the wildlife profession concerning the buck law for deer management, in which hunters were first encouraged to harvest only bucks and later to harvest either sex. He perceived the fisheries profession as having gone through a similar sequence of convincing anglers that closed seasons and restrictive harvest were essential for managing reservoir fisheries. This was followed by a period in which anglers were told that harvest regulations were not needed. Currently, the pendulum is swinging back to more restrictive creel and size restrictions. He stressed a need for much stronger efforts toward communication between fisheries scientists and the angling community. He suggested that fisheries biologists do not mingle enough with the angling public. Participation in civic groups and other organizations at the community level could do much to make the fisheries scientist better known within the community. Participation in the sport is a way to obtain a greater appreciation of the angler's concerns with fishing. He recommended that reservoir biologists be required to fish at least two or three times a month, to obtain a better perspective of how the fisherman perceives a reservoir fishery.

Mr. Gresham felt that other people around the country had similar viewpoints concerning cooperation between fisheries professionals and the angling community and were willing to become more involved in implementing management actions. He said that many leaders in the angling community wanted the profession to take tough management actions to improve sport fishing and would be willing to help in seeing that appropriate management actions were implemented. He concluded that these interests applied to a broad range of management options rather than specific management tools, and that anglers could offer a source of help.

Mr. Richard Kotis followed with comments related to the views of the tackle industry, which represents one of the strongest free enterprise systems in the U.S. Mr. Kotis drew the analogy of the customer who used fishing tackle on a daily basis, and stressed that anglers fish primarily to catch fish. Citing statistics from the 1980 Survey of Hunting, Fishing and Related Activities, he pointed out that about 32 million anglers catch catfish and panfish. However, a disproportionate amount of reservoir management activity is directed toward black bass, striped bass, and walleye. He recommended that attention should be focused on the needs of the casual angler. Increased emphasis on bluegill, catfish, and crappie fisheries would attract more people to fishing, but other sporting goods industries have out-promoted the fishing industry. He expressed a deep concern about the future of hunting and fishing.

Relating specifically to reservoirs, Mr. Kotis offered several useful suggestions to meet needs of different angling groups. He recommended developing more designated, easily accessible areas to provide youngsters and families greater opportunities for panfishing. He considered this primarily a problem of developing access, and recommended that the profession use available management background to accomplish needed actions. Specific suggestions included the construction of fishing piers on points and developing clearly marked artificial structures to concentrate sport fish and fishermen. Other management actions should attempt to accommodate a wide range of user needs, including designating some reservoirs as trophy fisheries. He was supportive of size limits from a psychological standpoint and felt that they had a positive influence on reservoir anglers.

Mr. Kotis suggested that every reservoir have some type of public relations effort. As reservoir managers, there is an obligation to provide a quality experience where anglers can enjoy the outdoors and also catch fish. Too often, the fisheries scientists becomes involved in managing the fishery to the extent that needs of the angler are overlooked. He perceived this function as providing an information source for people who want to fish a reservoir. He cited Ohio's program of recognizing anglers who catch a particular species of fish as positive action. In addition, a public relations person could create a recognition program, conduct fishing clinics, and communicate fisheries information to the public. He closed by stating "We have all the knowledge and techniques needed to do a good job. We may be overexploiting the fishery and we aren't using the fisherman. Let's communicate together to improve fishing."

Mr. Jantzen's keynote address is printed in the introductory section of this volume. As chief policy maker for the U.S. Fish and Wildlife Service, Mr. Jantzen reviewed current trends in the operation and perceived responsibilities of that organization. These included the decentralization of Federal responsibility and the transfer of several operations to state fishery programs. He stressed the importance of the Dingell-Johnson (Wallop-Breaux) expansion and the need for a strong continued partnership between the state and Federal governments in implementing this program. He reiterated the need for a greater product orientation and improved communications between fishery agencies and user interests.

User-Group Panel Discussion

An evening panel session provided an opportunity for representative user groups and biologists to communicate interests and to expand on concepts included in the opening session. After brief introductory comments by Mr. Spencer, the discussion opened with comments by Mr. Kotis, who suggested that fishing was undergoing a "convenience revolution" which is influencing fishermen's attitudes toward angling opportunities. He recommended that fishing should be made more convenient for the fisherman. Shorter trips and urban fishing programs were cited as examples of the changing attitudes of anglers. He suggested that management agencies give special consideration to actions which make it possible for the angler to gain easy access to fishing opportunities.

Mr. Aron Pass, Senior Associate Editor for Game and Fish Publications, stated that his company had mastered the art of encouraging biologists to write articles for their publications. The concept of improving the exchange of information between the biologist and the public remains an important consideration for successful management. Effective communication remains a cornerstone of society and an integral part of its workings. Most professional meetings stress more effective communication, and most professionals agree with the concept. However, finding time to accomplish the task remains an important problem. Mr. Pass stressed that the search for more effective communication should be made an important part of reservoir management because "today, effective communications is an increasingly critical resource management need, and the communication function must be recognized as part and parcel of the total management effort. So long as effective communication is seen as an optional duty or as someone in the Information and Education Division's job, the professional fisheries biologist is not effectively managing the total resource."

Mr. Pass suggested that fisheries biologists were usually less effective communicators than workers in game management and law enforcement. On the average, these workers have a much closer working relation with the public. The enforcement officer is in almost constant contact with the public. Similarly, wildlife biologists must interact with active groups of hunters. Effective communication may become essential to these individuals for self-defense. He perceived the fisheries biologist as having the refuge of his work to avoid public contact. The mesh of scientific study and management of fisheries resources could afford the fisheries manager a fulltime job unless a special effort were made to accommodate the public. Shortcomings in communications skills of many fisheries biologists develop as much from lack of practice as the result of using complex terms and scientific concepts.

He suggested that changes to make the responsibility for effective information exchange bear more directly on the resource manager. The development of more complex regulations has made it necessary for biologists to participate in public hearings on regulations, often as the "guest of dishonor." In addition, the rise of special interest groups seeking technical information has made it necessary to communicate more effectively. Organizations such as Trout Unlimited, Bass Anglers Sportman's Society, Stripers Unlimited, and many others have developed the expertise to ask pointed technical questions and the ability to critically analyze the answers. Many of these entities bypass Information and Education Divisions by design or accident. They seek more detailed information than that normally presented in a standard news release. Many times the information sought is too technical to be handled by workers in these sections.

Mr. Stu Tinney, Publisher of Striper Magazine, also stated that anglers fish primarily to catch fish. As a spokesperson for striped bass fishermen, he reflected on the controversy that has surrounded the species regarding potential negative effects of introducing the species into reservoirs. Concerns have centered on the striped bass competing with other predators for food and the perception by some anglers that the species preys heavily on other reservoir sport fishes. He commended the research efforts by fisheries scientists to help clarify biological relations between striped bass and other reservoir sport fish, and stressed the need for fishery managers to improve communications with publishers of popular magazines who might, in turn, communicate findings to the fishing public.

Dr. David Philipp discussed the incorporation of genetic principles into management decisions. He proposed that management efforts which reduce genetic variability in fish populations or disrupt specific gene combinations in native populations would most likely produce negative results. Existing stocks should be identified and characterized with respect to genetic features. He reported that new technologies offer the biologist powerful tools for addressing certain problems previously considered insurmountable, and that these new techniques should be incorporated into fisheries management. They should be considered long-term commitments, but with rewards being worth the investment of resources and time.

Mr. Bennie Fontenot reviewed relations between commercial and sport fishermen and reported that conflicts inevitably arose when sport and commercial fishermen operated in close proximity. The problem was particularly acute in reservoirs. The introduction of striped bass has added to the conflict because the species is particularly susceptible to gill netting. His agency believed that it was necessary to limit commercial fishing in many instances in order to benefit the greatest number of users. Solutions of these types of problems were considered difficult because they involved socioeconomic, political, and biological aspects. He stressed that managers must take actions that are not popular to all elements of their constituency to be effective.

He further suggested that conflicts between sport and commercial anglers differ at each reservoir, and solutions should be based on the composition of the fish community. Commercial fishing should be prohibited in reservoirs with very low populations of commercially valuable fish. He reported that commercial fishing was seldom prohibited in Louisiana, but that special regulations were placed on reservoirs to protect sport fish populations. Effective regulations are based on good enforcement. Conflicts were reduced when both commercial and sport fishermen were aware of regulations.

Mr. Darryl Lewis reviewed U.S. Army Corps of Engineers public use facilities. He considered the Corps of Engineers to be a facilitator, providing services to a variety of reservoir user groups. About 25% of these users are fishermen. He reported that recreational use of Corps of Engineers reservoirs has continued to increase since the mid-1970's, and this has resulted in increased stress on public use facilities. Total visitation to Corps of Engineers projects was 385 million recreation days in 1975. Visitation increased to 457 million days in 1980, and 480 million recreation days in 1982. Crime, littering, and maintenance demands have increased with the higher visitation levels. In addition, new types of recreation such as jet skiing and hang gliding have appeared. Construction of new reservoirs has slowed, and consequently, existing facilities receive higher use.

Problems at Corps facilities were classified as management or conflicting use. Conflicts between user groups are the most difficult to solve. The objective of the Corps of Engineers has been to provide the best mix of recreational opportunities while recognizing that there are certain quality thresholds. This involves developing optimum mixes of a wide variety of recreational uses.

Behavior of the user is an important consideration in developing management strategies within the Corps of Engineers. For example, weekdays present an entirely different set of problems than weekend days. Public feedback and input is sought, and this is important at the lowest organizational level.

Mr. Lewis stated that the Corps of Engineers maintains a high level of public involvement. This provides an excellent pathway for providing information to fishermen, and the Corps of Engineers might be able to help in communicating management objectives to anglers.

Mr. Holcomb completed the panel presentations with a discussion of user group needs and fisheries management for reservoirs and lakes. As Director of the Fisheries Division of the Florida Game and Fresh Water Fish Commission, he characterized two important reservoir user groups as the fisherman and the professional biologist; he then posed questions to stimulate discussion by the audience and panel. Speaking as a fisheries administrator and professional biologist, he offered the following comments concerning fishermen and the professional fishery scientist:

> "To the experienced fisheries biologist the term 'fisherman' broadly defines the largest, most important 'user group' we as professionals have the privilege of working for—a most complex group—at least as diverse as the fisheries they happily exploit and we attempt to manage. They are, collectively, the essential reason for your professional existence. In your day-to-day contact with recreational fishermen please consider the following.
>
> They need to know and understand who you are, so treat them with respect and consideration. They need to know that you understand who they are, and what they want, so communicate with them at every level. They need to know that you have special knowledge to apply toward those aquatic resources they use and you are charged with preserving and managing. So, before you sit down at the microcomputer to model the fishery, be absolutely sure you know the fishery and that your data gaps are as narrow as time and money allow. They need to know that you have personal and professional integrity—it's the only way to gain and keep their trust, sell your programs and get needed dollars. They need to know you are willing to go to bat for them in the face of public or political pressure and that you are steadfastly committed to spending their dollars wisely, according to plans that make good biological sense and which yield a product for them which is evident, worth their dollar, and easily measurable. These needs I've just mentioned are simple to state, often difficult to meet, but vital to keep in mind when you deal with your biggest 'user group'—the fishing

public. You can, I'm sure, add to the needs list and I hope you do when we open discussion.

I want to define another 'user group' which unlike the fisherman is small in number, often considers itself underrepresented, often is, and is the group partly responsible for this symposium. It is the group consisting of those professional biologists who must confront, on a daily basis, the difficult questions concerning productive reservoir management. A few of their needs are clearly defined as the five objectives of this Symposium. Another of their needs just may be the opportunity to discuss in an informal, open forum their personal and organization concerns on a subject being well debated in the halls, but not at all evident in the printed program."

Mr. Holcomb then discussed comments made by Mr. Jantzen during the Opening Session concerning changes in cooperative relations between the U.S. Fish and Wildlife Service and the state conservation agencies, and cited the closure of the National Reservoir Research Program. A principle reason for the closure of the Program was given as the judgement that the states had the ability to develop and carry out reservoir research and management within their boundaries. He then reported that 13 Southern states had 587 reservoirs larger than 500 surface acres with a total area of 4,776,000 acres, representing 37% of the total number and 48% of the total acreage of U.S. reservoirs having surface areas larger than 500 acres. He posed the following questions for consideration by the audience:

1. "As representative administrators or biologists of states wherein these resources reside, do you indeed have the capability to plan, research, and manage well in your respective area of responsibility?"
2. "Are there clearly needs which, by their nature, invite a coalition of interest and a necessary interstate cooperation, for example holistic river basin management?"
3. "If so, by what best formal mechanism should such shared problems be addressed? Why should Uncle Sam do it?"
4. "When last did the Reservoir Committee review comprehensively and prioritize the collective needs of member states?"

In regard to his last question, he requested that Mr. Fred Harris, then Chairman of the Reservoir Committee, Southern Division, American Fisheries Society, discuss reservoir needs with his committee and present a report at the next Southern Division annual meeting. In addition, he requested that Mr. Harris assist Mr. Gordon Hall in the development of a "White Paper" on reservoir management needs and policies under the sponsorship of an interagency committee of Federal and private conservation agencies and chaired by the American Fisheries Society.

ABSTRACTS OF POSTER PRESENTATIONS

Harvest Regulation

The Use of a "Slot" Length Limit Regulation for Largemouth Bass on a New Georgia Reservoir

In 1980, a slot length limit largemouth bass regulation was placed into effect on Lake Oconee one year after it was filled. An 11–14 in protected "slot" was selected after problems developed in the initial year classes of fish. We believed this regulation would (1) reduce the size of a rapidly growing initial year-class of gizzard shad; (2) alter the largemouth bass population structure; (3) improve the growth of largemouth and other game fishes; and (4) increase catch rates for largemouth bass.

Seining, electrofishing, and rotenone studies were carried out to monitor relative survival and growth of bass in 1979–1982. Shoreline electrofishing was carried out quarterly, and three cove rotenone studies were conducted each summer. Approximately 4,500 bass were tagged on five discrete occasions from 1980 to 1982. A $5 reward was offered for each tag returned by anglers, with an additional $100 drawing held at the end of each year. Roving and access creel surveys were run concurrently from September 1979 through September 1982.

There was tremendous variability in growth of the large initial year class of largemouth bass. Young-of-the-year ranged in total length in August 1979 from 64 mm to 229 mm. Subsequent year classes were much smaller, but survival was relatively good. The standing crop of largemouth bass was 29 lb/acre, 23 lb/acre, 28 lb/acre, 37 lb/acre in 1979, 1980, 1981, and 1982, respectively. Gizzard shad biomass peaked in 1980 at 131 lb/acre and declined to 88 lb/acre in 1982. Threadfin shad standing crops increased from 0.03 lb/acre in 1979 to 108 lb/acre in 1982.

From 1980 through 1982, 855 (19%) of the tags were returned. Information from anglers indicated that 57% of the tagged fish were released. Harvest rates averaged 0.11 bass per hour throughout the study period although actual catch rates (including fish caught and released) were double or triple that.

The size variability in the 1979 young-of-the-year largemouth bass in Lake Oconee was similar to that seen in West Point Reservoir during its first year. A mild winter, which aided in the establishment of a threadfin shad food base, led to a relatively high survival of this initial group. Standing crops of largemouth have remained two to three times that of other Georgia reservoirs with another large year class being spawned in 1982. The largemouth population continues to be dynamic with the prey source continuing to be more than adequate.

It appears that objectives 1 and 3 have been met. Gizzard shad biomass has declined and the average size of largemouth creeled has increased over the past 2 years. Objective 2 has not been successfully met. One possible cause was the apparent disinterest in anglers of creeling bass 152 mm to 254 mm long. Tag return data indicated that 55–64% of those caught are not being harvested. With standing crops of largemouth remaining high, the slot length regulation will be retained despite relatively low harvest rates.

An interesting result of the tagging study was the high percentage of legal largemouth caught and released. This information indicates that results from previous tagging studies where the assumption that all legal fish caught were harvested may be in error. This would bias total mortality estimates upward by assuming greater fishing mortality than was actually occurring.

Christopher Martin
Tim B. Hess
Department of Natural Resources
Game and Fish Division
Atlanta, Georgia 30334

The Effect of a Modified-Slot Regulation on Largemouth Bass Harvest at Lake Norman

North Carolina adopted new creel and minimum-size regulations for largemouth bass, effective 1 January 1982. The minimum size was increased from 30 cm to 35 cm, however, two bass smaller than 35 cm could be included in the daily creel limit of eight largemouth bass. The regulation was designed to permit a limited harvest of small bass (TL < 30 cm), and to reduce the harvest of intermediate-size bass (30–35 cm TL).

Lake Norman, a 13,156-hectare multiple-use reservoir on the Catawba River, has been characterized in recent years by having an abundance of small largemouth bass. Consequently, the impact of the new regulations on the harvest of this population is being evaluated through an exploitation study.

The Lake Norman project is a cooperative study involving Duke Power Company and the North Carolina Wildlife Resources Commission. In five nights of shoreline electrofishing during 24 March–5 April 1982, 3,906 largemouth bass (TL < 20 cm) were tagged with disc-dangler tags. An additional 4,324 largemouth bass (TL < 20 cm) were finclipped.

Rewards ranging in value from $1 to $100 are being paid to anglers for returning the disc-dangler tags. Through 1 March 1983, a total of 1,480 tags (37.9%) had been returned and $4,056 paid in rewards. A postcard survey to determine the harvest rate of angler-creeled bass (based on 1,131 card returns through 1 March 1983) showed that only 28.6% of the largemouth bass creeled actually were harvested.

Preliminary results revealed that 53.2% of the bass harvested were <30 cm TL, indicating extensive angler utilization of fish that could not have been harvested legally prior to passage of the new regulations. The size distribution of creeled bass approximated the size distribution of the total tagged population.

WILLIAM B. SMITH
WAYNE R. CHAPMAN
KIMBERLY E. GARNER
North Carolina Wildlife Resources Commission
512 North Salisbury Street
Raleigh, North Carolina 27611

Size Structures and Catch Rates of Unexploited Largemouth Bass Populations

Unharvested fish populations provide opportunities to study the basic interactions of fish assemblages without the confounding influence of exploitation. The objectives of this study were to determine the size structures of three unexploited largemouth bass populations, and to determine if a characteristic size structure existed in these populations.

Three unfished impoundments of varying physical and chemical characteristics were selected as study sites; these were Thomas Lake (8.0 hectares at Ft. Gordon, Georgia), Pond B, and Par Pond. The latter impoundments, on the Savannah River Plant near Aiken, South Carolina, are 106 hectares and 1,000 hectares, respectively. Largemouth bass were captured by angling and electrofishing at randomly selected sites or over the entire impoundment. Angling effort was 8.0 hours/hectare, 0.89 hours/hectare, and 0.32 hours/hectare for Thomas Lake, Pond B, and Par Pond, respectively. Impoundments were sampled quarterly from March through August, 1982. A Kolmogorov-Smirnov non-parametric test was used to analyze length frequency distributions.

These unexploited populations had a characteristic size structure, with the majority of the catchable population concentrated within a narrow size range. In Thomas Lake, this concentration was between 240 and 300 mm (all lengths refer to total length), in Pond B, between 280 and 340 mm, and in Par Pond, between 360 and 460 mm. Within lakes, there were significant differences ($P < 0.05$) between length frequency distributions from samples collected by angling and electrofishing.

The modes of length frequencies of samples collected by angling were 240 to 260 mm for Thomas Lake, 320 to 340 mm for Pond B, and 420 to 440 mm for Par Pond. All distributions were significantly different ($P < 0.05$). The modes of the size distributions from electrofishing samples for Thomas Lake were 160 to 180 mm, for Pond B, 300 to 320 mm, and for Par Pond, 380 to 400 mm. All distributions were significantly different ($P < 0.05$).

Catch rates during the study were higher than those typically reported for other southeastern reservoirs. The overall catch rates for Thomas Lake, Pond B, and Par Pond were 1.12, 7.02, 3.45 bass/angler hour, respectively. The highest individual sampling interval catch rate was 26.4 bass/angler hour, which was recorded at Pond B.

Our results are not directly comparable to angling catch data from exploited populations since data are usually unavailable for fish below minimum size limits. However, the results of this study suggest that a largemouth bass fishery maintained under restrictive harvest regulations would be characterized by high catch rates and a concentration of bass within a narrow size range.

DAVID S. JAKES
RONNIE J. GILBERT
Georgia Cooperative Fishery Research Unit
School of Forest Resources
University of Georgia
Athens, Georgia 30602

Management of Smallmouth Buffalo in Rough River Lake, Kentucky, During the First Twenty Years of Impoundment

The smallmouth buffalo population in Rough River Lake increased rapidly from impoundment in 1961 to 1967, when an experimental commercial harvest was first conducted. In 1961, three months after the lake reached seasonal pool, rotenone studies indicated 12 fish per hectare weighing an average of 0.7 kg. By 1963, the population had increased to 17 fish per hectare averaging 0.85 kg each. The contracted commercial harvest in 1967–1968 removed 2.5 fish/hectare averaging 4 kg each. From 1969 to 1976, commercial fishing was opened to all commercial fishermen with minimal gear restrictions of 10.1 cm bar mesh gill and trammel nets and 7.6 cm bar mesh hoop nets. During this time: 1) the average weight of buffalo harvested increased to 5.8 kg in 1972 and 4.8 kg in 1973, 2) the total number of fish harvested decreased to 121 in 1972, 169 in 1973, and was less than 50 per year from 1974–1976, 3) the number of fish per hectare increased to 267 weighing an average of 1 kg in 1975, and 4) the number of commercial fishermen decreased from 16 in 1972 to 7 in 1973 and 2 in 1976. Experimental netting studies in 1974–1976 identified a large population of buffalo averaging 1.8 kg, just under the 2.2–2.7 kg minimum weight for capture in 10.1-cm mesh netting. Commercial fishing was closed on January 1, 1977, and opened on a permit-only basis from December 1977 through June 1979. During the first 10 months of the season, a minimum mesh restriction of 10.1 cm was used, which produced an average fish weight of 2.8 kg. In the remaining nine months of the fishing period, the minimum mesh size was lowered to 8.9 cm, which resulted in an average fish weight of 2.4 kg, but with an eight times greater harvest than with the larger mesh. Based upon these catch data and the greater number of fish taken during the winter and spring months, commercial fishing regulations were rewritten in 1980 providing for a five-month, November through March, permit commercial fishing period utilizing minimum 8.9 cm mesh. From 1977, when permit commercial fishing was implemented, until 1983, the average weight of buffalo in the lake increased from 1.3 kg to 2.1 kg, and the number of fish per hectare increased from 113 to 244. The average weight of fish harvested since 1977 has ranged from 2.3 kg to 3.2 kg.

ROBERT D. HOYT
Department of Biology
Western Kentucky University
Bowling Green, Kentucky 42101

BONNY D. LAFLIN
Kentucky Department of Fish and Wildlife Resources
Frankfort, Kentucky 40601

Effects of a 15-Inch Minimum Length Limit on Largemouth Bass and Spotted Bass Fisheries in Table Rock Lake, Missouri

The largemouth bass and spotted bass fisheries of 43,000-acre Table Rock Lake were studied from 1973 through 1980 to evaluate the effects of a 15-in minimum length limit initiated January 1, 1976. Changes in the bass fisheries were monitored annually in the James River Arm of the lake with nighttime electrofishing during April and May and with a roving creel survey on a 1,700-acre portion of the arm. We intended for the length limit to improve catch rates for bass anglers and increase the abundance of bass longer than 15 in.

Total annual survival of protected-size largemouth and spotted bass increased after implementation of the length limit. However, poor recruitment of largemouth bass, unrelated to the size limit, kept the actual abundance of 15-in and longer largemouth bass at a lower level than during pre-length limit years. Bass population models indicated that if recruitment had been equal before and after the length limit, the abundance of 15-in and longer largemouth and spotted bass would have increased by 70% and 35%, respectively.

Angler harvest rates and the estimated numbers and pounds of bass harvested were lower after implementation of the length limit. Angler catch rates and the estimated numbers of bass caught increased due to improved survival rates of bass and repeat catches of individual bass. A higher proportion of the anglers who were fishing for bass were successful in catching at least one bass, although the proportion able to keep a bass declined.

The 15-in minimum length limit was considered a successful management technique for largemouth bass. Spotted bass at Table Rock Lake grow more slowly than largemouth bass, do not live as long, and could be better managed with a lower minimum length limit.

GARY D. NOVINGER
Missouri Department of Conservation
1110 College Avenue
Columbia, Missouri 65201

Evaluation of a 15-Inch Minimum Length Limit for Black Bass at Lake of the Ozarks, Missouri

On January 1, 1976, a 15-in minimum length limit on black bass was initiated at Lake of the Ozarks, a 55,000-acre reservoir in west central Missouri. Results of creel census and electrofishing surveys in the early 1970's indicated a low density black bass population that provided only a limited fishery. The purpose of the length limit was to control overharvest of black bass. The effects of overharvest combined with poor year class production of black bass led to the low density population. The desired results of the length limit included improved black bass population structure and abundance, black bass catch rates, and increased black bass harvest by weight.

Relative abundance of largemouth bass, as measured by electrofishing catch rates, has shown a sustained improvement since the implementation of the 15-in length limit. In 1974 only 13 largemouth bass (8 in or larger) were captured per hour of electrofishing. Since 1976, capture rates have ranged from 59 to 129 largemouth bass per hour with an average of 83. This increase in relative abundance resulted in improved angler catch rates. In 1973 and 1974 the total black bass catch rate was 0.07, meaning bass anglers expended an average of 14 hours to catch a bass of any size. From 1976 through 1981 the total bass catch rate improved to an average value of 0.79, indicating that the average bass angler caught nearly one bass per hour. During 1976 and 1977 the black bass harvest was reduced in terms of number and poundage, as expected. From 1978 through 1981 the harvest of legal bass averaged 2.2 pounds per acre, an increase as compared to the 1973 and 1974 average.

The 15-in length limit at Lake of the Ozarks has resulted in an improved quality of bass fishing, which has been recognized locally and nationally by anglers. Catch-and-release has become standard practice for many anglers and has added a dimension to fishing quality that cannot be measured through creel census surveys or biological sampling.

KEVIN RICHARDS
Missouri Department of Conservation
1104 South Grand
Sedalia, Missouri 65301

Results of a Black Bass 15-Inch Minimum Length Limit on Pomme de Terre Lake, Missouri

On January 1, 1976 a 15-in minimum length limit on black bass was implemented on Pomme de Terre Lake, a 7,800-acre Corps of Engineers flood control impoundment. The objectives of the regulation were to: 1) double the bass angler catch rate for all sizes of bass, including catch and release; 2) maintain black bass harvest at 4 lb/acre; 3) improve the overall size structure of the bass population. A six-year evaluation of the regulation was conducted in 1976–1982.

Bass were collected each spring with an electrofishing boat. No significant changes in the percent composition of largemouth or spotted bass were evident. Largemouth bass comprised an average of 88% of the total black bass population. Improvements in the size structure of largemouth and spotted bass were apparent with the regulation. Scale analysis of a representative portion of the electrofishing samples showed that virtually all largemouth bass were legal size at Age 5. No significant changes in the largemouth or spotted bass growth rate were apparent.

A random creel survey showed that bass angler catch rates for all sizes of bass increased from 0.22 bass/hour to 0.57 bass/hour after implementation of this regulation. The catch rate by bass anglers for legal bass increased from 0.04

bass/hour to 0.05 bass/hour after the regulation was imposed. However, the bass harvest decreased from 4.6 lb/acre before to an average of 1.3 lb/acre after the regulation. No major changes in total or bass fishing pressure occurred.

Total annual mortality estimates for Age 2 and older bass averaged 45% when calculated between successive years of individual age classes. The low bass harvest was a result of high annual mortality rates that reduced recruitment of Age 3 and 4 bass to 15 in. The low harvest caused angler dissatisfaction with the regulation. In the spring of 1982, a survey conducted to quantify anglers' opinions regarding the importance of bass catch versus harvest showed that 82% of the anglers wanted a higher harvest.

Modeling of the largemouth bass population indicated that a 13-in length limit would double the pounds per acre of bass harvested without significantly reducing angler catch rates. Because of the predicted results on the fishery and no significant changes anticipated in the bass population size structure, a 13-in minimum length limit was implemented on January 1, 1983.

RONALD J. DENT, JR.
Missouri Department of Conservation
1104 South Grand
Sedalia, Missouri 65301

The Effects of Commercial Fish Removal by Netting on Sport Fish Populations in Two Arkansas Reservoirs

Commercial netting occurred October through April, 1971–1976 on Nimrod Lake and from October through April, 1973–1977 on Blue Mountain Lake in west central Arkansas. Using a 7.6 cm or larger mesh gill and trammel nets, commercial fishermen harvested commercial fishes including buffalofishes (*Ictiobus* spp.), common carp (*Cyprinus carpio*), carpsuckers (*Carpiodes* spp.), drum (*Aplodinotus grunniens*), gars (*Lepisosteus* spp.), suckers (Catostomidae), and catfishes (*Ictalurus* spp. and *Pylodictis olivaris*).

During the study, cove rotenone samples were conducted on an annual basis. Fishes collected were placed into age classes and enumerated. Data were than analyzed for several general categories, including black basses (*Micropterus* spp.), crappie (*Pomoxis* spp.), sunfishes (*Lepomis* spp.), clupeids (*Dorosoma* spp.), commercial fishes, and catfishes.

Substantial reductions in standing crops of commercial fishes were noted in both reservoirs. However, in Blue Mountain Lake, reductions were temporary, and commercial rough fish biomasses had begun to increase markedly by the end of the study period. As catfishes were the species principally sought by commercial fishermen, it was somewhat surprising that total catfish biomasses increased in Blue Mountain Lake following the start of netting. For three of the four years prior to netting, catfish spawns were not recorded from Nimrod Lake. However, following the initiation of netting, spawns were recorded for the following four years.

Sport fish populations were for the most part unharmed by the commercial netting. In some instances, sport fish populations appeared to improve during the study period. Visual analysis revealed that the Blue Mountain Lake increases in numerical standing crops of smaller sized sport fish appeared to correspond to decreases in rough fish biomasses. Significant ($P \leq 0.05$) increases in sport fish young-of-the-year numbers were recorded for crappie in Blue Mountain Lake and sunfishes in Nimrod Lake and Blue Mountain Lake. Increases in numerical standing crops were also observed for intermediate black bass and crappie populations of both study lakes. Adult black bass mean numerical standing crops were virtually unchanged in both study lakes. Although not significant, adult crappie and sunfish populations exhibited some variation in the study. A small increase in the adult crappie populations was observed in Nimrod Lake, while a significant increase ($P \leq 0.05$) in the adult sunfish population was noted in Nimrod Lake.

In Blue Mountain Lake, an expansion of the forage base occurred, which included an increase in the numbers of adult and intermediate clupeid fishes present. However, adult shad populations decreased in Nimrod Lake while numbers of young of the year shad were lower in both study lakes.

TOMMIE CRAWFORD
Arkansas Game and Fish Commission
Natural Resources Drive
Little Rock, Arkansas 72205

RESERVOIR STOCKING

Evaluation of Supplemental Stocking of Yearling Largemouth Bass in Chatfield Reservoir, Colorado

Largemouth bass (*Micropterus salmoides*) were hatchery-reared to one year of age using intensive and extensive culture from 1978 to 1981. Each cohort was identified by a fluorescent pigment applied to the fish using compressed air and was stocked into Chatfield Reservoir at a length of approximately 125 mm. Stocking rates, which were based on a "habitat-acre" (H.A.) and took into account only littoral bass habitat, varied from 10 to 54/H.A. dependent on supply. The contribution of stocked bass, as a percentage of all bass in the same year class, was determined by electrofishing 1 year after stocking.

Stocked largemouth bass composed 12%, 59%, and 59% of all Age 2 bass collected during 1980, 1981, and 1982, respectively. Growth of stocked bass exceeded that of natural bass, which may have been due to selection of aggressive feeders during intensive culturing of stocked bass. Bass marked in 1981 and held as a control group showed mark retention of 75% for one year. Only small amounts of pigment were retained after one year with granules located under lateral and ventral scales and at the bases of pectoral fins.

DOUGLAS A. KRIEGER
STEVE PUTTMAN
Colorado Division of Wildlife
317 West Prospect
Fort Collins, Colorado 80526

The Electrophoretic and Morphometric Comparison of Lake and Stream Populations of the Troutperch

Fisheries management decisions have become increasingly dependent upon experimental research. One such decision is the selection of forage fish stock for newly impounded reservoirs. The troutperch, *Percopsis omiscomaycus* (Walbaum), has the desirable stocking characteristics of relatively long life expectancy, good reproductive potential, and food value for game fish. In the process of selecting troutperch for stocking, three populations of the species were sampled and analyzed for inter- and intra-population variation on the basis of electrophoretic and morphometric data. Both data types were interpreted as showing that the lake population displayed the greatest within-group variability. This leads to the assumption that lake populations should be most adaptable when placed in a new environment. In addition, lake and stream populations were found to be electrophoretically and morphometrically distinct. Biochemical and morphological markers were identified which could be used to separate individuals originating from the two habitat types.

ELIZABETH PITROLO
U.S. Army Corps of Engineers
210 Tucker Boulevard, North
St. Louis, Missouri 63101

Evaluation of the Striped Bass × White Bass Hybrid in Montrose Lake, Missouri

Hybrid striped bass (*Morone saxatilis* ♀ × *Morone chrysops* ♂) fingerlings were initially stocked in Montrose Lake, a 1,500-acre cooling reservoir in June 1980 at a rate of 15 per acre. An additional 13.3 fingerlings were stocked per acre in 1981 and 10 fingerlings per acre in 1982.

The objective of this study was to evaluate the introduction of hybrid striped bass by determining their incidence in the creel, growth rates, feeding habits, factors influencing abundance, and the effect of this introduction on the existing sport fishery. The evaluation involved the use of gill nets, electrofishing gear, and a one-quarter time angler survey initiated in December 1981.

Hybrid striped bass were caught in large numbers during the 1980 winter months (December–February). A 20-in minimum length limit and a daily creel limit of four was imposed in March 1981 to prevent overharvest and create a trophy fishery. Occurrence of legal size hybrid striped bass in the creel within 20 months of stocking is evidence of their excellent growth rate in Montrose Lake. No adverse effects of this introduction have been observed on the growth rates or population structure of other game species.

A major problem in sustaining the hybrid striped bass fishery has been the loss of fish through the spillway following periods of heavy runoff. These losses have been documented by observations of hybrid striped bass below Montrose spillway and in electrofishing samples and angler survey contacts in the lower reaches of Harry S. Truman Lake, a large reservoir located directly below Montrose Lake that was not stocked with the hybrid striped bass. The Montrose Dam water release procedure has been altered to minimize this loss. Releases now involve simultaneous use of all four tainter gates held at the smallest openings possible, which will still allow the desired volume of flow. Previous procedures utilized each gate to its maximum opening (4 ft) before use of the next gate was begun. Although relatively large water releases have been made in the 6-month period following initiation of the new procedure, no significant loss of hybrid striped bass has been observed.

WILLIAM M. TURNER
Missouri Department of Conservation
P.O. Box 180
Jefferson City, Missouri 65101

Establishment of a Walleye Population by Stocking in Gaston Reservoir

The Roanoke River, located on the Virginia and North Carolina line, once supported a substantial walleye (*Stizostedion vitreum vitreum*) fishery. However, this fishery disappeared following the construction of Kerr, Gaston, and Roanoke Rapids Reservoirs. Reservoir construction destroyed traditional riverine spawning habitat, and walleye numbers behind the new impoundments may have been too low to sustain reproducing populations.

The objective of this study was to determine if a self-sustaining population of walleye could be established in Gaston Reservoir. A population dynamics similator model was used in 1979 to evaluate Gaston Reservoir's potential for supporting walleye. The model evaluates three major parameters in determining site suitability for walleye introductions: water temperature during spawning seasons, amount of potential spawning area, and standing crop of potential predators. The model predicted that a self-sustaining population of walleye could be established from fry and fingerling introductions if the mature fish utilized available reservoir spawning sites.

Gaston Reservoir was stocked annually from 1980 through 1982 with 425 fry/hectare and 25 fingerlings/hectare. These fish were obtained from reservoirs where walleye were known to reproduce.

Walleye were collected in experimental gill nets fished annually since 1980. Growth rates compared favorably with those of walleye taken from other sites in Virginia and North Carolina. First year growth averaged 253 mm. Stomach samples were taken from all walleye collected, and food items were analyzed to the lowest possible taxon. These data were compared with standing crop estimates of available walleye prey from cove rotenone samples taken in 1979 and 1981.

Collections of young of the year walleye in 1983 or after will indicate that the introduced fish have reproduced. Annual sampling will be continued until 1986 to determine if a self-sustaining walleye population has been created.

T. WAYNE JONES
Route 3, Box 214
Asheville, North Carolina 28806

Utilization of the Forage Fish Population by Predators at Lake Rathbun, Iowa

Total estimated food consumption by crappie, white bass, walleye, largemouth bass, and channel catfish was examined at Lake Rathbun, Iowa in 1977–1979. Relationships between consumption and available food showed the greatest potential for severe competition existed between crappie-walleye, walleye-white bass, and crappie-white bass. Autumn diet overlap indices for these species combinations averaged 0.85, 0.74, and 0.69, respectively. Diet overlap in1978 was minimal. Associated August AP/P ratios showed fish-prey species, primarily gizzard shad, were more than adequate in 1977 and 1978 but insufficient in 1979. This was particularly true for predators in the 5–8 in size class. Inferences from a combination of AP/P, diet overlap, and total consumption showed severe competition between small walleye and intermediate crappie and white bass in 1979. Spring and late autumn showed greatest competition potential regardless of sample year.

LARRY R. MITZNER
Iowa Conservation Commission
Rural Route 1, Box 209
Chariton, Iowa 50049

The Culture of Fingerling Striped Bass Over Blue Catfish in the Lake Catherine Nursery Pond

Nursery ponds adjacent to large reservoirs are an important fisheries management tool in Arkansas. The direct discharge of the Lake Catherine Nursery Pond into Lake Catherine reduces the potential for mortality due to handling and transport during the culture of striped bass (*Morone saxatilis*) together with blue catfish (*Ictalurus furcatus*).

The nursery pond is prepared similar to a warmwater hatchery pond, with fescue hay, alfalfa pellets, and inorganic fertilizer being added upon filling. Intermediate size blue catfish (6–20 fish per lb) and fingerling striped bass (1.0–1.5 in) are stocked into the pond in early June. Growth and condition of fish are monitored bimonthly throughout the culture period by angling, seining, and electrofishing. Adult fathead minnows (*Pimephales notatus*) and young-of-year threadfin shad (*Dorosoma petenense*) are stocked to provide forage. Catfish are supplementally fed a commercial pellet diet. Plankton blooms are maintained throughout the culture period to inhibit the growth of filamentous algae.

Fish are released when the catfish have grown to a size that can be immediately harvested by the fishing public (around 1 lb per fish) and striped bass are 5–6 in TL, allowing them a greater chance of survival in the lake environment. The pond is drained directly into Lake Catherine via a pipe and requires about 18 hours to drain completely. Estimating the number of striped bass released and interference by vegetation (primarily *Pithophora* spp.) during sampling and draining remain problems.

The Lake Catherine Nursery Pond is an effective management tool which allows double cropping while reducing the dependence on hatchery pond space and the problems associated with handling and transporting the harvested fish.

Harry Tommey
Michael Armstrong
Les Claybrook
Arkansas Game and Fish Commission
Route 1, Box 477
Hot Springs, Arkansas 71901

Evaluation of Supplemental Stocking of Fingerling Largemouth Bass in Lake Coronado, Arkansas

Hatchery spawned largemouth bass (*Micropterus salmoides*) 1.5–2.0 in total length are stocked in selected Arkansas reservoirs in the spring to supplement poor survival of natural bass spawns. To evaluate this management technique, 20,000 fingerling largemouth bass with a mean total length of 1.96 in were marked with fluorescent pigment sprayed into the epidermis of the fish. These fish were stocked into a 384 acre reservoir at a rate of 50 bass per acre. It had earlier been determined that this lake experienced a poor natural spawn of bass. Seining, electrofishing, and cove rotenone sampling were used to monitor the total young-of-year size class made up of hatchery stocked bass. After stocking in June, these fish initially composed 17.2% of the 1981 year class. In August, hatchery bass comprised 90% of the 1981 year class. By September, hatchery bass made up 36% of the year class, and they comprised 10% of the young bass crop in October.

After October, mark retention in a pond of control fish dropped off to approximately 50%, and the lake sampling was discontinued. Black bass smaller than 3 in total length apparently do not retain the fluorescent pigment as long as larger fish due to incomplete scale formation at that age. During the June-October time period, hatchery bass were longer than their lake-spawned counterparts. Hatchery bass were also slightly larger than bass in a similar nearby control lake that had experienced an adequate bass spawn and higher forage spawns. Intermediate bass ranging from 6 to 8 in total length were 3.6 times more abundant in fall rotenone samples approximately 15 months after stocking, and almost one-third of these were hatchery stocked bass.

Stephen P. Filipek
Michael D. Gibson
Arkansas Game and Fish Commission
Route 1, Box 477
Hot Springs, Arkansas 71901

Pages 314–327 *in* G. E. Hall and M. J. Van Den Avyle, editors.
Reservoir Fisheries Management: Strategies for the 80's.
Reservoir Committee, Southern Division American Fisheries Society, Bethesda, Maryland. 1986.

SUMMARIZATION OF THE SYMPOSIUM

ROBERT C. SUMMERFELT
Department of Animal Ecology
Iowa State University
Ames, Iowa 50011

Man-made lakes, or reservoirs as we traditionally have called them, are significant components of the surface water resources of North America, and their number and area continues to grow. In 1980 there were 1,608 large (>200 hectares) reservoirs having a total surface area of 4 million hectares (9.9 milliion acres) at average annual pool level (Ploskey and Jenkins 1980). The number of large reservoirs constructed during 1970–1980 was similar to the number constructed during 1960–1970, but the increase in area of reservoirs was only 36% of the rate of increase in area (100,000 hectares per year) in 1960–1970.

The U.S. Army Corps of Engineers, the largest public reservoir constructing agency, has about 450 projects comprising about half of the total surface area of man-made lakes in the U.S. Recreational use on Corps of Engineers projects, where data were available, increased 27% in 7 years (1975 to 1982) to about 480 million days. In total, 40% of all freshwater recreational fishing in 1980 was on reservoirs and this activity involved user expenditures of $5.4 billion (USFWS 1982).

The magnitude and importance of public demand for reservoir fishery resources has led to the development and implementation of many State and Federal programs intended to conduct research on and manage these fisheries. In the early 1980's, the Reservoir Committee of the Southern Division of the American Fisheries Society recognized the need for an in-depth assessment of reservoir fishery management practices, and the Committee sponsored a symposium to fulfill this need. Specific objectives of the symposium were:

1) to provide a critical assessment and current practices for managing reservoir fisheries;
2) to provide for the exchange of information and ideas among users and managers to the resources;
3) to stimulate new ideas and approaches for managing reservoir fishery resources;
4) to identify critical research and management needs and goals; and
5) to clarify the role of fishery management relative to reservoir planning, operation, and other recreational uses.

The Committee achieved a desirable balance between review papers and case studies ("experience papers"). The former generally provided a critical assessment of current practices (objective 1) and the role of fishery management relative to reservoir planning, operation, and other recreational uses (objective 5). Critical research and management needs (objective 4) have been identified by some authors, and I have incorporated some in my recommendations at the end of this paper. Objectives 2 and 3 were accomplished by the interactions between speakers and participants in the symposium. The published Proceedings will advance the state-of-the-art of reservoir fishery management.

OVERVIEW OF THE TECHNICAL PAPERS

I shall try to add an independent perspective to the objective of "critical assessment" by attempting to synthesize the papers given in each of the seven technical sessions and then collect a list of critical research and management needs and goals. To save space, papers cited from the technical sessons are not listed in the references, and scientific names of fishes are rarely given here unless a necessity. Common names of fishes conform to those listed by Robins et al. (1980).

Management Goals and Economic Values

Measures of recreational use and economic value of recreation are needed in planning water development projects. A favorable benefit/cost ratio for public expenditures on water development projects requires evaluations of recreational resources in an economic context. Recreational use and expenditures do not provide the best measure of "market prices" for this type of recreation. Weithman gives alternative methods for "valuation" of a fishery; a prerequisite to the benefit-cost analysis, "so their value can be compared to other potential uses for the water and to determine the value of specific management programs."

Although fishing involves expenditures for tackle, other equipment, travel, meals, lodging, guides, and fees, a measure of "consumer surplus," not expenditures, is a better measure of economic value (Fisher et al.). The economic data in the 1980 National Survey of Fishing, Hunting and Wildlife-Associated Recreation are of the expenditure type. The 1980 estimate of $5.4 billion in expenditures "coincidentally" equaled an approximation of consumer surplus.

The value of the trout fishery in Lake Taneycomo (Weithman and Haas 1982) serves as a useful example. Also, Horvath (1974) has shown that when people were asked what they were willing to spend to give up a day of fishing in the southeastern United States in 1973, the average recreational fishermen would have required a payment of $55.09.

Recreational fisheries provide social benefits defined in terms of human experiences. Belusz and Witter reported details of a social-economic survey of entrants at the 1981 Pommede Terre Lake (Missouri) muskellunge tournament. The socioeconomic parameters provided more impressive statistics of the fishery than conventional catch rate or angler harvest. The lure of trophy fishing is suggested as a means to assist agencies in meeting increasing angler demand for "something new and challenging."

Simulation modeling is a process where relationships are used to predict future events—a simulation model is a mathematical statement that attempts to describe relationships among variables in a real system. The accuracy of the predictions reflect the nature of the modeling process, the quality and quantity of the input variables, and our understanding of the relationships. Cole and his colleagues used biological and economic consequences of altered management of river flow through six reservoirs on the Rio Grande River to develop a model to predict impacts on the ecosystem and various human uses of the water. Their "management model" will include measures of the consequences of redistribution of water and changes from altered delivery times or changed amounts of water storage—issues of importance everywhere, but crucial where water is scarce.

Recreational fishery management is always difficult in multipurpose water projects, but it is even more delicate in a legal-economic environment where water is a property right and ownership involves interstate compacts and international treaties. In this environment, fishery management is often constrained to the extent that management activities will not affect downstream water uses. The modeling approach described by Cole and his colleagues should be a useful tool for obtaining insight into effects of resource management alternatives involving a complex of socioeconomic, legal, and biological variables.

Assessment of Fish Populations and Measurement of Angler Harvest

Methods for assessing fish populations in large reservoirs need to be cost conscious, sensitive to public opinion, and adapted to the needs of specific types of management evaluations. There is a need for improvement in application of traditional methods and exploratory gear development. Because crappie are not easily captured with electrofishing gear, Colvin and Vasey used fall trap netting to assess white crappie populations in reservoirs, whereas Dent used nighttime electrofishing for assessment of largemouth bass. Siler evaluated a bow-fished Tucker trawl for estimating threadfin shad populations. His trawl estimates had exceptionally low (11%) coefficients of variability compared to cove rotenone samples (91 to 197%). The trawling estimates required only one-third the annual manpower as the cove rotenone samples. There is a continuing need to develop a better "mousetrap."

Quantitative estimates of fish abundance (fish standing crop, primarily derived from cove rotenone samples), when carefully related by correlation, regression, and modeling techniques to environmental variables such as climate, reservoir area, age, depth, relative shoreline length, water level fluctuation, water exchange rate, and water chemistry, have contributed immensely to an understanding of factors that affect fish production and harvest. Jenkins (1982) derived over 100 regressions to estimate standing crop and harvest of various species and groups of species of reservoir fishes. Jenkins stated, "Repeated analyses of (standing) crop versus environmental variables have consistently shown that total dissolved solids content is (at this time) the most useful predictor of fish biomass," and, "the best three-variable model developed to predict sport fish harvest (kg/hectare) included mean depth, area, and growing season." Carline summarized other literature relating fish yields and standing crops to environmental variables. Predictive indices began as simple linear regression analysis using an array of abiotic parameters, or ratios of variables, such as the morphoedaphic index, as the independent variable for estimating standing crop and yield. Multiple regression models incorporate quantitative description of more variables and their interactions. This research has added significantly to understanding the factors of reservoir morphometry and hydrology that influence fish standing stock.

Ultimately, the success of management programs must be described in terms of fishing success (catch/hour), quality, measures of angler satisfaction, other sociological and economic factors. Van Den Avyle reviews techniques for measuring angling effort, catch/hour, and harvest in reservoirs, summarizing attributes of the most common methods. He identifies areas where further research is needed: "to monitor resource trends, evaluate management tools, and document fishermen satisfaction associated with specific fishery resources." There is also a need for more emphasis and measurement of fishing quality such as described by Weithman and Anderson (1978). Papers cited by them showed that anglers go fishing to seek satisfaction unrelated to the catching and eating of fish. The outdoor experience, companionship, water quality, natural beauty of the area, privacy while fishing, and other variables may be important than catch in determining fishing quality.

Colvin and Vasey derived a population rating procedure for assessment of management problems for white crappie populations in Missouri's large reservoirs using relative catch rate, age structure, and three population rate functions (recruitment, growth, mortality) derived from trap net samples. The five parameters are given numerical values (0 to 10), and the sum is used for an overall population assessment.

Management of the Physical and Chemical Environment

Man-made lakes differ from natural lakes in many ways: man-made lakes are younger, generally have a larger (by a factor of 3) watershed to surface-area ratio, higher sedimentation and nutrient loading rates, conspicuous longitudinal gradients in many physical and chemical parameters, a higher shoreline development ratio, faster water exchange ratios, more extensive water level fluctuation, and often deep penstock rather than surface discharge (Benson 1982). Many key biological processes, including primary and secondary production, fish recruitment, mortality, and growth, and fish harvest are impacted by abiotic environmental factors. Variations in some of these factors are predictable and subject to manipulation. Fishery goals, however, are not often given sufficient status to exercise primary control over reservoir operations. Legislative authorization for the impoundment has generally focused on maximizing hydropower generation, navigation, flood control, irrigation, and other purposes.

Water level fluctuation (seasonal and annual) is often found to be a critically important environmental variable. Diurnal fluctuations are most extreme in pump-storage reservoirs. Inasmuch as induced fluctuations of water levels in man-made lakes are far more extreme on an annual basis than those observed in natural lakes, water level fluctuations have long been suspect as having detrimental impact on reservoir fishery resources. Several reviews have been written to summarize the literature since the 1930's (Triplett et al. 1980; Ploskey 1982).

Recruitment and growth of many fishes are responsive to the hydrological regime of reservoirs. Ploskey's regression analysis of fish standing crop in August with hydrological variables derived from monthly changes in reservoir area and storage ratio indicates that spring surface area, annual change in area, summer area, and spring flooding provided positive influences in over 90% of all correlation analyses. The increased year-class strength of many fishes after spring flooding usually was nullified when summer drawdown was equally extensive. Maintaining above-average water levels for as long as possible after successful spawning is important. Some researchers, however, have questioned whether the biological response would be as large if managed water levels are obtained for several consecutive years.

Water level management plans for reservoir fishery management objectives are reviewed by Willis. In Kansas, water level management plans to meet fishery management objectives typically have four major components: 1) a spring rise to flood terrestrial vegetation; 2) a summer drawdown of approximately 4 ft to allow regrowth of vegetation and concentrate predators and prey; 3) an autumn rise of approximately 2 ft to flood some terrestrial vegetation and attract waterfowl; and 4) a winter drawdown to concentrate predators and prey and protect remaining vegetation from water damage. In impoundments where 20% or more of the basin area was exposed during winter drawdown, turbidity decreased, and densities of walleye, white crappie, and white bass increased. Spring periods with exceptionally high discharges may cause a population "crash" of walleye from loss of young-of-the-year and adults.

Groen and Shroeder (1978) indicated that the Kansas water level management plan improved walleye growth and harvest and enhanced survival of stocked walleye fry. In Kansas reservoirs, a shortage of suitable spawning habitat limits walleye recruitment. Thus, a spring rise in water level is essential to increase spawning sites and flood terrestrial vegetation. Largemouth bass, however, were adversely affected by the plan. Aggus and Elliott (1975) found that the amount and duration of flooded shoreline vegetation strongly influenced survival of young largemouth bass in Bull Shoals Lake, Arkansas. Flooded vegetation reduced predation during the first three months of life by providing cover not normally available. The summer drawdown phase of the Kansas plan probably increases predation on young-of-the-year largemouth bass. Supplemental stocking with hatchery-reared or "nursery pond" fish may ameliorate the largemouth recruitment problem. Olmsted and Clugston described use of "nursery ponds" (subimpoundment strategy) for bass production by an electric utility on Clinton Lake, Illinois.

Brown reviews use of brush piles, tires, stake beds, and other structures for attracting fish. Artificial structures have been used in freshwater lakes to attract fish since at least 1931. The major purpose is to increase catch rate. It has been assumed that the need for artificial structure is lessened when segments of standing timber are saved from being cleared before impoundment. Burress (1961), and Davis and Hughes (1971), however, reported that catch rates in timbered areas were not statistically better, and may have been smaller than catch rates in non-timbered coves. The success rate of fishermen was better in the areas with standing timber. Is success rate more important than catch rate in the quality of the fishing experience? Brown suggests that fish attracting structures may enhance harvest of "underutilized" sport fish species which he identified as a "problem" in large (>20,000 acres) reservoirs. Do we really have an abundance of "underutilized sport fish" that can be attracted to artificial structures?

It has been shown that periphyton production, benthos, or fish standing stock has been increased in the area in which artificial structures have been placed. Yet we cannot expect artificial structures to increase overall biological production in a large lake because the area of impact is too small relative to the area of the lake. Attractors may be used by a few species for spawning, but it has not been shown that the spawning of those species was a limiting factor or that the small area of artificial spawning habitat added significantly to recruitment.

The potential value of artificial destratification or hypolimnetic aeration as a fish habitat management technique to overcome the substantial loss in fish habitat from hypolimnetic anoxia have been described in several major symposia and comprehensive reviews (Toetz et al. 1972; Dunst et al. 1974; Lorenzen and Fast 197; ASCE 1979; Pastorok et al. 1982). Most fish, and especially game fishes, will not occupy anoxic hypolimnia except for short forays in search of food. Artificial destratification has potential for expanding fish habitat and affecting primary and secondary production, fish growth, and predator-prey interactions.

Kimmel and Groeger consider reservoir aging over the life of a reservoir rather than the shorter-term "trophic upsurge-trophic depression" phenomenon observed in the initial 10 years of life. Paleolimnological evidence suggests that changes in lake trophic status are not necessarily gradual, unidirectional, and natural. Undisturbed natural lakes commonly remain in a state of trophic equilibrium for thousands of years. Natural and man-made lakes age by the same basic processes but at very different rates. Changes in productivity with reservoir aging per se are not significant compared to the effects of man-induced alterations of the watershed.

Reproductive success, growth and production of reservoir fishes are closely tied to fluxes in envirionmental variables. Miranda and Durocher related variation in growth of largemouth bass in Texas reservoirs to several environmental variables. The best multiple regression model (log of reservoir age, average reservoir depth, reservoir standing stock and a prey-predator ratio) accounted for 49% of the inter-reservoir variability in growth of largemouth bass. Given that over half of the variability was unexplained by the model, it is obvious that there still remains much room for fundamental studies of density dependent and other factors controlling growth, mortality, and recruitment of even our most intensively studied species.

Empirical information on the fish populations of a reservoir is derived by sampling the lake or the fishermen. Too often, sampling designs for fisheries and limnological purposes have neglected the significance of longitudinal gradients. Siler and Foris examined spatial distribution of the fish community and limnological characteristics. Heterogeneity was related to the longitudinal gradient from the dam upstream ("uplake") to the head of reservoir.

Management of Reservoir Fish Communities by Influencing Species Interactions

Species interactions can involve symbiosis, competition, and predator-prey relationships, Few cases of symbiosis have been observed in man-made lakes albeit Spall (1970) described a case of cleaning behavior. Fish stocking of large reservoirs have usually been limited to 1) establishing the initial population during or immediately following reservoir filling, 2) introducing new species, 3) maintaining important game or prey species that are unable to reproduce, and 4) special management purposes. Food partitioning and predator-prey relationships have received the most attention because fish community composition may be affected by fishing and stocking of predators or prey. The fishing public often regards fishery management as synonymous with stocking. Admittedly, hatchery-produced fish are indispensable for the larger part of trout (generally rainbow trout) management, given the inability of the species to reproduce in many situations where managers would like to have a sport fishery. Two-story fisheries and cold tailwaters have substantially expanded trout fishing opportunities and the need for hatchery production. Although maintenance stocking of large fingerling channel catfish may be a common practice in some small and intermediate sized lakes, stocking of channel catfish is rarely needed in large (>200 hectares) reservoirs.

Introductions discussed in this symposium have focused on piscivorous predators or prey species. Natural reproduction of striped bass in man-made lakes has been relatively rare; thus, there is a growing demand for fingerling production of striped bass to maintain fisheries. Recently, hybrid striped bass × white bass have been stocked in preference to striped bass. An *ad hoc* opinion survey of fishery agency personnel by Keith indicated that management of warmwater fisheries of large reservoirs may be shifting from habitat and population structure modification to more intensive stocking of non-native or diminishing species of native fishes. Keith recommended an expanded hatchery involvement in reservoir management, citing the use of stocking to supplement year-class strength as one justification.

Case histories in this symposium provide reasons to question the merits of certain stocking programs, including stocking to "fill a vacant niche." Stocking of large predatory fish, such as striped bass, was once proposed to "put predatory pressure" on "rough fish"—carp, carpsucker, gar, buffalo, and others. It is very evident, however, that nearly all predatory game fish prefer shad (*Dorosoma* spp.), rarely eating the so-called rough fishes. The flathead catfish is the only large piscivorus fish whose diet regularly contains large portions of rough fishes (Turner and Summerfelt 1971).

Recently, more attention has been given to competitive food partitioning between exotic and native fish and the addition of non-native prey fish and invertebrates. In this regard, we need to retain awareness of the vulnerability concept as it goes far to explain predator-prey relationships. Lewis (1967) demonstrated that vulnerability relates more to characteristics of the prey than to the predator. To optimize survival and growth or largemouth bass, prey must be the right size, located in the same habitat type, and be vulnerable to being captured by the predator.

Boxrucker defines the characteristics of an ideal prey: prolific, stable in abundance, trophically efficient, vulnerable to predation, non-migratory, and innocuous. The inland silversides seemed to have these characteristics; therefore, Boxrucker evaluated the introduction of this species in Thunderbird Reservoir, Oklahoma, with a focus on measuring utilization by largemouth bass and white crappie. Silversides replaced sunfish in the diet of small (≤200 mm) largemouth bass. However, sunfish and gizzard shad remained the primary prey of larger bass. The overall effect on growth rates of the sport fish population appeared to be minimal. The sunfish populations declined following the introduction of silversides, but cause-and-effect was not established. Silversides may not have been completely innocuous, and utilization of silversides by largemouth bass was less than what would have been desirable.

When introductions of striped bass were undertaken by state agencies, some biologists and fishermen were concerned that striped bass would eat or compete with black bass and crappie. Harper and Namminga describe population trends in several sport fishes in Lake Texoma, Oklahoma-Texas, during and after establishment of striped bass. After a 10-year (1965–1974) introductory period, a self-reproducing population of striped bass was established, and it became the most abundant sport species collected in the lake. Abundance (angler catch-hour and kg/hectare) of largemough bass and spotted bass did not decline in relation to expansion of striped bass. Harvest of black bass by anglers did decline, but this was related to a shift in fishing for striped bass that reduced fishing effort for largemouth bass. Neither populations nor fishing pressure for crappie changed following development of a strong striped bass population. The only distinct population change was a decrease in the abundance of white bass as the striped bass population expanded. Striped bass have shown almost no predation on native sport species, preferring primarily a shad diet. There was no evidence indicating that the striped bass over-exploited the prey. Fluctuations in abundance of gizzard shad and particularly threadfin shad were not directly related to population size of striped bass. The major fluctuations in the threadfin shad population were caused by severe winterkill rather than predation.

The Texoma case study provides a very positive overall view of a striped bass introduction, but in Lake McConaughy, Nebraska, striped bass grew beyond their food supply, and then growth rates declined sharply and mass starvation of striped bass occurred (Ellison and Eichner 1982). After a 6-year period of stocking, biomass of striped bass reached 13 kg/hectare; thereafter, heavy predation by striped bass and poor overwinter survival decimated gizzard shad populations. In Lake McConaughy, striped bass also preyed on rainbow trout, and a decline in the rainbow trout paralleled the expanding striped bass population. An overlapping thermal niche in summer of adult striped bass (>5 kg) and rainbow trout increased the vulnerability of trout to striped bass. As a result, further striped bass stocking has been terminated in Nebraska. In Lake Powell, Utah-Arizona, Pettengill and Gustaveson (1983) described a sharp decline in the condition of striped bass attributed to depletion of the threadfin shad population.

There is a widespread misconception that the limnetic or pelagic zone of a large reservoir has an "unoccupied niche" that needs "filling." On the contrary, the pelagic habitat generally contains one to two species of shad, silversides, two to three species of temperate bass, crappie, and many other epipelagic fishes such as mooneye. Also, except during times of hypolimnetic anoxia, there can be a diverse assemblage of benthopelagic fishes such as walleye, freshwater drum, several catfishes, many kinds of suckers, and carp. Thus, neither the hypolimnetic nor the epilimnetic habitats are "unoccupied." The pelagic zone contains a complex biotic community where habitat and resource partitioning are not always what they seem. The Claytor Lake, Virginia analysis of alewife introductions by Kohler et al., the *Menida* evaluation in Oklahoma by Boxrucker, the *Menidia* studies in California cited by Boxrucker, some *Mysis* introductions, and the redfin shiner problems of western trout lakes serve as examples to show that exotic prey introductions have had undesirable effects on other prey and the game fishes the prey were supposed to feed. Introductions of "pelagic" prey fish have negative impacts on the littoral zone communities as well. Kohler et al. criticized trial-and-error introductions of exotic prey fishes, and the Claytor Lake study serves as a demonstration that prey are not always innocuous.

Although managers focus on predator-prey relations among piscivorous game fishes and their prey, predator-prey interactions occur between planktivorous fish, including the young of most game fishes, and zooplankton (O'Brian 1979). Noble stressed the importance of understanding highly complex species interactions, stating that man-made lakes are too young to have the "highly-evolved" predator-prey relationships, that "typify our temperate lakes of glacial origin." The "highly-evolved" relationships include: trophic and spatial niches that provide for efficient utilization of resources; seasonal chronology of phytoplankton blooms, pulses of zooplankton grazers; and reproduction of fishes timed such that their young will be able to first utilize the zooplankton and then young of other fishes. Seasonal chronologies of biological events may be disrupted by many man-made variables. Fishery biology has long held that year-class strength of certain fishes is related to availability of food when the larval fish shifts from endogenous to exogenous food supplies. Noble says that the timing of largemouth bass spawning is often too late for their young-of-the-year to utilize larval shad unless a fall spawning of shad occurs. Food supply definitely affects growth. When an Oklahoma reservoir refilled following several years of drought, flooding innudated terrestrial vegetation creating new reservoir conditions, young largemouth bass growth was exceptionally fast, allowing a second growth stanza in the first summer (Summerfelt 1975).

Management of Reservoir Fish Populations by Harvest Regulation

Maximum sustained yield (MSY) persists as the dominant goal used in recreational fisheries management (Bennett et al. 1978). Yet, while the MSY management goal seeks to maximize harvest, it ignores measures of fishing quality, economic return, and social values. Most people want to catch fish, but not necessarily to keep or eat them (Ruckelshaus 1975). Redmond provides an excellent historical perspective on the evolution of fishing regulations and the beneficial effects of these regulations over a period coinciding with increased fishing pressure on sport fisheries. Since the 1960's, research across the U.S. has shown that predatory fishes, especially largemouth bass, but even white crappie, can be over-harvested with relatively little fishing effort. Over-harvest adversely affects fish population balance and fishing quality. Size limits function to prevent overharvest of predatory fish, and when this occurs, regulation of predatory game fish harvest may improve the size composition of the prey fish populations. In Stockton Lake, Missouri, Goddard and Redmond used size-limit regulation in combination with staged filling, stocking prey, and other predatory game fishes to prevent the typical "boom and bust" phenomenon of fishing in newly filled reservoirs.

Zagar and Orth generated length limit guidelines for largemouth bass from computer simulations of hypothetical reservoir populations under different harvest regulations. The computer models helped to explain variability in population response to harvest regulations and demonstrated the inadequacy of short-term field studies to describe population changes. They found that minimum length limits provide the most balanced harvest, whereas the slot limit will give a higher proportion of quality-size fish. Small size limits may be ineffective to produce quality angling opportunities.

Management of fish populations has been attempted by "rough fish" removal, but benefits have been inconclusive in many cases. Moyle et al. (1950) found that rough fish removal had little effect on total biomass of game fish in southern Minnesota lakes. The large biomass of non-game, food fishes ("rough fish") has encouraged a variety of control measures, but most management actions have not had much impact. Fritz and Wight reviewed the many facets of using a commercial fishery to help attain management objectives, including the "wise use" motive to provide a food fishery. Their survey of state fisheries agencies indicated that only six of 22 states that permitted commercial fishing used it for the purpose of manipulating fish stocks. The other states responded that the value of commercial fishing was mainly to utilize food resources.

Because many previous commercial activities did not achieve predictable biological or economic benefits and often provoked adverse public reaction or poor citizen reception, most agencies now have little interest in use of commercial fishing for management purposes. A successful commercial fishing program requries proper selection and adequate removal of target species over an extended period, a good market for fish, economic incentives if non-saleable species are to be removed, proper selection of removal operation, and selling the public and one's resource agency of the need and benefits of such a program. Stephen's description of a commercial fish harvest (4-in bar mesh trammel net) on Lovewell Reservoir, Kansas, provides an example of how a carefully implemented commercial fishery was used to enhance the abundance of game fishes without generating adverse public reaction. This was a limited-entry fishery awarded on a bid basis and was restricted to weekdays during daylight hours only. They concurrently stocked predatory fishes.

Management Implications of Energy Development

Human impacts from electric power generation—hydropower, pump storage, fossil fuel, and nuclear—cause large scale perturbations of aquatic ecosystems. Leaving aside the impacts of emissions of sulfur and nitrogen oxide precursors of acid rain, the technical papers in this session considered the multifaceted environmental impacts of electrical power generation. Thermal (fossil fuel and nuclear) power plants discharge a large volume of heated water (heat loading for a 2,000 MW nuclear plant can involve an 11 C rise in ambient temperature with a flow rate of >85 m^3/second). "Once-through" cooling systems at thermal plants also cause mortality of fish by entrainment (carry organisms through the cooling sytem of a power plant) and impingement (impact fish on intake screens). A "wet" cooling tower will use about 18,500 m^3 of water per year per MW at a 85% plant factor (Kaufmann and Schaefer 1977). Because of negative environmental impacts, regulation of once-through cooling systems forced the use of cooling towers and cooling impoundments. Lower comparative costs encouraged construction of cooling impoundments at a rapid rate. Cooling impoundments (designed to dissipate the entire heat load to the atmosphere via a lake or pond) rapidly expanded fishery habitat in the 1970's due to aggressive construction programs by utlities when growth in electrical demand was high. There are now existing or planned 135 cooling impoundments with an area of 0.5 million hectares.

Olmsted and Clugston described the nature of cooling impoundments, their fishery potential, and special problems. In addition to entrainment, impingement, and elevated temperatures, they have substantial horizontal water movement and vertical physical and chemical gradients. They also have a tendency for aquatic weed problems, poor reproduction of largemouth bass, overharvest, and the development of toxic substance problems (heavy metals and selenium) in cases where retention time is excessively long or where ash basin effluent discharge is not properly planned.

In spite of the significant number of problems, Olmsted and Clugston presented a positive impression of the value of cooling impoundments as fish habitat. Siting, design, and construction features may be employed to enhance fish production when planning reflects an understanding of reservoir ecology and ecological requirements of the fish to be managed. Pumping and heating of cooling lakes may be used to increase yield and enhance management options. At latitudes where threadfin shad would otherwise be killed at normal winter temperatures, cooling lakes can provide an overwinter environment. Although the thermal regimes of cooling impoundments may be too stressful for survival and propagation of some desirable native fishes, some exotic predatory and prey species may be used successfully.

Smithson et al. described problems of largemouth bass management in a "perched cooling pond"—constructed by enclosure of an area by dikes. The continuous recirculation of water and the elevated temperatures extended the growing season and resulted in a higher than average growth rate. However, the straight shoreline limited spawning sites, and exposure to wind resulted in poor recruitment. They attempted to overcome the poor recruitment problem by using supplemental stocking and high size limits to maintain a larger spawning population; however, they gave no evidence for a stocking benefit. Attraction of largemouth bass into the discharge canal in the winter encouraged overharvest of the concentrated fish, but this problem was prevented by designating the discharge canal a refuge.

Effective fishery management of cooling impoundments often is inhibited by certain regulatory constraints (discharge limits, prohibitions against use of sewage effluents for fertilization, and "onerous" monitoring requirements), the lack of economic incentives to carryout fishery management activities, liability problems, and the lack of cooperation between utilities and state management agencies (Olmsted and Clugston).

Selenium toxicity may adversely impact fisheries resources in cooling impoundments when ash disposal is not carefully planned. Olmsted et al. observed a drastic decline in centrarchid reproduction within a few years of impounding Belews Lake in north-central North Carolina. Young-of-the-year centrarchids were virtually absent, and the main body of the lake was dominated by fathead minnows, mosquitofish, and carp. Bioaccumulation of selenium was assumed to be the "most likely" factor. This toxic problem occurred because the ash basin effluent (sluice water) was discharged into the lake which has a long retention time (1,500 days).

Hydropower and pump-storage developments are important sources for peaking power demand. Applications received by the Federal Energy Regulatory Commission for hydropower development increased 8.7-fold between 1977 and 1981. Therefore, a proliferation of hydropower facilities at new and existing sites is underway (Prosser). This is visible in the form of new hydro projects and retrofitting of existing impoundments built for flood control and other purposes to provide hydropower. Barnhart (1982) summarized impacts of hydropower development on fisheries resources. Hydropower development is accompanied by streamflow alteration that has potential adverse effects on water levels in the impoundment and instream flow needs downstream of the dam. Within the reservoir, an altered hydrological regime may lead to modification of many physical and chemical variables, including sedimentation, erosion, eutrophication, stratification, gas supersaturation, oxygen depletion, and assimilative capacity (Walker 1982). These factors can affect standing crop and species diversity of aquatic life (Sale 1982).

A hydropower dam, like any other dam, can block upstream migration of adult fish, cut off access to upstream spawning grounds, inundate prime spawning sites, and change a riverine habitat to a slack-water impoundment. Even with fishways, anadromous fish are adversely impacted because of problems related to fluctuating flows, wandering, migration delays, gas supersaturation, physical fallback, and oxygen depletion (Gray and Rondorf). Mortality during downstream passage of fish has been identified as the major factor responsible for the decline of the chinook salmon and steelhead trout runs on the Columbia River. Both the riverine and reservoir fish communities are affected by the nature of the hydropower releases, and fish are killed by passage through turbines (Richards et al.). Fisheries agencies that have adjusted to a given set of operational impacts of an impoundment will find the need to monitor and re-evaluate the impacts of a hydropower modification.

Management of Reservoir Releases

Peters reviews the nature of tailwater fisheries, including physical, legal, biological, and political constraints and opportunities and some case histories of successful tailwater fisheries. Added to the comprehensive, interdisciplinary literature review by Walburg et al. (1981), we now have two comprehensive sources of information on a very important category of fishery habitat.

Jernejcic provides "fresh" documentation that fisheries in tailwaters may be better and harvest larger than in the reservoir as a whole. At Tygart Lake, West Virginia, fishing pressure (hours per unit area) in the tailwater was 24 times greater than in the lake, and angler catch rate was 75% higher than in the lake. Peters, however, states that "It may be quite by accident that quality tailwater fisheries develop for many existing projects."

To obtain quality fisheries, considerations must be interjected into project planning and operational considerations prior to construction. If it takes 10 to 30 years to plan and construct a large reservoir, there is certainly time enough to incorporate what has been learned from trial-and-error and recent ecological and fisheries research. If this is not now being done, then fisheries agencies and the fisheries profession should explain the problem(s) to their constituents (the fishermen) and use the political process to rectify the constraining factors.

Obviously, the tailrace environment is a function of the characteristics of the releases. Each project will have somewhat unique effects, but the most common problems are related to fluctuation in the volume of the discharge and the temperature and oxygen content. Tailwater temperature and oxygen are dictated by the depth of the penstock intake and limnological characteristics of the reservoir, particularly the occurrence of stratification. Water quality problems downstream from stratified reservoirs are numerous (Ruane et al.). Low dissolved oxygen, as well as occurrence of chemically reduced irion, manganese, hydrogen sulfide, and the undissociated ammonium ion, adversely impact fish and benthic invertebrates in tailwaters that have hypolimnetic anoxia.

Streamflow is obviously affected by the on-and-off again flow pattern of hydropower generation. During shutoff periods, dissolved oxygen can change from above saturation to almost complete depletion within a matter of minutes. Total dissolved gas pressure is the primary factor in gas bubble disease, a serious cause of mortality of young salmonids for many years and recently found to be a problem for warmwater fish as well

(Richards et al.). Retrofitting hydropower capabilities to dams may induce problems where none existed before. Water quality considerations of reservoir discharges must be mandated, monitored, and enforced to protect the value of tailrace fisheries. Forty years ago, Eschmeyer and Smith (1943) found that a cold hypolimnetic discharge eliminated reproduction of warmwater species in the area immediately below Norris Dam, Tennessee. In this symposium, Hickman and Hevel have shown that this coldwater discharge from Norris Dam also severely depressed reservoir water temperatures during spring and summer in Melton Hill Reservoir, 20 km downstream. The lowered water temperatures had a direct effect on recruitment and growth of young sport fish. The impacts of the coldwater discharge on the downstream reservoir are related to the rate of hypolimnetic discharge, distance separating headwaters of the receiving reservoir from the upstream dam, and the size of the receiving reservoir.

Cold, hypolimnetic releases from stratified reservoirs, although negatively impacting warmwater fisheries, have in a few fortuitous cases, created unique and highly valuable coldwater (trout) fisheries. Lake Taneycomo shifted from a warmwater to an outstanding coldwater (put-and-take rainbow trout) fishery after construction of Table Rock Dam, 3.3 km upstream of the headwaters. There are case histories, however, where a reservoir discharge can be too cold, even for trout. The Green River below Flaming Gorge Reservoir was too cold for rainbow trout before retrofiting the dam with a multilevel outlet structure to include warmer water in the discharge during certain times of the year (Sartoris 1976). Tailwaters may not be sufficiently cold or they may be inadequate in quantity or quality to provide habitat for a coldwater fishery.

Pump storage reservoirs offer significant challenges for fisheries management, and no reports were cited giving any pump storage facility with redeeming fishery values. The best that can be said for pump storage facilities is that in one case, Leesville Reservoir, a pump-back, storage impoundment on the Roanoke River, Virginia, the fishery was not adversely affected by water level fluctuations of 6 to 10 ft per week (Prosser). Richards et al., however, described extreme negative impacts of the pump storage hydroelectric facility located on the H. S. Truman Dam in Missouri.

Recommendations

Recommendations listed below were derived from the panel discussion to identify critical research and management needs, the technical sessions, and personal opinions.

A Federal reservoir research unit, with responsibility for protecting and fostering wise use of our reservoir fishery resources is in the best interests of users of the resource and the American public in general. It is needed to focus on fisheries resources of interstate waters, ecological problems related to the operation of Federal reservoirs, and to provide a centralized respository for data on reservoirs in relation to fish and fisheries. At present, Federal expertise in this area is scattered in several agencies, but most of it serving the development-oriented interests of the Corps of Engineers, Bureau of Reclamation, and the Tennessee Valley Authority. In the absence of a formal, centralized Federal reservoir program, an interagency team should be organized to facilitate coordination and cooperation for research and management of interstate waters. Incentives for the latter could be provided through the Wallop-Breaux program of federal aid.

More socioeconomic studies of fishery values should be encouraged, and quantification of fishery values needs more emphasis. Such information is needed for decision making on alternative uses for the water, such as development of commodity resources. Economic valuations of fisheries in the Pacific Northwest have affected land allocation decisions by the U.S. Forest Service and have established sport fishing as a major economic resource of the west; this is also true in certain southeastern states (Everest 1979). The value of specific management programs should be expressed in economic terms to obtain perspective on management directions. On the other hand, we must not lose sight of social benefits in terms of angler use and satisfaction—parameters that measure the value of fisheries programs in terms of human experiences. Socioeconomic values need consideration along with the traditional biological measures of success and catch rate.

Written fishery management plans should be developed for each large (>200 hectare) reservoir, and these plans should be available to the public. We might use for an example, the EPA administered "Clean Lakes Program," section 314 of the Clean Water Act (Public Law 92-500). This program requires each state to submit a classification and identification of publicly-owned freshwater lakes based on their eutrophic condition. It also requires procedures, processes, and methods to control pollution of the lakes, and techniques to restore their quality (Keup 1979). Financial assistance requirements for a state priority list would serve as a model for this type of planning requirement. Management plans should be carefully evaluated and fine-tuned by long-term monitoring.

Fishery management plans should incorporate and be integrated with reservoir water level and water discharge operational plans (hydrological plans). Inasmuch as hydrological manipulation involves entire water basin operational plans, state fishery management aspirations will require involvement in the complex hydrological and water quality simulation modeling that has been developed and is used by the water regulatory agencies. Fishery scientists and limnologists have come to understand some impacts of reservoir hydrology on aquatic communities, but we have not seen much application of this information. The "Kansas plan" of reservoir water level management has promise for improving certain types of sport fisheries, but given the highly variable water flow in the Great Plains, long term studies should also be undertaken to obtain a perspective of the plan over the 10- to 20-year, wet-dry cycle typical of the region and to assess whether the benefits and impacts of the plan are altered by continuous implementation.

Mitigation agreements developed during planning and included in authorization and environmental legislation should be monitored. Many sportsmen and the informed public have come to view the word "mitigation" as the means whereby the fishery agency is "bought off" so the water management agency can do what it wants. Even implementation of mitigation directives does not guarantee a satisfactory outcome. Although hatcheries are often the "mitigation carrot," the state agency generally assumes the operational costs. Hatchery fish are not the ultimate answer to habitat alteration from hydropower dams. One does not need to go further than the Columbia River to see that increased hatchery production alone cannot succeed in maintaining adult populations of anadromous salmonids stocks upriver of Bonneville Dam when impounded rivers provide habitat for native and exotic predators of juvenile salmonids.

Impoundments constructed primarily for the dissipation of waste heat are unique aquatic habitats with many inherent fishery problems. However, with the right type of regulatory revisions and economic incentives, they have potential for making a significant contribution to recreational fisheries. Some cooling impoundments with long retention times need careful monitoring for environmental contaminants (e.g., selenium toxicity).

Construction of new hydropower projects and retrofitting of many existing dams for electrical generation will likely take place at an increasing rate for the balance of this century. Bio-engineering for fisheries during the planning stages is needed to provide suitable habitat and minimize adverse impacts. These projects have the potential for seriously impacting stream and reservoir fishery resources. Pump-storage projects with large daily fluctuations in water level seem to be the most detrimental. Further study is needed to evaluate their impacts, ways to manage around the problems, and develop management methods to make constructive use of the pumping activity and water level fluctuations. The benefits of experience papers and research findings contained herein, and that of other recently published symposia and workshop proceedings, should be assimilated into planning documents.

Size regulations are a popular management tool for manipulating structure of reservoir fish populations. The impacts of size regulations on community composition may take longer than expected; therefore, regulations should not be abandoned after only a 2- or 3-year trial. Size regulation issues have served as a stage for the larger issue of measuring fishery management by maximum sustained yield, harvest-focused measure, versus optimum sustained yield, which uses catch, fishing quality, and socioeconomic considerations. Professional differences of opinion run deep; therefore, agreement will require continuing educational programs and further dialog before there is anything like unanimity on this issue. Given that largemouth bass are easily overharvested, more attention should be given to measuring the degree of angler compliance with respect to success or failure of size limit regulations.

Reservoir fishery habitat management should not be narrowly conceptualized and limited to fish attracting structures. Fishery habitat management must include a combination of activities that influence basic productivity, structure of the fish populations, and important population rate functions. The types of habitat improvement that are most needed are hydrological manipulations, erosion control, vegetation control or enhancement, manipulation of stratification and water quality characteristics, use of subimpoundments, addition of spawning structure, and other activities. Electrofishing is the most widely used and versatile fish sampling method; however, it is the least standardized sampling technique. A cooperative, interstate project is needed to standardize parameters for reservoir electrofishing, especially in relation to use of electrofishing for obtaining measures of fish population structure and rate functions.

User-group conflict was identified by Darrel Lewis (see the Management Needs of Reservoir User Groups section of this volume) of the Corps of Engineers as a major public relations problem. A better understanding of user (clientele) interests can lead to improved management of human behavior, but statistically reliable and economical methods for sampling clientele opinion are needed. Reservoir fishery management must be carried out with the reality of the reservoir operation plan and the entire mix of recreational and other uses in mind. Lewis stated that fishermen are 25% of the users of land and water on reservoir projects; therefore, management strategies of the Corps of Engineers are intended not only to benefit fishermen, but also to provide the "best mix." State fisheries agencies must represent their clientele, "lobby" for their interests among competing recreational groups, and protect fishery resources from negative impacts of energy development and consumptive uses of water.

Federal support of reservoir fisheries programs should be increased. Most state fishery agencies depend on random spot checks and troubleshooting approaches rather than continuous monitoring. At this symposium, one fisheries chief acknowledged that his division had one fishery manager for every 100,000 acres of water. Habitat management is the key to improved yield, but if a state agency were to spend on a per acre basis the same amount that is considered reasonable for small pond management, just one large reservoir could generally exceed a state fishery management budget (50,000 acres × $30/acre = $1.5 million). There has been an obvious need for many years for additional ear-marked funds for reservoir fishery management. In April 1972, I drafted a proposal for a "National Reservoir Fisheries Management Act" to provide federal funding to state fisheries agencies for management purposes. That proposal contained a recommendation for a minimum state allocation of $25,000 plus $1.00 per acre of federal reservoirs over 500 acres. At that time, I projected annual costs of $10 million: $1.25 million for the basic allocation and approximately $8 million for the pro-rata state payments on the basis of reservoir area. Eleven years later, adjusted for inflation and additional area, the program would cost $20 million. Yet, I still feel it is fully justified—most reservoirs over 500 acres have been constructed by federal agencies, but with rare exception, federal agencies do not contribute to costs of monitoring or management. Mitigation funds have been few and far between, and such funds are rarely provided for annual management costs. Unless new funds are available, a new reservoir may be more of a management liability than an asset, and worth less than the riverine fishery the reservoir destroyed or eliminated. A mitigation principle may be derived from the premise that states deserve compensation for a federal action that destroyed a riverine fishery resource and imposed a new financial burden.

References

Aggus, L. R., and G. V. Elliott. 1975. Effects of food and cover on year-class strength of largemouth bass in Bull Shoals Lake. Pages 317–322 *in* R. H. Stroud and H. Clepper, editors. Black bass biology and management. Sport Fishing Institute, Washington, D.C., USA.

ASCE (American Society of Civil Engineers). 1979. Symposium on reaeration research. Hydraulics Division American Society of Civil Engineers, New York, New York, USA.

Barnhart, G. A., editor. 1982. Hydropower development and fisheries: impacts and opportunities. Northeast Division American Fisheries Society, Bethesda, Maryland, USA.

Bennett, D. H., E. L. Hampton, and R. T. Lackey. 1978. Current and future fisheries management goals. Implications for future management. Fisheries 3(1):10–14.

Benson, N. G. 1982. Some observations on the ecology and fish management of reservoirs in the United States. Canadian Water Resources Journal 7:2–25.

Burress, R. M. 1961. Fishing pressure and success in areas of flooded standing timber in Bull Shoals Reservoir, Missouri. Proceedings of the Southeastern Association of Game and Fish Commissioners 15:296–298.

DAVIS, J. T., AND J. S. HUGHES. 1971. Effects of standing timber on fish populations and fisherman success in Bussey Lake, Louisiana. Pages 255–264 *in* G. E. Hall, editor. Reservoir fisheries and limnology. Special Publication 8, American Fisheries Society, Bethesda, Maryland, USA.

DUNST, R. C., S. M. BORN, P. D. UTTORMARK, S. S. SMITH, S. A. NICHOLS, J. O. PETERSON, D. R. KNAUER, S. L. SERNS, D. R. WINTER, T. WIRTH. 1974. Survey of lake rehabilitation techniques and experiences. Technical Bulletin 75, Wisconsin Department of Natural Resources, Madison, Wisconsin, USA.

ELLISON, D. G., AND D. L. EICHNER. 1982. Biological, environmental, and social risks of striped bass in Lake McConaughy, Nebraska. 44th Midwest Fish and Wildlife Conference, Des Moines, Iowa, USA. (Abstract).

ESCHMEYER, R. W., AND C. G. SMITH. 1943. Fish spawning below Norris Dam. Journal of the Tennessee Academy of Science 23(1):4–5.

EVEREST, F. H. 1979. How to demonstrate the importance of fishery resources to interdisciplinary planning teams. Fisheries 4(1):15–20.

GROEN, C. L., AND T. A. SCHROEDER. 1978. Effects of water level management on walleye and other coolwater fishes in Kansas reservoirs. Pages 278–283 *in* R. L. Kendall, editor. Selected coolwater fishes of North America. Special Publication 11, American Fisheries Society, Bethesda, Maryland, USA.

HORVATH, J. C. 1974. Economic survey of southeastern wildlife and wildlife-orientated recreation. Transactions of the North American Wildlife Conference 39:187–194.

Jenkins, R. M. 1982. Predictive methodology for reservoir fish production-standing crop, harvest, angler-use, and predator-prey relations. 44th Midwest Fish and Wildlife Conference, Des Moines, Iowa, USA. (Abstract).

KAUFMANN, K. O., AND R. R. SCHAEFER. 1977. Water for energy—western United States. Fisheries 2(1): 6–13.

KEUP, L. E. 1979. Fisheries in lake restoration. Fisheries 4(1):7–9, 20.

KING, D. L. 1970. Reaeration of streams and reservoirs—analysis and bibliography. U.S. Department of Interior, Bureau of Reclamation, REC-OCE-70-55, Denver, Colorado, USA.

LEWIS, W. M. 1967. Predation as a factor in fish populations. Pages 386–390 *in* Reservoir fishery resources symposium. Reservoir Committee, Southern Division American Fisheries Society, Bethesda, Maryland, USA.

LORENZEN, M. W., AND A. W. FAST. 1977. A guide to aeration/circulation techniques for lake management. U.S. Envirionmental Protection Agency, EPA-600/3-77-004, Washington, D.C., USA.

MOYLE, J. B., J. H. KUEHN, AND C. R. BURROWS. 1950. Fish-population and catch data from Minnesota lakes. Transactions of the American Fisheries Society 78:163–175.

NEY, J. J. 1981. Evolution of forage-fish management in lakes and reservoirs. Transactions of the American Fisheries Society 110:725–728.

O'BRIEN, W. J. 1979. The predator-prey interaction of planktivorous fish and zooplankton. American Scientist 67:572–581.

PASTOROK, R. A., M. W. LORENZEN, AND T. G. GINN. 1982. Environmental aspects of artificial aeration and oxygenation of reservoirs: a review of theory, techniques, and experiences. Technical report E-82-3. U.S. Army Corps of Engineers Waterways Experiment Station, Vicksburg, Mississippi, USA.

PETTENGILL, T. D., AND A. W. GUSTAVESON. 1983. Population dynamics of threadfin shad in Lake Powell, Utah-Arizona, 1977–1982. Pages 117-123 *in* C. R. Berry, editor. Proceedings of the 1983 Annual Meeting of the Bonneville Chapter, American Fisheries Society, Salt Lake City, Utah, USA.

PLOSKEY, G. R. 1982. Fluctuating water levels in reservoirs: an annotated bibliography of environmental effects and management for fisheries. Technical report E-82-5. U.S. Army Corps of Engineers Waterways Experiment Station, Vicksburg, Mississippi, USA.

PLOSKEY, G. R., AND R. M. JENKINS. 1980. Inventory of U.S. reservoirs, January 1, 1980. Unpublished report, National Reservoir Research Program, U.S. Fish and Wildlife Service, Fayetteville, Arkansas, USA.

ROBINS, C. R., R. M. BAILEY, C. E. BOND, J. R. BROOKER, E. A. LACHNR, R. N. LEA, AND W. B. SCOTT. 1980. A list of common and scientific names of fishes from the United States and Canada, 4th edition. Special Publication 12, American Fisheries Society, Bethesda, Maryland, USA.

RUCKELSHAUS, W. D. 1975. Keynote address. Pages 5–9 *in* R. H. Stroud and H. Clepper, editors. Black bass biology and management. Sport Fishing Institute, Washington, D.C., USA.

SALE, M. J. 1982. Assessing the impacts of regulated flows at hydropower projects. Pages 11–24 *in* G. A. Barnhart, editor. Hydropower development and fisheries: impacts and opportunities. Northeast Division American Fisheries Society, Bethesda, Maryland, USA.

SARTORIS, J. J. 1976. A mathematical model for predicting river temperatures—application to the Green River below Flaming Gorge Dam. U.S. Bureau of Reclamation REC-ERC-76-7, Denver, Colorado, USA.

SPALL, R. D. 1970. Possible cases of cleaning symbiosis among freshwater fishes. Transactions of the American Fisheries Society 99:599–600.

SUMMERFELT, R. C. 1975. Relationship between weather and year-class strength of largemouth bass. Pages 166–182 *in* R. H. Stroud and H. E. Clepper, editors. Black bass biology and management. Sport Fishing Institute, Washington, D.C., USA.

(Richards et al.). Retrofitting hydropower capabilities to dams may induce problems where none existed before. Water quality considerations of reservoir discharges must be mandated, monitored, and enforced to protect the value of tailrace fisheries. Forty years ago, Eschmeyer and Smith (1943) found that a cold hypolimnetic discharge eliminated reproduction of warmwater species in the area immediately below Norris Dam, Tennessee. In this symposium, Hickman and Hevel have shown that this coldwater discharge from Norris Dam also severely depressed reservoir water temperatures during spring and summer in Melton Hill Reservoir, 20 km downstream. The lowered water temperatures had a direct effect on recruitment and growth of young sport fish. The impacts of the coldwater discharge on the downstream reservoir are related to the rate of hypolimnetic discharge, distance separating headwaters of the receiving reservoir from the upstream dam, and the size of the receiving reservoir.

Cold, hypolimnetic releases from stratified reservoirs, although negatively impacting warmwater fisheries, have in a few fortuitous cases, created unique and highly valuable coldwater (trout) fisheries. Lake Taneycomo shifted from a warmwater to an outstanding coldwater (put-and-take rainbow trout) fishery after construction of Table Rock Dam, 3.3 km upstream of the headwaters. There are case histories, however, where a reservoir discharge can be too cold, even for trout. The Green River below Flaming Gorge Reservoir was too cold for rainbow trout before retrofiting the dam with a multilevel outlet structure to include warmer water in the discharge during certain times of the year (Sartoris 1976). Tailwaters may not be sufficiently cold or they may be inadequate in quantity or quality to provide habitat for a coldwater fishery.

Pump storage reservoirs offer significant challenges for fisheries management, and no reports were cited giving any pump storage facility with redeeming fishery values. The best that can be said for pump storage facilities is that in one case, Leesville Reservoir, a pump-back, storage impoundment on the Roanoke River, Virginia, the fishery was not adversely affected by water level fluctuations of 6 to 10 ft per week (Prosser). Richards et al., however, described extreme negative impacts of the pump storage hydroelectric facility located on the H. S. Truman Dam in Missouri.

Recommendations

Recommendations listed below were derived from the panel discussion to identify critical research and management needs, the technical sessions, and personal opinions.

A Federal reservoir research unit, with responsibility for protecting and fostering wise use of our reservoir fishery resources is in the best interests of users of the resource and the American public in general. It is needed to focus on fisheries resources of interstate waters, ecological problems related to the operation of Federal reservoirs, and to provide a centralized respository for data on reservoirs in relation to fish and fisheries. At present, Federal expertise in this area is scattered in several agencies, but most of it serving the development-oriented interests of the Corps of Engineers, Bureau of Reclamation, and the Tennessee Valley Authority. In the absence of a formal, centralized Federal reservoir program, an interagency team should be organized to facilitate coordination and cooperation for research and management of interstate waters. Incentives for the latter could be provided through the Wallop-Breaux program of federal aid.

More socioeconomic studies of fishery values should be encouraged, and quantification of fishery values needs more emphasis. Such information is needed for decision making on alternative uses for the water, such as development of commodity resources. Economic valuations of fisheries in the Pacific Northwest have affected land allocation decisions by the U.S. Forest Service and have established sport fishing as a major economic resource of the west; this is also true in certain southeastern states (Everest 1979). The value of specific management programs should be expressed in economic terms to obtain perspective on management directions. On the other hand, we must not lose sight of social benefits in terms of angler use and satisfaction—parameters that measure the value of fisheries programs in terms of human experiences. Socioeconomic values need consideration along with the traditional biological measures of success and catch rate.

Written fishery management plans should be developed for each large (>200 hectare) reservoir, and these plans should be available to the public. We might use for an example, the EPA administered "Clean Lakes Program," section 314 of the Clean Water Act (Public Law 92-500). This program requires each state to submit a classification and identification of publicly-owned freshwater lakes based on their eutrophic condition. It also requires procedures, processes, and methods to control pollution of the lakes, and techniques to restore their quality (Keup 1979). Financial assistance requirements for a state priority list would serve as a model for this type of planning requirement. Management plans should be carefully evaluated and fine-tuned by long-term monitoring.

Fishery management plans should incorporate and be integrated with reservoir water level and water discharge operational plans (hydrological plans). Inasmuch as hydrological manipulation involves entire water basin operational plans, state fishery management aspirations will require involvement in the complex hydrological and water quality simulation modeling that has been developed and is used by the water regulatory agencies. Fishery scientists and limnologists have come to understand some impacts of reservoir hydrology on aquatic communities, but we have not seen much application of this information. The "Kansas plan" of reservoir water level management has promise for improving certain types of sport fisheries, but given the highly variable water flow in the Great Plains, long term studies should also be undertaken to obtain a perspective of the plan over the 10- to 20-year, wet-dry cycle typical of the region and to assess whether the benefits and impacts of the plan are altered by continuous implementation.

Mitigation agreements developed during planning and included in authorization and environmental legislation should be monitored. Many sportsmen and the informed public have come to view the word "mitigation" as the means whereby the fishery agency is "bought off" so the water management agency can do what it wants. Even implementation of mitigation directives does not guarantee a satisfactory outcome. Although hatcheries are often the "mitigation carrot," the state agency generally assumes the operational costs. Hatchery fish are not the ultimate answer to habitat alteration from hydropower dams. One does not need to go further than the Columbia River to see that increased hatchery production alone cannot succeed in maintaining adult populations of anadromous salmonids stocks upriver of Bonneville Dam when impounded rivers provide habitat for native and exotic predators of juvenile salmonids.

Impoundments constructed primarily for the dissipation of waste heat are unique aquatic habitats with many inherent fishery problems. However, with the right type of regulatory revisions and economic incentives, they have potential for making a significant contribution to recreational fisheries. Some cooling impoundments with long retention times need careful monitoring for environmental contaminants (e.g., selenium toxicity).

Construction of new hydropower projects and retrofitting of many existing dams for electrical generation will likely take place at an increasing rate for the balance of this century. Bio-engineering for fisheries during the planning stages is needed to provide suitable habitat and minimize adverse impacts. These projects have the potential for seriously impacting stream and reservoir fishery resources. Pump-storage projects with large daily fluctuations in water level seem to be the most detrimental. Further study is needed to evaluate their impacts, ways to manage around the problems, and develop management methods to make constructive use of the pumping activity and water level fluctuations. The benefits of experience papers and research findings contained herein, and that of other recently published symposia and workshop proceedings, should be assimilated into planning documents.

Size regulations are a popular management tool for manipulating structure of reservoir fish populations. The impacts of size regulations on community composition may take longer than expected; therefore, regulations should not be abandoned after only a 2- or 3-year trial. Size regulation issues have served as a stage for the larger issue of measuring fishery management by maximum sustained yield, harvest-focused measure, versus optimum sustained yield, which uses catch, fishing quality, and socioeconomic considerations. Professional differences of opinion run deep; therefore, agreement will require continuing educational programs and further dialog before there is anything like unanimity on this issue. Given that largemouth bass are easily overharvested, more attention should be given to measuring the degree of angler compliance with respect to success or failure of size limit regulations.

Reservoir fishery habitat management should not be narrowly conceptualized and limited to fish attracting structures. Fishery habitat management must include a combination of activities that influence basic productivity, structure of the

TOETZ, D., R. C. SUMMERFELT, AND J. WILHM. 1972. Biological effects of artificial aeration in lakes and reservoirs—analyses and bibliography. U.S. Bureau of Reclamation REC-ERC-72-33, Denver, Colorado, USA.

TRIPLETT, J. R., D. A. CULVER, AND G. B. WATERFIELD. 1980. An annotated bibliography on the effects of water-level manipulation on lakes and reservoirs. Final report, Federal Aid Project F-57-R. Ohio Department of Natural Resources, Columbus, Ohio, USA.

TURNER, P. R., AND R. C. SUMMERFELT. 1971. Food habits of adult flathead catfish, *Pylodictis olivaris* (Rafinesque) in six Oklahoma reservoirs. Proceedings of the Southeastern Association of Game and Fish Commissioners 24:387–401.

USFWS (U.S. DEPARTMENT OF THE INTERIOR, FISH AND WILDLIFE SERVICE, AND U.S. DEPARTMENT OF COMMERCE, BUREAU OF THE CENSUS). 1982. 1980 national survey of fishing, hunting, and wildlife-associated recreation. U.S. Government Printing Office, Washington, D.C., USA.

WALBURG, C. H., J. F. NOVOTNY, K. E. JACOBS, W. D. SWINK, T. M. CAMPBELL, J. NESTLER, AND G. E. SAUL. 1981. Effects of reservoir releases on tailwater ecology: a literature review. Technical report E-81-12. U.S. Army Corps of Engineers Waterways Experiment Station, Vicksburg, Mississippi, USA.

WALKER, W. W. 1982. Water quality impacts and assessment procedures for hydropower projects. Pages 53–71 *in* G. A. Barnhart, editor. Hydropower and fisheries: impacts and opportunities. Northeast Division American Fisheries Society, Bethesda, Maryland, USA.

WEITHMAN, A. S., AND R. O. ANDERSON. 1978. An analysis of memorable fishing trips by Missouri anglers. Fisheries 3(1):19–22.

WEITHMAN, A. S., AND M. A. HAAS. 1982. Socioeconomic value of the trout fishery in Lake Taneycomo, Missouri. Transactions of the American Fisheries Society 111:223–230.